EMULSIONS AND EMULSION STABILITY

SURFACTANT SCIENCE SERIES

CONSULTING EDITORS

MARTIN J. SCHICK
Consultant
New York, New York

FREDERICK M. FOWKES
(1915–1990)

1. Nonionic Surfactants, *edited by Martin J. Schick* (see also Volumes 19, 23, and 60)
2. Solvent Properties of Surfactant Solutions, *edited by Kozo Shinoda* (see Volume 55)
3. Surfactant Biodegradation, *R. D. Swisher* (see Volume 18)
4. Cationic Surfactants, *edited by Eric Jungermann* (see also Volumes 34, 37, and 53)
5. Detergency: Theory and Test Methods (in three parts), *edited by W. G. Cutler and R. C. Davis* (see also Volume 20)
6. Emulsions and Emulsion Technology (in three parts), *edited by Kenneth J. Lissant*
7. Anionic Surfactants (in two parts), *edited by Warner M. Linfield* (see Volume 56)
8. Anionic Surfactants: Chemical Analysis, *edited by John Cross* (out of print)
9. Stabilization of Colloidal Dispersions by Polymer Adsorption, *Tatsuo Sato and Richard Ruch* (out of print)
10. Anionic Surfactants: Biochemistry, Toxicology, Dermatology, *edited by Christian Gloxhuber* (see Volume 43)
11. Anionic Surfactants: Physical Chemistry of Surfactant Action, *edited by E. H. Lucassen-Reynders* (out of print)
12. Amphoteric Surfactants, *edited by B. R. Bluestein and Clifford L. Hilton* (see Volume 59)
13. Demulsification: Industrial Applications, *Kenneth J. Lissant* (out of print)
14. Surfactants in Textile Processing, *Arved Datyner*
15. Electrical Phenomena at Interfaces: Fundamentals, Measurements, and Applications, *edited by Ayao Kitahara and Akira Watanabe*
16. Surfactants in Cosmetics, *edited by Martin M. Rieger* (out of print)
17. Interfacial Phenomena: Equilibrium and Dynamic Effects, *Clarence A. Miller and P. Neogi*
18. Surfactant Biodegradation: Second Edition, Revised and Expanded, *R. D. Swisher*
19. Nonionic Surfactants: Chemical Analysis, *edited by John Cross*
20. Detergency: Theory and Technology, *edited by W. Gale Cutler and Erik Kissa*

EMULSIONS AND EMULSION STABILITY

edited by
Johan Sjöblom
University of Bergen
Bergen, Norway

Marcel Dekker, Inc.   New York • Basel • Hong Kong

Library of Congress Cataloging-in-Publication Data

Emulsions and emulsion stability / edited by Johan Sjöblom.
    p. cm. – (Surfactant science series ; v. 61)
    Includes index.
    ISBN 0-8247-9689-6 (hardcover: alk. paper)
    1. Emulsions. 2. Stability. I. Sjöblom, Johan.
  II. Series.
  TP156.E6E614   1996
  541.3'4514–dc20   96-1662
                                           CIP

The publisher offers discounts on this book when ordered in bulk quantities. For more information, write to Special Sales/Professional Marketing at the address below.

This book is printed on acid-free paper.

Marcel Dekker, Inc.
270 Madison Avenue, New York, New York 10016

Current printing (last digit):
10 9 8 7 6 5 4 3 2 1

PRINTED IN THE UNITED STATES OF AMERICA

Preface

Emulsions are among the most important colloids in everyday life. In food emulsions such as margarine, butter, milk, dressings, and so forth, it is beneficial for the customer that the emulsified system retain high stability. On the other hand, in the process industry a stable emulsion can often cause a problem. Good examples are crude-oil-based emulsions, wash emulsions, and wastewater emulsions. The multiplicity of technical applications has naturally inspired in-depth studies that are needed in order to understand stabilization and destabilization processes in emulsified systems.

In this book, our intention is to give a fundamental presentation of emulsion science. The book gives a basic discussion of such topics as sedimentation, creaming, flocculation, and coalescence whereas the technical part describes crude-oil-based emulsions, food emulsions, and alkyd emulsions. The last mentioned topic is essential for the paint industry. In addition, the book also focuses on experimental techniques that can be conveniently applied to characterize emulsified systems.

In the first chapter, Friberg and Yang analyze emulsion stability in terms of third-phase stabilization (liquid crystal or solid) and necessary wetting/spreading conditions at equilibrium. The authors also describe processes in thin liquid films and surfactant association structures of immediate relevance for emulsion stabilization. Other topics covered are flocculation/coalescence kinetics, polymeric stabilizers and their surfactant complexes, and evaporation processes in emulsions and resulting equilibria.

Dukhin and Sjöblom give an analysis of flocculation/coagulation processes in dilute emulsions in the following chapter. A central topic in this analysis is to mathematically characterize the flocculation kinetics in dilute emulsions, both oil/water (o/w) and water/oil (w/o). In doing so, parallels are drawn between pure coagulation in solid particle systems and the flocculation process in emulsified systems.

In Chapter 3 on flexible surfactant films, Bellocq gives an extensive overview of the most recent achievements in the field of microemulsions. This chapter expresses the fundamental differences between microemulsions and macroemulsions. The author presents recent detailed phase diagram studies and theories for transition between droplet and bicontinuous structures together with experimental verifications from light-scattering experiments. Bellocq also applies the theories about flexible surfactant films to explain the differences in stability between bicontinuous microemulsions, lamellar liquid-crystalline phases, and sponge phases. At the end of the chapter, a summary of pertinent technical applications of microemulsions is presented.

Breen et al. give, in Chapter 4, a detailed discussion of destabilization processes in emulsions in general and in crude-oil-based emulsions in particular. The authors analyze thin-film rheology, Marangoni–Gibbs effects, and the occurrence of microstructures in ultrathin films. The authors also report on the effect of solid particles located at the w/o interface on the stability of w/o emulsions and especially crude-oil emulsions. In addition to this, pH effects and the influence of cosurfactants on crude-oil emulsion stability are also discussed.

Dalgleish gives an extensive analysis of fundamental mechanisms occurring in food emulsions. The author discusses in detail processes taking place at the w/o interface of different food preparations. These processes involve absorption of small surfactant molecules and proteins, competitive adsorption, interaction forms, and conformations at the interface. Dalgleish presents formation mechanisms in these kinds of emulsions together with the importance of accurate droplet size and droplet size distribution measurements. At the end of Chapter 5, the author correlates microscopic interfacial properties and macroscopic stability of food emulsions.

Bergenståhl and Östberg give a review of alkyd-based emulsions in the following chapter. As the authors point out, there is an ever-increasing interest in these kind of formulations rather than in organic solvent-based ones mainly due to health and environmental aspects. Alkyd emulsions are difficult to stabilize because of the presence of their many components. Besides the alkyd phase, the system contains solid pigments, dispersing agents, thickeners, wetting agents, biocides, and so forth. The authors discuss these components, their interaction forms, and their influence on alkyd emulsion stability. The efficiency of different emulsifiers is also compared from a mechanistic point of view.

Perfluorocarbon-based emulsions as red cell substitutes is the subject of the chapter by Kaufman. The author reviews the use of PFCs in medicine, including such topics as temporary oxygen transport, elimination of gaseous microemboli during cardiopulmonary bypass surgery, percutaneous transluminal coronary angioplasty, myocardial infarction, and cancer therapy. The chapter gives results from recent promising tests in these fields, together with prospects and challenges.

The last three chapters deal with NMR, dielectric spectroscopy, and ultrasound measurements as techniques to characterize emulsified systems. Söderman and Balinov update the reader on the use of NMR self-diffusion measurements on emulsions. The technique gives valuable insight into structures occurring in the continuous phase of the emulsion and how this association affects stability. The authors also present NMR as a nonintrusive method for studies of droplet sizes and size distributions. The technique has been successfully applied to both food and crude-oil emulsions. The NMR technique can also be used to analyze very concentrated emulsions, the so-called aphrons.

Dielectric spectroscopy is presented by Sjöblom et al. It is shown that the time-domain dielectric spectroscopy (TDS) method is very sensitive for monitoring flocculation processes in emulsions. It is shown that this process attains different levels, depending on the type of stabilizer used. The technique is very convenient for on-line analysis of emulsions. When equipped with an external high-voltage unit, the TDS method can be utilized to follow the coalescence in different emulsified systems. Results from model emulsions, food emulsions, and crude-oil emulsions are discussed.

In the last chapter, Frøysa and Nesse present ultrasonic techniques to characterize emulsions. These nonintrusive acoustical methods are used to determine the droplet sizes and the concentration of both the dispersed and continuous phases. An obvious advantage of the technique is its applicability to optically opaque media. The authors discuss different theoretical models for sound propagation in emulsified systems and multiple scattering models, together with the pros and cons of a variety of ultrasound-based experimental techniques. Finally, the authors report on studies of both o/w and w/o emulsions containing disperse phases in the range 30–80% by volume.

When editing a volume such as *Emulsions and Emulsion Stability* it is necessary to ask your colleagues to give priority to completing their contribution to the book. This volume has been very fortunate in that the contributors gave it that priority. Therefore, I would like to express my deepest gratitude to all the renowned contributors for their time and effort. I hope that our efforts will be well received. Finally, I hope that we have been able to contribute to a deeper understanding of the different kinds of complex emulsified systems.

Johan Sjöblom

Contents

Contributors

Balin Balinov Department of Physical and Analytical Chemistry, Exploratory Research, Nycomed Imaging AS, Oslo, Norway

Anne-Marie Bellocq Centre de Recherche Paul Pascal, Centre National de la Recherche Scientifique (CNRS), Pessac, France

Björn Bergenståhl Institute for Surface Chemistry, Stockholm, Sweden

Patrick J. Breen Fluid Separation Research, Baker Performance Chemicals, Inc., Houston, Texas

Douglas G. Dalgleish Department of Food Science, University of Guelph, Guelph, Ontario, Canada

Stanislav Dukhin Department of Theoretical Electrochemistry of Membranes and Colloids, Institute of Colloid and Water Chemistry, Ukrainian Academy of Science, Kiev, Ukraine

Harald Førdedal Department of Chemistry, University of Bergen, Bergen, Norway

Stig E. Friberg Department of Chemistry and Center for Advanced Materials Processing, Clarkson University, Potsdam, New York

Kjell-Eivind Frøysa Department of Industrial Instrumentation, Christian Michelsen Research AS, Bergen, Norway

Robert J. Kaufman HemaGen/PFC, St. Louis, Missouri

Young-Ho Kim   Department of Chemical Engineering, Illinois Institute of Technology, Chicago, Illinois

Øyvind Nesse   Department of Industrial Instrumentation, Christian Michelsen Research AS, Bergen, Norway

Alex D. Nikolov   Department of Chemical Engineering, Illinois Institute of Technology, Chicago, Illinois

Gunilla Östberg   Institute for Surface Chemistry, Stockholm, Sweden

C. S. Shetty   Nalco Chemical Company, Naperville, Illinois

Johan Sjöblom   Department of Chemistry, University of Bergen, Bergen, Norway

Tore Skodvin   Department of Chemistry, University of Bergen, Bergen, Norway

Olle Söderman   Physical Chemistry 1, University of Lund, Lund, Sweden

Darsh T. Wasan   Department of Chemical Engineering, Illinois Institute of Technology, Chicago, Illinois

Jiang Yang   Department of Chemistry and Center for Advanced Materials Processing, Clarkson University, Potsdam, New York

1
Emulsion Stability

STIG E. FRIBERG and JIANG YANG   Department of Chemistry and Center for Advanced Materials Processing, Clarkson University, Potsdam, New York

The stable state of an emulsion is in the form of its phases in layers separated by interfaces, which are as small as allowed by the form and size of the container. However, in an emulsion, several phases are dispersed in the continuous medium in the form of droplets. In the early approaches of emulsion science, only one dispersed phase, a liquid, was considered, and emulsions were defined as macrodispersions of one liquid in another. Such a state is common in commercial emulsions and its properties are

1

attractive from the fundamental point of view because of the simplicity. In the destabilization, such a two-phase emulsion (Fig. 1) goes through several consecutive and parallel steps before the final stage of separated layers is reached. As a first step, the droplets move due to diffusion or stirring, and if the repulsion potential is too weak, they become aggregated to each other; flocculation has taken place. The single droplets are now replaced by twins (or multiplets) separated by a thin film. This step is important because the destabilization process now passes from the realm of particles in random motion in a medium to describe the phenomena in a thin liquid film of colloidal dimensions. The thickness of the thin film is reduced due to the van der Waals attraction, and when a critical value of its dimension is reached, the film bursts and the two droplets unite to a single droplet. Coalescence has occurred.

In parallel with these phenomena, the droplets rise through the medium (creaming) or sink to the bottom (sedimentation) due to differences in density of the dispersed and continuous phase. This process is enhanced by the fact that larger droplets or aggregates move faster through the medium; in dilute suspensions, the velocity is proportional to the square

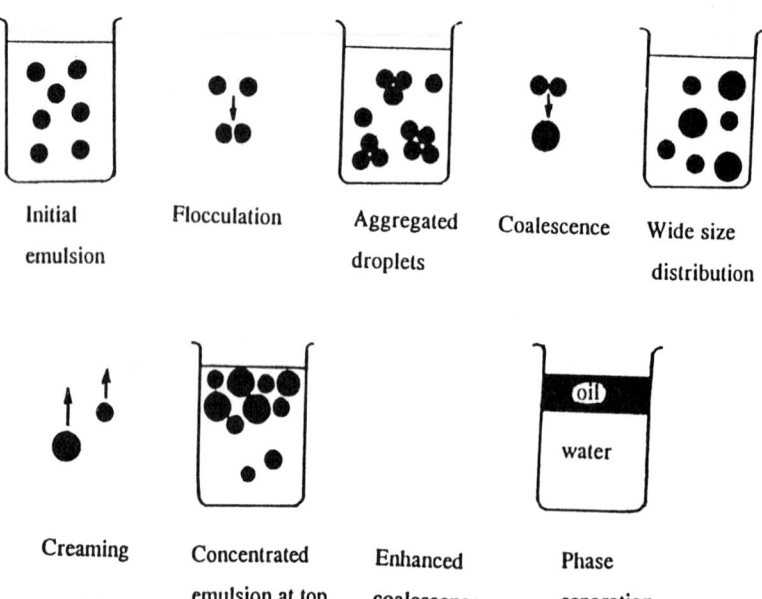

Initial emulsion   Flocculation   Aggregated droplets   Coalescence   Wide size distribution

Creaming   Concentrated emulsion at top   Enhanced coalescence   Phase separation

FIG. 1   The destabilization of an emulsion passes through several stages; flocculation, coalescence, and creaming before the final phase separation occurs.

of the radius. As a consequence, larger droplets pass the smaller ones, causing sedimentation-induced flocculation. The final result is a highly concentrated emulsion at the top or bottom of the container, and the increased number of droplets per volume increases the flocculation rate in a most decisive manner. The flocculation and coalescence processes lead to larger and larger droplets until, finally, a phase separation has occurred.

The understanding of flocculation rates using the well-known basis of colloidal behavior for solid particles has, by now, been well developed and reached a mature state as exemplified by the second chapter in this book. With this in mind this chapter will concentrate on two remaining phenomena. Of these, the first is obvious; the events leading to coalescence of the flocculated droplets needs a description. Second, and more important, a large number of commercial emulsions do not consist of two liquid phases only. In addition to these two phases, other phases of liquid, liquid crystalline, or solid state may also be present, as reflected in the International Union of Pure and Applied Chemistry (IUPAC) definition of an emulsions [1].

With this fact in mind, one essential phenomenon of emulsion stability is the question of to what extent the presence of a third phase will influence the different stages of the destabilization process. The influence of the third phase depends on its location and it will be briefly discussed before the general destabilization processes are examined.

One important factor for the location of the third phase at a local equilibrium is the ratio between the surface tensions (Fig. 2). The droplets may exist as individual droplets or they may be flocculated with each other. In the latter case, one droplet (Fig. 3a) may engulf the other (Fig. 3b) if the total surface free energy is reduced during that process.

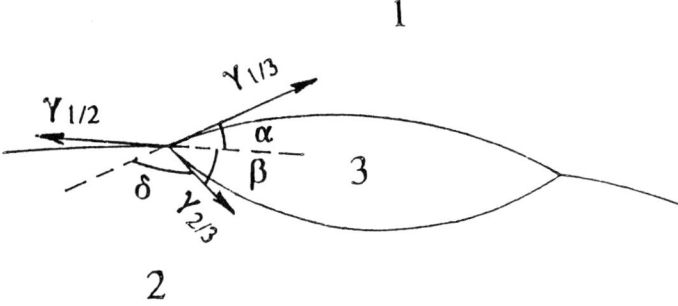

FIG. 2 With two different liquids, 2 and 3, dispersed in a third liquid, 1, the interfacial tension decides the shape of the droplets.

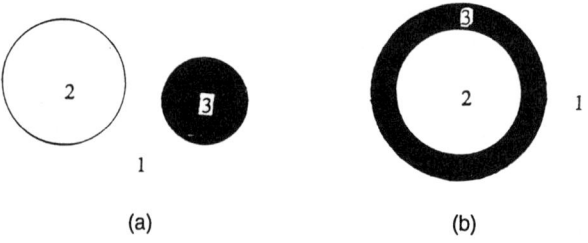

(a)   (b)

FIG. 3 With two liquids, 2 and 3, dispersed in a third liquid, 1, one of the dispersed liquids, 3, may entirely engulf the other, 2.

Taking the geometry into consideration [2], one finds the condition

$$\gamma_{2/3} < \gamma_{1/2} + \gamma_{3/1}[\psi^2 - (\psi^3 + 1)^{2/3}] \tag{1}$$

for phase 3 spreading on phase 2 with phase 1 as the continuous medium. The $\gamma_{v/\delta}$ are given in Fig. 2 and $\psi = r_3/r_2$ with r_v the radius of the droplet of radius v. The results are as expected. For $\psi \to \infty$, there are two infinite half-spheres of phases 3 and 1 and the location of phase 2 is decided by Eq. (1) for $\psi \to \infty$:

$$\gamma_{2/3} < \gamma_{1/2}. \tag{2}$$

$\psi \to 0$ leads to a situation in which an infinitely small droplet 3 covers droplet 2 if

$$\gamma_{1/2} > \gamma_{2/3} + \gamma_{1/3}. \tag{3}$$

If Eqs. (1)–(3) are not satisfied, a finite contact angle is obtained (Fig. 2).

Mechanical equilibrium along the direction of the three surface tensions give

$$\gamma_{1/2} = \gamma_{2/3} \cos \beta + \gamma_{1/3} \cos \alpha, \tag{4}$$

$$\gamma_{1/3} = \gamma_{1/2} \cos \alpha + \gamma_{2/3} \cos \delta, \tag{5}$$

$$\gamma_{2/3} = \gamma_{1/3} \cos \delta + \gamma_{1/2} \cos \beta. \tag{6}$$

Solving for $\cos \beta$ and $\cos \alpha$,

$$\cos \alpha = \frac{(\gamma_{1/2}^2 + \gamma_{1/3}^2 - \gamma_{2/3}^2)}{2\gamma_{1/3}\gamma_{1/2}}, \tag{7}$$

$$\cos \beta = \frac{(\gamma_{1/2}^2 + \gamma_{2/3}^2 - \gamma_{1/3}^2)}{2\gamma_{1/2}\gamma_{2/3}}. \tag{8}$$

Figure 4 shows the form of the aggregated droplets for different ratios of surface tensions. A certain ratio of surface tensions leads to the detachment of droplet 3 from droplet 2. The condition for the droplets to remain unattached is given by the following expression. A small area δ of attachment is visualized and the work of adhesion calculated. Putting phase 1 as the continuous phase, the condition for nonattachment is

$$A_2\gamma_{1/2} + A_3\gamma_{1/3} < (A_2 - \delta)\gamma_{1/2} + (A_3 - \delta)\gamma_{1/3} + \delta\gamma_{2/3}, \qquad (9)$$

$$\gamma_{1/2} + \gamma_{1/3} < \gamma_{2/3}. \qquad (10)$$

The special case of a liquid crystal as a third phase has an interest because of the extremely low interfacial tension $\gamma_{2/3}$ and $\gamma_{1/3}$ [3]. It is obvious from Eq. (10) that attachment will occur, and with both $\gamma_{2/3}$ and $\gamma_{1/3}$ small, the condition in Eq. (1) is also met and total coverage of the dispersed droplet is expected (*vide infra*).

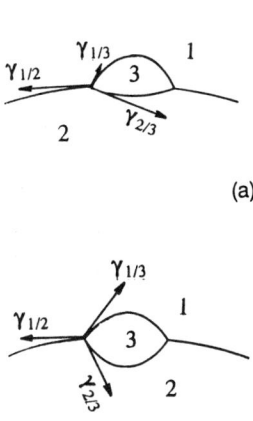

(a)

(b)

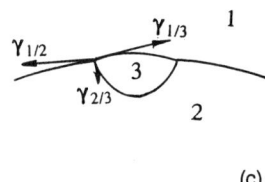

(c)

FIG. 4 The droplet shape for different interfacial tension ratios. (a) $\gamma_{2/3}/\gamma_{1/2}/\gamma_{1/3}$; $1:1:1/4$; (b) $\gamma_{2/3}/\gamma_{1/2}/\gamma_{1/3}$; $1:1:1$; (c) $\gamma_{2/3}/\gamma_{1/2}/\gamma_{1/3}$; $1/4:1:1$.

The location of small particles at an interface in an emulsion is characterized by Eqs. (1)–(8) except that the solidity of the particle simplifies Eqs. (4)–(6) to the well-known Young's condition for equilibrium at a solid surface

$$\gamma_{1/3} = \gamma_{2/3} + \gamma_{1/2} \cos \theta. \tag{11}$$

The location of a solid particle at the droplet–continuous phase interface is of decisive importance for its stabilizing or destabilizing action on an emulsion.

It is essential to realize the difference between these equilibrium conditions as described and the contact angle between two droplets of identical composition. For this case, the reduction in free energy when the thin film is formed by the continuous phase between the flocculated droplets is the decisive feature [4]. Hence, before the total process in the combined flocculation/coalescence may be discussed, a short review is given of the recent progress in the stability of thin liquid films.

I. CREAMING AND STABILITY OF THIN LIQUID FILMS

The stability of thin liquid films is one essential factor for a flocculated emulsion, but its importance is more directly visualized for foam bubbles. With this fact in mind, a large quantity of research is available on the conditions in a film per se. This research was characterized by its focus on the thin liquid film as an entity, consisting of a liquid layer limited by two interfaces toward a surrounding phase. For such a model system, the key factor for the stabilization is the thermal vibrations giving fluctuations in the film thickness. In short, a fluctuation wave pattern dampened with time results in stability, whereas growth of the amplitude leads to instability. It is immediately evident that fluctuations in phase cause local thinning and that when the thickness at a point is less than a critical value, the van der Waals forces will prevail and the film is destabilized.

Light scattering was established early as a useful tool for analyzing the wave pattern [5], and the predictions of stability (reduced fluctuations with time) or instability based on thermodynamic evaluations [6] could be verified. The agreement between theoretical predictions and measured values for the critical thickness showed excellent agreement [7].

For flocculated emulsion droplets, valuable information is obtained from the contact angle between them. A two-drop approach [8] shows the condition of two flocculated droplets to be different from that of the equilibrium Young equation. Now a decrease in free energy caused by the thinning of the interdroplet film decides the contact angle [8,9]:

$$-\Delta F = 2\gamma_{o/w} (1 - \cos \theta). \tag{12}$$

ΔF is the reduction in free energy of the formation of the semiequilibrium thin film between the droplets, $\gamma_{o/w}$ is the interfacial tension between the droplets and the continuous phase, and the contact angle, θ, is shown in Fig. 5. For a system in which the interaction free energy is low, the contact angle will be low [Eq. (12)]. Experimentally, it was found that the relative volume of the dispersed phase returned to a value close to 0.74 after emulsion centrifugation, followed by relaxation under gravity. For emulsions with substantial interaction energies or low interfacial tension, significantly higher volumes of dispersed phase were found. In addition, and more importantly, the time for clearing droplets from the low concentration part (Fig. 1) was significantly shortened.

However, the conditions are not static in a creamed emulsion, the most important stage for the coalescence of the flocculated droplets. Hence, it is valuable to observe the dynamics of creaming (or sedimentation depending on the relative density). The dynamics is exactly identical for the two cases except for the sign of the velocity and gradient vectors. In the following section, only creaming will be considered.

The process leads to a gradual division of the emulsion into two zones as exemplified by Fig. 1. The lower zone becomes depleted of droplets, whereas the top one develops into a highly concentrated emulsion. This division necessitates the treatment of different destabilization mechanisms; one of which is concerned with the conditions in the fast transport in the lower zone of the emulsion and the second one with the structure and dynamics in the space of the highly concentrated emulsion at the top.

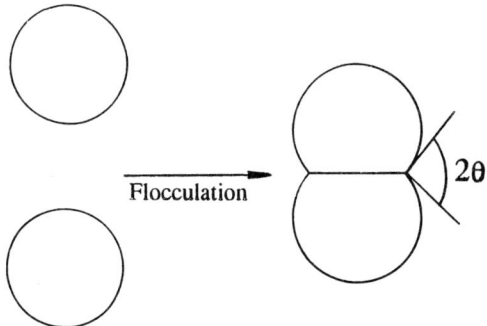

FIG. 5 The contact angle for flocculated droplets depends on the free-energy reduction during aggregation, Eq. (12). (From Ref. 4, with permission.)

The fast transport zone has been amply described by Dr. Dukhin in the second chapter of this book. The essential mechanisms are concerned with the balance between the gravitational flow versus the diffusional disordering. The analysis based on the early contributions by Fogler et al. [10,11] distinctly separated the conditions for convection- and diffusion-induced flocculation. This part of the emulsion is exceedingly amenable to an examination including only two droplet interactions. Such is not the case for the more concentrated part in which zone the network of droplets with the translational restrictions necessitates a different approach.

The conditions in the creamed and/or flocculated emulsion are not static; the movement of liquid is an essential factor. The dynamics leads to a surface tension gradient as examined by several authors [12] and surface rheological properties [13–16]. Good agreement was obtained between experiment and model evaluations when all the factors, including flow in plateau borders, were taken into account [17]. Hartland [18] also examined the combination effect of transversal interfacial tension gradients and drainage-induced circulation in the adjacent phase for the stability of a single film between flocculated droplets. Assuming a horizontal film, sheer stress balance at the upper and lower interface gave

$$\mu \left(\lambda \frac{\partial w}{\partial y} \right)_i - \mu_d \left(\lambda \frac{\partial w}{\partial y} \right)_i \tag{13}$$

in which μ is the viscosity in the thin film, μ_d is that in the adjacent phase, the derivatives are the shear gradients, and the i's indicate the lower or upper interface with $\lambda = 1$ for the lower phase border and -1 for the upper. The stress relation at the interfaces was used to calculate the mobilities of the film and adjacent phase interface. Setting the stresses equal and introducing the differential form of the drainage equation of reduced thickness h with time t,

$$-\frac{dh}{dt} = \left(\frac{h}{3\pi\mu r_f^4} \right) \left[2fh^2 + 3\pi\mu r_f^3 (w_{0f} + w_{hf}) \right]; \tag{14}$$

f is normal force across the film, μ the viscosity of the film, r_f the radius of the film, w_f the velocity at film peripheri at $v = 0$ lower and $v = h$ upper surfaces. The coalescence time t becomes

$$t = \frac{3\pi\mu r_f^4}{4fh_r^2 (1 + m_0 + m_h)} \tag{15}$$

in which h_r is the critical thickness at which the film breaks, m_v the mobility at the interfaces with $v = 0$ lower and $v = h$ upper surfaces. The results (Fig. 6) show the velocity profiles without an interfacial gradient

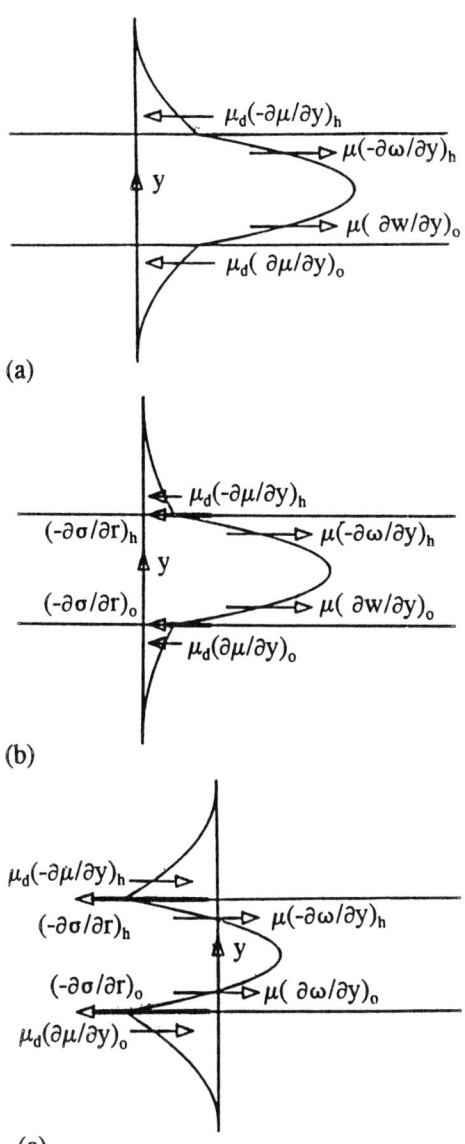

FIG. 6 The velocity profiles for a thin liquid film with a surface tension gradient. For sufficiently strong surface tension gradient (c), the net outflow from the film becomes zero. (From Ref. 18 with permission.)

(Fig. 6a), with a small negative gradient (Fig. 6b), and a strong negative gradient (Fig. 6c). The conditions in Fig. 6c are such that the counterflow induced by the gradient is sufficient to render the total outflow of the thin film equal to zero.

The overall conditions in a creaming emulsion stabilized by a polymer have been evaluated by Dickinson and collaborators [19–25].

The investigations confirmed the importance of the polymer critical overlap concentration [26] and related the significant increase of creaming below that polymer concentration to depletion flocculation [27,28]. The variation in creaming rates were covered for emulsions with different amounts of an added commercial Xanthan® gum. The influence of gum concentration is well illustrated by the results covering Xanthan concentrations from 0 to 0.173% at 30°C for emulsions containing 2% Tween 20 as stabilizer. The original emulsion without gum showed good stability against creaming, whereas 0.017% of the gum led to the separation of a serum layer at the bottom but no creamed layer. Such a layer was found for 0.035% gum, at which three layers are formed, namely, a serum layer, an emulsion layer, and a creamed layer. Higher concentrations led to emulsions separated into a cream layer and a serum one.

These results prompted the search for a reliable model for the conditions during creaming has been explored by Dickinson et al. [29]. A combination of the Stokes velocity and the diffusion-controlled movements of droplets was used. The Stokes velocity for particles of different sizes was calculated relative to the container by taking the fluid velocity, U_f, into consideration

$$U_f = \frac{\sum_j U_j \phi_j}{1 - \sum_j \phi_j}, \tag{16}$$

where ϕ_j is the volume fraction of component j and U_j is the particle velocity relative to the container

$$U_j = U_{\phi j} - U_f \tag{17}$$

in which $U_{\phi j}$ is the particle j velocity relative to the fluid. The U_ϕ is calculated from the velocity of an isolated particle U_0

$$U_0 = \frac{2r^2(\rho_1 - \rho_2)g}{9\eta} \tag{18}$$

in which r is the droplet radius, the ρ_1 and ρ_2 are the densities of the dispersed and the continuous phases, g is the gravitational constant, and η is the viscosity of the continuous phase.

For a collection of spherical particles, the velocity is related to the Stokes velocity using the model by Barneo and Mizrahi [30] to obtain the velocity for spherical particles of uniform size for a volume fraction ϕ,

$$\frac{U_\phi}{U_0} = \frac{1 - \phi}{(1 + \phi^{1/3}) \exp[5\phi/3(1 - \phi)]}. \tag{19}$$

The diffusion coefficient is expressed [31] as

$$D = D_0 B(\phi) S(\phi), \tag{20}$$

in which $B(\phi)$ are the hydrodynamic effects from the creaming [32]:

$$B(\phi) = \frac{(1 - \phi)U_\phi}{U_0} \tag{21}$$

and $S(\phi)$ represents the thermodynamic factor directing the flow of particles toward sites with less a value of ϕ:

$$S(\phi) = 1 + 8\phi + 30\phi^2. \tag{22}$$

It is essential to realize that both the creaming velocity and the normalized diffusion coefficient are reduced to zero at a volume fraction in the range 0.6–0.7.

The results of calculations for a sample 25 cm in height divided into 500 layers gave results (Fig. 7) showing both the serum layer (the intermediate layer with minor modification) and the creamed layer. The particle size distribution within the creamed layer revealed the larger particles to be more prevalent toward the higher parts of the layer (Fig. 8). The publication also gave the applicability of the Urick equation [33] to evaluate ultrasound analysis.

All these investigations use a model with no surfactant association structures, which have not been considered significant per se for the stability of a thin film. This neglect has left an important question unanswered. Why should surfactant concentrations in excess of the critical micellization concentration be beneficial for emulsions stability if only the properties of the interface are relevant? The fact is that the monolayer adsorption becomes saturated at the critical micellization concentration, and except for a minor increase of surfactant concentration in excess of that value to cover demands for extra availability of surfactant, there should be no need for higher amounts of surfactant, certainly not at magnitudes greater than the critical micellization concentration. This question went unanswered for a long time, delaying the understanding of emulsion stability.

Recently, Wasan and collaborators provided an elegant answer by revealing the stratification in last stages of the thinning of liquid films [34–36] to be due to layerwise removal of micelles [37–39]. Wasan et al. could

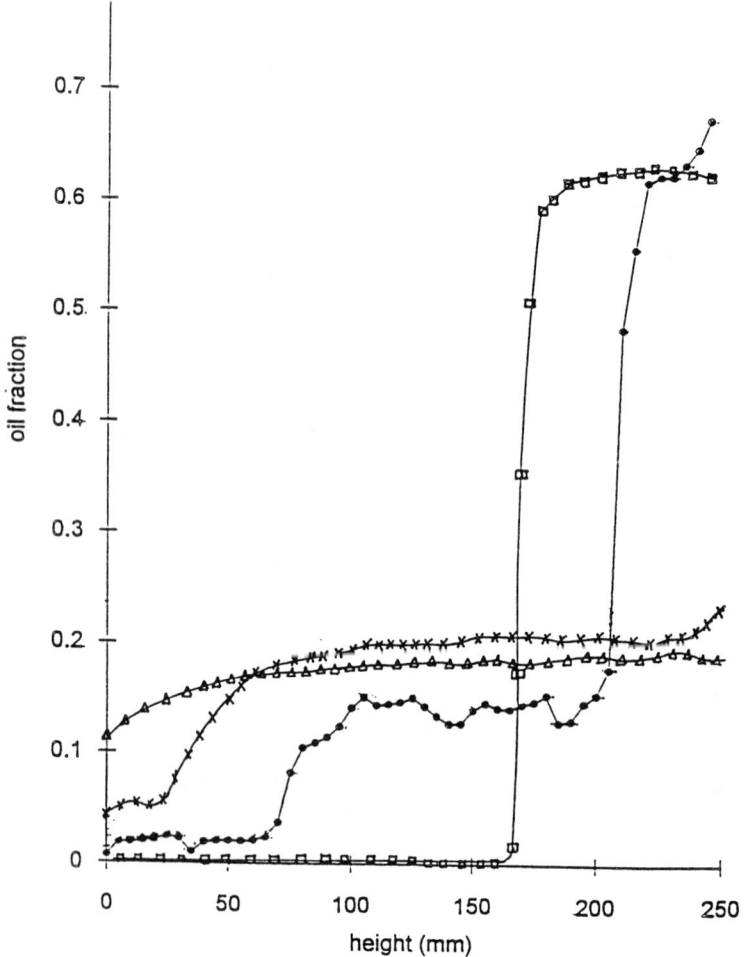

FIG. 7 Creaming profile of an emulsion, 18 vol% oil, stabilized by 2% Tween 20 after addition of different amounts of Xanthan gum and storage for 150–200 h. % Xanthan gum: △—0; ×—0.017; ●—0.035; □—0.170. (From Ref. 24 with permission.)

prove that the stratification in the thinning of films was due to stepwise removal of layers from close-packed micelles. They found that micelles within a thin film give a strong contribution to its stability by the mechanical influence by the close-packed micelles. It should be noted that the ordering of particles within a thin film is significantly different from the

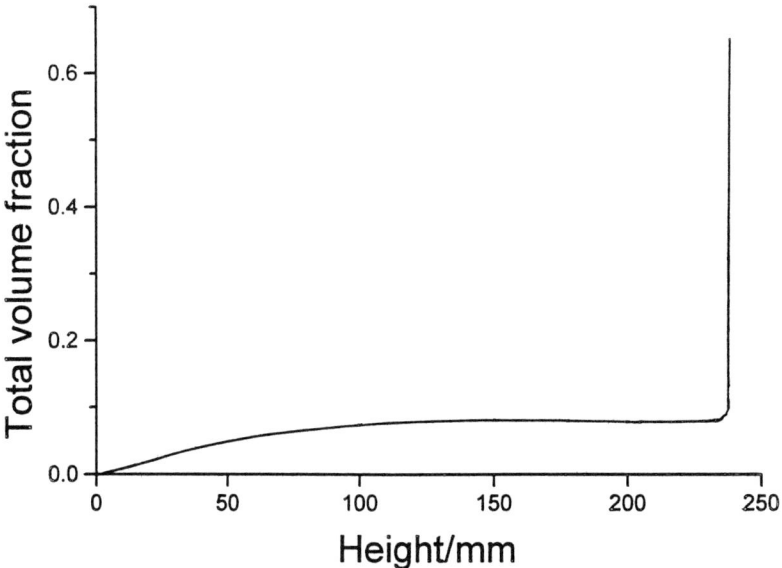

FIG. 8 Creaming of an emulsion leads to a highly concentrated emulsion at the top (240 mm) and total depletion in the bottom part (10 mm). (From Ref. 29, with permission.)

ordering in a bulk phase with infinite dimensions [40] because it is influenced by the presence of the thin-film walls. Recent calculations demonstrate this fact in detail [41].

The calculations used the 12–6 Lennard–Jones potential, which, when integrated over a thin film, gives a 10–4 potential.

$$V = 2\pi\epsilon\sigma^6\rho_s \left(\frac{0.4\sigma^6}{z^{10}} - 1/z^4 \right) \tag{23}$$

for the interaction between particles at distance z from the surface of the film. σ is the effective diameter of the particle, ρ_s the material density of the surface, and ϵ the depth of the potential well in the 12–6 potential curve. The numerical procedure by Valleau and Wittington [42] was used with a p value of 0.15 [43].

The results demonstrated the importance of the surface; ordering (Fig. 9) was pronounced at 35% volume fraction instead of 50% for the bulk phase. The calculated and experimentally determined disjoining pressure [44] were in complete agreement [45,46].

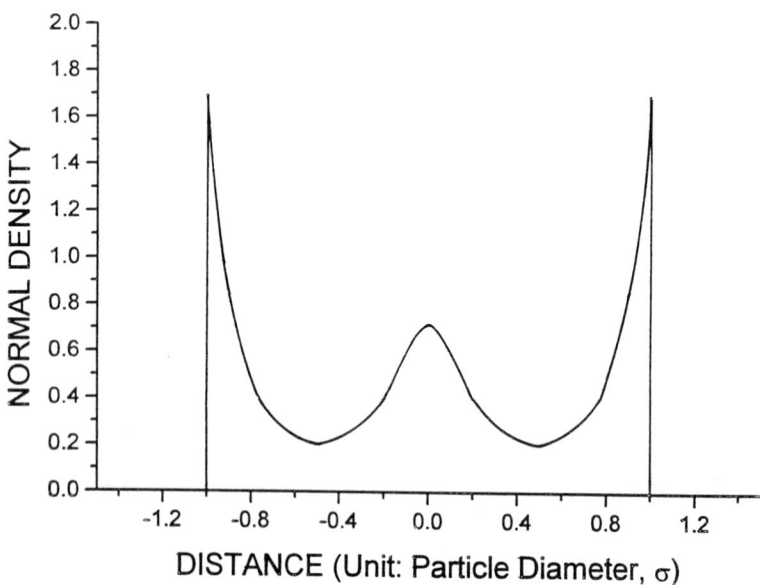

FIG. 9 Normal density distribution of particles in a thin film with particle volume fraction 0.35. (From Ref. 41, with permission.)

There is no doubt that this new mechanism answers the question of the enhanced stabilizing action of a surfactant when present in excess of its critical micellization concentration; an extremely important contribution to the knowledge of emulsion stability.

This mechanism brings the phenomenon of surfactant associations into the discussion of emulsion stabilization. The micellar association is but one step in a series of surfactant agglomerations, and a brief review of amphiphilic supramolecular structures is justified.

A. Surfactant Association Structures

The fact that surfactants show concentration-dependent associations has been known for more than 80 years and the criticality of the phenomenon was realized some 60 years ago. The studies in the area were concerned with phase diagrams and attempts were made to establish the thermodynamic driving force causing all the different structures to form.

The success of these efforts was rather limited until Israelachivili et al. [47] revealed the importance of steric factors to form specific structures.

They built their approach on the ratio between the volume of the surfactant hydrocarbon chain, v_H, and that of the area of the polar group, a_0, multiplied by the chain length, l:

$$R = \frac{v_H}{a_0 l}. \tag{24}$$

For a ratio R equal to 1.0, it is obvious that a lamellar structure is the only possible association form, but what is more important for emulsion science is the fact that, geometrically, the lamellar structure is the only possible in the R range 1.0–0.5. Lower ratios give cylinders $\frac{1}{3} < R < \frac{1}{2}$ and spheres $0 < R < \frac{1}{3}$.

The conditions for spheres to give micelles in the solution and their stabilizing action has been discovered and analyzed by Wasan et al. The influence of cylindrical micelles ($\frac{1}{3} < R < \frac{1}{2}$) on emulsions has not been treated at all, but the influence of lamellar association structures has been known for more than 25 years. The main action of lamellar liquid crystals will be discussed later in this section, but one aspect is closely related to the micellar stabilization and will be treated in this section.

In the lower part of the range 1.0–0.5 vesicles form during mechanical treatment in the emulsification process. They were first detected in an emulsion system a decade ago [48], but no evaluation of their specific stabilizing action has been made so far. However, some stabilizing features are self-evident from the structure of the vesicle. It consists of a bilayer or multilayer of the surfactant forming an w/w emulsion (Fig. 10). When present between two emulsion droplets, the vesicle serves to stabilize the emulsion with two mechanisms.

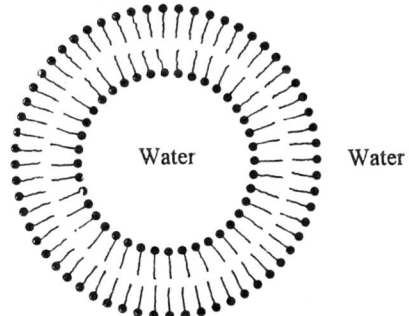

FIG. 10 A liposome forms a lipid bilayer around a water droplet in water.

At first, when the vesicle is deformed to an oblate, its surface area is increased, giving a free energy increase of

$$\Delta G = \int_{A_1}^{A_2} \gamma(A) \, dA \tag{25}$$

or, if γ is constant,

$$\Delta G = \gamma \Delta A. \tag{26}$$

The energy increase is huge, as revealed by a conservative example.

A liposome with a radius of 500 Å has an area of 3.14×10^{-10} cm^2. Assuming a low interfacial energy of 0.1 dynes/cm^2 and an area increase of 10%, one finds an increase in free energy of 3×10^{-12} ergs, which corresponds to 76 kT. Realizing that 20 kT gives a sufficient barrier maximum to reduce the flocculation rate to acceptable levels for practical applications, the importance of liposomes for the emulsion stabilization is easily perceived. As a second mechanism, the bending energy of the liposomal structure must also be taken into account.

Finally, if the forces during the approach of two droplets become sufficiently large to compress the vesicle to two bilayers, the dramatic influence on the distance dependence of the van der Waals forces (*vide infra*) remains as a stabilizing mechanism.

With the stability of thin films briefly reviewed, the kinetics of the combined flocculation/coalescence may be evaluated.

II. FLOCCULATION–COALESCENCE KINETICS

The time dependent changes in the number of droplets for aggregation (flocculation) is treated in Chapter 2 using the theoretical framework for solid particles. These equations show the state of aggregation of solid particles and, hence, cover the first part of the total destabilization process (Fig. 1). For emulsions of liquid droplets dispersed in another liquid, the flocculation process is followed by coalescence with the relative importance of the two processes varying with time, and an analysis of the relative rates is useful.

Van den Tempel [49] modeled the combined process using linear aggregates of droplets, obtaining the number of films between the droplets equal to $m - 1$ in which m is the average number of droplets in the aggregates. The approach leads to an overestimation of the rate of increase in aggregate size because, although the increase of aggregate size properly accounts for the flocculation of aggregates of different sizes, the fact that coalescence has taken place in the individual aggregates prior to their aggregation is not taken into account. Borwarkar et al. [50] have recently examined the problem with the following approach.

The aggregate of droplets in three dimensions and the number of films in each aggregate is equal to p per drop. Hence, the number of films n_f becomes

$$n_f = pm \tag{27}$$

and the rate of coalescence

$$-\frac{dn}{dt} = Kpmn_v, \tag{28}$$

where n_v is the number of aggregates.

The number of droplets, n, at time t with both flocculation and coalescence taken into account now becomes

$$\frac{n}{n_0} = \exp[-G'(\theta - 1)]$$

$$\times \left(1 + \frac{G^1(1 - 1/\theta)^{-1}}{\exp G^1} + \frac{G^{12} \ln \theta}{\exp G^1} - \frac{(G^{13})(1 - \theta)}{2! \exp G^1} \right. \tag{29}$$

$$- \frac{(G^{14})(1 - \theta^2)}{\exp G^1(2 \times 3!)} \cdots$$

$$\left. - \frac{(G^{1\,n+1})(1 - \theta^{n-1})}{(n - 1)n! \exp G^1} \cdots \right),$$

where $\theta = 1 + an_0t$, $1 < \theta < \infty$, $0 < n/n_0 < 1$, and $G^1 = Kp/an_0$.

The number of particles per aggregate shows a linear increase for short times (Fig. 11) but is drastically reduced at longer times. This latter result is an artifact due to the fact that as the coalescence proceeds, the value of m is strongly reduced and the number of film changes from pm to $m - 1$ for $m = 2$.

With this fact in mind, Borwankar et al. [50] decided to use $m - 1$ for the number of films within an aggregate of m droplets. The comparison between best-fit equation and experimental values is illustrated in Fig. 12. The fit is acceptable, giving similar rate constants for flocculation 2.6 $\times$ 10^{-8} cm^3 min^{-1} and 1 $\times$ 10^{-8} cm^3 min^{-1}, but a significantly reduced rate constant for coalescence, 4.4 $\times$ 10^{-4} and 1.5 $\times$ 10^{-2} min^{-1}. The first emulsion was considerably more stable and the results reveal the cause of the instability of the second emulsion. The coalescence rate within the aggregates obviously was the determining factor, and in this case, the stability of the thin films [51] is the essential factor to improve.

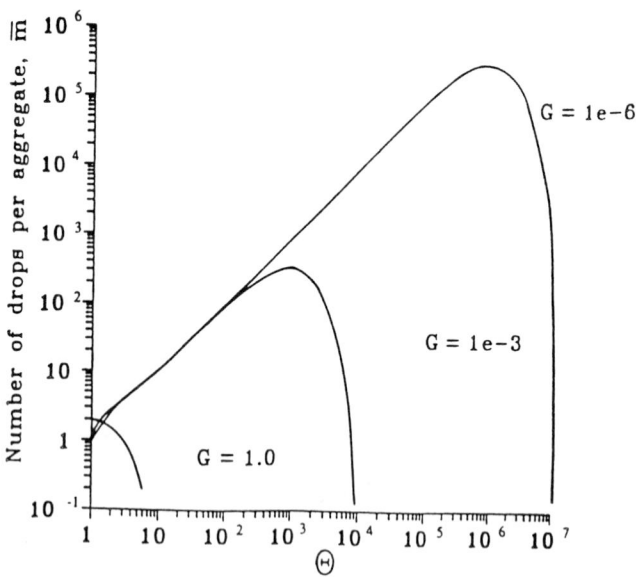

FIG. 11 Variation of $\bar{m}$ with Θ for concentrated emulsions. (From Ref. 50, with permission.)

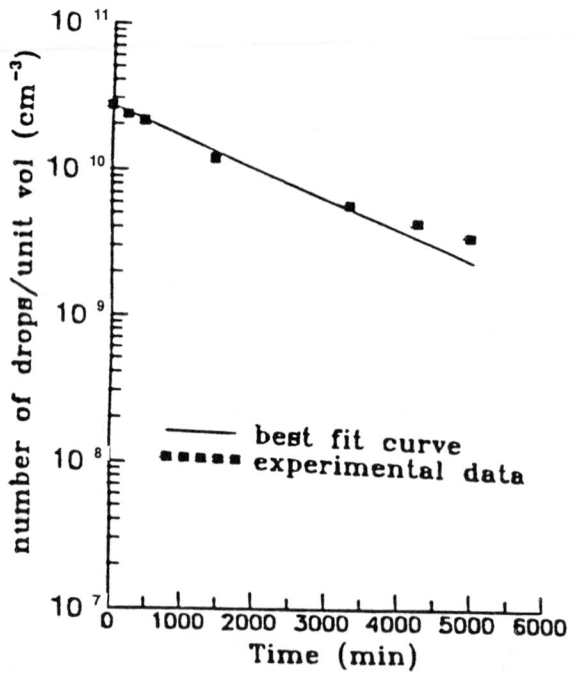

FIG. 12 Experimental determination of the number of drops per volume in an emulsion and calculations from the combined flocculation/coalescence rates agreed well. (From Ref. 50, with permission.)

The rate of coalescence is decisive for the fate of an emulsion during shear, the orthokinetic coalescence. The assumption is that in a partially flocculated system, subject to shear, two processes are monitoring the droplet size distribution. The shear motion will lead either to a separation of individual droplets (or smaller agglomerates) from the aggregates or to coalescence within the aggregate.

The disruption process is realistically assumed to be first order with regard to the number of aggregates per volume n and the coalescence process is postulated as of second order. The Bernoulli equation [52] may now be written [53]

$$\frac{da}{dt} = k_d n - k_c^1 (d_i^{3/D} + d_j^{3/D})^3 n^2, \tag{30}$$

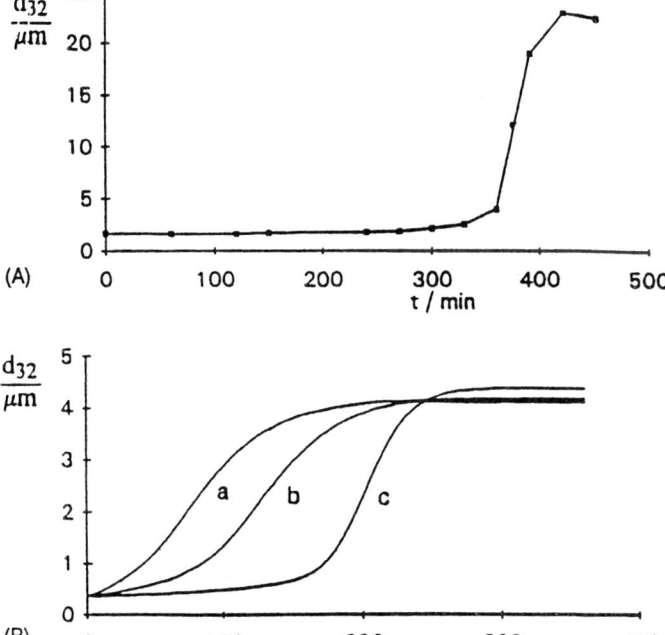

FIG. 13 Experimental determination (A) of droplet size in a flocculating coalescing emulsion under shear gave optimal agreement with calculated values assuming a fractal dimension of 2 (B). The fractal dimensions are as follows: (a) 3; (b) 2.5; (c) 2. (From Ref. 53, with permission.)

where k_d is the disruption constant, k_c^1 is the coalescence rate constant independent of droplet sizes d_i and d_j, and D is the fractal dimensionality. Furthermore, $k_d^1 = 0$ for $d_i < d_c$, reflecting the fact that turbulent flow shows a minimum size of droplet for description [54].

A comparison of experimental and calculated values (Figs. 13A and 13B) clearly indicates a fractal dimension of 2 which better reflects the experimental results. The relative importance of the disruption and aggregation rate constant was evaluated, showing the latter to be the obvious essential factor.

With these short sections, the fundamentals of the entire process in Fig. 1 have been covered and the specific action of different stabilizing molecules may be reviewed.

III. POLYMERIC STABILIZERS

The fundamentals of polymer emulsion stabilization was early established by Clayfield, Napper, Vincent, and others, who clarified the contributors from entropy and other factors [55–60].

Following this early analysis of the physics of dispersion stabilization, a number of more chemically oriented studies of the stabilizing action of block copolymers were made [61–63]. Special interest was attracted by the "star-shaped" polymers [64] because of their stabilization of both oil–water (o/w) and water–oil (w/o) emulsions [65].

The recent contributions by Lochhead [66,67] analyzed the conditions for polyacrylic acid (PAA), hydroxyethylcellulose (HEC), and their hydrophobically modified varieties hydrophobically modified hydroxycellulose (HMHEC) and hydrophobically modified polyacrylic acid (HMPAA).

The key factor for the evaluation of the potential for stabilization is the critical overlap concentration C^* value of the polymer [68]. At concentrations in excess of this value, the steric stabilization becomes ineffective because the basis for osmotic stabilization is removed when there is no polymer concentration difference between the bulk phase and the space between two approaching droplets. In addition, the polymers in this concentration range are entropically restricted throughout the continuous phase, and entropic stabilization through conformational restriction of the polymer between two droplets is not possible.

The critical overlap concentration C^* is defined as the concentration at which the product of concentration and the intrinsic viscosity reaches 1, but scaling arguments show specific viscosity below the C^* value to vary as the 1.4th power of the product of concentration, whereas for concentrations above that value, the scaling factor should be 3.3. Hence, a convenient determination of C^* is by a Huggins plot of specific viscosity versus polymer concentration. C^* is marked by an abrupt change in slope of the Huggins plot.

As a result, steric stabilization may only be expected for concentrations of polymer below the value of C^*. Lochhead [69] found the following results for HMPAA to illustrate these conditions. For $C > C_{un}^*$ (C_{un}^* is C^* for unneutralized polymer), the presence of a polymer stabilized the emulsions for all pH values through "trapping" of the emulsion droplets in hydrophobic sites of the polymer mesh, whereas for $C < C_{un}^*$, stabilization was achieved by the traditional steric stabilization for pH < 6.5. A pH value in excess of 6.5 gave instability for $C < C_{un}^*$ because of a combination of insufficient adsorption of the polymer, rending steric stabilization ineffective and $C < C_{un}^*$ preventing network formation in the bulk of the continuous phase. HEC was completely ineffective as a stabilizer.

The addition of PAA and hydrophobically modified PAA (HMPAA) gave stabilization for pH < 6.5, and interfacial tension measurements showed adsorption to be insignificant for pH values in excess of this value, in good agreement with the general rules found by Clayfield.

A. Polymer–Surfactant Combinations

The stabilizing action of polymer–surfactant combinations has not received the attention it deserves. A fundamental contribution to this kind of stabilization is provided by the investigations by Sjöblom et al. on the stability of w/o emulsions from the Norwegian continental shelf [70–79]. The interfacial fraction which stabilized the emulsions was separated [80] and a Langmuir–Blodget study of its monolayer properties against an aqueous subphase was made [81].

The results gave a direct correlation between the specific monolayer areas and the stability of the emulsion in question, and a fundamental study of the interfacial relations between an interfacially active polymer and conventional surfactants was made [76]. The results clarified the complex interaction between the surfactant and the polymer at the interface, demonstrating the importance of surfactant association concentration with the polymer. For surfactant concentrations less than the association concentration, it penetrates the polymer monomolecular film and the increased surface pressure leads to conformational relaxation of the polymer without giving rise to transfer of the polymer into the bulk phase. This is not the case at concentrations above the polymer/surfactant association concentration. The relaxation by the polymer monolayer is the dissolution of a polymer/surfactant association complex into the bulk phase.

IV. EMULSIFIER SELECTION

The fundamental factors in emulsion stability as discussed in the preceding
sections of this chapter are necessary for understanding emulsion stability
and are useful for identifying situations in which emulsion stability may not
be expected, but are only of limited value in the choice between different
emulsifiers for a specific oil–water combination. Several empirical rules
for the selection have been established.

These rules are built on the primary condition for an emulsifier to be
efficient in a two-phase, two-liquid emulsion. It has to be localized at the
oil/water interface to a maximal extent. This criterion is exemplified in
the HLB number introduced by Griffin in 1949 [82]. In passing, it may be
emphasized that the HLB number for a certain molecule denotes the balance between its hydrophilic/lipophilic properties only; the number per
se does not convey any information about the stabilizing efficacy of the
molecule. Formally, acetic acid, for example, according to the group number approach [83] has an HLB number of approximately 8.6 and should,
hence, be useful as an emulsifier for an o/w emulsion. Such is certainly
not the case.

The HLB number has been related to different thermodynamic entities
[84–90], but the most useful instrument to estimate the HLB number for
an individual molecule is the group number approach by Davies and Rideal

TABLE 1 HLB Group Numbers

Groups	Group number
Hydrophilic	
$-SO_4Na^+$	38.7
$-COO^-H^+$	21.2
$-COO^-Na^+$	19.1
N (tertiary amine)	9.4
Ester (sorbitan ring)	6.8
Ester (free)	2.4
$-COOH$	2.1
$-O-$	1.3
CH (sorbitan ring)	0.5
Lipophilic	
$-CH-)$	
$-CH_2-)$	-0.475
$-CH_3-)$	
$-CH-)$	

[83]. The HLB number for a molecule is calculated according to the sum with sign of the numbers for individual groups in the molecule according to Table 1.

The HLB number approach means that the emulsifier is optimal in a water–oil system in which the properties of the oil matches the surfactant. Hence, each water–oil combination is characterized by an HLB number.

The phase inversion temperature (PIT) or HLB temperature concept relates the emulsifier selection to the temperature at which an emulsion stabilized by a nonionic emulsifier of the polyethylene glycol type changed from oil-in-water to water-in-oil with rising temperature [91–94]. The studies showed emulsions to be maximally stable with an HLB temperature 20–65°C higher than the storage temperature.

In this system, the selection of emulsifier is experimentally connected to a certain oil–water combination with the advantage that it automatically takes into consideration changes brought forward by compounds added to the oil or the water. The colloidal aspects of the HLB temperature system are discussed in the section on emulsification.

V. OSTWALD RIPENING

An emulsion of droplets with different sizes may be stable against flocculation and coalescence but, with time, show a change of the droplet size distribution toward greater dimensions. This phenomenon is related to the dependence of solubility on the droplet radius, the Kelvin effect. The smaller the radius of the droplet, the higher the solubility [95]. The fundamental aspects of the process were established by Lifshitz [96], revealing decisive aspects: first, the ratio of particle sizes to the mean radius remains constant and, second, the cube of the mean radius grows linearly with time. The Ostwold ripening rate dr^3/dt is a constant depending on the interfacial tension, $\gamma_{o/w}$, the diffusion coefficient D, and, C, the solubility of the disperse phase in the continuous medium:

$$\frac{dr^3}{dt} = \frac{8\gamma_{o/w}V_mCD}{9RT}, \tag{31}$$

where V_m is the molar volume of the dispersed compound.

Early investigations by Davis et al. [97,98] demonstrated the expected strong dependence on the nature of the hydrocarbon structure [C and D in Eq. (31)] as well as the more important fact that small additions of the less soluble and less diffusive hydrocarbon significantly increased the stability against the ripening changes.

Later studies by Kabol'nov [99,100] have examined the conditions in a dispersed system of two compounds mutually soluble into an ideal solution. The simplest case is that one of the two components, the major one, is virtually insoluble into the continuous medium. Equilibrium [99] now gives

$$X_{e2}^1 - X_{e2}^{11} = \frac{2\gamma_{o/w} V_m (1/a_e^1 - 1/a_e^{11})}{RT},\tag{32}$$

where the X's and the a's are molar fractions and droplet radii after equilibrium, respective. Combining this equation with the one for material balance gives

$$C_{i1} + \left(\frac{4\pi V_i}{3}\right) \int_0^\infty a^3 f_i(a)\, da = C_{e1} + \left(\frac{4\pi v_e}{3}\right) \int_0^\infty a^3 t_e(a)\, da,\tag{33}$$

where i and e mark initial and equilibrium conditions, respectively, C_1 marks the concentration of the first (soluble) component in the continuous phase, v is the number concentrations of particles, and f is the size distribution of particles. In addition,

$$C_{e1} = C_{o1} \exp\left(\frac{\Delta_{e1}}{RT}\right).\tag{34}$$

The results are illustrative of the importance of the presence of the second component. For a high content of the second component in the dispersed phase, $X_{i2} \gg \gamma_{o/w} V_m / a_i RT$, the equilibrium is rapidly obtained with only a slight change in size distribution. In fact, the condition to obtain equilibrium is for a narrow initial droplet size distribution:

$$X_{i2} > 2\gamma_{o/w} V_m / 3RT \bar{a}_i.\tag{35}$$

The opposite condition means that because

$$\frac{\partial \mu_1}{\partial a(a_i)} < 0,\tag{36}$$

the Ostwald ripening will increase the difference in the chemical potential of component 1 in the initial droplets of different sizes, and a bimodal distribution develops. Actually, the size distribution of the coarse fraction is described by the Lifshitz theory.

An interesting consequence for emulsions is pointed out by Kabol'nov. If the second component is surface active and its concentration lower than its association concentration in any of the phases, its increased concentration in the small droplets will counteract the Ostwald ripening by reducing the value of $\gamma_{o/w}$.

VI. THREE-PHASE EMULSIONS

A. Two Liquids Plus Solid Particles

The presence of small particles in addition to the common two liquids may be used to assist in the stabilization of emulsions. The key to this stabilization is the location of sufficiently small particles at the oil–water interface.

Particles at the interface resist being forced into the dispersed droplet; the resistance (E) energy depends on the contact angle [101]

$$E = \pi r_p^2 \, \gamma_{o/w}(1 - \cos \theta)^2, \tag{37}$$

where r_p is the droplet radius, $\gamma_{o/w}$ is the interfacial tension between water and oil, and θ is the contact angle into the dispersed droplet (Fig. 14). The stabilization is excellent; calculations of the force taking into consideration both van der Waals attraction forces and the wetting force from the particle show a total repulsive force

$$F = 2\pi\gamma_{o/w}(h - r \cos \theta), \tag{38}$$

where h is the distance the particle has moved from its equilibrium position. The repulsive force may be significant for a contact angle of 90° (Fig. 15). These conditions are valid for spherical, solid particles such as latexes, clays, and other mineral particles that completely cover the oil–water interface. However, a large number of particles are formed in the emulsion and to a large degree consist of amphiphilic compounds. A typical example is fat particles in food emulsions [102–106].

The conditions are now complicated by structural changes due to the presence of surfactants [107] on the interface and by competitive adsorp-

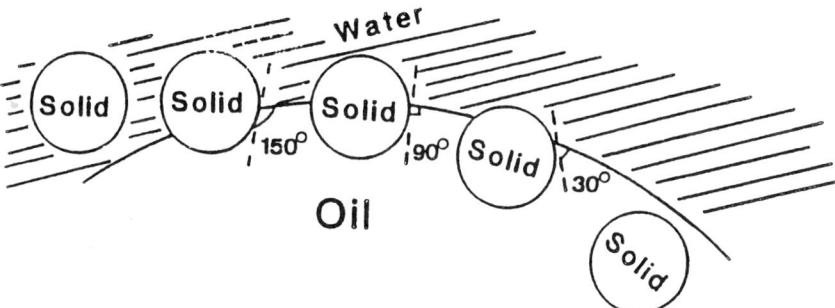

FIG. 14 The location of a solid particle at the oil–water interface is determined by the contact angle. (From Ref. 101, with permission.)

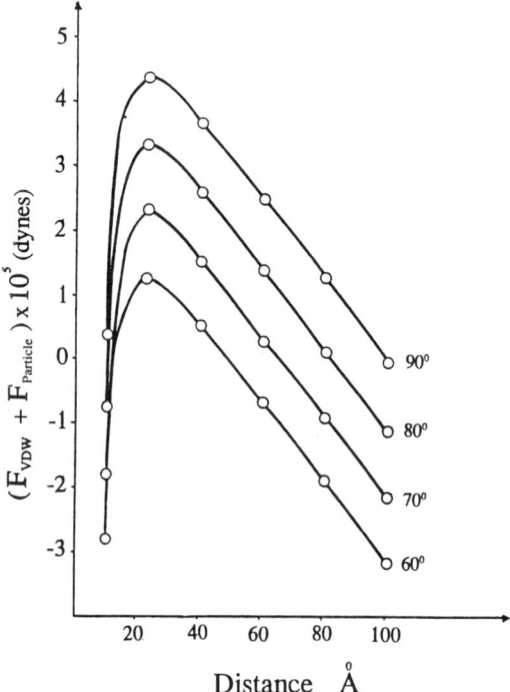

FIG. 15   The combined van der Waals attractive forces and the wetting repulsive forces from small particles ($r = 100$ Å) will reach positive values for high contact angles. (From Pharmaceutical Dosage Forms with permission.)

tion of surfactants onto the oil–water, oil–solid, and water–solid interfaces [108–110]. Boode and Walstra [105] have evaluated these phenomena, emphasizing the temperature dependence of the behavior. The results are not entirely without ambiguity, but some general trends were observed. First, the preferential wetting, the Lonza effect [111] may have a most pronounced effect on emulsion droplets with solidified fat as demonstrated in Fig. 16. One day after the addition of a water-soluble surfactant, sodium dodecylsulfate, the oil was dispersed in small droplets into the aqueous phase and the solid fat was separated into a broken network of flocculated particles.

The second factor of importance is the orientation of the fat crystals at the interface. With incomplete coverage of the oil–water interface, a protruding crystal (Figs. 17a and 17b) will have more of a tendency to bridge into another droplet than a crystal oriented in parallel to the inter-

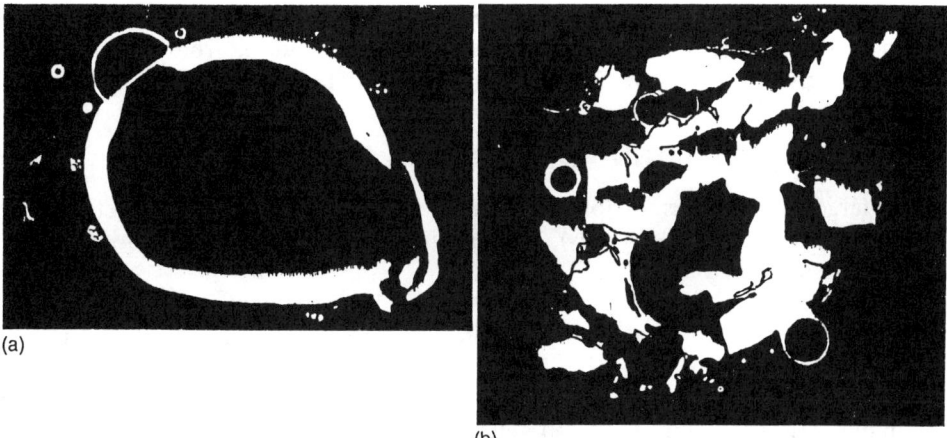

(a)

(b)

FIG. 16   Addition of a surfactant to an oil droplet with a solid lipid film (a) leads
to breakage and dispersion of both the oil and the solid (b). (From Ref. 105, with
permission.)

face according to Fig. 17b. As pointed out by Boode and Walstra [105],
the conditions are complicated by recrystallization, crystal growth, flocculation of crystals, and surface modification.

The crystallization of oils in droplets is related to this phenomenon,
and crystallinity as such has a significant influence on the stability
[102,112]. The systematic clarification of the influence is complicated by
the pronounced supercooling found in oil-in-water emulsions [113]. The
crystallization may take place by surface or homogeneous nucleation [114]

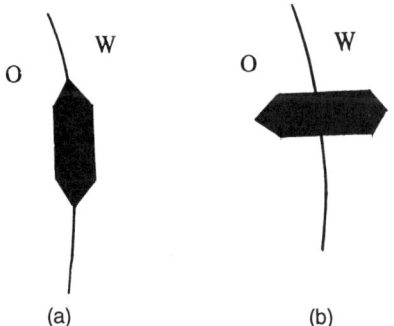

(a)   (b)

FIG. 17   The orientation of a solid crystal at the oil–water interface will influence
the coalescence rate in a most significant manner. (From Ref. 105.)

or by transfer between liquid and solid droplets [115]. A reasonable explanation is transfer of small crystalline aggregates resulting from the influence between the surfactant and the solid particle [105].

VII. EVAPORATION AND DILUTION OF EMULSIONS

A large number of emulsion applications involve evaporation (cosmetics,
pharmaceutics, and coatings), whereas in other cases (foods), dilution is
an essential factor.

The evaporation of the continuous phase from an emulsion shows the
expected fast vaporization of a liquid, but, more unexpectedly, the dispersed phase also evaporates rapidly. This is illustrated by evaporation
rates from emulsions in comparatively thick layers with convertion movement [116] (>0.5 cm) (Fig. 18). The evaporation curve is from a water–
decane emulsion stabilized by an isostearic acid–triethanolamine combination. The evaporation of water is similar to that of decane and the inversion from o/w to w/o in the range 70–80 vol% of the decane gave no
difference in evaporation rate. This fact has found an interesting applica-

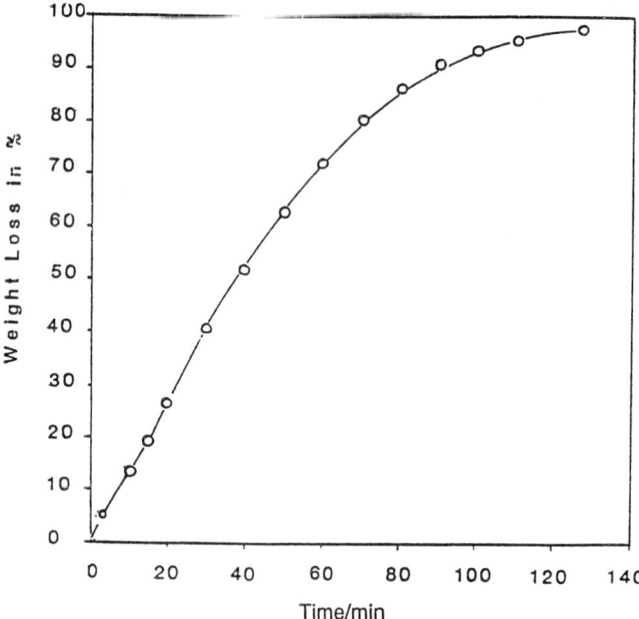

FIG. 18 The evaporation rate from an oil-in-water emulsion with stirring showed
no change at the phase inversion. (From Ref. 116.)

tion in the preparation of small drug particles from the evaporation of the drug solution dispersed as oil droplets in water [117]. The results demonstrated the evaporation from the emulsified droplets take place during the encounter with the emulsion surface.

The conditions are entirely different when the stagnant conditions in a thin layer of an emulsion are encountered. Now the concentration gradients are essential for the behavior. The results from such evaporation are at a first glance surprising, as demonstrated by the evaporation path in a typical phase diagram (Fig. 19). In quite a number of cases, water is considered the only substance to evaporate because the vapor pressure of typical oils is low, at the level of 0.25 mm Hg. In the example given, a "reasonable" (but mistaken) conclusion is an evaporation path according to line 1 in the diagram, resulting in a residue of an oil solution.

This is not the case. The coordinates on a diagram of this kind are in weight units and the weight fraction of oil (F_0) in the vapor becomes equal to

$$F_0 = \frac{P_0 M_0}{P_0 M_0 + 18 P_w}. \tag{39}$$

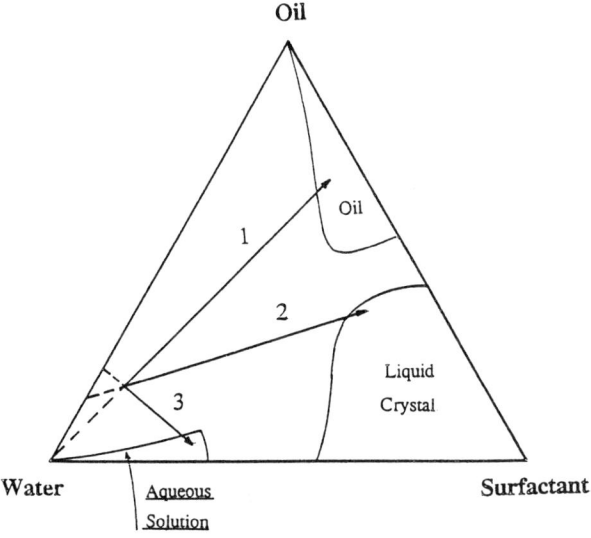

FIG. 19 The final structure after evaporation of an emulsion of 75% water, 20% oil, and 5% surfactant is sensitive to the oil vapor pressure line 1: 0 mm Hg; line 2: 2.5 mm Hg; line 3: 0.40 mm Hg.

With a typical molecular weight of 250 of the oil, the weight fraction of oil in the vapor becomes 0.15 and the residue in the example is not an oil but a liquid crystal (line 2, Fig. 19). With a modest vapor pressure of 0.4 mm Hg, the residue becomes an aqueous solution of the surfactant (line 3, Fig. 19). These evaluations are important for the behavior of fragrances in an emulsion system.

Because of the tendency to preferential evaporation, the oil, even for low vapor pressures, becomes even more pronounced at lower water contents as demonstrated for a microemulsion system [114,118]. The evaporation path bends toward the surfactant corner, causing the oil weight fraction in the vapor to increase from an original value of 0.27 to 0.76 for the last point measured (Fig. 20).

In addition, evaporation under stagnant conditions leads to a high concentration of phases with low water content toward the top of the layer. If this top layer constitutes a barrier to water or oil transport, the effect

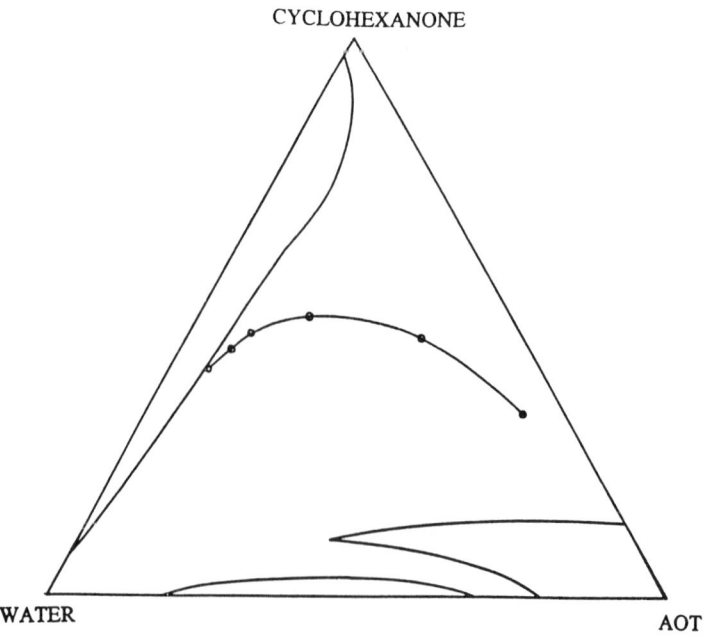

CYCLOHEXANONE

WATER   AOT

FIG. 20 The evaporation path in a microemulsion curves toward high surfactant concentrations for low water content. (From Ref. 118.)

may be very pronounced. As an example, Fig. 21 shows a strong reduction in evaporation for a certain composition. This change of rate is related to the first appearance of the lamellar liquid crystal. The liquid-crystalline layer formed on top of the film is an efficient barrier to evaporation because of the low diffusion coefficient of water and solvents through it [119]. It should be noted that due to the stagnant conditions, there is no necessity for the entire emulsion to be changed to a liquid crystal; the formation of a liquid crystal within the emulsion is sufficient. The emulsion changes to a liquid crystal when approximately 50% of the water is evaporated (Fig. 22), but the three-phase region appears much earlier (Fig. 23).

The opposite of evaporation, dilution, also presents stability problems, sometimes of an unexpected nature. As an example, a common flavor, vanillin, in a simple model food system of water, soybean oil, lecithin, and vanillin is chosen. The most common approach to such an emulsion is to consider it as a two-phase system with vanillin mostly in the oil phase with an oil–water partition coefficient of approximately 3. This approach

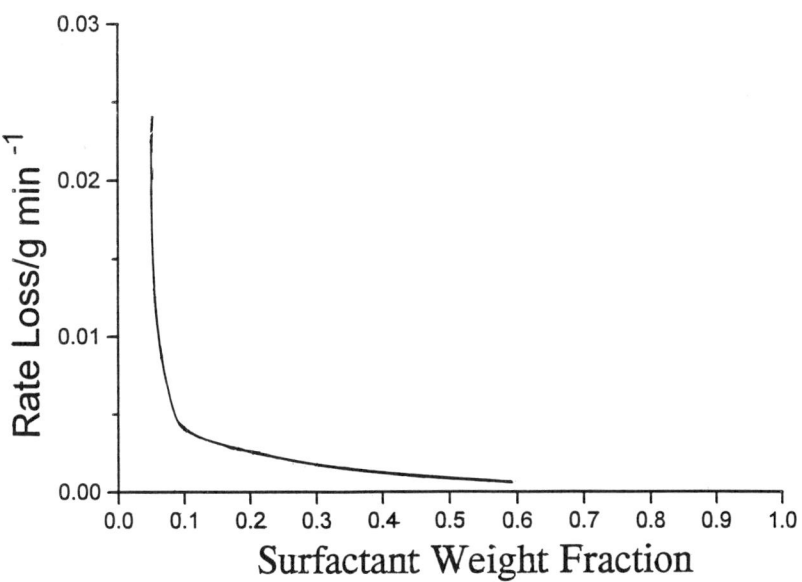

FIG. 21 The evaporation rate from a stagnant emulsion layer shows a strong reduction when a lamellar liquid crystal is formed at the surface.

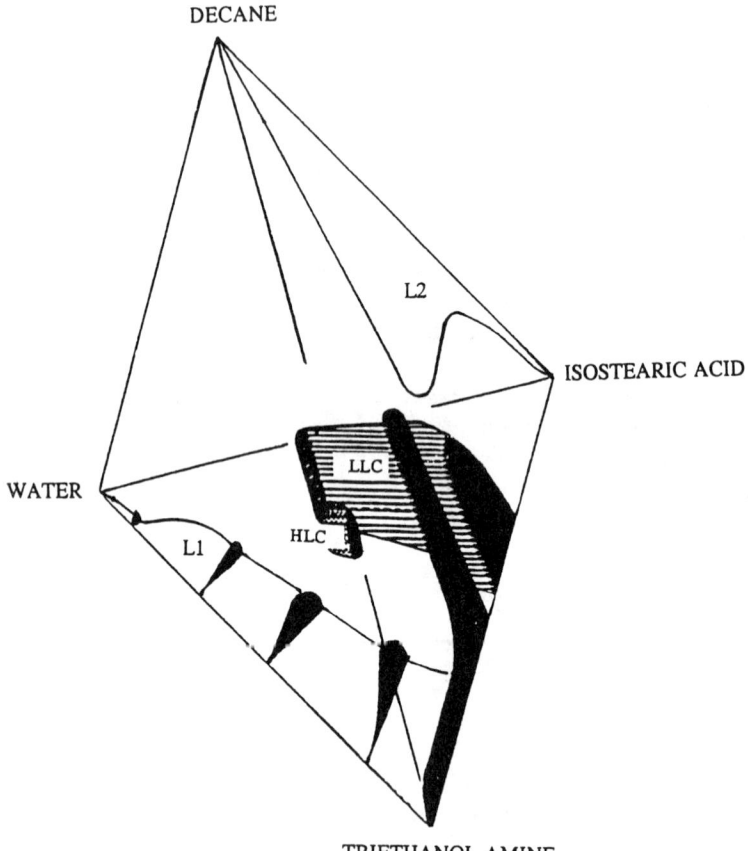

DECANE

L2

ISOSTEARIC ACID

WATER

LLC

L1

HLC

TRIETHANOL AMINE

FIG. 22 The phase diagram for emulsions in Fig. 20.

leads to an erroneous opinion about the true partition of vanillin. In fact, for an emulsion with typical water/oil/lecithin ratios, most of the vanillin is located in the aqueous part of the emulsion as demonstrated by the phase diagram (Fig. 24).

The maximum solubility of vanillin is determined by the four-phase space (solid lines, Fig. 25). The maximum solubility of vanillin is 1.0% by weight in water, 2.5% in the oil, and 8.8% in the liquid crystal that in addition contains 64% lecithin, 30% water, and 6% oil. With these values,

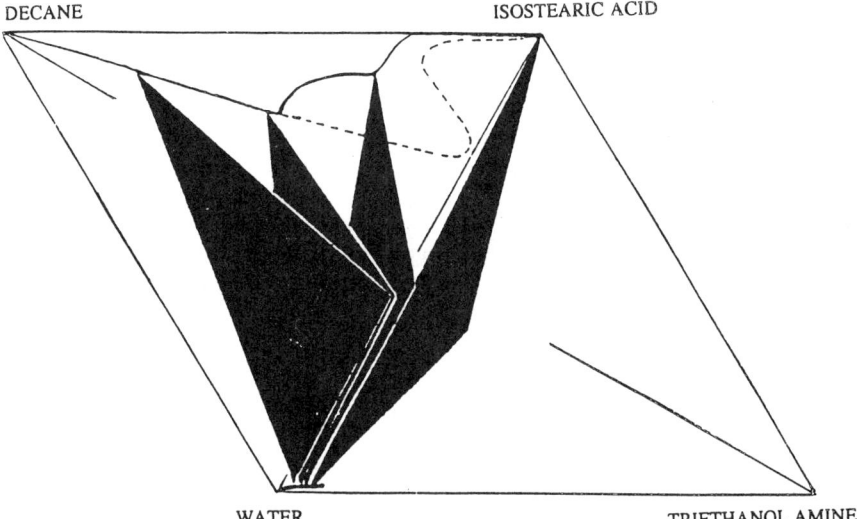

DECANE   ISOSTEARIC ACID

WATER   TRIETHANOL AMINE

FIG. 23   In the phase diagram (Fig. 22), the liquid crystal enters the three-phase region at low concentrations of surfactant.

the total solubility of vanillin may be written

$$S_{\text{tot}}^{\text{van}} = 0.010w + 0.025\sigma + 0.134l, \tag{40}$$

where w, σ, l are weight fractions of water, oil, and lecithin.

The high solubility in the lamellar liquid crystal should be noted because this phase is dispersed in the aqueous phase in the form of liposomes during emulsification. For a typical emulsion with 75% water, 21% oil, and 4% lecithin, the water–oil partition coefficient now is not 0.33 as expected but 1.6. In short, the presence of the lamellar liquid crystal has a most drastic effect on the location of the flavor substance.

The solubility of the vanillin has an effect on the dissolution of vanilla during the dilution of the emulsion in the mouth cavity. Starting the dilution with 50 wt% water and 9 wt% vanillin, the equations show the water constant brings the emulsion to the phase of maximum solubility (Fig. 26).

The equation for the vanillin content during dilution becomes

$$S_{\text{dil}}^{\text{van}} = -0.18w + 0.18. \tag{41}$$

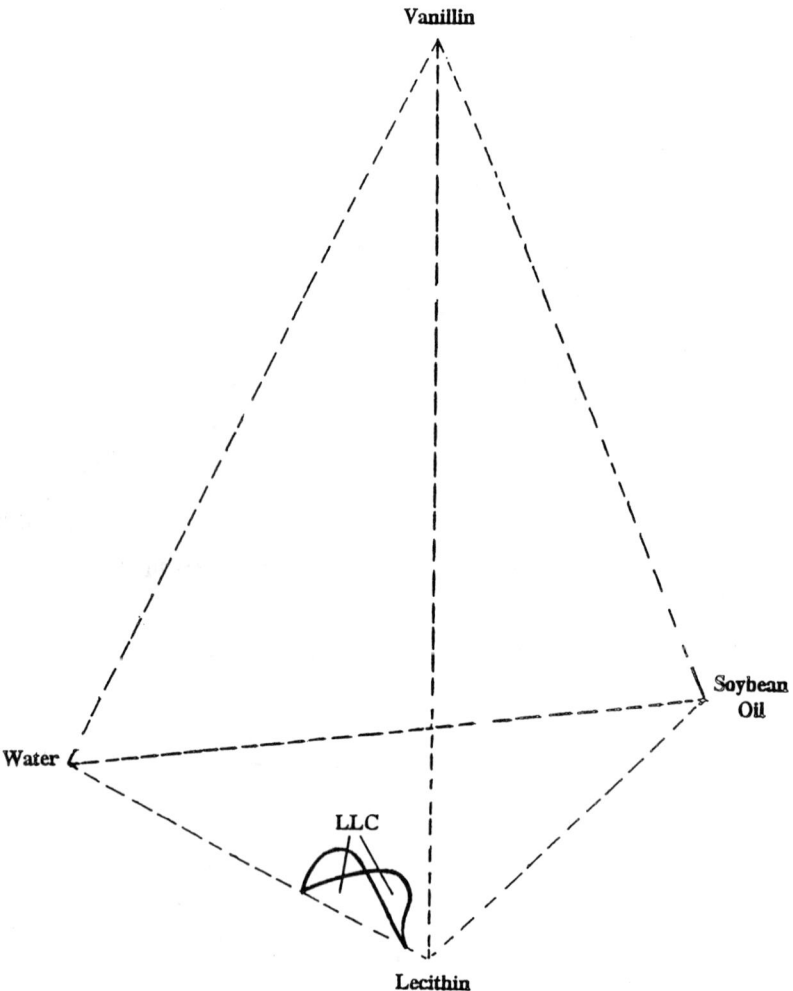

FIG. 24 The phases in the water, lecithin, soybean oil, and vanillin system show four phases of interest: water with dissolved vanillin, oil with dissolved vanillin, and a lamellar liquid crystal with dissolved vanillin, and solid vanillin.

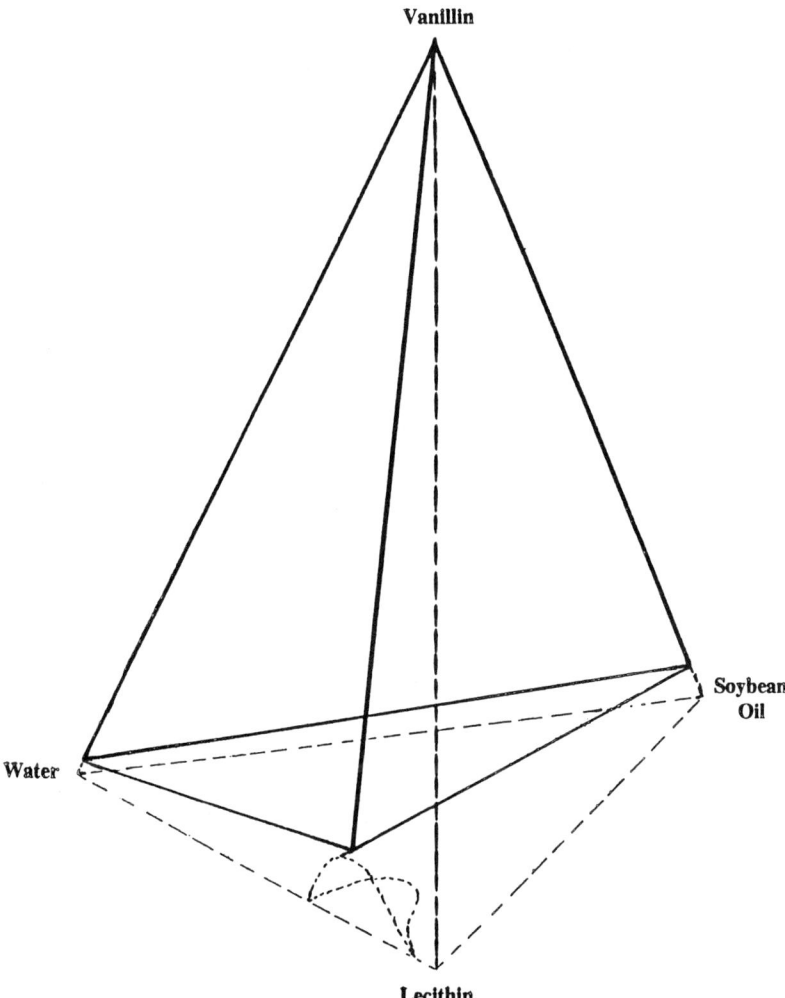

FIG. 25 With the vanillin exceeding its solubility, four phases are present. The solid vanillin is in equilibrium with an aqueous solution, an oil solution, and a lamellar liquid crystal, all with maximally dissolved vanillin.

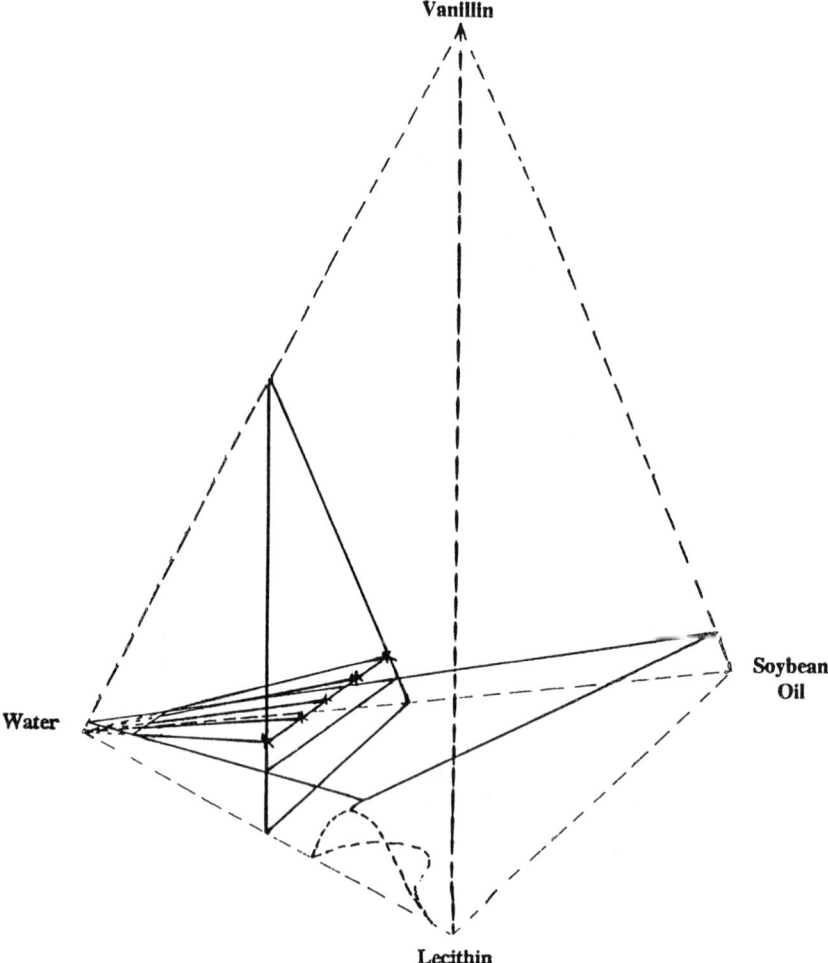

FIG. 26 Dilution with water leads to dissolution of vanillin.

Furthermore, if the weight ratio of oil to lecithin is R, the water content at which the vanillin dissolves becomes

$$W = \frac{R + 0.563}{1.06R + 0.626}. \tag{42}$$

Excellent agreement between experimental and calculated values was found [120].

REFERENCES

1. International Union of Pure and Applied Chemistry, *Manual on Colloid and Surface Science*, Butterworths, London, 1972.
2. K. Madany and S. E. Friberg, Prog. Colloid Polym. Sci. *65*:164 (1978).
3. L. Ghosh and C. A. Miller, J. Phys. Chem. *91*:4528 (1987).
4. H. M. Princen, J. Colloid Interface Sci. *71*:55 (1979).
5. A. Vrij, J. Colloid Sci. *19*:1 (1964).
6. A. Vrij and J. Th. G. Overbeck, J. Am. Chem. Soc. *90*:3074 (1964).
7. B. R. Radoev, A. D. Scheludko, and E. D. Manav, J. Colloid Interface Sci. *95*:254 (1983).
8. M. P. Aronson and H. M. Princen, Colloids Surf. *4*:173 (1982).
9. H. M. Princen, J. Phys. Chem. *72*:3342 (1968).
10. S. R. Reddy and H. S. Fogler, J. Phys. Chem. *84*:1570 (1980).
11. S. R. Reddy and H. S. Fogler, J. Colloid Interface Sci. *79*:105 (1981).
12. D. D. Huang, A. D. Nikolov, and D. T. Wasan, Langmuir *2*:672 (1986).
13. I. B. Ivanov and D. J. Dimitrov, Colloid Polym. Sci. *252*:982 (1974).
14. R. K. Jain and E. R. Ruckenstein, J. Colloid Interface Sci. *54*:1 (1976).
15. Z. Zapryanov, A. K. Malheha, N. Aderangi, and D. T. Wasan, Int. J. Multiphase Flow *9*:105 (1983).
16. I. B. Ivanov, D. S. Dimitrov, P. Somasundaran, and R. K. Jain, Chem. Eng. Sci. *40*:137 (1985).
17. A. K. Mahlhatra and D. T. Wasan, AIChE. J. *33*:1533 (1987).
18. S. Hartland and S. A. K. Jeelani, Colloids Surf. *88*:289 (1994).
19. J. Chen, E. Dickinson, and G. Iveson, Food Structure *12*:135 (1993).
20. E. Dickinson, D. J. McClements, and M. J. W. Povey, J. Colloid Interface Sci. *142*:103 (1991).
21. Y. Cao, E. Dickinson, and D. J. Wedlock, Food Hydrocolloids *5*:443 (1991).
22. E. Dickinson, J. Chem. Soc. Faraday Trans. *88*:2973 (1992).
23. E. Dickson, M. I. Goller, and D. J. Wedlock, Colloids Surf. A *75*:195 (1993).
24. E. Dickinson, J. Ma, and M. J. W. Powey, Food Hydrocolloids *8*:481 (1994).
25. E. Dickinson, R. K. Owusu, and A. Williams, J. Chem. Soc. Faraday Trans. *89*:865 (1993).
26. Y. Cao, E. Dickinson, and D. J. Wedyork, Food Hydrocolloids *4*:185 (1990).
27. B. Vincent, J. Edwards, and R. Croot, Colloids Surf. *31*:267 (1988).
28. H. N. W. Lekkerkerker and A. Strobants, Physica A *195*:387 (1993).
29. V. J. Pinfield, E. Dickinson, and M. J. W. Poney, J. Colloid Interface Sci. *166*:363 (1994).
30. E. Barneo and J. Mizrahi, Chem. Eng. J. *5*:171 (1973).
31. G. K. Batchelor, J. Fluid Mech. *131*:155 (1983).
32. G. K. Batchelor, J. Fluid Mech. *74*:1 (1976).
33. R. J. Ruick, J. Appl. Phys. *18*:983 (1947).
34. E. Manev, S. V. Sazdanova, and D. T. Wasan, J. Disper. Sci. Technol. *3*:435 (1982).
35. D. T. Wasan, Chem. Eng. Ed. *26*:104 (1992).
36. A. D. Nikolov, D. T. Wasan, N. D. Durkov, P. A. Kralchevsky, and I. B.

Ivanov, in *Ordering and Organization in Ionic Solutions* (N. Ike and I. Soganic, eds.), World Scientific, Singapore, 1988, p. 26.
37. A. D. Nikolov, P. A. Kralchevsky, I. B. Ivanov, and D. T. Wasan, J. Colloid Interf. Sci. *133*:13 (1989).
38. A. D. Nikolov and D. T. Wasan, Langmuir *8*:2985 (1992).
39. D. T. Wasan, K. Koczo, and A. D. Nikolov, in *Foams: Fundamentals and Applications in the Petroleum Industry* (L. L. Schramm, ed.), ACS Symposium Series 242, American Chemical Society, Washington, DC, 1994, p. 47.
40. J. G. Kirkwood, E. D. Maun, and B. J. Alder, J. Chem. Phys. *18*:1040 (1950).
41. X. L. Chu, A. D. Nikolov, and D. T. Wasan, Langmuir *10*:4403 (1994).
42. J. P. Valleau and S. G. Wittington, in *Modern Theoretical Chemistry* (B. J. Berne, ed.), Plenum Press, New York, 1977, p. 128.
43. M. S. Wertheim, L. Blum, and D. Beatko, in *Micellar Solutions and Microemulsions* (S. H. Chen and R. Rajagapalon, eds.), Springer-Verlag, New York, 1990, p. 73.
44. B. V. Derjaguin, J. Colloid Interf. Sci. *56*:492 (1976).
45. P. Kékicheff and P. Richetti, Prog. Colloid Polym. Sci. *88*:8 (1992).
46. A. D. Nikolov and D. T. Wasan, J. Colloid Interf. Sci. *133*:1 (1989).
47. J. N. Israelachivili, J. Mitchell, and B. W. Ninham, J. Chem. Soc. Faraday Trans. II *76*:1525 (1976).
48. M. J. Groves, M. Wineberg, and A. P. R. Brain, J. Disper. Sci. Technol. *6*:237 (1985).
49. M. Van den Tempel, Rec. Trav. Chim. *72*:419, 433 (1953).
50. R. P. Borwankar, L. A. Lobo, and D. T. Wasan, Colloids Surf. *69*:135 (1992).
51. A. K. Malhotra and D. T. Wasan, Chem. Eng. Commun. *55*:95 (1987).
52. P. Becher and M. J. McCann, Langmuir *7*:1325 (1991).
53. E. Dickinson and A. Williams, Colloids Surf. *88*:317 (1994).
54. J. T. Davies, *Turbulence Phenomena*, Academic Press, New York, 1972, Chaps. 8 and 10.
55. E. J. Clayfield and E. C. Lumb, Macromolecules *1*:133 (1968).
56. D. H. Napper, J. Colloid Interface Sci. *29*:168 (1969).
57. D. H. Napper, *Polymeric Stabilization of Colloidal Dispersions*, Academic Press, London, 1983.
58. ThF. Tadros and B. Vincent, J. Colloid Interface Sci. *72*:505 (1979).
59. B. Vincent, Adv. Colloid Interface Sci. *4*:193 (1974).
60. Th.F. Tadros (ed.) *The Effect of Polymers on Dispersion Properties*, Academic Press, London, 1982.
61. S. Marti, J. Nervo, and G. Riess, Prog. Colloid Polym. Sci. *58*:114 (1975).
62. P. Marie, Y. Herrenschmidt, and Y. Gallot, Macromol. Chem. *177*:2773 (1976).
63. R. Reeb and G. Riess, C.R. Acad. Sci. Paris Ser. C *283*:663 (1976).
64. Huynh-Ba-Gia, R. Jérôme, and Ph. Teyssié, J. Polym. Sci. Chem. Ed. *18*:3483 (1980).

65. Huynh-Ba-Gia, R. Jérôme, and Ph. Teyssie, J. Appl. Polym. Sci. 26:343 (1981).
66. R. Y. Lochhead, J. A. Davidson, and G. M. Thomas, Adv. Chem. Ser. 223 (1989), Chap. 7.
67. R. Y. Lochhead, Cosmetics Toiletries 109:93 (1994).
68. B. Vincent and S. Whittington, Colloid and Surface Science, Vol. 12. (E. Matijević, ed.), Plenum Press, New York, 1982, p. 1.
69. R. Y. Lochhead and C. J. Rulison, Colloids Surf. 88:27 (1994).
70. J. Sjöblom, L. Mingyuan, H. Hoiland, and E. J. Johansen, Colloids Surf. 46:127 (1990).
71. J. Sjöblom, H. Soderlund, S. Lindblad, E. J. Johansen, and I. M. Skjärvö, Colloid Polym. Sci. 268:389 (1990).
72. E. J. Johansen, I. M. Skjärvö, T. Lund, J. Sjöblom, H. Soderlund, and G. Boström, Colloid Surf. 34:353 (1989).
73. B. Gestblom, H. Fordedal, and J. Sjöblom, J. Disper. Sci. Technol. 15:449 (1994).
74. J. Sjöblom, O. Urdahl, K. Grete, K. Nordli Borve, L. Mingyuan, J. O. Saeten, A. A. Christy, and T. Gu., Adv. Colloid Interface Sci. 41:241 (1992).
75. K. Nordli Borve, H. Ebeltoft, and J. Sjöblom, Kjemi 8:6 (1993).
76. H. Ebeltoft, K. Nordli Borve, J. Sjöblom, and P. Stenius, Prog. Colloid Polym. Sci. 88:131 (1992).
77. O. Urdahl, A. E. Movik, and J. Sjöblom, Colloids Surf. 74:293 (1993).
78. J. Sjöblom, L. Mingyuan, A. A. Christy, and T. Gu, Colloids Surf. 66:55 (1992).
79. B. Balinov, O. Urdahl, O. Soderman, and J. Sjöblom, Colloids Surf. 82: 173 (1994).
80. J. Sjöblom, O. Urdahl, H. Hoiland, A. A. Christy, and E. J. Johansen, Prog. Colloid Polym. Sci. 82:131, 1990.
81. K. Nordli Borve, J. Sjöblom, J. Kizling, and P. Stenius, Colloids Surf. 57: 83 (1991).
82. W. C. Griffin, J. Soc. Cosmetic Chem. 1:311 (1949).
83. J. T. Davies and E. K. Rideal, Interfacial Phenomena, Academic Press, New York, 1961.
84. P. Becher, J. Colloid Interface Sci. 18:665 (1963).
85. P. Becher, in Surfactants in Solution, Vol. 3 (K. L. Mittal, ed.), Plenus Press, New York, 1984, p. 1925.
86. J. T. Davies Proc. Int. Congr. Surface Activity, 2nd ed., Vol. 1, Butterworths, London, 1959, p. 426.
87. L. Marszall, Fette Seifen Anstrichmittel 79:41 (1977).
88. L. Marszall, J. Colloid Interface Sci. 59:376 (1977).
89. H. Lange, J. Soc. Cosmetic Chem. 16:697 (1965).
90. L. Richter and D. Vollhardt, Colloids Surf. 95:113 (1995).
91. K. Shinoda and H. Arai, J. Phys. Chem. 68:3485 (1964).
92. K. Shinoda and H. J. Saito, J. Colloid Interface Sci. 30:258 (1969).
93. K. Shinoda, H. Saito, and H. Arai, J. Colloid Interface Sci. 35:624, (1971).
94. K. Shinoda and H. Sagitani, J. Colloid Interface Sci. 64:68 (1978).

95. L. M. Skinner and J. R. Sambles, J. Aerosol Sci. *3*:199 (1972).
96. I. M. Lifshitz and V. V. Slösob, Zh. Exp. Teor. Fiz. *35*:479 (1958).
97. R. Buscall, S. S. Davis, and D. S. Potts, Colloid Polym. Sci. *257*:636 (1979).
98. S. S. Davis, H. P. Round, and T. S. Turewal, J. Colloid Interface Sci. *80*: 508 (1980).
99. A. S. Kabol'nov, A. V. Perzov, and E. D. Shchukin, Colloids Surf. *24*:19 (1987).
100. A. S. Kabol'nov, K. N. Makarov, A. V. Pertzov, and E. D. Shchukin, J. Colloid Interface Sci. *138*:98 (1990).
101. S. E. Friberg, in *Food Emulsions* (S. E. Friberg, ed.), Marcel Dekker, Inc., New York, 1976, p. 1.
102. M. A. J. S. van Bockel and P. Walstra, Colloids Surf. *3*:99 (1981).
103. M. van den Tempel, J. Colloid Sci. *16*:284 (1961).
104. K. Boode, C. Bisperink, and P. Walsstra, Colloids Surf. *61*:55 (1991).
105. K. Boode and P. Walstra, Colloids Surf. *81*:121 (1993).
106. K. Boode, P. Walstra, and A. E. A. de Groot-Mostert, Colloids Surf. *81*: 139 (1993).
107. E. H. Lucassen-Reynders and M. Van den Tempel, J. Phys. Chem. *67*:731 (1963).
108. A. G. Gaonkar and R. P. Boowankar, Colloids Surf. *59*:331 (1991).
109. H. Oortwijn and P. Walstra, Neth. Milk Dairy J. *33*:134 (1979).
110. J. A. de Feijter, J. Benjamins, and M. Tamhoer, Colloids Surf. *27*:243 (1987).
111. F. Lonza, German Patent 191238 (1905).
112. D. F. Darling, J. Dairy Res. *49*:695 (1982).
113. E. Dickinson, M. I. Goller, D. J. McClements, S. Peasgood, and M. J. W. Povey, J. Chem. Soc. Faraday Trans. *86*:1147 (1990).
114. S. E. Friberg, T. Young, R. Mackay, J. Oliver, and M. Breton, Colloids Surf. *100*:83 (1995).
115. D. J. McClements, E. Dickinson, and M. J. W. Povey, Chem. Phys. Lett. *172*:449 (1990).
116. S. E. Friberg and B. Langlois, J. Disper. Sci. Technol. *13*:2 (1992).
117. B. Sjöström, K. Westesen, and B. Bergenståhl, Int. J. Pharm. *94*:89 (1993).
118. S. E. Friberg, B. Yu, J. Lin, E. Barni, and T. Young, Colloids Polym. Sci. *271*:152 (1993).
119. S. E. Friberg and I. Kayali, J. Pharm. Sci. *78*:639 (1989).
120. I. Kayali, C. Heisig, and S. E. Friberg, in *Physical Chemistry of Flavors* (C. T. Ho, C. T. Tan, and C. H. Tong, eds.), ACS Symp. Series (in press).

2
Kinetics of Brownian and Gravitational Coagulation in Dilute Emulsions

STANISLAV DUKHIN   Department of Theoretical Electrochemistry of Membranes and Colloids, Institute of Colloid and Water Chemistry, Ukrainian Academy of Science, Kiev, Ukraine

JOHAN SJÖBLOM   Department of Chemistry, University of Bergen, Bergen, Norway

I. GENERAL

A. Stability Mechanisms

In emulsions, the water droplets in oil or oil droplets in water, are large (of the order of microns). Because of the large surface area per drop, the excess Gibbs energy per drop is high and cannot be compensated for by entropy contributions as in microemulsions.

The stability of a disperse system is characterized by a constant behavior in time of its basic parameters, namely, the dispersity and the uniform distribution of the dispersed phase in the medium. The problem of stability is one that is most important and complicated in colloid chemistry.

Notwithstanding their thermodynamic instability, many emulsions are kinetically stable and do not change appreciably for a prolonged period (sometimes for decades). These systems exist in the metastable state, that is, the potential barrier preventing aggregation of the particles is sufficiently high.

To understand the reasons for the relative stability of such systems, it is necessary to first determine the stability [1] and mechanisms of destabilization. *Sedimentation stability* distinguishes the stability of the disperse phase with respect to the force of gravity. Phase separation due to sedimentation is a typical phenomenon for droplets in coarsely dispersed emulsions resulting in settling (or floating) of the drops. Highly dispersed emulsions are *kinetically* stable. They are characterized by diffusion–sedimentation equilibrium. Isothermal distillation of fine drops in coarser ones causes subsequent sedimentation. The loss of *aggregative* stability is due to a combination of drops.

Such forces of molecular attraction may result in the formation of a continuously structurized system with *phase* stability. *Coagulation or flocculation* is a process of particle cohesion, the formation of larger aggregates with a loss of sedimentation and phase stability and the subsequent phase separation, that is, a destruction of the emulsion. Hence, *aggregative stability* can be defined as the ability of emulsion to retain the dispersity and individuality of drops. In aggregates, not withstanding the change of their mobility, the drops still remain as such for a certain time (the "lifetime") after which they can merge spontaneously with diminishing phase interface. The merging of droplets is called *coalescence*.

The aggregative *stability* or *instability* is most characteristic for colloid systems. This chapter is devoted to this type of instability.

Two types of the aggregative stability can be discriminated: a very low coagulation rate or an equilibrium in the aggregation and disaggregation processes (reversible coagulation).

The theoretical and experimental investigations of flocculation are focused on monodisperse systems because it simplifies to quantify the flocculation kinetics. Polydisperse emulsions are not usually chosen as a model system for the flocculation kinetics. Another reason for not choosing emulsions is the deformation of droplet surfaces upon interaction, which further complicates a quantitative treatment. In addition, the mentioned coupling of flocculation and coalescence creates difficulties in the modeling of the flocculation kinetics.

B. Hydrodynamics of Flocculation. Main Notions

When two particles approach each other, several types of interaction patterns may arise, affecting the flocculation process. There are two different but related ways in which colloid interactions influence flocculation. First, they have a direct effect on the *collision efficiency*, which is the probability that a pair of colliding particles will form a permanent aggregate. A strong long-range repulsion between the particles will reduce the chance of aggregate formation, and flocculation will occur very slowly, if at all. The other aspect of colloid interactions is their effect on the *strength* of aggregates, which is much less well understood but of great practical importance [2].

Flocculation occurs only if particles collide with each other and adhere when brought together by collision (i.e., the particles have low colloidal stability). To a large extent, these two processes, which could be termed *transport* and *attachment* steps, may be regarded as independent and can be treated separately.

Practically all colloidal interactions are of a short range, almost never extending over distances greater than the size of the particles. Hence, they have little influence over the transport of particles, although they are crucial in determining the collision efficiency. To a large extent, this justifies the treatment of transport and attachment as separate steps. There is an important type of interaction to which this conclusion does not apply, the so-called *viscous* or *hydrodynamic* interaction, which arises during the approach of particles in a viscous fluid. The effect reduces the rate of approach and gives a lower collision efficiency.

The main qualitative difference in flocculation kinetics is caused by the sign of the surface forces. Naturally the attractive forces enhance the flocculation and repulsive forces retard it. This difference is reflected (expressed) in the notions of rapid and slow flocculation. If attractive forces predominate, the flocculation is called "rapid," whereas "slow" coagulation relates to the systems in which the repulsive interaction between particles predominates.

In addition, the surface force dependence on the interparticle distance is important. As is known, the van der Waals attractive forces can predominate at small and large distances, whereas the repulsive forces cause the electrostatic barrier at intermediate distances. The flocculation/coagulation in the primary minimum is slow because it is retarded by the electrostatic barrier. It is not valid for the flocculation/coagulation in secondary minimum, which is rapid.

The formation of the drop doublet is the basic process in the flocculation kinetics. In a polydisperse emulsion, a statistical ensemble of the pairs of the drops with the radius a_1 and a_2 is considered [3,4]. One of the drops is chosen as the reference sphere, and the flux of all drops with another radius to the surface of the reference drop is considered. Thus, the flux represents the result of the averaging over all drop pairs of the statistical ensemble. The flux determines the loss of the particle drop number during the flocculation and the rate of the doublet formation.

The drop loss rates can be computed from the distribution of drops around the reference sphere. In the case of coagulation, the steady-state capture rate is given by the net inward flux of drops through an arbitrary surface enclosing the test sphere. The net flux depends on the drop distribution as well as the relative sphere velocities, which are known functions of the relative drop positions. Once the drop loss rates are known, the overall stability of an emulsion can be determined by solving the governing population balance equations which incorporate these drop loss rates. The solution of this system of coupled nonlinear partial differential equations gives the drop size distribution as a function of time and position.

C. Importance of Flocculation Kinetics

Tadros and Vincent [5] classified emulsion-breakdown processes. The transport processes (Brownian diffusion and differential settling) can manifest themselves in any kind of breakdown processes. For instance, it is clear that the first step in Ostwald ripening [6] is due to the molecular diffusion. However, the mutual approach of the droplets caused by the differential settling or Brownian movement influences on the molecular diffusion and kinetics of the Ostwald ripening. Thus, one concludes that the role of the transport processes can be understood more widely than in this chapter.

In the flocculation case, two final states have to be distinguished: creaming and emulsion gelling [7,8], which occupies the emulsion volume as a whole. The larger the difference in liquid densities, the larger the droplet dimensions and the lower the volume part, which is favorable for creaming.

The quantifying of the conditions of the two final states in flocculated emulsions is, in principle, possible on the basis of the flocculation kinetic theory.

Flocculation kinetics is important because it is accompanied by the change of emulsion properties and finally leads to the creaming and phase separation. The coalescence process is usually follows the flocculation; that is, the *coalescence kinetics are coupled with flocculation kinetics*. Thus, the flocculated state and flocculation kinetics have independent significance [9–11].

D. Scope of the Chapter

In monographs and reviews, much more attention is paid to surface forces [12,13] and stable colloids than to flocculation kinetics. Exceptions are Refs. 14 and 15 and literature on aerosols [16,17] and flotation [18–20].

In emulsion science, the main attention is focused on interparticle forces or colloid interactions and, correspondingly, to the emulsion stability or instability, in terms of deemulsification. The preceding chapter treats these problems. Here, we turn to the question of transport mechanism, the coagulation kinetics, and to a lesser degree to the floc properties.

Now, due to the success of the investigations of flocculation in disperse model systems, the prerequisites are created for the investigation of flocculation in emulsions. In this review, we try to couple the studies of coagulation in disperse systems with flocculation in emulsions.

In a manner similar to Tadros and Vincent [5] and Melik and Fogler [15], we are concerned solely with "dilute" emulsions and suspensions in which the particle behavior is dominated by two-body interactions. Consequently, for emulsions, the present analysis is valid only for conditions predominated by singlets and doublets. However, the true application of this chapter is with emulsion systems which are characterized by coalescence times much faster than the corresponding flocculation times. The present analysis remains rigorous for predicting the flocculation and creaming behavior of a new larger particle, which is the result of a coalescence process, in relation to the remaining particles in the system. At faster coalescence times, an emulsion consists of the singlet and temporary doublet. At faster flocculation times, the emulsion consists of flocs. The structure of these flocs will not be considered in this chapter.

Dilute emulsions play an important role in, for instance, oil-dewatering processes [21]. At the high initial volume fraction, the dewatering rate is high too. The smaller the volume fraction, the lower the dewatering rate. Thus, the intensification of the oil-dewatering technology is connected with the intensification of the flocculation kinetics. The same is true with respect to water purification from oil in environmental protection [22]. These two examples prove the importance of the flocculation process in dilute emulsions.

Tadros and Vincent [5] classified the wide range of emulsion systems. The classification subdivides the different types into (1) the nature of the "stabilizing moieties" and (2) the basic "structure" of the system. In both cases, the list represents a hierarchy of increasing complexity. In this hierarchy, the simplest systems will be considered here, which is justified by the limited knowledge of emulsion flocculation. The systems with steric stabilization [23,24], polymer bridging [25], and depletion stabilization [26] are beyond the scope of this chapter. This does not mean that the Brownian and gravitational coagulations described earlier are not important for the more complicated emulsion systems.

A general mathematical formulation of the problem of evolution of liquid drops spectrum is given in Ref. 15. A deductive method is used for the presentation. Equations of doublet formation kinetics and population balance equations (BBE) are written in the most general form. Then the stability problem is being considered as a purely mathematical one. In principle, such an approach is very efficient if reliable information is available about surface forces, kinetic coefficients, and subprocesses affecting the coagulation kinetics. Unfortunately, available information about surface forces in emulsions is incomplete and often inaccurate with a clear discrepancy between theory and experiment.

Along with a general approach characterized in Refs. 15, 27, and 28, it is very important at the present stage to discriminate between facts established reliably and the links in the complex process of coagulation in emulsions which need a systematic investigation. As applied to definite systems and conditions, the general model of coagulation kinetics in emulsions described in Ref. 15 can be used at present; however, without guarantee of the required accuracy. Generally, the use of potentially very wide possibilities of the general coagulation theory can lead to unreliable results because of incomplete information about surface forces and subprocesses. There is a possibility to provide efficiency of the general theory with regard to a wider range of problems and systems which can result in a qualitatively new level of the emulsions science. The objective of this review is to provide a motion in this direction. This has also determined the accepted inductive method of presentation from a simple to a complicated matter, from reliably established facts to those insufficiently studied.

During the few last years, the influence of the droplet surface mobility on the collision efficiency and the droplet deformation on the colloidal interaction in emulsions is quantified [29]. This also stimulated the preparation of this chapter.

II. SURFACE FORCES

Surface forces are described in all monographs and reviews concerning emulsion stability [5,15,30–32]. In the following section we deal with some general notions and some new results.

A. Van der Waals Interaction

1. Macroscopic Approach

The universal attractive forces between atoms and molecules, known as van der Waals forces, also operate between macroscopic objects and play a very important role in the interaction of colloidal particles. Indeed, without these forces, aggregation of particles would usually be prevented by the hydrodynamic interaction.

The interaction between macroscopic bodies arises from spontaneous electric and magnetic polarizations, giving a fluctuating electromagnetic field within the media and in the gap between them. In order to calculate the force, the variation in electromagnetic wave energy with the separation distance has to be determined. Lifshitz [12,30] derived an expression for the force between two semi-infinite media separated by a plane-parallel gap. His treatment was later extended by Dzyaloshinskii et al. [12,30] to deal with the case of two bodies separated by a third medium. In principle, these methods should enable the interaction between systems of interest to be calculated, and direct measurements of van der Waals forces between mica sheets [40] have confirmed the essential correctness of the

Lifshitz or *macroscopic* approach. However, proper application of this approach requires detailed knowledge of the dielectric responses of the interacting media over a very wide frequency range.

2. Hamaker Expressions

Because of difficulties in applying the macroscopic theory, an older approach, due mainly to Hamaker [33] is still widely used. This is based on the assumption of pairwise additivity of intermolecular forces. The interaction between two particles is calculated simply by summing the interactions of all molecules in one particle with all of the molecules in the other particle. Hamaker replaced the summation by a double integration procedure, which leads to very simple expressions, especially when the separation distance is small. For two spheres, radii a_1 and a_2, separated by a distance h, the interaction energy at close approach ($h \ll a$) is given by

$$V_A = -\frac{A_{12}}{6h}\frac{a_1 a_2}{a_1 + a_2},\tag{1}$$

where V_A is the attraction energy between the two spheres and A_{12} is the *Hamaker constant* for media 1 and 2, of which the spheres are composed.

The results given above apply to the interaction of media across a vacuum.

A useful approximation for Hamaker constants of different media is the geometric mean assumption

$$A_{12} \cong (A_{11}A_{22})^{1/2}.\tag{2}$$

For similar materials, medium 1 interacting across medium 3,

$$A_{131} \cong (A_{11}^{1/2} - A_{33}^{1/2})^2.\tag{3}$$

Equation (3) led Hamaker to the conclusion that the van der Waals interaction between similar materials in a liquid would always be attractive (positive Hamaker constant) whatever the values of A_{11} and A_{33}.

Oil phases are characterized by fairly low dielectric constants ranging between 2 and 5, which makes the pure entropic term fairly constant. The nonretarded dispersion contribution is the major cause of differences in Hamaker constants between different emulsions [34]. The Hamaker constant can be expected to vary between 3×10^{-21} and 10×10^{-21} J in most food emulsions [32].

3. Retardation

Because dispersion forces are electromagnetic in character, they are subject to a *retardation* effect. The finite time of propagation causes a reduced correlation between oscillations in the interacting bodies and a smaller interaction. Pailtorpe and Russel [35] considered the effect of retardation and found that the total Hamaker constant decreases by approximately 70% for $a = 0.1$ μm and 87% for $a = 0.25$ μm over the range $2.001 < S - 2 < 2.5$. Melik and Fogler [15] conclude that the equation proposed in Ref. 36 cannot be considered a reasonable quantitative approximation to the retarded interparticle potential. However, in light of the tedious calculations required to account rigorously for any spatial variation of the Hamaker constant, they used this equation in their calculations of flocculation in secondary minimum. Its distance to the surface can be equal to 5-7 Debye radius, that is, 10-100 nm (Sec. V.A.1). According to Lyklema [37], the van der Waals energy dependence on distance can always be expressed through the Hamaker function $A(h)$. At a small separation, $A(h)$ becomes independent of h and identical to the *Hamaker constant*. At large h,

$$A(h) \sim h^{-1}. \tag{4}$$

However, the systematic investigations of Rabinovich and Churaev [38–40] led them to the conclusion that for the systems including water and a dielectric, the asymptotic equation [Eq. (4)] is not valid at any distance. In Ref. 40, they carried out a numerical calculation of $A(h)$ for 36 systems.

B. Electrical Interaction

1. Electrical Double Layer

Most particles in aqueous media are charged due to various reasons, such as the ionization of surface groups, specific adsorption of ions, and so forth. In an electrolyte solution, the distribution of ions around a charged particle is not uniform and gives rise to an *electrical double layer* (DL). This topic has been the subject of several reviews [34,41,42] and details are not given here. The essential point is that the charge on a particle surface is balanced by an equivalent number of oppositely charged *counterions* in solution. These counterions are subject to two opposing effects: electrostatic attraction tending to localize the counterions close to the particles and the tendency of ions to diffuse randomly throughout the solution due to their thermal energy. The surface charge on a particle and the associated counterion charge together constitute the electrical double layer. A widely accepted model for the double layer is that of Stern, later modified by Graham (see Hunter [41]), in which part of the counterion charge is located close to the particle surface (the so-called *Stern layer*) and the remainder is distributed more broadly in the *diffuse layer*.

The interaction between charged particles is governed predominantly by the overlap of diffuse layers, so the potential most relevant to the interaction is that at the boundary between the Stern and diffuse layers (the Stern potential, Ψ_δ), rather than the potential at the particle surface. This boundary (the *Stern plane*) is generally considered to be at a distance of about 0.3–0.5 nm from the particle surface, corresponding to the diameter of a hydrated counterion.

There is no direct experimental method for determining the Stern potential. The two major influences on electrical interaction between particles are the magnitude of the effective "surface potential" (generally assumed to be Ψ_δ or ζ) and the extent of the diffuse layer because the latter governs the range of the interactions. Surface potentials can be modified in two distinct ways. If the ionic strength is raised, then a greater proportion of the potential drop occurs across the Stern layer, giving a smaller Stern potential. This effect can be produced by adding salt, and those which act in this way only are known as *indifferent electrolytes*. A more dramatic effect can be produced by the addition of salts with *specifically adsorbing counterions*. These adsorb on the particles because of some specific, non-electrostatic affinity and can be regarded as located in the Stern layer. In many cases, such ions can adsorb to such an extent that they reverse the sign of the Stern potential.

The extent of the diffuse layer is also dependent on the ionic strength and is best seen through the variation of the potential as a function of distance from the Stern plane, for which the Poisson–Boltzmann approach is most commonly employed [41,42]. For fairly low potentials the linear form of the Poisson–Boltzmann expression is appropriate and a very simple result is obtained:

$$\Psi = \Psi_\delta \exp(-\kappa x), \tag{5}$$

where Ψ is the potential at a distance x from the Stern plane and κ is the *Debye–Hückel* reciprocal length [41,42], which is of great importance in colloid stability. For aqueous solution at 25°C, κ is given by

$$\kappa = 2.3 \times 10^9 \left(\sum c_i z_i^2 \right)^{1/2} \quad (\text{m}^{-1}), \tag{6}$$

where c_i is the molar concentration and z_i is the valence of ion i. The sum is taken over all ions in solution.

The Debye–Hückel parameter has the dimensions of reciprocal length and $1/\kappa$ is a characteristic length which determines the extent of the diffuse layer.

2. Double-Layer Interaction

When two charged particles approach each other in an electrolyte solution, their diffuse layers overlap and, in the case of identical particles, a repulsion is experienced between them. The precise way in which the double layers respond to each other depends on a number of factors which cannot be treated in detail here. One distinction is between interaction at constant surface potential or constant surface charge. It will be considered in Sec. 3.7.

The repulsion energy of identical spherical particles at constant potential is [41], according to DLVO theory [1]

$$V_R = \frac{4\pi a_1 a_2}{a_1 + a_2} \epsilon \Psi^2 \ln(1 + e^{-\kappa h}). \tag{7}$$

When deriving Eq. (7), it was assumed that the Stern potential is low, that is,

$$\tilde{\Psi} = \frac{e\Psi}{kT} < 1, \tag{8}$$

where e is the electron charge, k is the Boltzmann constant, T is the absolute temperature, and the double layer is thin, that is,

$$\kappa a \gg 1. \tag{9}$$

For larger potentials, the analytic equations for large separation and thin layers [43,44],

$$\kappa h > 2, \tag{10}$$

is

$$V_R = 32\pi\epsilon \left(\frac{kT}{e}\right)^2 \tanh^2 \left(\frac{\tilde{\Psi}}{4}\right) a e^{-\kappa h}. \tag{11}$$

3. The Problem of the Determination of the Stern Potential

The electrostatic interaction is very sensitive to the value of the surface potential. Its identification with the ζ potential is correct for the surfaces which can be described by the so-called standard electrokinetic model [41,42]. In this model, the surfaces are described as molecularly smooth, impermeable for ions, and under no conditions porous and rough. The surface of an emulsion droplet is usually covered by the adsorption layer of an organic substance and does not satisfy this condition. In Ref. 32, values concerning the size and penetration of the different polar groups of the adsorbed organic molecular into water are given as 0.3–1.6 nm. The ions of the diffuse layer are distributed partially outside the adsorption layer and partially inside it. The latter part is not small if the electrolyte concentration is not low. The counterions inside the adsorption layer participate in the electrostatic interaction of particles also. The ζ potential yields information only about the counterions outside the adsorption layer. Thus, using the ζ potential instead of the Stern potential leads to an underestimation of the electrostatic interaction.

The ζ potential manifests itself in Bickerman surface conductivity. The counterions distributed between the slipping plane and the interface are mobile and produce the additional surface conductivity. Measurements of the surface conductivity enables the evaluation of the total charge of the diffuse layer and the efficient Stern potential [45]. Low-frequency dielectric dispersion is the most exact method for those measurements [46]. High-frequency conductometry is also useful for the Stern potential evaluation [47].

C. Hydrophobic Interaction

In some cases, surfaces may have significant areas with a hydrophobic (nonwettable) character, such as polymer latex particles with a low density of surface ionic groups, or negative particles with adsorbed cationic surfactant. The possibility then arises of another type of interaction which can give appreciable attraction—the so-called *hydrophobic interaction*. The extensive hydrogen bonding present in ordinary water is responsible for a considerable degree of association between water molecules and a significant "structuring" effect. A hydrophobic surface offers no possibility of hydrogen bonding or ion hydration, so there is no inherent affinity for water. However, the presence of such a surface tends to limit the "structuring" tendency of water molecules simply by reducing the possibility of hydrogen bonding in certain directions. Consequently, water avoids contact with such surfaces as far as possible. In aqueous solutions, molecules with hydrophobic segments can associate with each other in such a way that contact between water and the hydrophobic regions is minimized. This is known as *hydrophobic bonding*. The best known examples of this are surfactant and proteins solutions.

It has recently been found [48] that the same type of interaction occurs between macroscopic hydrophobic surfaces. The resulting attractive force can be surprisingly large and of quite long range [49]. Measurements of the force between mica surfaces, modified by adsorbed surfactants, show that the hydrophobic attraction is much stronger than the van der Waals force and extends to a distance of about 80 nm in pure water.

With increasing hydrophilicity with the introduction of charged head-groups onto the surface, the measurable range of the hydrophobic attraction decreases rapidly [50]. No long-range hydrophobic interaction is observed for surfaces having contact angles below about 40–70° (water/air). It, thus, appears that a low density of hydrophilic groups on a hydrophobic surface is sufficient to considerably reduce the magnitude and range of the hydrophobic force. The hydrophobic interactions between macroscopic surfaces seem to be important only in rather pure systems [32] and not in technical emulsions.

D. Hydration Effects

The nature of water close to a particle surface is usually very different from bulk water. The major consequence of hydration at a particle surface is an increased repulsion between approaching particles because of the need for ions to lose their water of hydration upon contact between particles. This involves work and, hence, an increase in free energy of the system.

The most direct evidence of hydration effects comes from measurements of the force between mica sheets in various electrolyte solutions [51]. At low ionic strengths, the repulsion follows the expected exponential form for the double-layer interaction (see Sec. II.C.2). At salt concentrations above about 1 mM, a monotonic short-range force is apparent, in addition to the double-layer repulsion, which is due to adsorbed hydrated cations. This extra force increases with the degree of hydration ($Li^+ \cong Na^+ > K^+ > Cs^+$) and is roughly exponential over the range 1.5–4 nm, with a decay length of the order of 1 nm. In their earlier studies, Derjaguin and Zorin [52] introduced the concept of a "structural component of disjoining pressure" to account for the anomalous behavior of thin films.

The range of these hydration forces is quite appreciable in relation to the range of double-layer repulsion, and they may be expected to have an effect on colloid stability, especially at high ionic strengths. Similarly, the way some flocculated colloids can be redispersed ("repeptized") simply by washing away the electrolyte [53,54] is a clear indication that aggregation is not occurring in a true primary minimum, but in a "hydration minimum," where the particles are prevented from coming into true contact by the presence of hydrated ions. In these cases, the van der Waals attraction is not sufficiently strong to prevent separation of particles when the salt is diluted and double-layer repulsion is reestablished.

Another type of hydration repulsion arises when adsorbed layers of hydrophilic material are present, but this is usually considered as a "steric" interaction. In Ref. 32, it is shown that the hydration forces acting between bilayers of different phosphatidylcholines have a comparable magnitude out to a bilayer separation of about 2–3 nm. The authors suggest the important role of the hydration forces in emulsion stabilization.

III. RAPID BROWNIAN COAGULATION

A. von Smoluchovski Theory

The classical understanding of coagulation kinetics is given by von Smoluchovski theory [3,4], which follows from the assumption that the collisions are binary and that fluctuations in density are sufficiently small. Computer simulations [55–58] serve as a means to test the validity of the mean field approach.

An aggregate formed from i identical particles is called an i-mer. The average number of i-mers per unit volume is the particle concentration z_i.

The coagulation of two clusters of the kind i and j is given by the following relation:

$$i\text{-mer} + j\text{-mer} \xrightarrow{\ k_{ij}\ } (i + j)\text{-mer} = k\text{-mer},$$

where k_{ij} is the concentration-independent coagulation constant or kernel. Physically, it means that the coagulation rate between all kinds of i-mers and j-mers is identical.

For dilute dispersions with volume fractions less than 1%, only two-particle collisions need to be considered because the probability of three-particle collisions is small.

The equation describing the temporal evolution of the cluster of kind k is as follows:

$$\frac{dz_k}{dt} = \frac{1}{2} \sum_{\substack{i=1 \\ j=k-1}}^{j=k-1} k_{ij} z_i z_j - z_k \sum_{i=1}^{\infty} k_{ik} z_i; \tag{12}$$

k_{ij} depends on details of the collision process between i-mers and j-mers. This kernel embodies the dependence on i and j of the meeting of an i-mer and a j-mer, including effects such as the volume dependence of the collision cross section and the diffusion constant.

The first term in Eq. (12) describes the increase in z_k owing to coagulation of an i-mer and a j-mer and the second term describes the decrease of z_k owing to the coagulation of a k-mer with other aggregates.

It is important to note that Eq. (12) is for *irreversible* aggregation—no account is taken of the breakup of aggregates, which would require a third term on the right-hand side. Also, it is assumed, for the present. That each collision results in the formation of an aggregate (i.e., the collision efficiency and stability ratio are both unity).

In principle, it would be possible to use Eq. (12) to derive the concentrations of all aggregate types at any time, but there are great difficulties, notably in assigning values to the rate coefficients, which depend not only on aggregate size and shape but also on the particle transport mechanism. By converting Eq. (12) into an integral expression and considering continuous aggregate size distributions rather than discrete numbers, it is possible to derive solutions in certain cases but only for specific forms of the rate coefficients (or kernels), which may not be physically realistic. Nevertheless, such approaches can give some insight into the way aggregate size distributions may evolve during flocculation processes.

Particles in suspensions are subject to Brownian motion [59,60]. As a result, particles will collide. The Smoluchowski approach is to imagine a stationary central particle and to calculate the number of particles colliding with it in unit time. Allowance can then be made for the fact that the central particle itself is one of many similar particles undergoing Brownian motion and so the appropriate collision frequency can be derived:

$$J_{ij} = 4\pi R_{ij}(D_i + D_j)n_i n_j; \tag{13}$$

J_{ij} is the total number of collisions occurring between particles of types i and j in unit volume per unit time. D_i and D_j, and n_i and n_j are the diffusion coefficients and number concentrations, respectively. The term R_{ij} is the *collision radius* for the pair of particles and represents the center-to-center distance at which the particles may be assumed to be in contact. In many cases, the collision radius can be taken simply as the sum of the particle radii, but if there is long-range attraction between the particles, the effective collision radius will be somewhat larger. In what follows, we will assume spherical particles, radii a_i and a_j and that $R_{ij} = a_i + a_j$. Also, for the diffusion coefficients, the Stokes–Einstein expression is used:

$$D_i = \frac{kT}{6\pi a_i \mu}, \tag{14}$$

where μ is the viscosity of the fluid, k is the Boltzmann constant, and T is the temperature.

Equation (13) then becomes

$$J_{ij} = \frac{(2kT/3\mu)n_i n_j(a_i + a_j)^2}{a_i a_j} \tag{15}$$

and so, by comparison with Eq. (12), the rate constant can be written as

$$k_{ij} = \frac{(2kT/3\mu)(a_i + a_j)^2}{a_i a_j}. \tag{16}$$

For an initially monodisperse suspension of particles, radius a_1, the initial collision rate can be calculated easily from Eq. (12) because only one type of collision (1–1) is involved. Also, the initial rate of decrease of the *total* particle concentration, n_T, follows directly from the collision rate because each collision reduces the number of particles by one (two primary particles lost, one aggregate gained). The result is

$$-\frac{dn_T}{dt} = \left(\frac{4kT}{3\mu}\right) n_T^2 = k_F n_T^2, \tag{17}$$

where $k_F (= 4kT/3\mu)$ is known as the flocculation rate constant and has a value of $6.13 \ 10^{-18} \ \text{m}^3 \ \text{s}^{-1}$ for aqueous dispersions at 25°C.

The most noteworthy feature of Eq. (17) is that it does not include the particle size because the size terms cancel from Eq. (16) when $a_i = a_j$. As the particle size increases, the diffusion coefficient decreases, but the collision radius increases. These have opposing effects on the collision rate and, for equal particles, they balance exactly. Even for spheres of different size, the term $(a_i + a_j)^2/a_i a_j \cong 4$, provided that the sizes do not differ too greatly (say, by no more than a factor of 3, in which case the size term has a value of about 5).

The simplest second-order rate expression of Eq. (17) can be integrated to give the total number of particles as a function of time:

$$n_T = \frac{n_0}{1 + k_F n_0 t} \tag{18}$$

where n_0 is the initial concentration of primary particles.

There is a characteristic flocculation time t_F, in which the number of particles is reduced to half of the initial value ($n_T = n_0/2$) and this follows immediately from Eq. (18):

$$t_F = \frac{1}{k_F n_0}. \tag{19}$$

The flocculation time can also be thought of as the average time in which a particle experiences one collision. From the value of k_F quoted above, t_F turns out to be about $1.6 \times 10^{17}/n_0$ s and so for an initial particle concentration of $10^{16} \ \text{m}^{-3}$, the flocculation time would be about 16 s. Particle concentrations are often much lower and the corresponding flocculation time would be greater.

As well as the total particle concentration, it is also possible to calculate the concentration of single particles and aggregates, assuming the rate constant is the same for all collisions. For single particles,

$$n_1 = \frac{n_0}{(1 + t/t_F)^2},$$ (20a)

and for doublets,

$$n_2 = \frac{n_0 t/t_F}{(1 + t/t_F)^3}.$$ (20b)

The general Smoluchowski expression for k-fold aggregates is

$$n_k = \frac{n_0 (t/t_F)^{k-1}}{(1 + t/t_F)^{k+1}}.$$ (21)

Results from these expressions are shown in dimensionless form (n_k/n_0 versus t/t_F) in Fig. 1, for aggregates up to threefold. For all aggregates, the concentration rises to a maximum at a characteristic time and then declines slowly. Note that at all times the concentration of singlets exceeds that of any other aggregate.

Experimental data for rapid flocculation show reasonable agreement with Smoluchowski predictions, at least in the early stages of the process (see Sec. III.D).

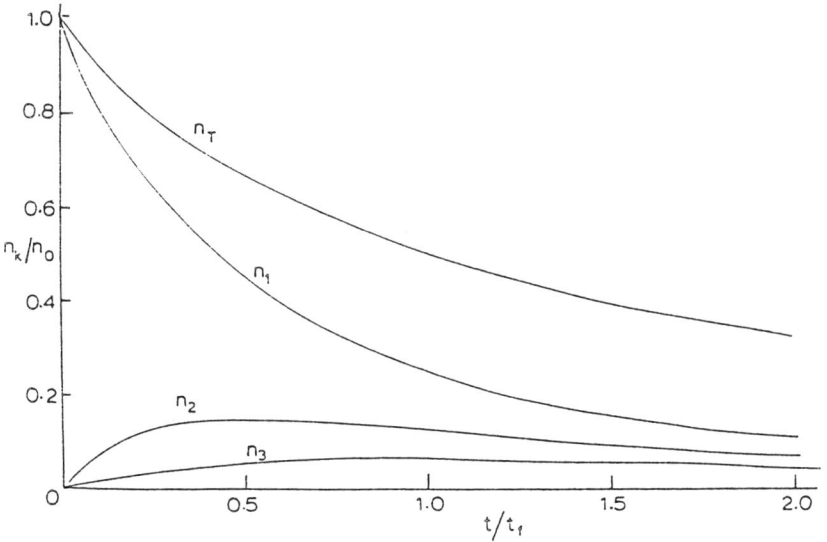

FIG. 1 Reduced concentration of k-fold aggregates plotted against reduced time for $k = 1, 2, 3$ and for the total concentration, n_T.

Rate constants determined for the rapid flocculation of latex particles are about half of the value given by Eq. (17) and this is now known to be a result of hydrodynamic interaction between approaching particles. This effect will be considered briefly in Secs. III.B and III.C.

B. Incorporation of Hydrodynamic Interaction and Attractive Surface Forces in the Theory of Rapid Coagulation

Theory of perikinetic coagulation continued to develop along the way as proposed by Smoluchowski. Oversimplifications were not allowed, which led step by step to an improvement in theory. In the Smoluchowski model, every collision is accompanied by coagulation. The possibility of the disaggregation is neglected. In other words, Smoluchowski proposed the theory of the irreversible coagulation, which corresponds to the case of strong attractive surface forces. However, the introduction of the notion of long-range surface forces necessitates the generalize of the model of the mutual free diffusion of the particles. At small interparticle distances, the additional flux caused by the particle attraction to each other has to be incorporated in the expression for the particles flux to the reference particle. This important generalization was made by Fuchs [61] with respect to repulsive forces. However, the attractive surface forces can be used in the Fuchs expression for particle flux too. Fuchs has not paid attention to another shortcoming in the Smoluchowski model which was noticed by Derjaguin [62].

The particle approach leads to the necessity of a liquid flow out from the gap between them. The thinner liquid interlayer dividing the particles, the higher is the hydrodynamic resistance to its thinning. It leads to the decrease of the coefficient of the particle mutual diffusion. Instead of the constant particle diffusivity in Smoluchowski–Fuchs theory, Derjaguin and Krotova [63] introduced in the Fuchs equation the particle diffusivity which is proportional to h at a small interparticle distance and consequently equals zero at particle contact.

The Einstein equation for interconnection of diffusivity and friction coefficient f,

$$fD = kT, \tag{22}$$

is valid for any interparticle distance. It allows one to obtain the dependence $D(h)$ using the dependence $f(h)$. The asymptotic expression for $f(h)$ at small h is given by Taylor [64] without a derivation. The derivation of this linear dependence is given by Derjaguin [62]. The diffusivity decreases to zero at $h \rightarrow 0$ and does not lead to the disappearance of particle flux because the attractive surface force causes the opposite influence. However the neglection of the long-range surface forces leads to an erroneous conclusion about when the coagulation takes place. The elimination of this paradoxial result is the main consequence of the incorporation [63] of the attractive surface forces in the theory of perikinetic coagulation. This phenomena has been considered in some articles [65,66]. To distinguish this from analytical result of Derjaguin and Krotova [63], the numerical calculation were used [65,66]. The results obtained in Refs. 65 and 66 are less convenient for application and confirm the original investigation [63].

The collision of droplets of intermediate dimensions are controlled by the joint action of the Brownian movement and sedimentation. In this respect, the Fuchs equation was generalized by Batchelor [67]. The Batchelor equation was used by many investigators and led to noticeable progress in the kinetic theory. Taking this into account, we shall represent the Fuchs equation in the next section in the notation used by Batchelor and his successors.

C. Brownian Collisions in Emulsions

There are few studies available of the coalescence of fluid drops, presumably because of the more complex interactions which involve fluid flow both inside and outside of the drops. Further, there is a possibility that the drops will deform as they collide. A notable exception is that Zinchenko [68–70] has calculated the rate of gravity-induced coalescence of spherical drops of different sizes numerically, using a trajectory analysis for pairs of drops, without considering the effects of interparticle forces. His results confirm that, in contrast to rigid spheres, drop collision is possible at finite rates under the action of a finite external force only. He also showed how the rate of drop collisions decreases with an increasing ratio of the drop fluid viscosity and the surrounding fluid viscosity.

Under conditions of low Reynolds numbers, the relative motion of two spherical drops may be decomposed into motions along and normal to their line of centers. Hetsroni and Haber [71] used the method of reflections to describe the hydrodynamic interaction for widely separated drops in both configurations. More accurate image techniques, which are valid unless the separation distance is small relative to the radius of the smaller drop, have been developed by Fuentes *et al.* [72,73]. Exact solutions based on bispherical coordinate methods have been developed by Haber et al. [74] and Rushton and Davies [75] for axisymmetric motion along the line of centers, and by Zinchenko [69] for asymmetric motion normal to the line of centers. Each of these solution methods yields an infinite series for the hydrodynamic force between the drops, which diverges when the distance between the drops tends to zero. Because coalescence phenomena depend critically on the near-contact interaction, these earlier solutions may be matched with the recent lubrication theories of Davis et al. [76]. In these, the nature of the hydrodynamic force resisting the near-contact relative motion of two spherical drops in the direction along their line of centres has been analyzed in detail.

1. Specification of the Fuchs Equation for Emulsions

Zhang and Davis [77] employs the above solutions for the hydrodynamic interactions of two spherical drops and calculates collision rates by extending the previous work by several authors for rigid particles and by Zinchenko [70] for spherical drops. According to Zhang and Davis [77], for creeping flow, the external driving forces on each drop balance the hydrodynamic forces, and the velocity V_{12} of drop 1 to drop 2 is linearly related to the sum of the external forces and depends only on the relative position of the two drops. An expression for this relative velocity has been presented by Bachelor [67] for rigid spheres.

Zhang and Davis [77] use Bachelor's expression for V_{12} and modify it for spherical drops:

$$V_{12} = -\frac{D_{12}^{(0)}}{kT} G \left(\frac{dU_{12}}{dr} + kT \frac{d \ln p_{12}}{dr} \right), \tag{23}$$

where r is the vector from the center of drop 2 to the center of drop 1. Similarly, the relative diffusivity due to Brownian motion for two widely separated drops is

$$D_{12}^{(0)} = \frac{kT(\hat{\mu} + 1)(1 + \lambda^{-1})}{2\pi\mu (3\hat{\mu} + 2)a_1}. \tag{24}$$

The pair-distribution function, p_{12}, represents the probability that drop 1 is at position r relative to drop 2, normalized such that $p_{12} \rightarrow 1$ as $r \rightarrow \infty$. The interparticle force is described by the potential function $U_{12}(r)$.

The relative mobility function for motion along the line of centers G describes the effects of hydrodynamic interactions between the two drops. The function depends on the size ratio of the two drops

$$\lambda = \frac{a_1}{a_2}, \tag{25}$$

the viscosity ratio of the drop fluid and the surrounding fluid

$$\hat{\mu} = \frac{\mu_1}{\mu_2}, \tag{26}$$

and the dimensionless distance between the drops, $s = 2r/(a_1 + a_2)$. It is unchanged when λ is replaced with λ^{-1}.

In the development of Davis et al. [76], the dimensionless lubrication force between two spherical drops in near contact is shown to depend on a single dimensionless parameter,

$$m = \hat{\mu} \left(\frac{a}{h_0} \right)^{1/2} \equiv \frac{a_1 a_2}{a_1 + a_2}, \tag{27}$$

where $a_1 a_2/(a_1 + a_2)$ is the reduced radius of the two drops and $h_0 = r - (a_1 + a_2)$ is the closest separation between two drop surfaces. This parameter describes the mobility of the interfaces: When $m \ll 1$, the drops behave as rigid spheres, whereas when $m \gg 1$, the drops have fully mobile interfaces and offer relatively little resistance to the drop relative motion. Note that the interface mobility, m, is not a property of the interfaces themselves but instead represents the viscous resistance of the fluid inside the drops to the flow exerted on their interfaces by the external fluid as it is squeeced out of the gap between the drops. Using this interface mobility, the lubrication forces acting on the drops in the direction along their line of centers can be simply expressed as

$$-F_{1,1} = F_{1,2} = 6\pi\mu a^2 \frac{V_{12}}{h_0} f(m), \tag{28}$$

where V_{12} is the component of $\mathbf{V}_{12}$ in the direction along the line of centers and $f(m)$ is a dimensionless function which is approximated by Davis et al. [76] using the following Pade-type expression:

$$f(m) = \frac{1 + 0.402m}{1 + 1.711m + 0.461m^2}. \tag{29}$$

Note that the lubrication force for drops with mobile interfaces ($m \gg 1$) is inversely proportional to $(h_0/a)^{1/2}$, indicating that spherical drops can come into contact in a finite time under the action of a finite force, in contrast to that for immobile interfaces ($m \ll 1$) for which the lubrication force is inversely proportional to h_0/a. This lubrication force dominates the hydrodynamic resistance unless the drop viscosity is very small [$\hat{\mu} < O(h_0/a)^{1/2}$], in which case the fluid slips out of the gap with little resistance.

When the drops are close to one another ($\zeta \rightarrow 0$), the lubrication force dominates the hydrodynamic force and directly balances the external force on each of the drops, that is,

$$G(\xi) = \frac{2 + 3\hat{\mu}}{3 + 3\hat{\mu}} \frac{(1 + \lambda)^2}{2\lambda} \frac{\xi}{f(m)}. \tag{30}$$

2. Expression for the Drop Collision Rate

The rate at which the drops of radius a_1 collide with the drops of radius a_2 per unit volume is equal to the flux of pairs into the contact surface $r = a_1 + a_2$ and is expressed in terms of the pair-distribution function $p_{12}(\mathbf{r})$ and the drop relative velocity V_{12} by

$$J_{12} = -n_1 n_2 \int_{r=a_1+a_2} p_{12} V_{12} \mathbf{n} \, dA, \tag{31}$$

where $\mathbf{n} = \mathbf{r}/r$ is the outward unit normal to the spherical surface represented by $r = a_1 + a_2$, and n_1 and n_2 are the number of drops at the given time in the size categories characterized by radius a_1 and radius a_2, respectively, per unit volume of the dispersion.

For a dilute dispersion, the pair-distribution function is governed by a quasisteady mass conservation equation for regions of space outside the contact surface:

$$\nabla \bullet (p_{12} V_{12}) = 0. \tag{32}$$

As the colliding drops come into contact, they are assumed to coalesce, and so $p_{12} = 0$ for $r = a_1 + a_2$:

$$p_{12}\big|_{r=a_1+a_2} = 0. \tag{33}$$

Provided that all of the drop–drop encounters originate at wide separations in a homogeneous dispersion, the other boundary condition is

$$p_{12} \to 1 \quad \text{as } r \to \infty. \tag{34}$$

Due to the spherical symmetry and quasisteady mass conservation equation, the expression for rate can be simplified:

$$J_{12} = 4\pi r^2 n_1 n_2 p_{12}(\mathbf{r}) V_{12}(\mathbf{r}); \tag{35}$$

it does not depend on r. The substitution in it $V_{12}(\mathbf{r})$ according Eq. (23) yields the linear differential equation of first order where J_{12} is an unknown constant. The integration of this equation and using the boundary conditions for p_{12}, Eqs. (33) and (34) yield the equation for the collision rate:

$$J_{12} = 4\pi n_1 n_2 D_{12}^{(0)} \left(\int_{a_1+a_2}^{\infty} \frac{\exp(U_{12}/kT)}{r^2 G} \, dr \right)^{-1}. \tag{36}$$

If the drops are assumed to move independently, that is, without any hydrodynamic interactions ($G = 1$) or interparticle forces ($U_{12} = 0$), other than a sticking force on contact, the collision rate is that obtained by Smoluchowski [Eq. (13)]. From Eq. (13), Eq. (12) for the total number of collisions occurring between particles of types i and j follows. We define the collision efficiency,

$$E_{12} \equiv \frac{J_{12}}{J_{12}^{(0)}}, \tag{37}$$

as the ratio of the predicted collision rate with hydrodynamic and interparticle interactions to that obtained in their absence. Using the dimensionless center-to-center distance, s, this is then

$$E_{12} = \left(2 \int_2^\infty \frac{\exp[U_{12}(s)/kT]}{s^2 G(s)} \, ds \right)^{-1}. \tag{38}$$

Note that the inverse of the collision efficiency is often called the "stability ratio." Of considerable interest is the influence of the viscosity ratio on the collision efficiency. For viscous drops, the relative mobility function G for near-contact relative motion is inversely proportional to the square root of the distance between the drops when the interface mobility is large. It leads to the integration in Eq. (38) being finite instead of being infinite as for rigid spheres. Furthermore, because G decreases with increasing viscosity ratio, Eq. (38) indicates that the collision efficiency will decrease monotonically as the ratio of the drop phase viscosity to the suspending phase viscosity is increased. On the other hand, comparing the effects of the hydrodynamic interactions and interparticle force which are represented by G and U_{12}, respectively, on E_{12} through Eq. (38), it is seen that the hydrodynamic interactions appear in a preexponential factor and, therefore, are subordinate, for moderate values of A/kT, to the interparticle forces, which appear in the argument of the exponential.

3. Brownian Collisions Without Interparticle Forces

Figure 2 shows the results for E_{12} as a function of λ for $\hat{\mu} = 0, 0.1, 1.0, 10, 100,$ and 1000. As expected, E_{12} decreases as $\hat{\mu}$ increases because this corresponds to decreasing the interface mobility and internal drop flow, which leads to a higher hydrodynamic resistance to the close approach. In the limit as $\hat{\mu} \to \infty$, corresponding to that of rigid spheres, $E_{12} \to 0$, although this limit is approached only slowly.

As λ decreases from unity, E_{12} increases and it tends to unity when λ tends to zero. One reason for this is that when λ decreases, the influence of the smaller drop on the Brownian diffusion of the larger one is decreased. More important is that the contribution of the smaller drop to the relative Brownian diffusivity increases as λ decreases. As $\lambda \to 0$, the hydrodynamic interactions become important only within an increasingly small boundary layer around the larger drop, and so $E_{12} \to 0$ as $\lambda \to 0$.

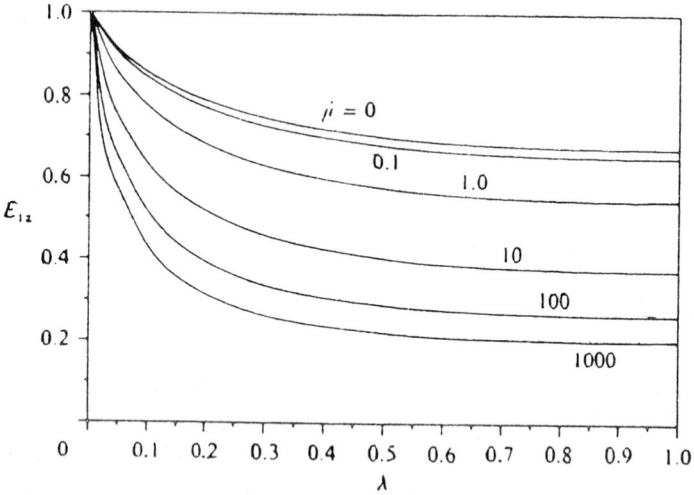

FIG. 2 The collision efficiency for Brownian drops as a function of the size ratio for various viscosity ratios without interparticle forces. (Redrawn from Ref. 77.)

4. Brownian Collisions with van der Waals Forces

In addition to λ and $\hat{\mu}$, the collision efficiency depends on A/kT. This dimensionless parameter is called the Hamaker group, and it provides a measure of the strength of the van der Waals forces relative to the Brownian motion. Typically, it attains a value of unity or less.

The effects of van der Waals attractions on the collision efficiency of Brownian drops are shown as a function of $A/6kT$ in Fig. 3 for different viscosity ratios with $\lambda = 1$. As expected, the attractive force increases the collision rate. In fact, the collision efficiency becomes larger than unity for $A \gg 6kT$, but attractive forces of this magnitude are not usually encountered in practice. Moreover, the van der Waals attraction plays an increasingly important role as $\hat{\mu}$ increases. In particular, the collision efficiency for $\hat{\mu} \gg 1$ is independent of $\hat{\mu}$ at large values of $A/6kT$ but not for small values of $A/6kT$. This is because van der Waals forces are too weak when $A \ll 6kT$ to become important until after the viscous drops have become sufficiently close that the squeeze flow in the gap between them causes their interfaces to become mobile so that the internal flow and viscosity affect the collision process. In contrast, if both $\hat{\mu}$ and $A/6kT$ are large, then the drops are pulled into contact rapidly by the attractive forces before they approach within this range and so the interface

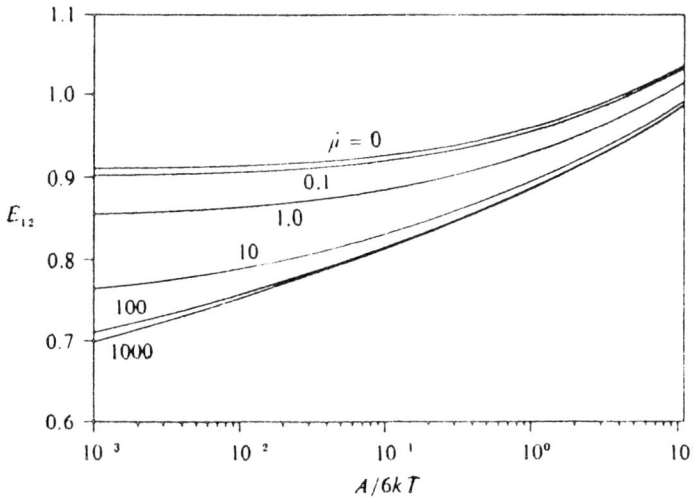

FIG. 3 The collision efficiency for Brownian drops as a function of the Hamaker group for $\lambda = 1.0$ and various ratios with unretarded van der Waals attraction. (Redrawn from Ref. 77.)

mobility and internal flow do not affect the process and the drops behave as rigid spheres. In this limit, results [76] agree to within 0.2% with the earlier calculations of Spielman [65], which are shown as filled circles in Fig. 3 for $\lambda = 1$ and $\hat{\mu} = \infty$.

Numerical results of Zhang [78] for E_{12} with retarded van der Waals' attraction show no qualitative change in the effects of van der Waals' attractions on the collision efficiency.

D. Experiments

The systematic consideration of the experimental investigation is given by Sonntag [79]. The measurements [80,81] confirm the equations for singlet diffusitivity and doublet diffusitivity in line of axis and perpendicular to it. In Ref. 79, several techniques are described for the measurement of the coagulation of colloidal particles. Some of these give only global information on the state of aggregation; others give a detailed picture of the particle and floc size distribution. Ideally, the monitoring technique should be suited to on-line application without a sample pretreatment such as dilution. All methods are divided into bulk techniques (turbidity, static light scattering) and single-particle techniques (Coulter counter, flow ultramicroscopy). First, quantitative experiments on rapid coagulation were performed with gold particles by slit ultramicroscopy and individual counting [82–84]. The average value of the coagulation constant of gold particles was found to be $k_{ij} = 12 \times 10^{-18} \pm 1$ m^3 s^{-1} at 298 K. These are the only values described in the literature that confirm the theoretical value.

The coagulation of gold particles in different electrolytes was also investigated later with streaming ultramicroscopy with visual counting by Derjaguin and Kudravtzeva [85]. For sodium chloride, $k_{ij} = 8.2 \times 10^{-18}$ m^3 s^{-1}; for magnesium sulfate, $k_{ij} = 8.9 \times 10^{-18}$ m^3 s^{-1}; and for lanthanum nitrate, $k_{ij} = 6.5 \times 10^{-18}$ m^3 s^{-1} was obtained. Selenium sols were investigated spectrophotometrically by Watillon et al. [86]. The rapid rate constant was 4×10^{-18} m^3 s^{-1}.

Silver iodide sols were coagulated with barium nitrate and lanthanum nitrate by Ottewill and Rastogi [87]. The coagulation constants were determined to 8.78 $\times$ 10^{-18} and 10.2 $\times$ 10^{-18} m^3 s^{-1}, respectively.

The influence of the particle size on rapid coagulation was investigated with hematite particles by Penners and Koopal [88].

In recent years, many experiments have been carried out using polystyrene lattices because of their monodispersity and their ideal spherical shape. The coagulation rate was lower than the theoretical value, even when the hydrodynamic interaction was taken into consideration.

Sonntag [79] considers two ways to explain this behavior. The first one is the introduction of reversibility into the coagulation process. This was suggested by Frens and Overbeek [53]. Under this assumption the coagulation kernel can remain constant or depend on the aggregate size.

The Smoluchowski treatment is based on the collision of spheres and this assumption is questionable in the case of aggregates. Except when coalescence occurs, aggregates cannot be truly spherical. Two colliding solid spheres must form an aggregate in the form of a dumbbell, and with higher aggregates, many different shapes become possible, as illustrated in Fig. 4. The collision rates of such aggregates are likely to differ from those for spheres. In particular, the effects of size on the collision radius and the diffusion coefficient cannot be expected to balance each other as for spheres, which led to the very simple, size-independent form of the collision rate constant in Eq. (16). As the size of an irregular aggregate increases, it is likely that the increase in collision radius more than compensates for the decrease in diffusion coefficient and the results of Cahill et al. [89] tend to support this conclusion. Gedan et al. [90], from their measurements of aggregate size distribution, assigned different rate constants to the various types of collision. For the collision of primary parti-

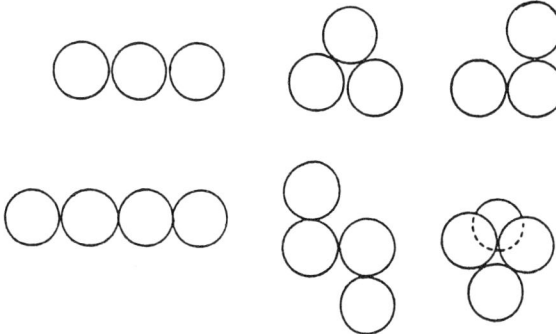

FIG. 4 Possible shapes of aggregates up to fourfold.

cles, k_{11} was found to be 3×10^{-18} m^3 s^{-1}. For other collisions, they found $k_{12} \cong 2k_{11}$ and $k_{13}, k_{22}, k_{12}, k_{33} \cong 4k_{11}$. These values show a more pronounced effect of size on collision rate than expected. However, the value found for k_{11} is only about 25% of the Smoluchowski value and there may be some doubt about the accuracy of the value.

IV. KINETICS OF RAPID GRAVITATIONAL COAGULATION IN THE ABSENCE OF REPULSIVE SURFACE FORCES

A. Orthokinetic Coagulation and Particle Capture from Flow

The role of Brownian diffusion in the coagulation process decreases with increasing particle size, and other factors begin to dominate. The difference in particle dimensions and, therefore, in their sedimentation rate play the determining role in the particle approach in a quiescent media. The "orthokinetic coagulation" term used in the case when a directed motion of particles is favorable for coagulation was initially connected just with this mechanism.

Particles of similar size also approach in a inhomogeneous hydrodynamic field. The term "orthokinetic coagulation" is now more frequently used as applied to coagulation in inhomogeneous hydrodynamic field. Coagulation in stationary medium is often called the "gravitational coagulation," "gravity-induced coagulation," or "differential settling."

The theory is mainly developed to describe an approach of particles with largely differing dimensions. The coordinate system is related to the larger particle, and the trajectory of the smaller particle is considered in the hydrodynamic field of the larger one. This hydrodynamic field arises at the cost of the motion of the larger particle with respect to the medium, which in the laboratory coordinate system is considered to be stationary. But in the coordinate system related to the particle, a liquid flow runs on the latter, which is homogeneous at infinity and carries particles of small size. In this problem, the large particle is called the collector because small particles settling from the flow accumulate on it. Only the relative velocity of the medium and collector is substantial for the settling mechanism.

The situation is fully identical to the following: A liquid comprised of small particles is at rest as a whole (in the laboratory coordinate system) and the collector moves with respect to it, for example, under the effect of gravity, electric, or centrifugal field; the collector rests (in the laboratory coordinate system) and liquid flow is present. It is precisely the setting of the experiment aimed at the study of the elementary orthokinetic coagulation act. Similar to that is the elementary act of particle capture from the flow in filtering with the use of porous media, for example, of granular filters. A particle from the flow is settled on individual elements of the filtering medium (on a granule of a granular filter or on fibers of a fibrous filter). Thus, considering the process of particle sedimentation from the flow on collector, we obtain, at the same time, information about the elementary act of orthokinetic coagulation.

Processes controlling the formation of cloud drop spectrum or emulsion drop spectrum are more complex, as it is much more difficult to describe coagulation of drops of almost equal size.

It is natural to first consider the gravitational coagulation of drops strongly differing in size (Secs. IV.A–IVH), which will facilitate the understanding of the more complex process of coagulation of drops of close dimensions (Secs. IV.I and IV.J).

Investigations of coagulation in aqueous aerosols [16,17,91] at flotation [18,19] and filtration [92,93]. were started earlier and have been conducted on a front wider than that in the case of emulsions. Therefore, in studying gravitational coagulation in emulsions, one should not ignore the experience accumulated in adjacent scientific leads. In addition, the specific nature of emulsions can be taken into consideration.

The first gravitational coagulation model was proposed by Smoluchowski [4]. The next important step was the investigation by Langmuir and Blodgett [94] in which the role of inertial and hydrodynamic forces was demonstrated apparently for the first time. Then investigations have been conducted mainly applied to aerosols and flotation. A systematic consideration of orthokinetic coagulation was given in Refs. 15 and 93.

B. Primitive Model of Gravitational Flocculation and Its Specification for Emulsions

Particles of different size or density will settle at different rates and the resulting relative motion can cause particle collisions and, hence, flocculation.

The first attempts to estimate the rate of gravity-induced coagulation of a dispersion were made by Smoluchowski [4].

The collision frequency can be calculated very simply, assuming that the Stokes or the Hadamard–Rybchynski law [95,96] applies and that particle motion is linear up to contact with another particle. The resulting single-particle collision rate was found to be

$$J_{Gr}^0 = \pi(u_{02} - u_{01})(a_1 + a_2)^2 n_1, \tag{39}$$

where u_{0i} is Stokes creaming velocity for particles of radius a_i. Because all resistances are ignored in this analysis, the Smoluchowski flocculation rate provides a useful scale on which other flocculation rates can be compared. The ratio of Smoluchowski flocculation rate to the actual flocculation rate in a system undergoing gravity-induced flocculation, G_{Gr}, is known as the stability ratio,

$$W_{Gr} = \frac{J_{Gr}^0}{J_{Gr}}. \tag{40}$$

The gravity-induced capture efficiency is defined as the reciprocal of the stability ratio:

$$E_{Gr} = \frac{1}{W_{Gr}}. \tag{41}$$

Some authors assume erroneously that this first primitive model of gravity-induced coagulation was proposed by Saffman and Turner [97].

Zhang and Davis [77] specified the Smoluchowski model for emulsions, taking into account that the drag coefficient of a drop is described by the Hadamard–Rybchynski equation [95,96].

In distinguishing this from the Stokes equation, the sedimentation velocity of a droplet expressed according to the Hadamard–Rybchynski theory is given by

$$\bar{u}_0 = \frac{2}{3} \frac{\Delta \rho g a^2}{\mu} \frac{1 + \hat{\mu}}{2 + 3\hat{\mu}}, \tag{42}$$

where g is the gravitational acceleration vector and $\Delta \rho = \rho' - \rho$, where ρ' and ρ are densities of oil and water, respectively. The internal circulation in a drop and the mobility of its surface reduces the media hydrodynamic resistance to a drop movement that causes the increase in its creaming velocity in comparison with a solid sphere. The relative velocity due to gravity for two widely separated drops is given by

$$u_{12}^{(0)} = u_0^1(\hat{\mu}) - u_0^2(\mu) = \frac{2(\hat{\mu} + 1)\Delta\rho a_1^2(1 - \lambda^2)g}{3(3\hat{\mu} + 2)\mu}. \tag{43}$$

It is seen that when $\hat{\mu} \to \infty$, Eq. (42) transforms into the equation for the Stokes settling velocity, and when $\hat{\mu} \ll 1$, the settling velocity increases 3/2 times.

In Ref. 98, the experimental data concerning the droplet settling velocity are analyzed. Some experimental data confirm the Stokes equation, whereas other are in agreement with the Hadamard–Rybchynski theory. The systematic experimental theoretical investigations of Frumkin and Levich described in Ref. 99 proved that the transition from the Hadamard-Rybchynsky regime to Stokesian behavior of a droplet occurs due to the retardation of the surface movement of a droplet caused by the adsorption layer of a surfactant. The mathematical representation of this result led to the notion of the retardation coefficient χ_r. The Hadamard–Rybchynski equation (42) was generalized by Levich and Frumkin by incorporation of the retardation coefficient χ_r:

$$u_0(\hat{\mu}) = \frac{2}{3} \frac{\Delta g a^2}{\mu} \frac{\mu + \mu' + \chi_r}{2\mu + 3\mu' + 3\chi_r}, \tag{44}$$

where

$$\chi_r = \frac{2}{3} \frac{RT}{D_i} \frac{\Gamma_0^2}{c_0}, \tag{45}$$

D_i is a surfactant molecule diffusion coefficient and c_0 and Γ_0 are its bulk and surface equilibrium concentration, respectively.

At a very low surfactant concentration, χ_r is small and Eq. (44) transforms into the Hadamard-Rybchynski equation (42). The surfactant concentration growth leads to the surface retardation and the decrease of settling velocity. For $\chi_r \to \infty$, Eq. (44) transforms into the Stokes equation.

The different equations for χ_r corresponding to the different types of the adsorption kinetics are given in Refs. 98 and 100. The level of impurities in water and their adsorption usually is sufficiently high to cause a strong or even complete retardation of droplet surface mobility in emulsions. The larger the droplet, the lower the retardation. The surface movement is possible for sufficiently large drops and by the special purification of water and oil. Thus, at Re $\ll$ 1, the deviation from the Stokesian drag coefficient is weak even under special experimental conditions. At the same dimension of a drop and a bubble the gravity force and correspondingly the viscous stresses which cause the surface movement are weaker in emulsions than in foams. Thus, neglecting the droplet surface mobility in the differential settling in emulsions is justified even in higher degrees than for bubbles in foam.

One underlying restriction is that the flow in and around the drops is sufficiently slow that inertia is small relative to viscous forces. This requires that the Reynolds number

$$\text{Re} = \frac{\rho \mu_0 a_2}{\mu} \tag{46}$$

is small compared to unity for all phases, where a_2 is the larger drop radius and μ_0 is its sedimentation velocity. Typical conditions are $\mu = 0.01$ g cm^{-1} s^{-1}, $\rho = 1$ g cm^{-3}, $\Delta\rho = 0.1$ g cm^{-3}, and $g = 10^3$ cm s^{-2}, this requires that $a_2 < 50$ μm.

As seen from Eqs. (44) and (46), Re is proportional to a_2^3. It means that condition (46) is strongly violated for $a_2 > 50$ μm. In addition, the experimental investigations of the collision efficiency are simplified for large drops.

The droplet velocity in centrifugal field can increase up to hundred times, which leads to a similar growth of Reynolds number. Meanwhile, the ultracentrifugation is important in the investigations of emulsion stability [101–103]. Hence, the case Re $\gg$ 1 becomes interesting too. For this condition, the empirical dependence of the settling velocity on the radius of a solid particle or a droplet with the retarded surface is known. It follows from the expression for the resistance coefficient for the spherical particle which is a function of the Reynolds number [104,105].

The relative motion of different sized particles can be induced in other ways. In this respect, one important method is the acoustic method, in which ultrasonic waves induce vibration of suspended particles. The smaller particles are better able to respond to the inducing frequency and to vibrate with greater amplitude than larger particles, facilitating collisions between particles of different size [106].

C. Long-Range and Short-Range Hydrodynamic Interaction

The process when two droplets of different size approach each other undergoes qualitative changes as the distance between their surfaces diminishes. At large distances, this process is determined by two parameters: forces of inertia and the long-range hydrodynamic interaction (LRHI).

A sufficiently large drop moves linearly under the effect of the forces of inertia until it collides with the bigger one, which takes place if the target distance $b < a_2 + a_1$ (Fig. 5), where R is the radius of the bubble.

The liquid flow envelops the surface of the big drop, and the small drops are entrained to a greater or a lesser extent by the liquid. The smaller the drops and their difference in density relative to the medium, the weaker the inertial forces acting on them and the more closely the drop

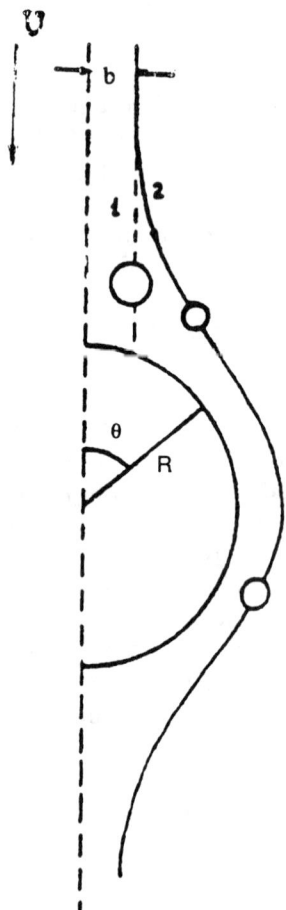

FIG. 5 The influence of the inertia of particles on their trajectory in the vicinity of the floating bubble. Trajectories of the great (inertial) (line 1) and the small (inertial-free) (line 2) particles at the same target distance b.

trajectory coincides with the liquid streamlines. Thus, at the same target distance, fairly large drops move almost linearly (Fig. 5, line 1), whereas fairly small drops move essentially along the corresponding liquid flow line (line 2). The trajectories of drops of intermediate size are distributed within lines 1 and 2; as the size of drops decreases, the trajectories shift from line 1 to line 2 and the probability of collision thus decreases.

The deviation of trajectory of small droplets from the rectilinear path to the surface of the biggest drop at distances of the same order as the biggest drop is caused by LRHI. The biggest drop causes a curving of liquid streamlines and thereby bends the trajectory of small drops (i.e., acts on them hydrodynamically due to the liquid velocity field). In the case of large drops, the forces of inertia considerably exceed the LRHI which is, therefore, not clearly manifested. In the case of small drops, the forces of inertia are small compared with the LRHI.

Thus, the process of the approach of large drops to the biggest drop is ensured by forces of inertia, whereas in the case of small drops, this process occurs in an inertia-free manner and is strongly hindered by the LRHI [107,108].

In addition, the hydrodynamic interaction at distances comparable to the drop radius has to be taken into account; the latter causes the drop's trajectory to deviate from the liquid flow line and should naturally be called the short-range hydrodynamic interaction (SRHI). Using Taylor's solution of the hydrodynamic problem involving the squeezing out of liquid from the gap as spherical particles approach the flat surface, Derjaguin and Dukhin [108] have shown that the SRHI may prevent drops from coming into contact.

The process of the approach of particles to the bubble surface can be described quantitatively by taking into account both the LRHI and the SRHI. One introduces a dimensionless parameter of the collision efficiency,

$$E = \frac{b^2}{a^2}, \tag{47}$$

where b is the maximum radius of the cylinder of flow around the bigger drop encompassing all particles deposited on the small droplets surface (Fig. 6). The particles moving along the streamline at a target distance b are deposited on the surface of a bigger drop (Fig. 6, as indicated by a dashed line). Otherwise the particle is carried off by the flow. From Fig. 6 it is evident that the calculation is essentially reduced to the so-called "limiting (grazing) trajectory" (continuous curve) and, correspondingly, the target distance.

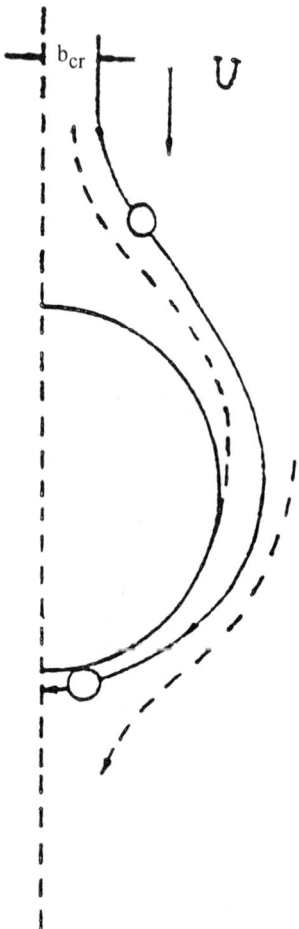

FIG. 6 Continuous line illustrating the concept of the limiting trajectory of particles. Dashed line indicates the trajectories of the particle at $b < b_{cr}$ and $b > b_{cr}$.

According to Taylor, at a gap thickness h, which is much smaller than a_1, the hydrodynamic resistance of the film to the thinning process is

$$F_h = \frac{u(h)\mu a_1^2}{h},\tag{48}$$

where $u(h)$ is the velocity at which the drops approach a certain surface area of the biggest drop, which may be considered to be flat because a_2

$\geqslant a_1$. If a constant pressing force F_h is applied to the smaller drop, then according to Eq. (48),

$$u(h) \equiv \frac{dh}{dt} = \frac{F_h h}{a_1^2 \mu}.$$

It may be inferred that complete removal of the liquid from the gap requires an infinitely long time:

$$t \sim -\int_h \frac{\mu a_1^2 \, dy}{F_h y} \sim -\frac{a_1^2 \mu}{F_h} \ln y \Big|_h \to \infty.$$

In the area above the equatorial plane, the liquid flow lines approach the surface of the biggest drop, which means that the radial component of the liquid velocity is directed toward the surface of the biggest drop. Because the motion of the particle toward the surface is obstructed within the zone of the SRHI, the radial velocity of the liquid is higher than that of the small drop. Thus, at a small gap thickness where the viscous resistance is high, the radial velocity of the liquid will be even higher. The radial flow of liquid envelops the small drop whose approach to the biggest drop has been retarded and presses it against the latter. As a first approximation, this hydrodynamic force can be estimated from the Stokes formula by substituting into it the radius of the small drop and the difference in the local values between the velocity of the liquid and that of the small droplet.

D. The Effect of Droplet Inertia

In order to understand the mechanism of inertial deposition of smaller drops on the biggest drop, one must introduce the notion of the particle inertial path, l. The latter is defined as the distance the particle with an initial velocity u_0 is able to cover in the presence of the viscous resistance of the liquid due to the initial velocity

$$l = \frac{2}{9} \frac{u_0 a_1^2 \rho'}{\mu}. \tag{49}$$

This result follows from the linear differential equation of the particle hydrodynamic relaxation which consists of two terms. The first term describes the inertial force and second one the hydrodynamic resistance.

Because the drop surface is impermeable to liquid, the normal component of the liquid velocity on the surface is zero. As the distance from the drop surface increases, the normal component of the liquid velocity also increases. The thickness of the liquid layer, in which the normal component of the liquid velocity decreases because to the effect of the drop, is of the order of a_2. The particle traverses this liquid layer due to the inertial path whereby the deposition of a smaller drop depends on the dimensionless parameter:

$$\tilde{l} = \frac{l}{a_2}. \tag{50}$$

When $\tilde{l} > 1$, the deposition is obviously possible; yet, calculations have shown that it can also take place at $\tilde{l} < 1$ as long as this value is not too small. This conclusion becomes apparent if it is considered that in a layer of thickness a_2, the smaller drop moves toward the surface not only due to inertia but also together with the liquid of the biggest drop. The motion component of the latter normal to the surface of the biggest drop becomes zero at its surface. Inertial deposition proves to be impossible if $\tilde{l}$ is smaller than some critical value $\tilde{l}_{cr}$. Neglecting the size of smaller drop, Levin [109] obtained

$$\tilde{l}_{cr} = \frac{1}{24}. \tag{51}$$

This is taken into account in the equation describing the hydrodynamic relaxation in the hydrodynamic field of the biggest droplet. The one-dimensional equation describes the small droplet movement along the symmetry axis crossing the centers of both droplets:

$$\text{St} \frac{d^2x}{dt^2} + \left(\frac{dx}{dt} - u(x)\right) = 0, \tag{52}$$

where the Stokes number $\text{St} = \tilde{l}$ characterizes the ratio of inertial and viscous forces when the Stokes mobility is valid for smaller droplets. It means its Reynolds number is less than 1. x is the distance to the surface of biggest droplet and $u(x)$ is the liquid velocity distribution along the symmetry axis. It is described by the Stokes equation for the hydrodynamic field if the Reynolds number for the biggest droplet is less than 1 also. The characteristic equation $\text{St } m^2 + m + 2 = 0$ has one real (negative) root and separation between the droplet surface decreases exponentially with time. However, the (point) particle never reaches the surface because the particle velocity and fluid velocity coincide. This conclusion holds for nonzero St value if St is smaller than a critical value,

$$\text{St} < \text{St}_{cr}. \tag{53}$$

Considerable attention was paid to the calculation of St_{cr}. The numerical calculations of Natanson [110] are in good agreement with the original Langmuir data [94] for E. Later, the general solution of the problem for St_{cr} was given by Levin [109] who derived the equation for St_{cr} for any hydrodynamic field. It is important because the condition (53) is often violated for the biggest drop.

The dependence $E(\text{St})$ can be obtained by the solution of a set of two differential equations for two components of particle velocity. The results of the numerical calculation of Fonda and Herne [111] for the potential and the Stokes flow are given in Ref. 112.

These results and the analytical results of Natanson [110] and Levine [109] are in good agreement [91].

The differential equation (53) describes the movement of a spherical particle center. Inertial deposition proves to be possible even at $\text{St} < \text{St}_{cr}$ if it is considered that at a distance equal to $a_1 + a_2$ between the smaller droplet center and the biggest droplet surface, the surfaces touch each other.

The possibility of the inertialess collision due to interception and the importance of the particle dimension was proved by Sutherland [107]. During some decades, there was no coupling between two directions in the theory of the collision efficiency. At $\text{St} < \text{St}_{cr}$, the collision efficiency was considered to be caused by the interception. The inertial forces were neglected and the dimension of a particle was important. At $\text{St} > \text{St}_{cr}$, the inertial forces were taken into account. However, the role of a particle dimension was neglected. These simplifications can be verified at extreme cases of

$$\text{St} \ll \text{St}_{cr} \quad \text{and} \quad \text{St} \gg \text{St}_{cr}. \tag{54}$$

In Ref. 113, it is emphasized that in the intermediate range of Stokes values [i.e., at the violation of the conditions (54)], both the interception mechanism and inertial forces are important.

Condition (46) is violated at the conditions which are necessary for the manifestation of inertial forces. It is violated even at smaller droplet dimensions when the inertial forces can be neglected. Thus, the extension of the droplet range in the theory of the collision efficiency can be achieved by the refusal of condition (46). The next and more difficult task is the incorporation of inertial forces in the theory of macroemulsions. The inertialess collision at the intermediate Reynolds number will be considered in the next section. Some additional remarks concerning the inertial collision in emulsions will be given in Sec. IV.F.

E. Collision Efficiency Caused by Long-Range Hydrodynamic Interaction

1. The Grazing Trajectory Method

Collision efficiency calculations were presented for the first time in Ref. 107. They were based on the consideration of liquid streamlines and of

the finite size of spherical particle. Functions characterizing liquid streamline are related to the velocity field by well-known differential relations,
and in the Stokes case

$$\Psi = a_2^2 U \left(\frac{1}{2} \rho^2 + \frac{1}{4} \rho^{-1} \right) \sin^2\theta, \tag{55}$$

where ρ denotes the radial position scaled on a_2 and θ denotes the angle
from the front stagnation point. As can be seen from Eq. (55), streamline
$\rho(\theta)$ is symmetrical with respect to equatorial plane of bubble: $\theta = \pi/2$.
This means that the droplets moving along the liquid streamline approach
the drop surface most closely at $\theta = \pi/2$. Because the center of a droplet
moves along the liquid streamline, it will touch the surface of bigger drops
only in the case when the distance between the streamline and the droplet
is equal to the droplet radius. According to Ref. 107, liquid streamline
passing through a point with coordinates

$$r = a_2 + a_1 \quad \text{and} \quad \theta = \frac{\pi}{2} \tag{56}$$

represents the limiting trajectory of the droplet ($\Delta\rho = 0$). Under conditions
(56), the droplet touches the drop. But if we increase the tangent distance,
then even the nearest distance between the streamline and the drop surface
($\theta = \pi/2$) proves to be larger than the droplet radius. By substituting
values of r and θ according to Eqs. (56) into Eq. (55), Ψ_c which characterizes grazing streamline giving the basis for the collision efficiency E_{0c} can
be determined:

$$\lim_{\rho \to \infty} (\rho^2 \sin^2\theta) = \frac{1}{a_2^2} \lim_{r \to \infty} (r^2 \sin^2\theta) = \frac{b^2}{a_2^2} = E_{0c} = \Psi_p.$$

After substitution of the expression for Ψ_p in which at, $a_1/a_2 \ll 1$, only
the term linear in small parameter is considered, we obtain

$$E_{0s} = \frac{3}{2} \frac{a_1^2}{a_2^2}. \tag{57}$$

We introduced the superscript zero to note that SHRI and surface forces
are not taken into account. According to it, when the size of particle
decreases, LRHI retards coagulation kinetics by a factor of hundreds and
even thousands.

2. Equation for the Smaller Drop Flux on the Surface of a Larger Drop

In the case of noninertia collision and without regard for SHRI, the velocity of smaller drop is well known at any distance from the bigger one:

$$\mathbf{u} = \mathbf{u}(r, \theta) + \mathbf{u}_g, \tag{58}$$

so that it is advisable to derive an expression for the radial component of smaller drop flux density

$$J_r = n(r, \theta)u_r, \tag{59}$$

where $n(r, \theta)$ is the smaller drop number concentration next to a surface of a bigger drop. Integrating the flux density over that part of the surface on which sedimentation takes place, we obtain the number of smaller drops colliding with the bigger one in unit time N, which relates to E_0 as

$$N = 4\pi a_2^2 E_0 u_0. \tag{60}$$

Earlier [114], a theorem was proven that the particle concentration remains constant if the velocity field is solenoidal (i.e., the condition is met). From this, it follows that

$$E_0 = \frac{1}{\pi a_1^2 u_0 n_1} \int_0^{\theta_c} n_1(u_r + u_g) 2\pi a_1^2 \sin \theta \, d\theta, \tag{61}$$

where θ_c characterizes the boundary of the region of deposition of smaller drops at the surface of the bigger drop. To consider the effect of a finite size of a smaller drop, integration over a concentric sphere of radius $a_1 + a_2$ has to be performed. From this condition, θ_c can be derived:

$$u_r(a_1 + a_2, \theta_c) = 0. \tag{62}$$

Substituting the Stokes velocity field into Eq. (61) and using Eq. (44) at $\hat{\mu} \to \infty$, we obtain [114]

$$E_{0s} = \frac{3}{2} \left(\frac{a_1}{a_2}\right)^2 - \frac{\Delta\rho}{\rho} \left(\frac{a_1}{a_2}\right)^2. \tag{63}$$

This method of calculation was proposed almost a quarter of a century later by Weber [115], who, probably, were not aware of the work of Dukhin and Derjaguin [114]. The usefulness of the method was well demonstrated, although the discussion was restricted by the consideration of systems in which the particle density differs only slightly from water density and allows one to ignore the role of sedimentation. A more general approach was demonstrated in the recent work of Nguen Van and Kmet [116].

It is important that Eq. (61) is applicable not only under Stokes and potential flow but also at any solenoidal hydrodynamic fields, which ensures its wide application.

3. Long-Range Hydrodynamic Interaction at Intermediate Reynolds Numbers

The important Eq. (63) is valid for condition (46). At higher Reynolds numbers, the hydrodynamic flow pattern around a drop changes and causes the change in the equation for collision efficiency. Flow conditions representative of very high Reynolds numbers can be approximated by the potential flow. It is not valid in the surface vicinity within the so-called hydrodynamic boundary layer. This case of very high Reynolds numbers will be not considered here because it corresponds to very big drop dimensions and the deviation of shape from sphericity.

The hydrodynamic at intermediate Reynolds numbers with respect to spherical particles is considered in Ref. 117.

There is a large qualitative distinction in flow pattern around a drop, depending on the ratio $\hat{\mu}$. If an oil viscosity strongly exceeds water viscosity, the flow pattern is similar to that of a solid sphere.

According to experimental investigations and numerical solutions (cf. Clift et al. [117]) qualitatively different hydrodynamic regimes exist at different Reynolds numbers. At $7 < Re < 20$, a so-called unseparated flow is observed. Flow separation is indicated by a change in the sign of the vorticity and first occurs at the rear stagnation point, approximately at $Re = 20$.

If Re increases beyond 20, the separation ring moves forward so that the attached recirculating wake widens and lengthens. A steady wake region appears at 20. The onset of wake instability corresponds to separation angle

$$Q_s = 180 - 42.5(\ln Re/20)^{0.48}.$$

To generalize Eq. (63), the theories of flow pattern around the bubble at intermediate Re can be used [116,118,119].

If an oil viscosity is less than water viscosity, the theory in Ref. 120 predicts the disappearance of flow separation. For this case, the solution of the Navier–Stokes equation can be accomplished according to numerical methods only [121].

Investigations of droplet collisions in emulsions at intermediate Reynolds numbers seem to be lacking. However, the emulsion specificity disappears in the presence of even traces of a surfactant with respect to long-range hydrodynamic interaction. This statement cannot be valid with respect to short-range hydrodynamic interaction, which will be discussed in next section.

(a) Complete Surface Retardation. As discussed in Sec. IV.A, a droplet surface can be substantially retarded even in distilled water due to the presence of surfactants.

Weber and Paddoek [122] applied a curve-fitting technique to the numerical solutions of the Navier–Stokes equations obtained by Masliyah [123] and Woo [124] and derived the expression for the interceptional collision efficiency:

$$E_0 = \frac{3}{2} \lambda^2 \left(1 + \frac{(3/16)\text{Re}}{1 + 0.25 \, \text{Re}^{0.56}} \right). \tag{64}$$

Yoon and Luttrel [118] empirically developed an analytical expression for the streamline function at intermediate Reynolds number by combination of the stream function for creeping and potential flow regimes and derived the equation for the interceptional collision efficiency. As it is seen from Fig. 4 of their article [118], the difference of their results from Eq. (64) is small.

The complete Navier–Stokes equation was solved numerically again by Nguen Van and Kmet [116] who obtained the interceptional collision efficiency. Its first term coincides with Eq. (64); the second one characterizes the particles sedimentation on the bubble surface. It is important in the condition of mineral flotation and can be neglected in emulsions.

Thus, the theoretical results of Refs. 116 and 118 do not contradict Eq. (64).

(b) Incomplete Surface Retardation. The term "emulsion" usually relates to the dispersion with drops less than 100 μm in diameter. At these dimensions, the droplet surface mobility is retarded completely or very strong by the traces of surfactants. At strong retardation, a small residual surface mobility is possible. However, under some special conditions, the participation of the droplet of bigger dimensions in collisions can be of interest. In this extreme case, the interceptional collision efficiency can be described by means of the theory [125] elaborated for the description of the flotation kinetics at intermediate Reynolds number and unretarded bubble surface. As surface retardation decreases, both the tangential liquid velocity and its normal component at the distance of order a_1 from the surface increased. Thus, the smaller drop flux increases, and this increase can be described by the Sutherland equation as the surface retardation decreases. Thus, the interceptional collision efficiency for free surface E_{0c} is less sensitive to the small value of the radius ratio λ than that for immobile surface E_{0r}. For the intermediate Reynolds number, Rulyov and Leshchov [125] used Hamieloc theory [126] and obtained

$$E_{0c} \sim \frac{a_1}{a_2^{0.9}}, \qquad E_{0r} \sim \frac{a_1^2}{a_2^{1.8}}. \tag{65}$$

For small values of the radius ratio surface retardation reduces the interceptional collision efficiency by magnitudes of hundred or thousand. Experimental investigation is easier for big drops and the discussed theory [125] can be useful for this purpose.

(c) Experiments. Experimental investigations of the droplet collisions in emulsions at intermediate Reynolds numbers are not reported. However, some information can be extracted from the experiments accomplished for the investigation of particle collisions with a small bubble at an intermediate Reynolds number. In experiments [116], a single bubble is generated and captured on a needle. Afterward, the small particles' trajectories in the vicinity of the captive bubble were observed and the grazing trajectory was determined. In these experiments [116], the size of a captive bubble was in the range of 0.5–2 mm, the Reynolds number in the range 30–300, and the particle size in the range 10–35 μm. In other experiments [118], the bubble size was in range 100–500 μm and the particle size in the range 1–40 μm. The experiments [116,118] confirmed the theoretical prediction of collision efficiency dependence on particle and bubble dimensions.

In these experiments, the influence of solid-particle sedimentation on bubble surface predominated due to big difference of densities of solid particles and water. It means that the manifestation of a second component of the particle transport to the bubble surface, caused by the normal component of liquid velocity, was weak. Thus, the exactness of the verification of the theory of the convective transport in these experiments was low. Meanwhile, the convective transport predominates in emulsions due to small difference in densities of water and oil.

Thus, Eq. (64) can be used for the description of the interceptional collision efficiency in emulsions to a first approximation. However, the exactness of its experimental verification with respect to emulsions is limited.

The experimental methods and the experimental installations, described in [116], can be applied for the experimental investigations of collision efficiency in emulsions. It is sufficient to replace the captive bubble by a big captive oil drop.

If a particle shape is isometrical, its behavior in the long-range hydrodynamic interaction is similar to the spherical particle with the same dimension. However, their behavior in short-range hydrodynamic interaction will be different. The drainage in the gap between a bubble (drop) and particle (small drop) is very sensitive to the microrelief of the surface. The resistance to thinning of the interfacial film can be drastically reduced if the angles between the facets are sufficiently sharp so that "intrusion" of the interface film by a sharpened section of a particle surface may take place. Anfruns and Kitchener [127] determined the efficiency of the accumulation of glass spheres and broken quartz particles and established that the manifestation of SHRI is weakened in the case of quartz particles.

Thus, the discrimination between LRHI and SRHI contributions to measured particle accumulation by a big bubble (drop) is possible. The broken particle accumulation yields information concerning LRHI. The decrease of small droplet accumulation yields information concerning SRHI. The exactness of this procedure can be decreased if SRHI preserves in some degree for the broken particles also.

F. Flocculation in Centrifugal Field and Dynamic Adsorption Layer (DAL)

Centrifugal fields are used for demulsification [128] and in the investigation of emulsion stability [101,102]. The increase of a droplet velocity in centrifugal field (for example, 100-fold) led to the corresponding increase of Reynolds and Stokes numbers. Thus, the movement of not very small droplets in the centrifugal field occurs in the intermediate range of Reynolds number (Sec. IV.E.3) and their flocculation can be complicated by the action of the inertial force (Sec. IV.D).

Larger drop velocities create stronger viscous stresses in the vicinity of its surface. Thus, the residual surface mobility increases as the Reynolds number increases and has to be taken into account in the theory of orthokinetic coagulation. Probably this question was not considered with respect to drops. Thus, the experience of the flotation theory can be used [18,19].

Steady-state motion of a drop induces adsorption–desorption exchange with the subsurface, with the amount of substance adsorbed on one part of its surface being equal to the amount desorbed from another part. Obviously, the surface concentration is lower on the part where adsorption takes place and it is higher on the part where desorption occurs. Thus, surface concentration varies along the surface of the moving drop, taking a maximum value at the rear pole and a minimum one at the front pole [99]. The adsorption difference between the poles of a drop causes the surface tension drop and Marangony–Gibbs phenomena; that is, the surface movement in the direction of the surface tension decreases. This secondary surface flow is directed opposite to the primary surface and flow; consequently, the surface velocity retards.

Therefore, the state of adsorption layer on a moving drop surface is qualitatively different from that on a resting one. Such an adsorption layer was called a dynamic adsorption layer [129].

With respect to the large Reynolds numbers, the retardation coefficient can be evaluated at the strong surface retardation by using the concept of the hydrodynamic and diffusion layers [99,129] having a thickness δ_G and δ_D, respectively, independent of angle θ. It follows from the boundary condition expressing the interconnection between adsorption–desorption flux and the surface divergence of the surface convective flux of the adsorbates [129]

$$\chi_b = \frac{RT\Gamma_0^2}{Dc_0}\frac{\delta_D}{a}\frac{\delta_G}{a} = \frac{RT\Gamma_0^2}{Dc_0}\frac{D}{u_0\,\text{Re}} \tag{66}$$

because

$$\frac{\delta_c}{a} \sim \text{Re}^{-1/2}, \qquad \frac{\delta_\delta}{a} \sim \text{Pe}^{-1/2} \tag{67}$$

and Peclet number $\text{Pe} = \dfrac{\mu a}{D}$

$$\text{Pe} = \frac{\nu}{D}\,\text{Re}. \tag{68}$$

Comparing Eqs. (66) and (45), one concludes that the decrease of the retardation coefficients is

$$\frac{\chi_B\big|_{\text{Re}\gg1}}{\chi_B\big|_{\text{Re}\ll1}} = \frac{D}{\nu\,\text{Re}}. \tag{69}$$

Taking into account that $\nu/D \sim 10^3$, one concludes that the transition from a small Reynolds number to a larger one is accompanied by a reduction in the retardation coefficient by a factor of 10^4–10^5. Thus, for big drops or under the action of centrifugal force, the state of the mobile surface is possible even at noticeable surfactant concentrations.

The dynamic state of the adsorption layer causes the change in all stages of the flocculation. It influences the degree of the drop surface retardation, its velocity and the relative velocity for two widely separated drops, the hydrodynamic velocity distribution around a drop, and, consequently, the LRHI. It modifies surface forces because they depend on the surface concentration. Thus, it is important at the last stage of flocculation too.

G. Incorporation of Short-Range Hydrodynamic Interaction and Attractive Surface Forces in the Theory of Inertialess Collision

In the process involving inertialess approach of a smaller drop to a bigger one, the size of the latter plays an important role. It is in the equatorial plane that the closest approach of a streamline to the surface of a bigger drop is attained. In Fig. 7, the broken line (curve 1) represents the liquid streamline whose distance from the surface of bigger drop in the equatorial plane is equal to the radius of the smaller drop. Some authors erroneously believe that this liquid streamline is limiting for the drops of that radius. The error consists in that the SRHI is disregarded in this case. Under the

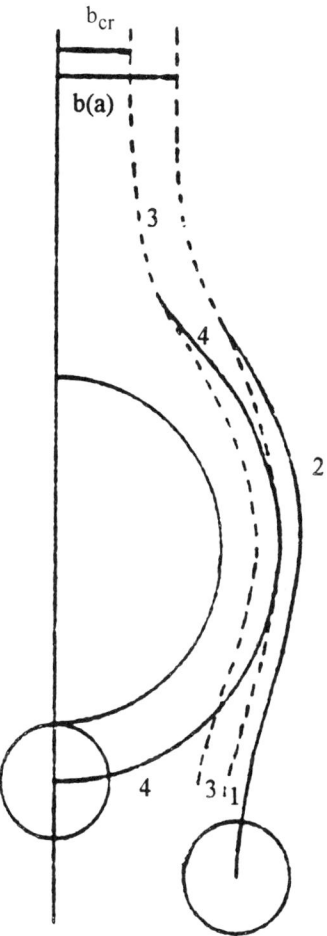

FIG. 7 The influence of the finite dimension of particles in the inertial-free flotation on their trajectory in the vicinity of the floating bubble. The liquid flow lines corresponding to target distances $b(a)$ and b_{cr} are indicated by dashed lines. The continuous lines are characteristic of the deviation of the trajectory of particles from the liquid flow lines under the influence of the short-range hydrodynamic interaction.

influence of the SRHI, the drop is displaced from liquid streamline 1 so
that its trajectory (curve 2) in the equatorial plane is shifted from the
surface by a separation larger than its radius. Therefore, no contact with
the surface occurs and, correspondingly, $b(a_1)$ is not a critical target distance.

Due to the SRHI, the distance from the smaller drop to the surface in
the equatorial plane is larger than the distance from the surface to the
liquid streamline with which the trajectory of the smaller drop coincides
at large distances from the bigger drop. It may thus be concluded that b_{cr}
$< b(a_1)$. The limiting liquid streamline (curve 3) is characterized by the
drop trajectory (curve 4), which, branching off under the influence of the
SRHI, runs in the equatorial plane at a distance a_1 from the surface of
the bigger drop.

The value of b_{cr} decreases, first, due to the deflection of the liquid
streamlines under the influence of the LRHI and, second, due to the deflection of the small drop trajectory from the liquid streamline under the
influence of the SRHI. Therefore, the collision efficiency is expressed as
the product of two factors, E_0 and f, both smaller than unity. The first
one represents the influence of the LRHI and the second, the influence
of the SRHI.

It is evident that SRHI, like LRHI, manifests itself in the gravitational
coagulation of drops, solid particles of suspensions, bubbles, and in their
sedimentation from a flow on large-size surfaces, with the SRHI mechanism being common in these systems. In essence, identical SRHI theories
are created as applied to flotation, aerosols, and suspension purification
from particles by filtration. Only lately has problem received further development as applied to emulsions [27,28].

The mechanism of SRHI is the same under conditions of perikinetic
and orthokinetic coagulation. Formulation and solution of this problem
as applied to perikinetic coagulation is usually associated with the work
by Derjaguin and Muller [23].

It is less known that it was preceded by a work by Derjaguin and Dukhin
[108] in which the problem of SRHI was formulated as applied to orthokinetic coagulation or, more properly, to the elementary act of flotation
of a small spherical particle by a bubble.

When the mineral particle comes closer than its own radius to the bubble viscous resistance forces should develop retarding thinning of the
intervening layer . . . assuming the trajectory of the particle centre to
be independent of its radius it is easier for a particle of larger radius
to contact the bubble surface. It can easily be shown that it is impossible
to predict the outcome of the competition between two opposite effects

without considering the forces of interaction between particle and bubble surfaces. When the radius of action of the attraction forces acting in zone 3 is much smaller than the particle radius the influence of viscous deflection of the particle trajectory will always outweigh the other effect and contact . . . will not occur [108].

If we replace "bubble" by "a big drop" and "a particle" by "small drops" in this quote, then we obtain the statement of the problem of the role of SRHI in emulsions. As can be seen from the quote, the main features of this stage of coagulation process were already clear at that time. When the size of the particle increases, the effect of LRHI on collision efficiency decreases and the negative effect of the SRHI increases. It is shown in the further development of quantitative theory that the effect of particle size prevails on LRHI. Along with this, the remark about flocculation of large particles not taking place even in the presence of attractive forces is not always correct, as seen below.

1. Quantitative Theory of Short-Range Hydrodynamic Interaction Given Attractive Surface Forces

A more general approach to the problem of small-particle deposition from laminar flow on the surface of a big particle (collector) was elaborated by Spielman [131] and Goren [132,133]. They introduced a local coordinate system for the description of a local hydrodynamic field around the small particle and in the interlayer between it and the collector. They presented the equations for the short-range hydrodynamic forces caused by the field. Their results is an important component of the Derjaguin–Dukhin–Rulyov (DDR) theory of SHRI in flotation [19]. The particles move along a bubble surface and the condition for particle–bubble interactions changes, (i.e., the short-range interaction is unsteady). However, in the local coordinate system which is bound to the particle and moves with it, the interaction can be considered as a quasisteady one. The cylindrical system of coordinates is introduced. Its center is lying on the bubble surface with the z-axis crossing the center of the moving particle.

Cylindrical symmetry in the chosen coordinate system simplifies the description of hydrodynamic interaction. Separate equations for the normal component $u_r = dh/dt$ and tangential component $u\theta = a_2(d\theta/dt)$ of a particle velocity are written respectively

$$\frac{dh}{dt} = \frac{F_n f_1(h)}{6\pi\mu a_1}, \tag{70}$$

$$a_2 \frac{d\theta}{dt} = u_\theta f_2(h). \tag{71}$$

The equations describe the stage of a particle movement when the liquid interlayer is thin and the distance between the centers of a particle and the bubble equal to $a_1 + a_2$.

The rates of interlayer thinning and particle movement to the bubble surface coincide and are determined by the action of pressing force F_n and the resistance force which is characterized by the product of the Stokes drag coefficient and the dimensionless function $f_1(H)$.

The particle trajectory equation follows from the set of Eqs. (70) and (71) after the introduction of H instead of h and the time exclusion

$$\frac{dH}{d\theta} = \frac{a_2 F_n(\theta, H) f_1(H)}{6\pi\eta a_1^2 u_\theta(\theta, H) f_2(H)}. \tag{72}$$

In the general case, the pressing force is a superposition of many forces:

$$F_n = F_A + F_\Psi + F_V + F_g, \tag{73}$$

where F_A, F_Ψ, F_V, and F_g are the attractive surface force, the repulsive surface forces, the hydrodynamic pressing force, and the gravity force, respectively. The semianalytical solution of Eq. (72) is possible if the radius of action of surface forces is small in comparison with the particle dimension.

The hydrodynamic pressing force is proportional to the local value of the normal component of the hydrodynamical velocity:

$$F_n = F_H = 6\pi a\eta u_r(H, \theta) f_3(H). \tag{74}$$

The function $f_3(H)$ yields its dependence of the film thickness. After the substitution of Eq. (74) into Eq. (73), one obtains

$$\frac{dH}{d\theta} = \frac{R u_r(\theta, H) f_1(H) f_3(H)}{a u_\theta(\theta, H) f_2(H)}. \tag{75}$$

The functions f_1, f_2, and f_3 were determined in the works of Goren [133], Goren and O'Neil [132], Goldman et al. [134], and Spielman and Fitzpatrick [92]. At a distance which is comparable with the particle dimension (i.e., at $H \geq 1$), a simplification occurs:

$$\frac{f_1 f_2}{f_3} \sim 1. \tag{76}$$

It means that at this distance, Eq. (72) transforms into an equation for the liquid streamline. Consequently, at this distance, the particle trajectory coincides with the liquid streamline.

The theory specifies how surfaces of the two models influence the drainage [19].

1. A critical thickness h_{cr} exists. Naturally, there is a correlation between h_{cr} and DL thickness. This is not taken into account and surface forces action beyond h_{cr} is neglected. In this model, a particle is attached after the film thinning is culminated, satisfying the boundary condition

$$H\left(\frac{\pi}{2}\right) = H_{cr} \qquad \left(H_{cr} = \frac{h_{cr}}{a_1}\right). \tag{77}$$

Naturally, a particle is attached when $h(\theta) = h_{cr}$ at any θ. However, condition (77) separates the grazing trajectory. The result for creeping flow is

$$f_s = 1.5(1 - 0.3 \ln H_{cr})^{-2}. \tag{78}$$

2. The notion of h_{cr} is neglected. The attachment is caused due to a predominating attraction [135]. In the case of the first model, neglecting the molecular attractive force we could not consider the particle attachment on the lower surface of a bubble because any particle departs there from the bubble surface under the gravity action. Neglecting h_{cr}, we consider the particle almost on the bubble surface under the action of the predominating attractive force which can exceed the gravity. Thus, the model enables one to consider the particle attachment on the lower surface of a bubble. The grazing trajectory finishes at the rear stagnant pole of the bubble. Due to cylindrical symmetry,

$$\left(\frac{dH}{d\theta}\right)_{\theta=\pi} = 0. \tag{79}$$

Comparing condition (39) and Eq. (72), one concludes that the final coordinates of the grazing trajectory satisfy conditions

$$\theta_0 = \pi, \qquad F_n(\pi, H_0) = 0. \tag{80}$$

2. Theory Specification for Different Models

First, the model function (78) was calculated. At all values under consideration, this function is smaller than unity; it decreases with H_{cr} and becomes zero at $H_{cr} = 0$. This confirms the aforementioned representation of the mechanism based on the influence of the SRHI on the particle deposition process.

As H_{cr} decreases from 10^{-1} to 10^{-3}, f decreases from 0.5 to 0.15; that is, the dependence of f (and of E) on the absolute value of h_{cr} is weak. Thus, the inclusion of the SRHI is important not only in considering the problem of flotation. This effect reduces the number of collisions by several times.

Rulyov [135] has developed the SRHI theory without considering the phenomenological parameter, h_{cr}, by directly taking into account the dependence of molecular forces on h, and obtained

$$E = E_0 f(W), \qquad W = \frac{A a_2^2}{27 u_s \pi \mu a_1^4}. \tag{81}$$

The function $f(W)$ is plotted in Fig. 8 and can be approximated [136] to

$$f = 1.2 W^{0.15} = \frac{1.2 A^{0.15}}{(2\Delta \rho g/9)^{1/6} a_1^{0.7}}. \tag{82}$$

The first model can be applied to unstable emulsions. The interdroplet lamellae is unstable and its critical thickness is h_{cr}. The second model corresponds to emulsions which are unstable with respect to flocculation and stable with respect to coalescence. The drop doublet forms under the action of attractive molecular forces. However, the coalescence is prevented due to the presence of a stable adsorption layer.

The second model was used by Spielman and Fitzpatric [92,93] who earlier considered the same problem as Rulyov [135]. It is interesting to compare the results of these investigations.

The dimensionless number N_A introduced in Ref. 92 differs from W by a factor of 4/3 only.

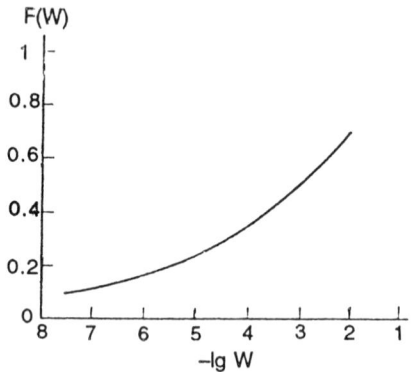

FIG. 8 The multiplier in Eq. (81) expressing the influence of hydrodynamic and molecular interaction between particle and bubble at inertialess flotation. (After Ref. 136.)

In Ref. 92, the dependence E/E_0 is presented as a function of N_A (Fig. 9). Taking into account that $E/E_0 = f$, one concludes that the ordinates in Figs. 8 and 9 coincide. Neglecting the difference in abscissas caused by the multiplier 4/3, one concludes that the functions $f(W)$ and $f(N_A)$ practically coincide. In Ref. 92, N_A is called the attraction number. Its physical sense is explained in Ref. 93.

Different velocity and time scales are appropriately close to the particle surface where viscous and colloidal forces become important. The characteristic fluid velocity is smaller [i.e., $O(a_1 u_0/a_2)$] due to the proximity of the surface, so the time spent in the neighborhood of collector, the dwell time, is $O(a_2^2/u_0 a_1)$. The importance of interparticle attraction is assessed by calculating the time required to capture a particle once it has been moved close to the surface and comparing this with the dwell time. A particle mobility of $O((\mu a_1)^{-1})$ and a dispersion force of A/a (see Sec. II.A) produce a velocity toward the surface of $O(A/\mu a_1^2)$. Therefore, a representative capture time, the time required to move a distance $O(a_1)$, is $O(\mu a_1^3/A)$. The ratio of the dwell time to the capture time is called the attraction or adhesion number, N_A.

As the dwell time exceeds the capture time, that is,

$$N_A > 1, \tag{83}$$

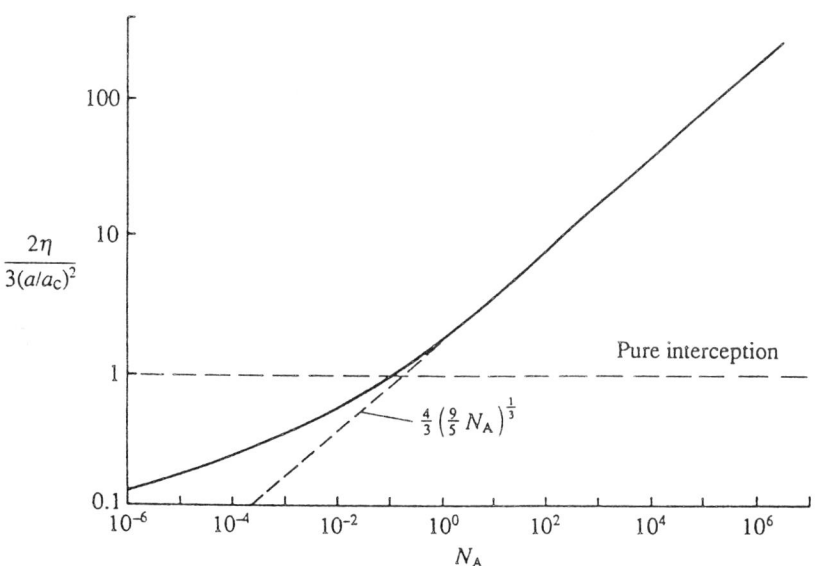

FIG. 9 Capture efficiencies for a spherical collector. Solid line from a numerical solution. (From Ref. 92.)

the smaller drop moves along the surface of bigger drop during a sufficiently long time and can be attracted to the surface from a distance of
the order of its dimension. It means that in condition (83), the collision
efficiency can increase due to attraction forces in comparison with its
value caused by the interception. It explains why at condition (83), $f >$
1, as it is can be seen from Fig. 9. The smaller the N_A, the shorter the
distance from the grazing trajectory to the surface of the bigger drop; that
is, the smaller the collision efficiency and f.

3. Peculiarities of Gravitational Coagulation at Small Aggregation Numbers

An interesting peculiarity of short-range hydrodynamic interaction is the
fact that according to Eq. (82), it depends to a larger degree on the size
of the smaller particle (power 0.7) than on the Hamaker constant (power
1.6). As can be seen from this formula and from Fig. 8, a decrease in the
Hamaker constant is accompanied by a decrease in collision efficiency.
But this dependence is very weak; and so, the impression arises that
coagulation is possible at any small (but finite) value of the Hamaker
constant.

This impression is most likely incorrect. With the increasing size of
the smaller drop, the stability of the doublet is attained at the cost of
the closest separation between two droplet surfaces, h. In this case, it is
assumed that the attractive force infinitely increases. The infinite growth
of attractive force with decreasing h cannot be proved due to difficulties
in measuring surface forces at very small distances [137].

The initial model of surface forces with an infinitely deep primary energy minimum in DLVO theory was replaced by a model with the potential
well of a finite depth in the works by Frens and Overbeek [53] and Martinov [130]. This has resulted in the abandonment of the notion of completely irreversible coagulation and in the development of important notions of reversible coagulation (XIII). In this context, the paradoxal result
can be discussed, which points to the possibility of gravitational coagulation at an arbitrarily small but finite value of the Hamaker constant. Hydration effects and/or structural components of disjoining pressure (Sec.
II.D) which will further restrict the possibility of gravitation coagulation
should be taken into consideration in the future.

The mentioned factors restricting gravitational coagulation manifest
themselves when we consider conditions (73) and (80), which in the absence of the repulsive force can take the form

$$\frac{Aa_1}{6} h_m^2 = 6\pi\mu a_1 \left(4.8 V_2 \frac{a_1^2}{a_2^2} - V_1\right) = (3.8)6\pi\eta a_1 V_2 \frac{a_1^2}{a_2^2} = 2.2\pi\Delta\rho g a_1^3.$$

$$(84)$$

The first term on the right-hand side of the equation is the detaching hydrodynamic force and the second one is the gravity force.

Equations for the absolute values of the detaching force and the pressing force are identical. The normal component of liquid flow has different directions in the vicinity of the leading and rear hemispheres and causes the pressing force on the leading hemisphere and the detaching force on the rear hemisphere. Consequences of Eq. (84) on emulsions will be discussed in the next section.

4. Estimation of the Size of Emulsion Drops Stable Against Flocculation with Larger Drops

As was pointed out in Sec. IV.6.2, the value of the Hamaker constant in oil–water (o/w) emulsions is approximately known. In the case of emulsion-stabilized surfactants, we can obtain the lower bound of h also, i.e., h_m. In this case, h_m exceeds the double thickness of the adsorption layer h_a. Then for the critical radius a_{cr} of the smaller of the drops, it follows from Eq. (84) that

$$a_{1cr} \le \left(\frac{A}{2.2\pi\Delta\rho g}\right)^{1/2} (2h_a)^{-2}. \tag{85}$$

Drops with a radius exceeding a_{1cr} cannot flocculate with drops of a much larger radius, $a_2 \gg a_{1cr}$.

Substituting the values $A = (3 \times 10^{-21} \div 10^{-20})$, $\Delta\rho = 0.1$ g cm^{-3}, $2h_a \cong 2 \times 10^{-7}$ cm into Eq. (85), we obtain $a_{1cr} \le 100$–300 μm.

It is apparent that h_a can be both larger and smaller than the estimated value when using emulsifiers of different nature. The condition of a small Reynolds numbers is poorly fulfilled at this value of a_{1cr} and is completely violated for a larger drop with radius $a_2 \gg a_{1cr}$. This limits the accuracy to estimate according to Eq. (85). Nevertheless, it is clear that the detaching force cannot prevent gravitational coagulation in emulsions for drops of Stokesian size if additional repulsion forces are absent because of the small value of $\Delta\rho$. In suspensions, $\Delta\rho$ can be 10–70 times higher and a_{1cr}, can be 3–8 times smaller. Thus, the statement [108] cited in Sec. IV.G.1 that molecular attractive forces cannot result in coagulation of particles of a larger size is to some extent confirmed. But at a radius $a_1 < a_{1cr}$ even very small molecular forces provide coagulation which was underestimated in Ref. 108.

H. Experiments

There is a lack of systematic experimental investigations of coagulation kinetics in emulsions. This drawback can be compensated in some degree by the consideration of the capture efficiency in adhesion experiments with packed beds and in flotation, because the mechanism of the collision is the same.

(a) Experiments with Packed Beds. Experiments with beds of solid spheres have been carried out to establish the applicability of theories based on single collector-to-collector arrays where several transport processes operate and the flow field is not known in detail. Here, filter coefficients are measured instead of single-particle capture efficiencies and a flow structure parameter is used.

An extensive study of non-Brownian particles was carried out by Spielman and Fitzpatrick [92,93] to test theories for attraction-dominated capture. Several hundred experiments were done wherein latex particles were filtered from aqueous suspensions using beds of glass particles. Particle diameters ranged from 0.7 μm to 21 μm, bead diameters from 0.1 mm to 4 mm, and velocities between 0.01 and 1.0 cm-s^{-1}. Electrolyte type and ionic strength were varied in order to control electrostatic repulsion and to avoid interference from particle deposits. Filter coefficients were calculated from particle concentrations measured with a Coulter counter.

In Ref. 93, data are shown for conditions in which the electrostatic repulsion and sedimentation are negligible. Although there is a fair amount of uncertainty in the data, the data cluster about the line derived from the theory given by Spielman and Fitzpatrick [92].

At low ionic strengths, where repulsion is strongest, agreement between theory and experiment was poor. This was attributed to the acute sensitivity of the capture rate to small changes in the surface potential. Upon making small adjustments (~1 mV) to the values of the ζ-potential of the film used in the theory, the agreement has improved.

(b) Experimental Investigation of Collision Efficiency at a Small Radius Ratio. Equation (63), which can be recommended for the collision efficiency in emulsions, was confirmed in systematic investigations devoted to microflotation kinetics. The monodisperse latex or glass spherical particles were used in the experiments. The radius ratio was very small. Thus, in Eqs. (63) and (82), a_1 corresponds to the latex particle dimension and a_2 to the bubble dimension. The dependence $E \sim a_2^{-2}$ was confirmed in Refs. 139 and 140. The dependence $E \sim a_1^{1.5}$ was established in Refs. 141 and 142. The combination of Eq. (63) describing LRHI and Eq. (82) describing the SRHI influence leads to the dependence $E \sim a_1^{1.4}$. Thus, the theory of short-range hydrodynamic interaction is confirmed by the investigations [137,139–142].

Recently, the direct observation of the grazing trajectory of latex particles (0.9 μm) in the vicinity of the rising bubble with radius 15 μm was accomplished [143,144]. The cell was attached to the microscope stage which is capable of moving vertically in the same velocity as that of the rising bubbles in cell.

The similar experiments in emulsion can be very important. The predicted particle trajectory agreed well with those obtained experimentally. It was found that the collision efficiency values obtained experimentally were at maximum when the absolute values of ζ-potentials of both the bubbles and particles were at minimum.

I. Long and Short Hydrodynamic Interaction and Collision Efficiency at any Radius Ratio

If the difference between particle dimensions is not large, the dynamics of the interacting spheres have to be represented in terms of the relative and center of mass motion by defining $dr/dt = u_2 - u_1$ and $d\bar{x}/dt = u_{12} = (u_1 + u_2)/2$, where u_1 and u_2 (see Fig. 10) are the velocities of particles 1 and 2. Every particle moves under the action of the external field and the interaction with the neighboring particle. Thus, its velocity can be represented by the sum

$$u_i = \sum_{i=1}^{2} \omega_{ij} F_j \qquad (86)$$

with $i = 1, 2$. The mobility tensors ω_{ij} express the response of the ith sphere to a force acting on the jth and depend on the separation. The superposition (86) is justified by the linearity of the Stokes equation. The

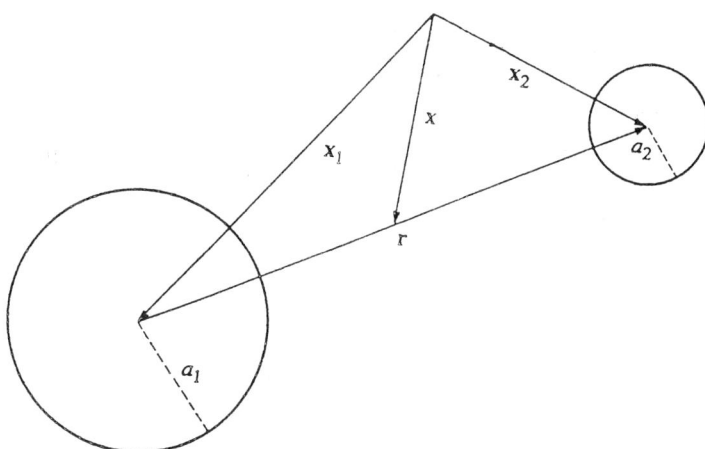

FIG. 10 Two interacting spheres.

set of Eqs. (86) can be transformed in the expression for the relative velocity of two sphere centers. Batchelor and colleagues [67,145,146] accomplished this transformation:

$$\mathbf{u}_{12}(\mathbf{r}) = \mathbf{u}_{12}^{(0)} \left[\frac{\mathbf{rr}}{\mathbf{r}^2} L(s) + \left(\mathbf{I} - \frac{\mathbf{rr}}{\mathbf{r}^2} \right) M(s) \right]. \tag{87}$$

The relative mobility functions for motion along the line of centers (L) and motion normal to the line of centers (M) describe the effects of hydrodynamic interactions between the two drops. These functions depend on λ, $\hat{\mu}$, and s.

When the spheres are very close ($\xi = s - 2 \ll \lambda$, $\xi \ll 1$), the mobility functions have the asymptotic forms given by Jeffrey and Onishi [147]. Using these asymptotic forms, Davis [148] incorporated them in the equation for the particles' trajectories. Its solution with the boundary condition (80) enabled one to determine the grazing trajectory and accomplished the numerical calculation of the collision efficiency. At $\lambda \ll 1$, his results agree with Rulyov's results (Sec. IV.G). Davis' data are plotted in Fig. 13 for $\hat{\mu} = 1000$.

The kinetics of the doublet formation were also analyzed in Ref. 145 by another method. Authors note the agreement with Ref. 148.

J. Collision Efficiency for Interacting Drops

Zhang and Davis [29] use the Batchelor expression [Eq. (87) for u_{12} and modify it for spherical drops.

(a) Mobility Functions for the Relative Motion of Two Drops. Their derivation [29] of expressions for the function L is based on the equation for the lubrication force on drops in the direction along their line of centers. The force balance enables one to obtain the functions $L(s)$ and $M(s)$ in Eq. (87). They consider two unequal drops which are nearly touching and which move together as a pair due to gravity. Superimposed on this is a small relative velocity of the larger drop approaching the smaller one. A force balance on each drop yields

$$F_{g,i} + F_{d,i} + F_{l,i} = 0, \tag{88}$$

where $F_{g,i}$ is the gravity force acting on the drop i, which can be described by the Hadamard–Rybczynski result, Eq. (42):

$$F_{g,i} = 6\pi\mu a_i U_i^{(0)} \frac{3\hat{\mu} + 2}{3\hat{\mu} + 3}. \tag{89}$$

$F_{d,i}$ is the drag force exerted on the drop i by the surrounding fluid and is defined as the total hydrodynamic force minus the lubrication force.

According to the analysis of Reed and Morrison [149] for two touching drops, it can be expressed as

$$F_{d,i} = -6\pi\mu a_i U_p \frac{3\hat{\mu} + 2}{3\hat{\mu} + 3} \beta_i, \tag{90}$$

where β_i is a correction factor of the Hadamard–Rybczynski formula for drop i to account for the presence of the second drop.

Equations (88)–(90) for $i = 1$ and $i = 2$ may be solved for the pair velocity U_p and the lubrication force, $F_{1,1} = -F_{1,2}$. It enables the determination of the coefficients L_1, M_0, and M_1 in asymptotic equations:

$$L\xi = L_1\xi, \qquad M(\xi) = M_0 + M_1(\xi). \tag{91}$$

The coefficient dependences on $\hat{\mu}$ and λ are given numerically in Table 1 of Ref. 29.

The trajectory equation follows from the definition of functions L and M,

$$\frac{ds}{d\theta} = s \frac{-L(s) \cos \theta}{M(s) \sin \theta} \tag{92}$$

In the absence of interparticle forces, the limiting trajectory is one with the final condition $s = 2$ (contact) when $\theta = \pi/2$, because by symmetry, the point of closest approach occurs at $\theta = \pi/2$. Using the result of the integration of Eq. (92), the collision efficiency is [29]

$$E_{12} = \exp\left(-2 \int_2^1 \frac{M - L}{sL} ds\right). \tag{93}$$

With interparticle forces considered, the trajectory equation can no longer solve L analytically for an explicit formula for the dimensionless critical impact parameter, and hence the collision efficiency. Instead, the determination of the collision efficiency has to be performed by integrating Eq. (92) numerically along the limiting trajectory from the infinite separation of two drops to the termination point.

The dimensionless critical impact parameter may be determined by integrating Eq. (92) backward along the limiting trajectory from the termination point $\theta = \pi$ and $\xi = \delta$, to a position $s = s_1$ and $\theta = \theta_f$, beyond which the van der Waals forces are negligible. This numerical solution may be matched with the solution in the outer region ($s > s_f$). Setting $s > s_f$ and $\theta = \theta_f$ as the matching condition reveals that

$$E_{12} = \left\{4\left[s_f \sin \theta_f \exp\left(\int_{s_f}^{x} \frac{L - M}{sL} ds\right)\right]^2\right\}^{-1}. \tag{94}$$

Figure 11 shows the results for E changing with λ for several different $\hat{\mu}$ [29]. The results for the collision efficiencies predicted by Zinchenko [70] are presented as solid circles for comparison. There is a very good agreement between the present results and Zinchenko's, with the relative difference between them being smaller than 3%. The *collision efficiency* approaches a finite value as the drops become equisized ($\lambda \to 1$). However, the *collision rate* goes to zero in this limit because the relative velocity of the two drops approaches zero. The collision rate may be nondimensionalized with a quantity not involving the size ratio: $J_{12}/J_{12}/(n_1 n_2 u_{01} \cdot \pi a_1^2) J_{12}/(n_1 n_2 u_{01} \pi a_1^2) = E_{12}(1 - \lambda^2)(1 + \lambda)^2$. This quantity is shown in Fig. 12. The collision rate is small for small size ratios because of the reduced collision cross section and collision efficiency (as discussed previously), achieves a maximum at moderate size ratio, and then decreases as the size ratio approaches unity because of the reduced relative velocity.

The unique peculiarity of results characterized by Fig. 12 is that the nonzero collision efficiency occurs in the absence of attractive molecular forces. This is possible due to the mobility of droplet surfaces. The mobility decrease caused by liquid viscosity increase leads to a drastic decrease in the collision efficiency. As $\hat{\mu}$ increases, E decreases and its value is unusually low for $\hat{\mu} = 100$ and 1000. The increase in the unrestricted $\hat{\mu}$

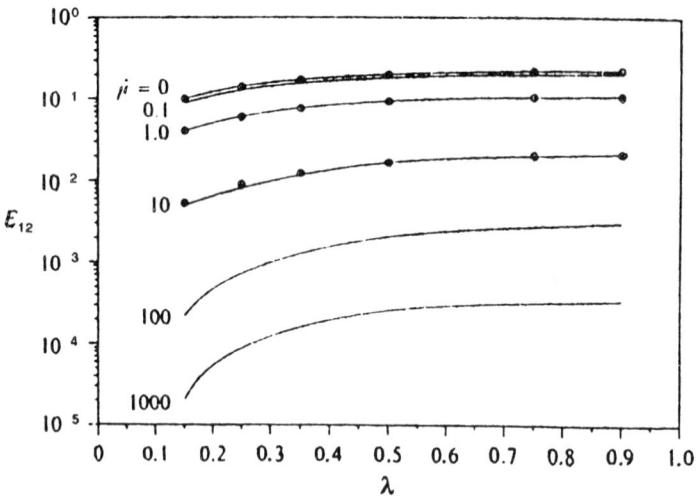

FIG. 11 The dimensionless collision rate, $J_{12}/(n_1 n_2 U_0 \pi a_1^2) = E_{12}(1 - \lambda^2)(1 + \lambda)^2$, for gravity sedimentation of drops as a function of the size ratio for various viscosity ratios without interparticle forces. (After Ref. 29.)

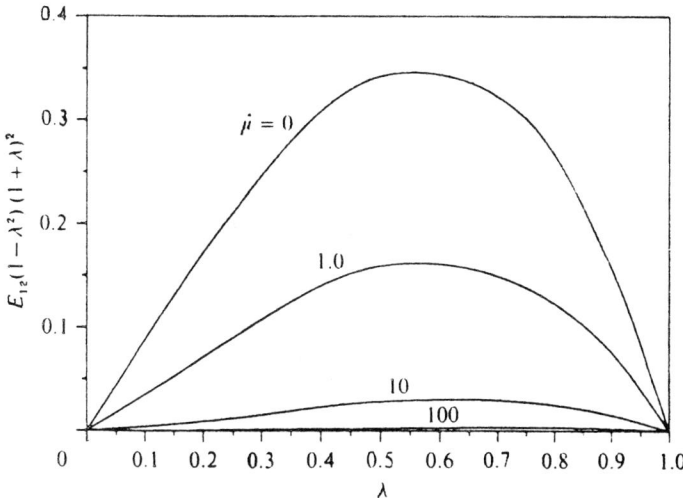

FIG. 12 The collision efficiency for gravity sedimentation of drops as a function of the size ratio of various viscosity ratios without interparticle forces. (After Ref. 29.)

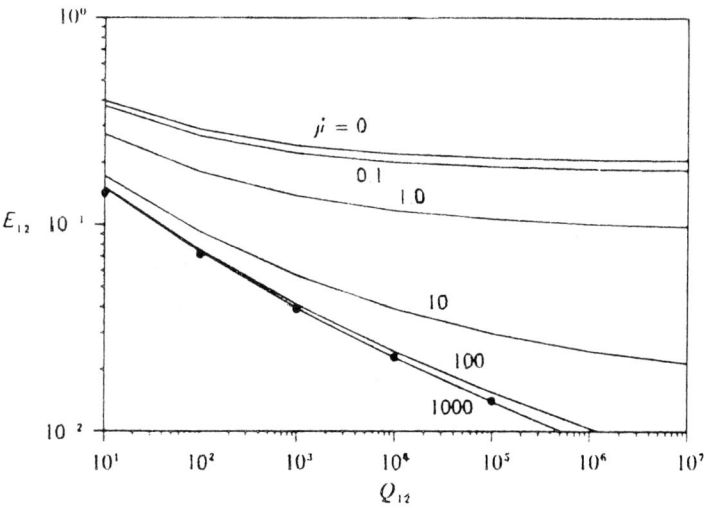

FIG. 13 The collision efficiency for gravity sedimentation of drops as a function of the interparticle force parameter for $\lambda = 0.90$ and various viscosity ratios with unretarded van der Waals attractions. The solid circles are the rigid sphere results of Davis (1984). (After Ref. 65.)

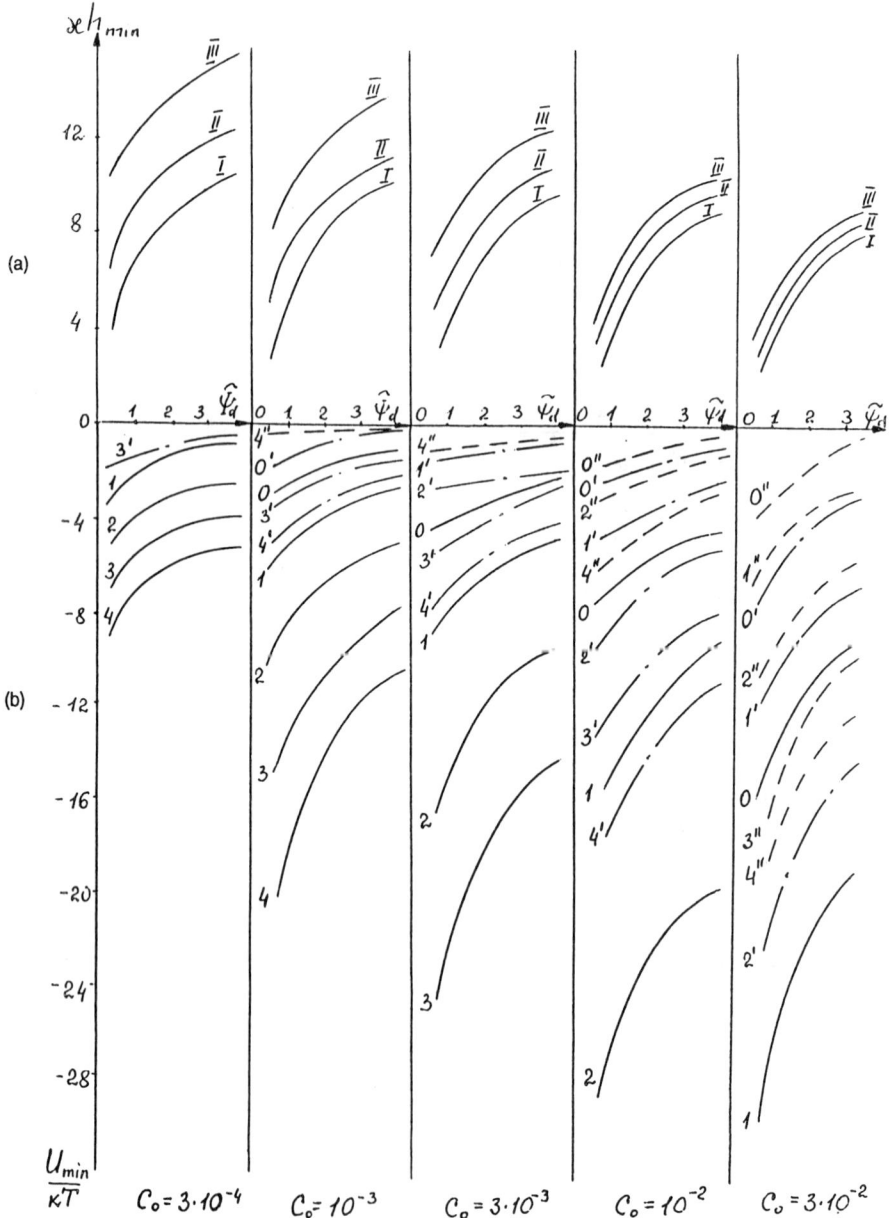

FIG. 14 Dependence of the Stern potential for the coordinate (a) and the depth (b) of the secondary minimum: I. 0–4 without the retardation and screening; II, 0'–4' with account of retardation; III, 0''–4'' with account of retardation of retardation and screening. (0—a = 0.5 μm; 1—a = 1 μm; 2—a = 2 μm; 3—a = 3 μm; 4—a = 4 μm). The electrolyte concentrations are shown at the bottom.

corresponds to the transition to the solid particle and to the transition to $E_{12} = 0$ in the absence of attractive surface forces.

Typical results for the collision efficiency as a function of the parameter Q_{12} are shown in Fig. 13 for $\lambda = 0.9$ for various $\hat{\mu}$, where the corresponding results of Davis [148] for rigid spheres are shown as solid circles. In the limit of $\hat{\mu} \rightarrow \infty$, there is an excellent agreement between our new results [29] and those of Davis [148]. As expected, the collision efficiency increases with increasing values of the Hamaker constant. The collision rate is more sensitive to the van der Waals attraction for drops with high viscosities relative to the surrounding fluid than for drops with moderate or low viscosities. Drops with high viscosities offer considerable resistance to near-contact relative motion. For collisions to occur, these resistance must be overcome by the van der Waals attraction. The internal flow for drops with low viscosities allows them to collide with relatively little resistance and without the aid of attractive forces.

V. GRAVITATIONAL COAGULATION IN SECONDARY MINIMUM

A. Secondary Minimum of Interaction Energy in Emulsions

Thermal motion energy of a colloidal particle has a value proportional to kT. The particle will easily escape a potential well whose depth has the same order of magnitude. An energy of the order of 10 kT is seldom acquired by a particle. Hence, a particle will stay rather long in a potential well of this magnitude. The depth of the secondary minimum can be fairly large (i.e., of the order of 10 kT and more) if two conditions are fulfilled simultaneously. The thickness of the double layer is small (of the order of nanometers) and the size of the particles is rather large.

The necessity of these conditions is clear without further calculations. The secondary minimum appears at the periphery of DL where electrostatic repulsive forces are weakened and molecular attractive forces prevail. Therefore, the thinner the DL, the closer the secondary minimum is to the surface; but the closer the secondary minimum to the surface, the higher the molecular attractive forces. A calculation shows that the secondary minimum is localized at a distance of several Debye radii (Fig. 14). Then its distance to the surface is of the order of 10 nm as applied to conditions of interest to us. These conditions are realized rather often in natural colloidal and technological processes. Indeed, electrolyte concentration usually is fairly high under technological conditions and it can be equal to several centimoles in 1 liter, which corresponds to a DL thickness of 3 nm and more.

The depth of the minimum strongly depends on the value of the Hamaker constant and on the size of particle. The energy due to attractive forces and electrostatic energy (and therefore energy as a whole) is proportional to particle size. As can be seen from Fig. 15, the depth of the secondary minimum expressed in kilotesla units quickly decreases with the size of particles. Long-range aggregation is possible for particles in colloidal dispersions only for very high values of the Hamaker constant (i.e., for metal particles). On the other hand, long-range aggregation for large particles in microheterogeneous systems ($a > 1$ μm) is carried out over a wide range of Hamaker constants.

When the charge increases, the electrostatic energy of repulsive forces increases, the primary minimum disappears, and the height of the repulsion barrier grows. The electrical contribution to primary coagulation is very substantial. It is practically impossible to prevent long-range coagulation by increasing electrostatic repulsion (at real high electrolyte concentrations). The electrostatic factor of stability is strong with respect to short-range aggregation and is weak with respect to the long-range one. In colloidal systems, it dominates over other factors because the depth of the secondary minimum is small and long-range aggregation is impossible. The electrostatic factor appears as a weak one in microheterogeneous systems when the depth of the secondary minimum is large and slightly depends on particle charge so that it does not prevent the long-range aggregation.

The depth of the secondary minimum was calculated from the expression for the total interaction energy between a spherical particle and a planar surface or (what is the same) with a particle of a much larger size. With regard to Eqs. (1) and (7), this expression has the form

$$y = de^{-\kappa h} - B\frac{\kappa a}{\kappa h}, \tag{95}$$

where

$$y = \frac{U}{kT}, \qquad U = U_A + U_\Psi, \qquad B = \frac{A}{6\,kT}, \qquad d = \frac{\epsilon\Psi^2 a}{kT}. \tag{96}$$

From $y' = 0$ we can find the equation for κh_m:

$$(\kappa h_{min})^2 \exp(-\kappa h_{min})^2 = \frac{B\kappa a}{d} = \beta. \tag{97}$$

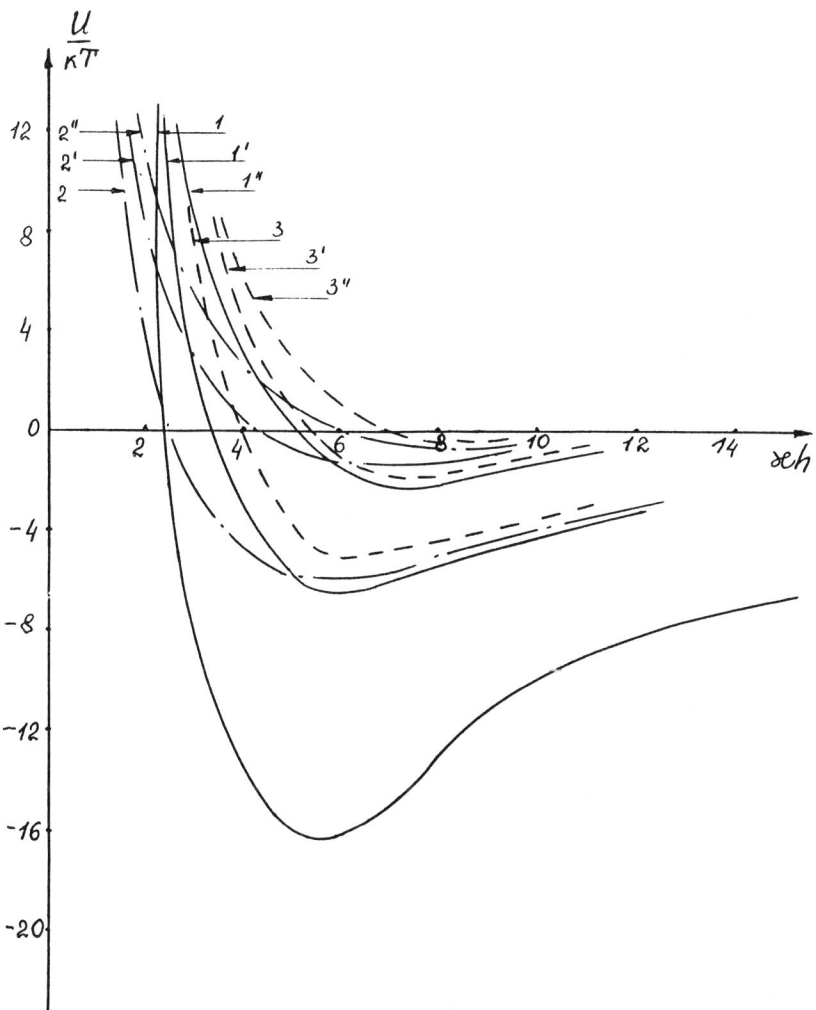

FIG. 15 Pair interaction energy versus κh: 1, 1', 1"—$\tilde{\Psi}_d = 1$; $c_0 = 10^{-2}$ mol L^{-1}, 2, 2', 2"—$\tilde{\Psi}_d = 0.5$; $c_0 = 10^{-3}$ mol L^{-1}, 3, 3', 3"—$\tilde{\Psi}_d = 1$; $c_0 = 10^{-3}$ mol L^{-1}, 1, 2, 3: without retardation and screening; 1', 2', 3': with account of retardation: 1", 2", 3" both screening and retardation are taken into account.

This equation has two solutions, one of which corresponds to the "bottom" of the distant well and the other to the "top" of the barrier. The expression for the depth U_{min} of the secondary minimum follows from Eqs. (95) and (96):

$$-\frac{U_{min}}{kT} = \frac{A\kappa a(\kappa h_{min} - 1)}{6kT(\kappa h_{min})^2}. \tag{98}$$

1. Secondary Minimum in Potential Curve of Charged Emulsions

The dimensionless parameter on right-hand side of Eq. (97) characterizes the balance between the attractive and repulsive forces:

$$\beta = \frac{1}{96\epsilon} \left(\frac{e}{kT}\right)^2 \frac{A\kappa}{\tanh^2\left(\frac{\tilde{\Psi}_d}{4}\right)}. \tag{99}$$

The left-hand side of Eq. (97) has a maximum at $\kappa h_m = 2$. Its height is $2^2 e^{-2} = 0.54$. It is convenient to introduce the boundary value $\beta_b = 0.54$. There is no solution of Eq. (97) for $\beta > \beta_b$. This corresponds to the predomination of the attractive forces and the absence of the energetic barrier. Accordingly, the barrier and the secondary minimum exist at

$$\beta < \beta_b. \tag{100}$$

Equation (100) is a necessary condition for secondary flocculation. The second condition is a sufficient depth of the secondary minimum. Note that Eq. (11) is valid for $\kappa h > 2$. This condition is always satisfied at the secondary flocculation.

The dependence of the depth of the secondary pit on the droplet dimension is plotted in Fig. 14. One can see that it is impossible to remove the secondary pit by the potential increase.

According to Eqs. (1) and (7), the influence of the droplet dimensions is identical. Thus, Eq. (95) can be generalized by multiplying by $\lambda/1 + \lambda$. The localization of the secondary minimum does not depend on λ. However, the equation for its depth of minimum has to be generalized:

$$\tilde{U}(\lambda) = \frac{\lambda}{1 + \lambda} U\big|_{\lambda = 0}, \tag{101}$$

where $\tilde{U}\big|_{\lambda = 0}$ is given by Eq. (95).

Note that the depth of the potential pit is a factor of 2 smaller for identical droplets in comparison with the case $\lambda \ll 1$.

At lower electrolyte concentrations and higher potentials, the distance of the secondary minimum from the surface is larger by 5–10 nm. Thus, the retardation of the dispersion forces (Sec. II) influences the localization of the secondary minimum and the depth of the potential pit (Figs. 14 and 15).

2. Repulsive Hydration Forces and the Secondary Minimum

Similar to the electrostatic energy, hydration forces decline (Sec. II) exponentially with the distance. Thus, the conditions for the formation of the secondary potential pit are identical. This identity exist even on the quantitative level because the same function describes the dependence on the distance for electrostatic and hydration forces. Due to this similarity, the introduction of the dimensionless ratio h/h_s instead of κh enables one to derive an equation similar to Eq. (98):

$$\tilde{U}_{\min} = A \frac{a}{h_s} \frac{h_{\min}/h_s - 1}{6kT(h_{\min}/h_s)^2}, \tag{102}$$

where $h_{\min}$ satisfies the equation similar to Eq. (97):

$$\left(\frac{h_{\min}}{h_s}\right)^2 e^{-(h_{\min}/h_s)} = \beta_s = \frac{A}{6\pi h_s^3 K_s}. \tag{103}$$

β_s characterizes the balance between the attractive and the hydration repulsive forces. Because Eqs. (97) and (103) are identical, we can conclude that the hydration forces can produce the barrier and the secondary minimum at the condition

$$\beta_s < \beta_{sb}, \tag{104}$$

which is identical with condition (100).

If K_s and h_s are not sufficiently large, the hydration forces cannot give rise to the barrier and the secondary minimum. The manifestation of the hydration repulsive forces was found in the investigation of emulsion stability in Ref. 150. The substitution of the values of K_s and A determined in Ref. 150 and $h_s = 1$ nm, assumed by authors, into Eq. (103) yields β_s value exceeding β_{sb}. However, Eq. (103) for β_s is written for an interaction at small λ, whereas the equation for any λ is used in Ref. 150. For $\lambda \ll 1$, the multiplier 0.5 has to be introduced into Eq. (103). After this correction, it turns out that the experimental data in Ref. 150 agree with condition (104). The information concerning K_s and h_s values in Ref. 150 agrees with condition (104). The data cited in Ref. 32 confirm the importance of repulsive hydration forces in the formation of the energy barrier and secondary potential pit. These forces were introduced to explain the forces observed between bilayers in lamellar phases [151,152]. In Ref. 32, the data concerning h_s for different phospholipids, lipids, and a few surfactant are summarized in the Table 14 (of Ref. 32). The table shows that $h_s =$ 2–3 nm for phosphatidylcholins [153], about 2 nm for phosphatidylethanolamine and approximately 1.5 nm for glycerol groups on alkoxyglycerol surfaces. The K_s values change in broad range.

In Ref. 32, the authors emphasize that the surface charge in food emulsions is low, electrolyte concentration is high, and, hence, the DL is not responsible for the emulsion stability. The stabilization can be caused by the repulsive hydration forces. However, the secondary flocculation preserves.

B. Estimation of the Size of Drops Not Coagulating in the Secondary Minimum. Sedimentation–Hydrodynamic Stability Mechanism of Microheterogeneous Dispersed Systems

Kinetics of primary gravitational coagulation as a joint action of attractive and repulsive forces will be considered in Sec. X. The important question of a critical size of drops not undergoing a coagulation in the secondary minimum can be considered to be very simply based on the balance of forces at the rear stagnant pole of larger drop (Sec. IV.G.4). Most likely, this effect was considered for the first time when applied to flotation [19].

It was shown in Ref. 19 that only particles of a rather small size can float, at the cost of coagulation in the secondary minimum. The larger the particle, the deeper the secondary minimum and the larger the gravity force. Gravity forces increase with increasing particle size faster than the depth of the secondary minimum. Therefore, coagulation of rather large particles in the secondary minimum is impossible. The failure of long-range aggregation for rather large particles (microheterogeneous disperse systems) is an experimentally established fact considered by Kruyt [154], although he provided no explanation for the phenomenon.

The notion of a sedimentation–hydrodynamic mechanism of aggregate stability (SHAS) of microheterogeneous systems was introduced in Ref. 155. The importance of this phenomenon for classification of powders is shown in Ref. 156.

The dependence of particle interaction force on h is obtained by differentiating the known curve of total interaction energy (Fig. 16). The decrease in the curve to the right of the minimum of total interaction force F is caused by the decrease of each of the forces with distance. The

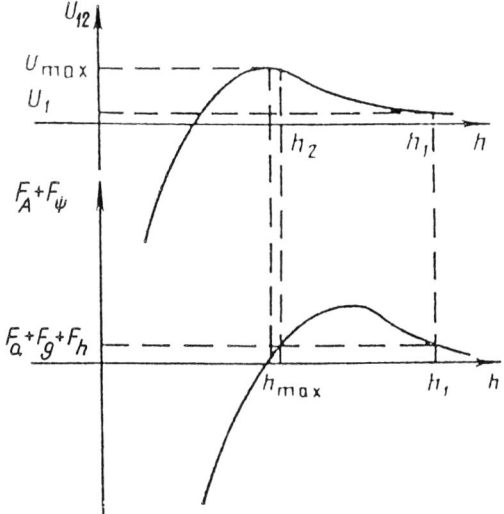

FIG. 16 Illustrative curves representing the dependencies of surface forces F_A + F_Ψ and corresponding energy U_{12} on h. The points h_1 and h_2 are characterized by the equality of the surface forces and the bulk forces $F_a + F_h + F_g$. (After Ref. 155.)

decrease at the left is caused by the faster increase in repulsive forces than in the attractive forces. The minimum size of the particle is determined from the condition of equilibrium between the pull force and the maximum force of attraction between particles in the floc:

$$F_A(h_{f\text{min}}) + F_\Psi(h_{f\text{min}}) = F_h + F_g + F_a. \qquad (105)$$

As mentioned in Sec. V.A, energy due to molecular attractive forces predominates in the secondary energy minimum. This is true to a still larger extent in the secondary force minimum because $\kappa h_{f\text{min}} > \kappa h_{\text{min}}$ (see Fig. 16). Therefore, the electrostatic repulsive force in Eq. (95) can be omitted. In view of a large distance from the surface, the molecular attractive force in the region of the secondary force minimum should be calculated with regard to electromagnetic delay, that is, by Eq. (4). F_g and F_a in Eq. (105) are the gravity and the Archimed force, respectively, for the smaller of the particles; F_h is the detaching hydrodynamic force. Taking into consideration the aforementioned, the formula follows from Eq. (105) [155]:

$$a_{1cr} = \sqrt{\frac{B}{2\Delta\rho g(h_{fmin})^3}} F(\lambda),$$

(106)

$$F(\lambda) = \sqrt{\frac{\lambda^2[\lambda\Lambda(\lambda) + \Lambda(\lambda^{-1})]}{(1 + \lambda)[\Lambda(\lambda) - \lambda^2\Lambda(\lambda^{-1})]}},$$

where κh_{fmin} is the greater of two possible positive roots of the analog to Eq. (97),

$$(\kappa h_{0min})^4 e^{-\kappa h_{0min}} = \frac{2\pi\kappa^2 B}{\Psi^2\epsilon},$$

(107)

and $B = \lim(A(h))_{h\to\infty}$.

A very complicated function $\Lambda(\lambda)$ was obtained in Ref. 157. Its graphical and asymptotic representations are given in Ref. 158. The plot of the function $F(\lambda)$ is given in Ref. 158. The calculation is performed for the values $B = 2.5 \times 10^{-20}$ J m, $\Psi = 25$ mV, $c = 10^{-3}$ M, and $\lambda = 0$ (i.e., approximately at $a_2 \gg a_1$. $a_{1cr} = 1$ μm corresponds to this case). If we increase λ to 0.9 and make it very close to a_2, $a_{1cr} = 4.2$ μm at $a_2 = 4.2$ μm. These calculations have been performed for titanium carbide powder.

Using formula (106), the graphical representation of function $F(\lambda)$, and the calculated value of $a_{1cr}|_{a_{1cr}\ll a_2}$, we obtain curve $a_{1cr}(a_2)$ as presented in Fig. 17. At any fixed a_2, particles with radius $a_1 > a_{1cr}(a_2)$ cannot aggregate with particles with radius a_2. Only for particles with radius $a_1 < a_{cr}(a_2)$ is this possible.

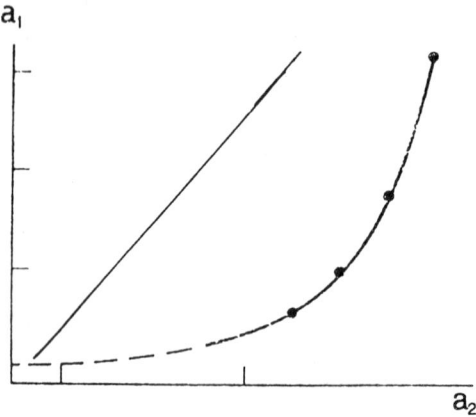

FIG. 17 Flocculation in the secondary minimum is impossible for a pair particles belonging to the region between bisectris $a_1 = a_2$ and curve $a_{1cr}(a_2)$.

Let us consider a region in Fig. 17 bounded below by curve $a_{1cr}(a_2)$ and by bisectors $a_1 = a_2$. Each point corresponds to a pair of particles with radii a_1 and a_2. For a pair of particles belonging to this region, secondary flocculation is impossible due to the sedimentation–hydrodynamic mechanism of stability. The position of the boundary of this region [i.e., of curve $a_{1cr}(a_2)$, depends on values of B] $\Delta\rho$, electrolyte concentration, and Stern potential according to Eqs. (106) and (107).

C. Wet Fractionation Method and Sedimentation–Hydrodynamic Mechanism of Aggregate Stability

It is apparent that aggregation prevents powders from separation. The sedimentation–hydrodynamic aggregative stability mechanism (SHAS) prevents particles of rather large size from aggregation and thus creates necessary conditions for separation of narrow fractions. This explains the substantial success in the classification of wet powders (159). The difference in sedimentation rate for particles of different size is used in wet classification. In the light of results obtained in the preceding section, pairs of particles belonging to region 1 (Fig. 17) can be separated in principle using this method. We will divide the region under curve $a_{1c}(a_2)$ into two zones. Zone 2 in which either a_1 or a_2 is reasonably small is characterized by a small depth of the remote potential well. Flocculation does not occur in zone 2. The boundary between zones 2 and 3 is fuzzy because the lifetime of a doublet due to secondary flocculation depends on the size of particles.

Pairs of particles corresponding to points in zone 2 are subjected to secondary aggregation. Liquid classification according to the method above is difficult. This difficulty is overcome by centrifuging. As can be seen from Eq. (106), an ordinate of curve $a_{1cr}(a_2)$ decreases by a factor of 10 when we apply centrifugal acceleration of $100 \times \mathbf{g}$.

D. Peculiarities of Secondary Flocculation in Emulsion and Their Fractionation

According to Eq. (106), a_{1cr} for the emulsion is $\sqrt{\Delta\rho_{TiO_2}/\Delta\rho_{em}}$ times higher than in the case of titanium carbide (i.e., approximately 7 μm at $a_2 \gg a_1$). At $\lambda = 0.9$ $a_{1c} \sim 30$ μm and $a_2 = 33$ μm.

Thus, due to the small $\Delta\rho$, emulsions are subjected to secondary flocculation to a substantially larger extent than microheterogeneous solid suspensions.

Nevertheless, the SHAS mechanism can presumably have a strong effect on secondary flocculation in emulsions and creaming and deserves systematic studies.

Considerations set forth in Secs. V.A–V.C relate to the initial stage of flocculation. In going from doublets to multiparticle aggregates, the mechanism behind SHAS becomes more complicated.

In accordance with the analysis of Fig. 17, after a lapse of time, the rear stagnant part of the surfaces of the big droplet will be covered by small droplets. It is apparent that applicability of Eq. (106) needs a revision as applied to paired hydrodynamic interaction of such large particle aggregates.

Differing from powders, a substantial difficulty arises in fractioning micron-size drops as applied to emulsions because they are aggregated so that SHAS does not function in this case. It is likely that fractioning can be achieved by applying a centrifugal field. A change to the regime of fairly large Reynolds numbers (Sec. IV.E.3) and supercritical Stokes numbers (Sec. IV.D) is possible in this case. DAL can have an effect on interaction of drops in this situation (Sec. IV.F). Theoretical analysis of emulsion fractioning in centrifugal field is thereby substantially complicated.

This range of questions is of fundamental interest for emulsion stability problem because, in this way, it will most likely be possible to obtain narrow fractions of drops. This will allows us to apply the flocculation investigation methods to emulsions which have proved to be advantageous in studying flocculation kinetics of monodispersed systems (Sec. III.D).

E. Gravitational Coagulation Kinetics in the Secondary Minimum

1. General

Flocculation of coarse emulsions in the secondary minimum proceeds at any electrokinetic potential and at fairly high electrolyte concentrations or even at low electrolyte concentrations when combined with thicker DL and small surface charge density (Sec. V.1). In other words, secondary coagulation of coarse emulsions is more a rule than an exception. Because of the large drop size, this process is mainly a gravitational coagulation. For some intermediate drop size, Brownian coagulation might not be excluded; but gravity coagulation predominates for large drop sizes.

If we consider a general case of emulsion containing also small drops, the posed question has great importance. Secondary flocculation will be of minor importance for small drops because the depth of the secondary potential well is small. But the coarse part of drops will be involved in the secondary coagulation process, and after a period of time, the system will be represented by doublets rather than by singlets.

The formulated problem, in fact, defines concretely the problem of the initial stage of flocculation. Of course, this problem is presented in the general formulation and in general equations considered in articles and in the review by Melek and Fogler [15], but it is not treated as a separate problem. In a series of works by Batchelor and Wen [146], the effect of surface forces on gravity coagulation was restricted to attractive forces. The same holds for works by Davies et al. [29,148]. However, Spielman and Cucor [160] pay attention to the gravitational coagulation in the secondary minimum.

Derjaguin et al. [19,135,136,158] raise the question about the importance of gravity coagulation in the secondary minimum. The necessary condition imposed on drop size is formulated [29]. The equation and the boundary condition are derived for finding the limiting trajectory but the equation is solved under the assumption of lack of repulsive forces, that is, in accordance with Refs. 29 and 146. Omitting electrostatic component, Rulyov (134) was able to obtain an exact analytical solution of the equation. The inclusion of this term along with attractive forces excludes the possibility of an exact solution.

However, an approximate method for analytical consideration of the problem can be proposed.

2. Similarity in Collision Efficiencies of Primary and Secondary Gravitational Coagulation

Secondary gravitational coagulation occurs in the force well of a larger drop and is correspondingly caused by the transport of the smaller droplet.

The larger the drops, the smaller the part of the well occupied by them. The drops of a finite dimension distribute symmetrically with respect to the maximum of the surface force h_{fm}. The maximum distance for the droplet of a finite dimension corresponds to the balance of the attaching surface force and detaching hydrodynamic force [see Eq. (105)].

The boundary condition for the grazing trajectory of the droplet is expressed by Eqs. (79) and (80), in which h_0 has to be substituted. The equation for the grazing trajectory in the case of the secondary flocculation differs from the treatment in Sec. IV.G due to the additional repulsive electrostatic force. However, the electrostatic force can be neglected, as seen below.

Take into account that the electrostatic force is κh_m times smaller at the force maximum than the attraction force and that $\kappa h_{fm} \gg 1$ (Sec. V.A). The distance to the surface increases along the grazing trajectory when the angle θ decreases. Thus, the maximum error in neglecting the electrostatic force corresponds to the point $\theta = \pi$, $h = h_0(a_1)$. Neglecting the electro-static force corresponds to an increase in the molecular force or in the Hamaker constant by the multiplier $(1 - \kappa h_{fm}^{-1})$. This approximation causes a negligible error in the collision efficiency because its dependence on the Hamaker constant is very weak. Taking into account Eq. (98), one concludes that neglecting the electrostatic force is equivalent to omitting the multiplier $(1 - \kappa h_{fm}^{-1})^{0.15}$. This estimation concerns the drop of the critical dimension a_{1cr} (Sec. V.B) which occupies only one point in the well corresponding to force maximum. At $a_1 < a_{1cr}$, $h_0(a_1)$ exceeds h_{fm} It means that the ratio of the electrostatic force to the attraction force for $a_1 < a_{1cr}$ is less than for a_{1cr}. Thus, for drops with radius less than a_{1cr}, the error decreases.

One concludes that Eq. (94) (and, correspondingly, Fig. 13) describes the collision efficiency of the gravitational coagulation in the remote well too. However, there is a large difference in the application of Eq. (94) to the collision efficiency in the case of uncharged droplets (Sec. IV) and charged droplets (Sec. V). This concerns the boundaries of the application of the equation. For the uncharged droplets, Eq. (94) is valid for any $a_2 \gg a_1$. For the charged droplets, Eq. (94) is valid for the case

$$a_1 < a_{1cr}(a_2). \tag{108}$$

VI. GRAVITATIONAL COAGULATION IN THE PRIMARY MINIMUM

The pressing hydrodynamic force (Sec. IV.7) produces the pressure excess in the gap between sedimenting drops and causes the liquid to flow from the gap between them.

The hydrodynamical pressing force can exceed the force barrier of the disjoining pressure, thereby allowing the possibility of coagulation in the primary minimum. This problem may be examined when considering the drop motion along their symmetry axis [19]. Under this condition, the hydrodynamic pressing force is maximum; hence, one can obtain the necessary and sufficient condition for the primary gravitational coagulation of drops.

Along the axis of a drop pair, the tangential flow velocity is equal to zero. Thus, the duration of the deposition process can be indefinitely long. This means, in turn, that the viscous resistance of the interface film can be neglected in the balance of acting forces. It leads to an equation similar to Eq. (105). The difference concerns the pressure hydrodynamic force instead of the detaching hydrodynamic force. Both forces are expressed by the same equation but of different sign.

Equation (105) indicates that for all values of h, the smaller droplet is subjected to a force directed toward the surface of bigger drop (front pole). Otherwise, the deposition and coagulation cannot occur. The same expression imposes limitations in the values of the parameters at which the disjoining pressure can be overcome. We represent it as the critical radius of the small drop for the gravitational primary coagulation. It occurs at

$$a_1 > a_{1cr}^{pc}(a_2), \tag{109}$$

where a_{1cr}^{pc} is the smaller root of Eq. (107).

Instead of only electrostatic forces, the repulsive hydration forces have to be considered too (Sec. V.A.2).

VII. CLASSIFICATION OF REGIMES OF GRAVITATIONAL COAGULATION

There are two regimes in the absence of repulsive forces. At condition (109),

$$a_1 < a_{1cr},$$

coagulation occurs. Due to SHAS (Scc. IV.G.4), there will be no repulsive forces in the opposite case.

There are three regimes in the presence of repulsive forces: the primary coagulation, the secondary coagulation, and lack of coagulation.

One can see in Fig. 18 that at $a_1 > a_{1c}^{det}(a_2)$ detachment occurs, at $a_1 < a_{1cr}^{pc}$ primary coagulation occurs, and the condition of the secondary coagulation is

$$a_{1cr}^{pc}(a_2) < a_1 < a_{1cr}^{det}(a_2). \tag{110}$$

Again, Spielman and his co-workers [131,160] were first to classify the regimes of gravitational coagulation in systems of dimensionless criteria. In order to characterize the role of the attractive and the repulsive forces, they used the attraction Eq. (99) and the repulsion numbers:

$$N_R = \frac{6\epsilon\Psi^2 a}{A} = \frac{\kappa a}{16\beta}\left(\frac{\bar{\Psi}}{th\bar{\Psi}/4}\right)^2, \tag{111}$$

where β is given by Eq. (99).

The result is displayed in Fig. 18 which is also presented in Ref. 112, (Fig. 11.8).

Naturally, the primary coagulation occurs at a sufficiently low Ψ potential when the electrostatic barrier is absent. It corresponds to low values

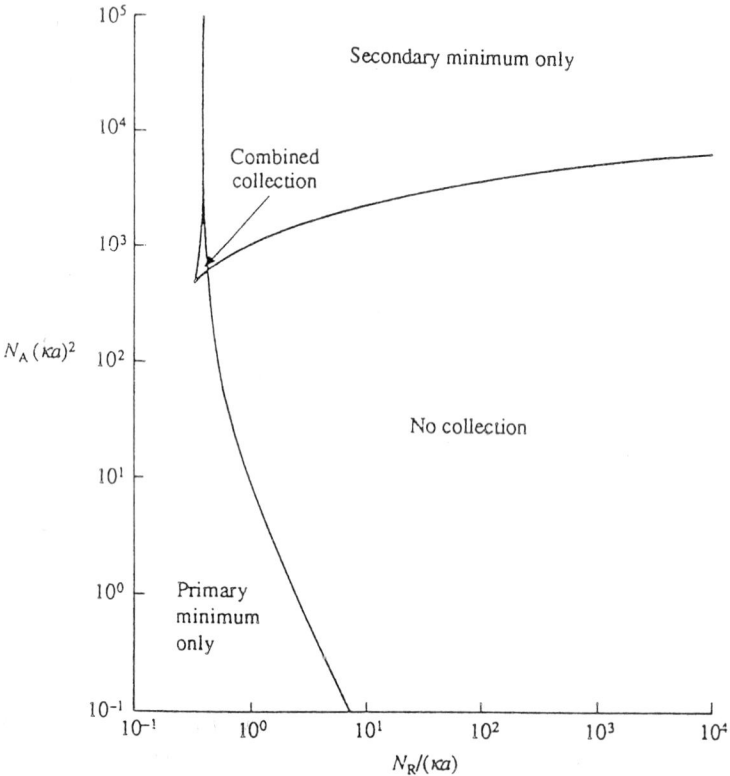

FIG. 18 Stability diagram for a spherical collector showing the influence of electrostatic repulsion at constant potential. (After Ref. 112.)

of the repulsive number and to the left domain in Fig. 18. The growth in Ψ and N_R leads to a growth in electrostatic barrier and excludes the possibility of primary coagulation. However, at any high value of the repulsive number, the secondary coagulation is possible, corresponding to the domain in the higher right corner of Fig. 18. The second alternative at high repulsive numbers is the absence of coagulation, which corresponds to the third domain in Fig. 18. Which alternative is realized depends on the value of attractive number. At a fixed N_R value (i.e., at fixed Ψ and a values), N_A increases the mean molecular force growth, and the relative velocity of droplet decreases. Drop coagulation in the secondary minimum results from a flow too weak to carry the droplet through the secondary well. It corresponds to high values of attractive numbers and to the domain in the upper right corner of Fig. 18.

The boundary between the domain of stability and of secondary coagulation is given by the curve of the critical condition of the sedimentation–hydrodynamic aggregative stability [SHAS (Sec. V.4) and Eq. (106)].

The slope of the boundary curve can be explained. At fixed values of the drop dimensions, the Hamaker constant, and the Ψ-potential, the relative velocity only changes along the boundary curve. The decrease in velocity corresponds to an increase in the attraction number and to a decrease in the pressing hydrodynamic force. Thus, the increase in the attraction number corresponds to the decrease in the pressing hydrodynamic force which causes the shrinkage of the domain of the primary coagulation in Fig. 18.

For an interpretation of the slope of the boundary curve of SHAS, fixed droplet dimension, relative velocity, and Hamaker constant can be assumed. For these conditions, N_A is invariant and the growth in $N_R(\kappa a)^{-1}$ corresponds to the growth in Ψ-potential and the decrease in attraction force between the droplets in a doublet. Thus, the increase in $N_R(\kappa a)^{-1}$ is favorable for the sedimentation–hydrodynamic stability. It explains the shrinkage of the secondary coagulation domain in Fig. 18 upon growing $N_R(\kappa a)^{-1}$ values.

The evaluation of the coagulation kinetics for a specific emulsion starts with the determination of the stability diagram, that is, the calculation of the boundary curves for the primary and secondary coagulation. The attraction number for emulsions can be specified by expressing the relative velocity by means of the Stokes equation and using the known value of the density difference $\Delta\rho$, that is,

$$N_A = \frac{A}{2\pi g \Delta \rho a^4}, \tag{112}$$

$$N_A(\kappa a)^2 = \frac{A\kappa^2}{2\pi g \Delta \rho a^2}, \tag{113}$$

$$N'_R = N_R(\kappa a)^{-1} = \frac{3\epsilon \Psi^2}{4\pi A \kappa}.$$

In emulsions, the value of A changes in a narrow range and the information concerning $\Delta\rho$, κ, and $\Psi \cong \zeta$ is usually available too. For a specific emulsion, $\Delta\rho$, κ, Ψ, and A are invariants. However, a can vary in the broad range.

The evaluation of N'_R enables one to discriminate between the gravitational stability and the gravitational coagulation. For sufficiently high ζ-potentials and sufficiently low electrolyte concentrations, the gravitational stability or secondary coagulation will take place. These two regimes can be discriminated by the N'_R evaluation. As the electrolyte concentration increases, the secondary minimum approaches the surface, which is accompanied by an increase in the molecular attraction. Thus, experimental evaluation of N_R and N_A criteria enables one to determine both boundary curves and identify the coagulation regime. Depending on the values of the parameters, the regimes of primary or secondary coagulation or gravitational stability can be identified. However, these simple regimes exist for emulsions with a narrow drop size distribution. Due to the strong dependence of N'_R on real emulsions with a broad droplet distribution, mixed regimes will occur. For the largest droplets, SHAS becomes of increasing importance and droplets satisfying condition Eq. (110) participate in secondary coagulation.

The stability diagram (Fig. 18) does not account for Brownian coagulation. In principle, Brownian coagulation manifests itself at any value of repulsion number.

For an application of the kinetic theory of emulsion stability, the stability diagram is extremely important. Unfortunately, the exactness of the boundary curves plotted in Fig. 18 is very low. The boundary curve of primary coagulation reflects the interaction constant of both charge and constant potential. The curve in Fig. 18 is plotted for an invariant potential. General criteria for the discrimination between the two cases are known (Sec. II) and formulated on the basis of Debora number [177]. However, a quantitative determination is difficult due to the lack of information concerning the charge relaxation.

The exactness of the boundary curve of SHAS is limited due to the restriction in knowledge concerning the retardation of the van der Waals attraction (Secs. II and V.D).

The boundary curve in Fig. 18 is calculated for the extreme case of λ = 0. The incorporation of the multiplier $\Lambda(\lambda)$ (Secs. IV.I and V.B) into the equation for the balance of the hydrodynamic detaching force and attractive surface force equation (110) enables a generalization of the theory. It leads to a splitting of the boundary curve into a family of curves.

VIII. SLOW BROWNIAN COAGULATION

A. Stability Ratio

Rapid Brownian coagulation was considered in Sec. III taking into account the role of the attractive van der Waals interaction and short-range hydrodynamic interaction and neglecting electrostatic interaction. The expression for the total energy of interaction, including molecular attraction and electrostatic repulsion, between a spherical particle and a particle of much greater size is given in Sec. V.A, which can easily be generalized for any λ value [see Eq. (101)]. A substitution of the expression for the total energy interaction according to Eq. (96) into Eq. (38) yields the stability ratio for the slow coagulation. This ratio can exceed unity in distinction to the rapid coagulation. This happens at sufficiently low electrolyte concentrations when the height of electrostatic barrier is sufficiently high. The calculations can be based on the extremes of the potential curve [i.e., according to Eq. (97)], which has two roots. The larger the value of the roots corresponding to the "bottom" of the distant well, the smaller the value to the barrier.

The major contribution to the integral in Eq. (38) comes from the expression for the exponent [163] that enables one to accomplish integration and leads to the simple result

$$ W \sim \left(-\frac{2}{\pi} \frac{\partial^2}{\partial r^2} \frac{U_{max}}{kT} \right)^{-1/2} \frac{\exp(-U_{max}/kT)}{h_{max}}. \tag{114} $$

As the electrolyte concentration decreases, the Stern potential as well as U_{max} increases according to Eq. (11). This causes an exponential growth of the stability ratio. The conditions causing the simplification of the expressions for the electrostatic repulsion energy and for the integral equation (38) are altered as the electrolyte concentration increases and the Stern potential diminishes.

Prieve and Ruckensten [164] calculated numerically the stability ratio with the Derjaguin approximation and the constant-charge boundary condition. They found that the approximation

$$ W = W_R + 0.25 \left[\exp\left(-\frac{U_{max}}{kT} - 1 \right) \right] \tag{115} $$

accurately represents their numerical results for a limited range of conditions.

B. Experiment

An increasing stability ratio with decreasing electrolyte concentration was confirmed in experiments by Ottewill and Shaw [165] and Watillon and Joseph-Petit [166]. At higher concentrations, the stability ratio remains constant because a rapid coagulation occurs. According to Eqs. (114) and (99), the slope of the log W–log c curve should depend on the particle size and inversely on z^2 (z is a counterion valence) at potential invariancy.

In fact, neither of these predictions is verified by experiment. The measurements by Ottewill and Show [165] on latex particles show that the slopes are nearly independent of particles size. The slopes for La^{3+}, Ba^{2+}, and K^+ did not significantly differ from the experiments of Frens and Heuts [167]. The stability ratio extracted from the experiments of Lips and Willis [168] and Zeichuer and Schowalter [169] shows a poor agreement with theory. In Ref. 160, the authors state that "the lack of agreement may stem from the extreme sensitivity of rate to the magnitude of the repulsion." The repulsion cannot be calculated with sufficient accuracy because the ζ-potential measurements are not sensitive enough and the Stern potential deviates from the ζ-potential. Wiese and Healy [170] paid attention to the discrimination between the slow primary minimum flocculation and the secondary minimum flocculation. In other words, by neglecting the secondary minimum flocculation, one can obtain an apparently low calculated stability ratio. However, some experimental data cannot be explained in this way. First, the experiments by Matijevic et al. [171–173] cannot be interpreted by the above approach. The large discrepancy between theory and experiment was established by measurements [171] of aqueous suspensions of monodisperse spherical particles of polystyrene latex of 65.5 nm. The secondary minimum flocculation of the latex is impossible if the particle dimension is so small. At low electrolyte concentration, the discrepancy between theory and experiment is so large that a possible misinterpretation in the ζ-potential measurement cannot be used for explanation.

Coagulation rates and electrophoretic mobilities in dispersions of uniform spherical particles of surfactant-free latex, silica, and cerium (hydrous) oxide particles were measured [171] as a function of the ionic strength by the low-angle scattering technique. The measured stability ratio as a function of the ionic strength was compared [171] with theoretically calculated curves of W by using the advanced Overbeek [174] procedure for the calculation of the double-layer interaction energy. Let us note that the distance of the electrostatic barrier to the surface can be very small under slow coagulation. It means that Eqs. (11) and (97) simplified by the assumption of sufficiently large κh values are not valid for slow coagulation. Usually, the more general Hogg–Healy–Furstenau (HHF) linear approximation is used. Overbeek [175] has developed more accurate nonlinear formulations for the calculation of the double-layer interaction energy, which is important especially for the interpretation of slow coagulation [172,173]. However, despite the application of the advanced Overbeek model, significant discrepancies are observed in the slow coagulation between the calculated and measured stability ratios. In order to account for observed discrepancies, the variations of the most common parameters were tested without success [171]. Alternating parameters characterizing hydrodynamic retardation, hydrophobic–hydrophilic interaction, and van der Waals attraction caused only a parallel shift in the log W–log c plot, and the large discrepancy in the slope was preserved. Both the experiment and theory led Kïhira et al. [171] to conclude that the discrepancy can be caused by the heterogeneity of the particle surface. In Ref. 171, the discreteness–charge effect and more extended surface charge segregation are mentioned. The latter model corresponds to a stronger decrease of electrostatic barrier. Thus, the patchwise heterogeneity [176] can cause the lower values of the observed stability ratio. If particles interact by weakly charged patches, the electrostatic barrier will decrease, which corresponds to a decrease in the stability ratio. In Ref. 171, the averaging procedure for the interaction at patchwise heterogeneity is proposed and agreement between calculated and measured stability ratios was achieved. The independent measurement of the heterogeneity degree is the most important for the verification of the proposed modification of the slow coagulation theory.

When the particles approach and their double layers start to overlap, the Stern potential will depend on the thickness of the liquid interlayer. Usually two extreme cases are considered when assuming constant Stern potential or surface charge. At the same value of the Stern potential (ζ-potential), the electrostatic barrier at constant charge can exceed substantially that at quasiequilibrium interaction. It can cause a very large difference in stability ratio. Lyklema and Dukhin [162,177] formulated this topic as the problem of the dynamic of colloid particle interaction. The dynamic of colloid particle interaction preserves its role at the surface heterogeneity of the interacting particle. The interaction energy of the patches can differ substantially depending on the relaxation properties of the double layer.

C. Specific Features of Slow Coagulation in Emulsions

In investigations of slow coagulation, the attention was paid to monodisperse suspensions (Sec. VIII.B). Probably this is the reason to the small interest to study slow coagulation in emulsions, which are polydisperse systems. The slow coagulation in emulsions should deserve special attention because the droplet surface can be homogeneous. The simple variant of the slow coagulation theory can be applied to emulsions. Folger and colleagues [27,28,178–180] investigated slow coagulation in emulsions. It is interesting that the authors do not notice the contradiction of their experimental data with Fuchs' theory. Usually, experimental data on slow coagulation in suspensions contradicts Fuchs' theory, whereas similar data on emulsions agree with theory [27,28,180]. This can be considered to be due to the nature of the surface's topology. The interpretation of the experiment in monodisperse suspensions is easier in comparison with polydisperse emulsions. In addition, the influence of many parameters (particle dimensions, electrolyte concentration, etc.) on slow coagulation in suspensions was investigated in comparison with the restricted investigations in emulsions [179,180]. The average droplet dimension and electrolyte concentration in Ref. 27 changed in a very narrow range. The Stern potential was very low (7 mV). The stability ratio calculated with this potential differed slightly from unity. The slow coagulation manifested itself [27] at low electrostatic barriers which does not cause a discrepancy, as discussed in the preceding section.

The influence of droplet dimension on the stability ratio is in line with Fuchs' theory because the increase of the initial droplet size caused a retardation of the coagulation [27]. In the preceding section, the measured stability ratio decreased mainly with an increase in particle dimension or is not sensitive to it. However, in some suspensions, the growth of stability ratio with the increase of particle dimensions is observed. For monodisperse gold sols, a decrease, an increase, and an independence of colloid stability on particle size has been reported [163]. An increase in stability with increasing particle size followed by an decrease has been reported for monodisperse lateces [181].

D. Importance of Slow Coagulation in Emulsions for Colloid Stability

Systematic investigations of the slow coagulation in emulsions are extremely important. First, they are important because of the lack of experimental data concerning this phenomenon in emulsions. Second, they are important for the development of the theory of slow coagulation theory, in general, because the droplet surface is homogeneous.

"It is not too difficult to predict conditions at which the onset of 'rapid deposition' will occur, but the deposition rate in the presence of a repulsive barrier is usually greater than the predicted (DLVO) rate . . . *this remains one of the major unsolved problems in colloid science*" [2].

The investigations by Lyklema, Dukhin, and Shulepov [162,177] clarified the importance of the dynamics of colloid particle interaction. The dynamics of the droplet interaction is characterized by the liquid interface mobility. The water drainage which accompanies the droplet approach can involve a lateral movement in the droplet surface which is impossible in suspensions. Due to the lateral surface movement in the gap dividing the adjacent droplets, the ion adsorption can decrease. This leads to a decrease in the Stern potential, in the electrostatic barrier, and in the stability ratio.

The problem of the drainage and the dynamic adsorption layer is very complicated [18,182] and deserves a separate consideration.

There are some features of emulsions which are favorable for studies of the slow coagulation in a broad range of droplet dimensions.

The discrepancy between the theory and experiment builds up with increasing characteristic flocculation times. The measurement of the high characteristic flocculation time in suspensions is usually restricted due to a rather rapid sedimentation. The density difference $\Delta\rho$ in emulsions is usually a factor of 10–100 smaller than in suspensions. Another restriction in the measurement of long coagulation time is the gravitational coagulation in the primary minimum (Sec. VI).

The larger the $\Delta\rho$, the larger the pressing gravitational force which enables one to overcome the disjoining pressure and causes the gravitational coagulation in primary minimum. In other words, the smallest droplet resulting in dimension a_{cr}^{pc}, the gravitational coagulation in primary minimum [see Eq. (109)], decreases with the increasing $\Delta\rho$ value. The boundary between the slow Brownian coagulation and the gravitational coagulation in the primary minimum is determined by the value a_{cr}^{pc}. It means the droplet dimension range for Brownian coagulation,

$$a_1 < a_{1c}^{pc}(a_2), \qquad (116)$$

in emulsions greatly exceeds that in suspensions. The slow Brownian coagulation is not favorable in suspensions for the increasing a_1 value at rather small a_{1c}^{pc} values and this is the reason for the discussed discrepancy in Sec. VIII.C. In emulsions, $a_{1c}^{pc}(a_2)$ exceeds that in suspension. Hence, it can be assumed that the slow Brownian coagulation in emulsions can be observed in the broad range of droplet dimensions.

However, the possibility of simultaneous slow Brownian coagulation and gravitational coagulation in the secondary minimum preserves in emulsion too. It means that for a manifestation of slow coagulation in the broad droplet dimension range, the coagulation in the secondary minimum has to be suppressed. This can be achieved by a decrease in the electrolyte concentration because the distance of the secondary minimum to the surface increases and its depth decreases (Sec. V.A.1).

It might be that a decrease in electrolyte concentration and a retardation of the coagulation in the secondary minimum will not be sufficient for measuring the slow coagulation. At low electrolyte concentrations, the Stern potential can be rather high, which leads to a high stability ratio. This can be avoided if ion adsorption and surface charge can be decreased in parallel with the increase of the diffusion layer thickness. It restricts the growth of the Stern potential and the stability ratio.

E. Influence of Surface Heterogeneity and Surface Roughness on Slow Coagulation

The influenace of surface roughness on interaction energy of colloid particles (and thus on colloid stability) is discussed in Charnecki's review [183].

It is usually assumed that the particles are smooth spheres and that the collectors are flat, smooth, and infinitely large as compared with particle sizes. Under these assumptions, the theory predicts forces normal to the surface of the interacting bodies only. However, the very fact that deposited particles become immobilized at the collector surfaces and are not washed out under the influence of hydrodynamic forces imply the existence of tangential forces as well [184].

The tangential forces may originate from surface roughness and inhomogeneities in the surface charge distribution. It should be noted that the surface irregularities are always present in real systems, even those being carefully processed.

In the model adopted, the rough collector surface was simulated as follows: On a perfectly smooth flat surface, a number of small spheres with randomly selected radii from 0 to 0.05 μm were randomly scattered. A large spherical particle (with a diameter of 1 μm) was placed above this surface and the interaction energy of the system was calculated, assuming additivity of the dispersive and double-layer contributions. Such a procedure allows the use of simple expressions for both sphere–sphere and sphere–half-space geometries.

The dispersive interactions are very short ranged and only those parts of the interacting bodies which are very close to each other make a significant contribution to the total interaction energy. Thus, if particles have irregular shapes or rough surfaces, this will influence only the closest adjacent particles. This is true for a small separation between interacting bodies. At larger distances, the effects of local surface irregularities are "smoothed out" and the interaction energy is determined by the total volumes of the interacting bodies rather than by details in their shapes.

To determine whether it is necessary to consider surface roughness, the ratio of the thickness of rough surface layers to the separation between the bodies may be used. If the thickness of the rough layers are of the same order of magnitude as the surface-to-surface separation, the effect of surface roughness must be taken into account. For a small separation, surface roughness may cause a decrease in the van der Waals attraction energy by a factor of 10 or more.

The interaction energy resulting from the overlapping of electric double layers may be expected to behave in a similar way.

In other words, the roughness can cause a decrease in the electrostatic barrier by a factor of 10 or more, which can explain the above-mentioned discrepancy between the theory and experiment for slow coagulation. At a sufficiently large separation, the repulsion energy is very low in comparison with the electrostatic barrier, corresponding to the interaction of homogeneous surfaces. At a smaller separation, the repulsion does not increase due to the surface heterogeneity. As an approximation, one can consider a sphere with a dimension comparable with the separation. Taking into account that interaction energy is proportional to the radius of the sphere [see Eq. (101)], one concludes that the surface inhomogeneity dimension determines the energy of the interaction of rough particles instead of their own dimensions. This very simple suggestion can be used to explain the observed low sensitivity of slow coagulation to the particle dimension.

IX. DELINEATION OF DIFFERENT PARTICLE LOSS MECHANISMS. RAPID COAGULATION

In emulsions, flocculation will be significant for smaller particles and creaming will be significant for larger particles. However, it is extremely important to have quantitative relationships that can predict the state of an emulsion or suspension; that is, whether the particles are creaming or flocculating, and if both are occurring, determine the predominant one. If they are flocculating, what type of flocculation dominates, if any? This information can be used to predict the behavior of the colloidal system from the solution of the convective–diffusion equation and population balance equations by including only those destabilizing factors which dominate the particle breakdown process. This type of analysis has the potential of greatly simplifying the process of predicting the stability of a given emulsion or suspension. Melic and Fogler [15] accomplished this for both uncharged and charged emulsions. Taking into account the results of the preceding section, a specification is necessary. In this section, the delineation concerns a weakly charged emulsion where the role of the electrostatic component of disjoining pressure can be neglected. As to emulsions stabilized by electrostatic interaction, the delineation procedure as a whole cannot be realistic due to the large discrepancy between theory of slow coagulation and experiment.

A. Gravity-Induced Flocculation Versus Brownian Flocculation

This section is primarily based on the review by Melic and Fogler [15]. As a first approximation, the relative strength of gravity-induced flocculation as compared to Brownian flocculation is given by the gravity number

$$Gr = \frac{(u_{02} - u_{01})(a_1 + a_2)}{2D_0} = \frac{2\pi g \Delta \rho a_2^4}{3kT} \lambda(1 - \lambda^2) \qquad (117)$$

and represents the relative strength of gravitational to Brownian forces.

In Fig. 19, the gravity number Gr is shown as a function of the particle size ratio λ for the case $\Delta\rho = 0.1$ g cm^{-3} at $T = 298$ K under normal gravity. As λ deviates from unity, the difference in particle size causes the differential creaming contribution to Gr to be enhanced at a rate faster than the increase of the Brownian motion of the smaller particle. A further decrease in λ only results in a small increase in differential creaming,

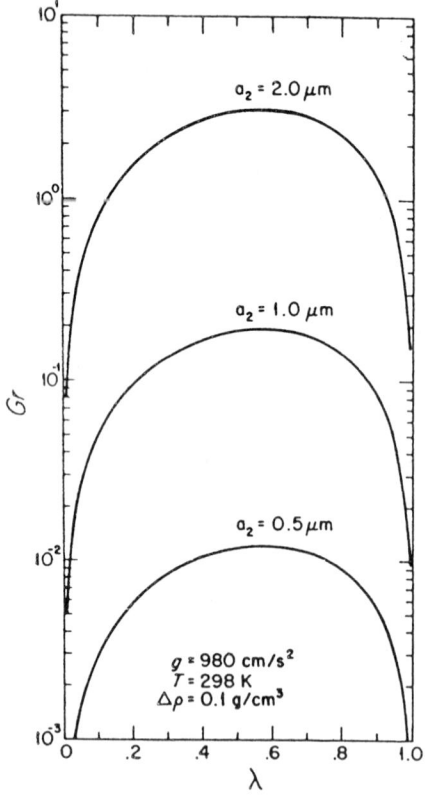

FIG. 19 Effect of particle size ratio on the gravity number Gr for various type 2 particles sizes. (After Ref. 15.)

whereas the diffusivity of the smaller particle starts becoming quite vigorous. The maximum value of Gr at $\lambda = (1/3)^{1/2}$ reflects the balance point between these two competing effects. That is, one can estimate the relative importance of gravity-induced flocculation and Brownian flocculation from a plot similar to Fig. 19. For example, if the colloidal particles are about 1.0 μm in diameter with $\Delta\rho = 0.1$ g cm^{-3} and $T = 298$ K, Brownian flocculation will be approximately 100 times more important than gravity-induced flocculation. On the other hand, for particles about 4.0 μm in diameter, gravity-induced flocculation will be about three times larger than Brownian flocculation. A plot of this type can provide useful qualitative information on the stability of a colloidal dispersion. This approach was used by Melek and Folger [15] to delineate the various particle loss mechanisms in quiescent media.

However, an underlying assumption that is invoked when using the parameter Gr is that the stability factor for the loss mechanism of each particle is identical. The gravity number implies that the particle loss mechanisms due to differential creaming and Brownian motion can be quantified independently. Although there are regimes for which the assumption of linear independence is quite good, one should be careful because we can only account for the differences in the gravity-induced and Brownian flocculation stability factors. The ratio of the particle loss due to gravity-induced flocculation to the particle loss due to Brownian flocculation is given by

$$R_{GB} = \frac{J_{Gr}}{J_{Br}} = \frac{Gr}{2} \frac{W_{Br}}{W_{Gr}}, \tag{118}$$

where J_{Br} is given by Eq. (13) or (14). As shown by Melek and Fogler [15] for Gr $\ll 1.0$, additivity is justified only in the absence of electrostatic repulsion. Consequently, we will restrict our discussion to conditions of rapid flocculation.

The effect of gravitational forces on the flocculation ratio R_{GB} is shown in Fig. 20 for various strengths of interparticle attraction. As expected, the effect of gravity-induced flocculation becomes more pronounced as the gravitational force is increased. It is interesting to note that the gravity number Gr does not give a good estimate of the relative importance of gravity-induced flocculation as compared to Brownian flocculation. For example, if Gr $= 10.0$, one would probably assume that the rate of particle loss due to gravity-induced flocculation is approximately 10 times higher than the particle loss due to Brownian flocculation. However, as can be seen from Fig. 20 for Gr $= 10.0$, the particle loss ratio is only approximately 1.5 for a dimensionless Hamaker constant of $A/kT = 1.0$, typical of colloidal dispersions. This result clearly shows that hydrodynamic inter-

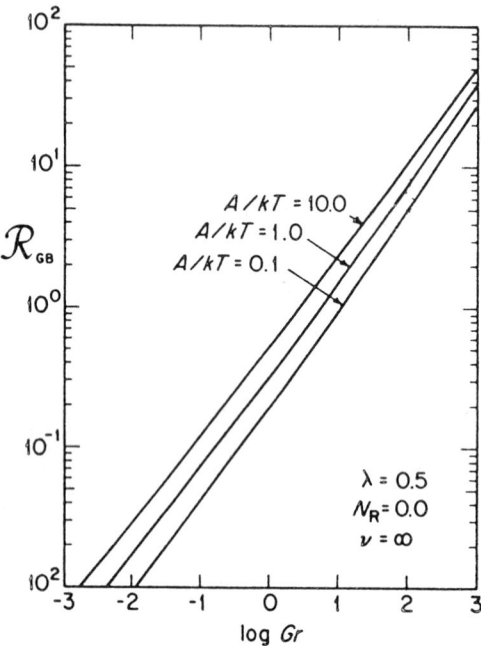

FIG. 20 The importance of gravity-induced flocculation and Brownian flocculation. (After Ref. 15.)

actions reduce the rate of gravity-induced flocculation more than the rate of Brownian flocculation. This effect is absent when using the gravity number to delineate the mechanisms of gravity-induced and Brownian flocculations.

B. Gravity-Induced Flocculation Versus Creaming

Whenever a colloidal system is destabilized due to gravity-induced flocculation, a breakdown due to particle creaming occurs simultaneously. The ratio of the total rate of particle loss due to creaming to the net rate of particle loss as a result of gravity-induced flocculation R_{CG} was determined by Melck and Fogler [15]. Their result is given by the following analytical expression:

$$R_{CG} = \frac{f(a_2, N_{02}, \lambda, R_N)}{\pi HE_{1,2}(1 - \lambda^2)(1 + \lambda)^2 R_N}, \tag{119}$$

where $f(a_2, N_{02}, \lambda, R_N)$ with $H = h/a_2$ is the dimensionless container height and $R_N = N_{01}/N_{02}$ is the particle concentration ratio.

Under conditions of negligible electrostatic repulsion, the particle loss ratio R_{CG} is most sensitive to changes in the total initial particle concentration N_{TOT} (where $N_{TOT} = N_{01} + N_{02}$), the size of the larger particle a_2, and the particle size ratio λ. The effect of each of these parameters is discussed next.

Figure 21 shows the effect of the total particle concentration N_{TOT} on the relative rates of creaming to gravity-induced flocculation. As expected, the higher the particle concentration, the more significant the process of gravity-induced flocculation becomes. The presence of a minimum in Fig. 20 is to be expected. For small values of R_N, $N_{02} \gg N_{01}$ and for large values of R_N, $N_{02} \ll N_{01}$. In each of these extreme cases, the effect of gravity-induced flocculation is reduced because there are either not enough collectors (a_1) for the given number of particles (a_2) or there are too many collectors. In either case, creaming begins to dominate the col-

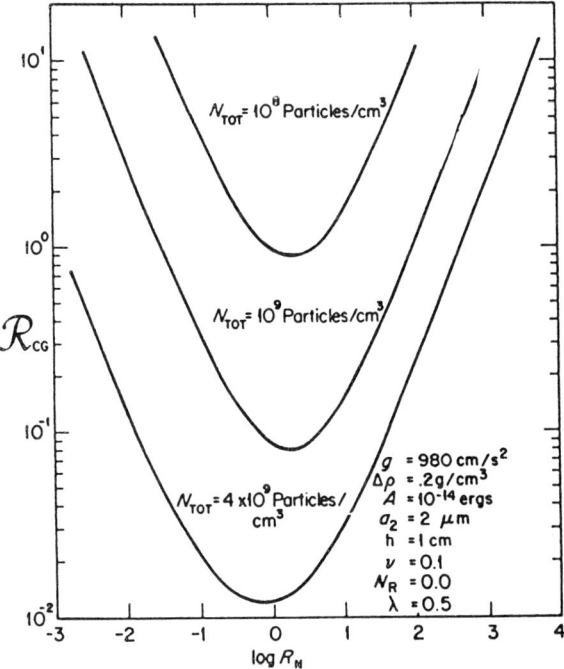

FIG. 21 Effect of particle concentration ratio on the relative particle loss rates of creaming and gravity-induced flocculation for various total particle concentrations. (After Ref. 15.)

loidal breaking process. When $N_{02} \cong N_{01}$, the effect of gravity-induced flocculation is at a maximum in the breaking process.

Figure 22 shows the effect of increasing particle size on the relative rates of creaming to gravity-induced flocculation. As the particle size increases, the rate of flocculation increases faster than the creaming rate of the particles.

Figure 23 shows how changes in the particle size ratio affect the relative rates of creaming and gravity-induced flocculation. When the concentration of smaller particles is less than the concentration of larger particles ($R_N < 1$), the larger particle size ratio reduces the rate of gravity-induced flocculation as compared to the creaming rate of the particles. On the other hand, when the concentration of the smaller particles is slightly higher than that of the larger particles ($R_N \geq \sim 3.2$), the situation is reversed. Smaller particle size ratios favor gravity-induced flocculation over creaming more than the larger particle size ratios.

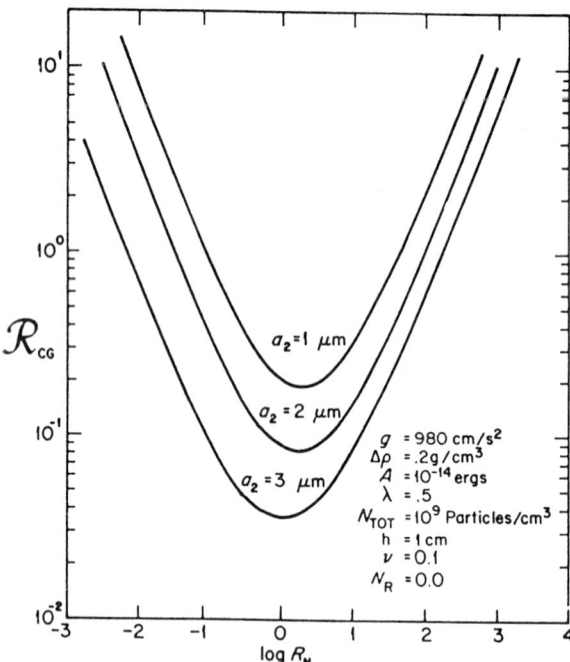

FIG. 22 Effect of particle concentration ratio on the relative particle loss rates of creaming and gravity-induced flocculation for various type 2 particle sizes. (After Ref. 15.)

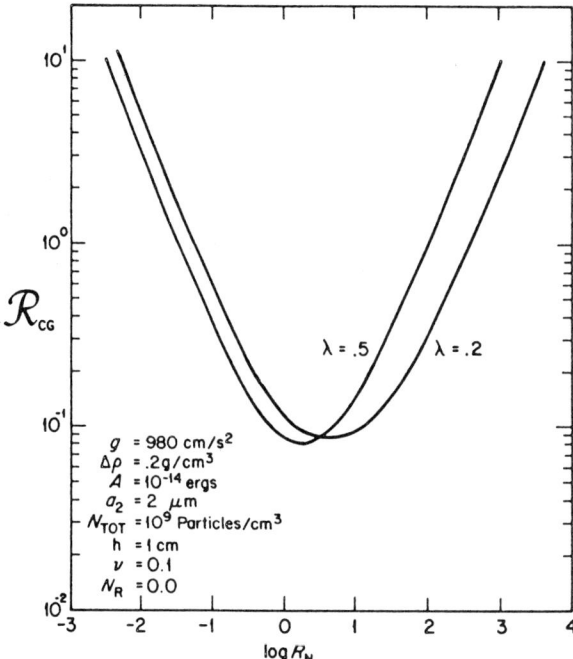

FIG. 23 Effect of particle concentration ratio on the relative particle loss rates of creaming and gravity-induced flocculation for various particle size ratios. (After Ref. 15.)

C. Domains of Dominant Particle Loss Mechanisms

Regimes of the various limiting cases of particle loss can be delineated quantitatively by manipulating the equations for the creaming rate and the Brownian and gravity-induced flocculation rates. For a given fluid, the creaming rate is strongly depending on particle size. For very small particles, the creaming velocity will be negligible as compared to the Brownian motion of the particles. It is generally accepted that the gravitational forces acting on a particle can be neglected for creaming velocities of 1.0 mm day^{-1} or less [15] and the case of negligible interactions (surface and hydrodynamic):

$$u_{0i} = \frac{2g\Delta\rho a_i^2}{9\mu_f} < 1.0 \text{ mm day}^{-1} \tag{120}$$

or

$$a_c = \left(\frac{(1.0 \text{ mm day}^{-1})9\mu_f}{2g\Delta\rho}\right)^{1/2}. \tag{121}$$

Particles with a radius smaller than a_c will disappear from the system primarily due to Brownian flocculation. For values larger than a_c, particles can disappear by any of the three particle loss mechanisms.

Brownian flocculation will be negligible for large values of the flocculation ratio R_{GB}, whereas gravity-induced flocculation will be negligible for small values of R_{GB}. Consequently, Eq. (IX.2) can be used to define quantitatively the regions where only one mechanism will be predominant. Melek and Fogler [15] define Brownian flocculation as predominant if the flocculation ratio R_{GB} is less than 0.1, and gravity-induced flocculation as predominant if R_{GB} is higher than 10.0. As discussed in Sec. IX.A, this rationale does not account for any possible coupling between differential creaming and Brownian motion:

$$\frac{\text{Gr}}{2}\frac{W_{Br}}{W_{Gr}} \leq 0.1 \quad \text{(Brownian flocculation),} \tag{122}$$

$$\frac{\text{Gr}}{2}\frac{W_{Br}}{W_{Gr}} \geq 10.0 \quad \text{(gravity-induced flocculation).} \tag{123}$$

The key parameters in Eqs. (121)–(123) are the particle radius a_2, the net gravitational force $g\Delta\rho$, and the particle size ratio λ. Consequently, these parameters can be used to map the various domains for the particle loss mechanisms.

In a fairly general manner, plots of a_2 versus λ for a constant value of $g\Delta\rho$ can be prepared by using Eqs. (121)–(123). An example of this analysis is shown in Fig. 24 for the case of negligible electrostatic repulsion ($N_R = 0$) and for a Hamaker constant $A/kT = 1.0$. Equation (121) was used to obtain the upper boundary of region I, and Eqs. (122) and (123) to obtain the lower and upper boundaries of region III. A plot of this type can also be used for centrifugal creaming instead of gravitational creaming. For centrifugal creaming, g is simply replaced with $\omega^2 X$, where ω is the angular velocity of the centrifuge rotor and X is the distance from the axis of rotation.

Various regions in Fig. 24 represent the possible dominant particle loss processes in quiescent media. For example, in region I, creaming and gravity-induced flocculation will be negligible and particles disappear primarily by Brownian flocculation. Brownian flocculation in region I will depend on the electrostatic properties of the particles and also the total

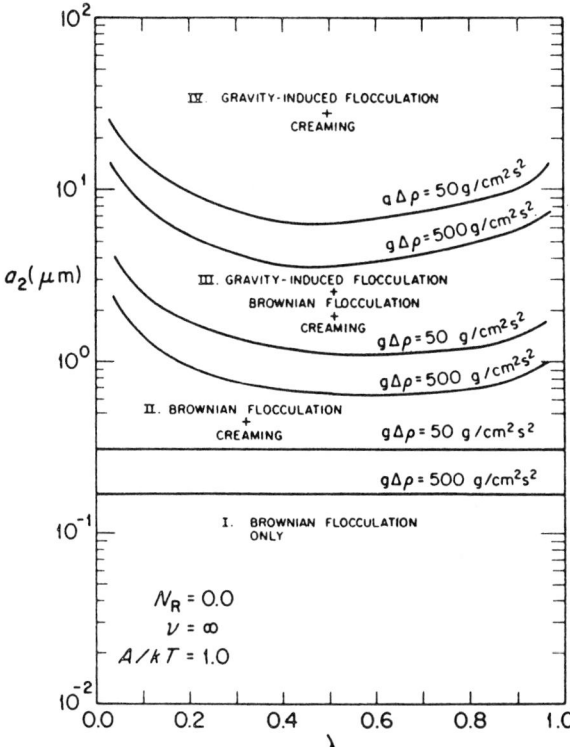

FIG. 24 Regimes of dominant particle loss mechanisms in quiescent media. (After Ref. 15.)

particle concentration. If the colloidal system is very dilute, Brownian flocculation will also be negligible.

For colloidal systems in region II, gravity-induced flocculation will be negligible as compared to Brownian flocculation. In region III, creaming is bound to occur and the rate of Brownian and gravitational flocculation will depend on the total particle concentration. Similarly, in region IV, Brownian flocculation will be negligible as compared to gravity-induced flocculation. Creaming will also occur in this region and the rate of gravity-induced flocculation will depend on particle concentration, and even on the polydispersity of the colloidal system.

Figure 24 is useful in determining the state of a colloidal system; that is, whether the colloidal particles are creaming or flocculating.

D. Experiment

A methodology of experimental investigation of the kinetic and aggregative instability of miniemulsions is elaborated by Ostrovsky and Good [185]. The methodology is illustrated by Figs. 25A and 25B. The invariant slope of section c of the experimental curve characterizes the rate of coalescence. Its value depends only on the composition of the system and is called the coalescence time parameter. More than 50 different composition were studied.

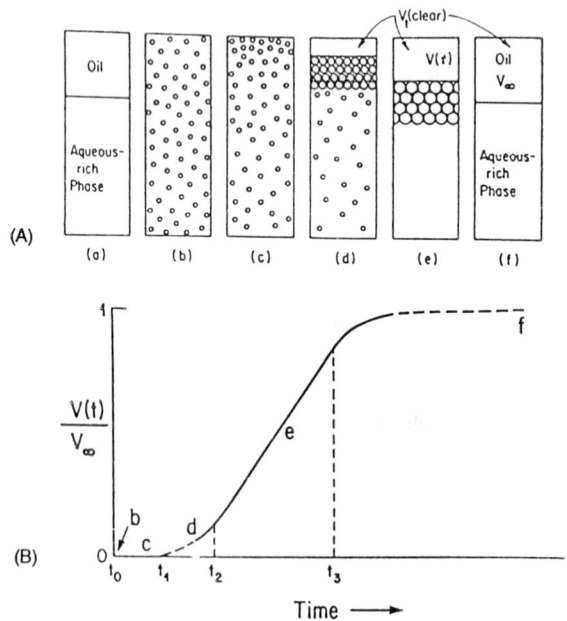

FIG. 25 (A) Commonly observed sequence of events in separation processes. Schematic sequency of direct observations: (a) Before shaking; equilibrated volumes. (b) Just after stopping shaking; $t = 0$. Macroemulsion of oil phase in aqueous phase. (c) Droplets of oil have migrated toward top of aqueous phase, but no appreciable coalescence is detectable. $t_0 < t < t_1$. (d) Appreciable coalescence of oil droplets; V_t increasing. Note larger droplet size just below clear layer, and sedimentation from lower part of aqueous phase still in progress. $t_1 < t < t_2$. (e) Sedimentation effectively complete. Coalescence of droplets continues in layer at top of aqueous phase. $t_2 < t < t_3$. (f) Separation complete. $t \gg t_3$. (B) Schematic kinetic curve of separation, corresponding to conditions in (A). (Parts A and B after Ref. 225.)

By means of the linear extrapolation, as shown in Fig. 25B, the authors introduced the notion of τ_{sed}. The authors neglect the flocculation and coalescence during stages b and c and at $t < \tau_{sed}$. They interpret the time τ_{sed} as being determined by the mean velocity of the translating droplet and identify averaged velocity with the velocity corresponding to averaged radius $\bar{r}$.

For the droplet velocity characterization, the radius, the density difference $\Delta\rho$, and the viscosity of the external phase μ were varied and measured in a wide range. For τ_{min}, a rather good correlation between the measured and calculated values is established. It means that the flocculation influence on the droplet dimension and velocity was weak.

Comparing this result with the delineation procedure characterized in Fig. 24, one concludes that the droplet dimensions in the experiment in Ref. 185 were large. They corresponded to the higher domain in Fig. 24.

Unfortunately a direct measurement of the droplet dimension was not undertaken in Ref. 185. However, the conclusion can be confirmed by the comparison of the Stokesian droplet velocity and velocity from the measured effective sedimentation time τ_{min}. Substituting the measured values of μ and $\Delta\rho$ in the Stokesian expression yields $\bar{r}$. Its value exceeds 5–10 μm; that is, it corresponds to the highest domain in Fig. 24. This is not surprising because the miniemulsions were prepared by gently shaking by hand. The authors recognize the serious restrictions with respect to droplet dimensions in their investigations.

Both the investigations by Melek and Fogler [29] and by Ostrovsky and Good [185] emphasize the interdependence of the kinetic and aggregative types of emulsion instability and the importance of this topic. Melek and Fogler introduce this topic in its theoretical aspect. Ostrovsky and Good make something similar in an experimental approach. Their "two-parameter methodology" is a step forward as compared to the traditional methodology [186–188]. It is common [186–188] to report the time for the appearance of a given fractional volume (e.g., 50% of the clear phase) without any attempt to distinguish between creaming and coalescence.

The experimental system in investigations [185] was a set of surfactant–brine–oil–cosurfactant mixtures that were relevant to the optimization of composition for enhanced oil recovery [189,190].

In this system, the lower phase was a microemulsion. This choice of external phase gave opportunity to change $\Delta\rho$ and μ in a wide range of compositions. However, the buoyant droplets can interact in another way in a microemulsion environment. The closest separation between two drop surfaces h_0 can be small and approach the dimension of the microemulsion droplet. The discrete nature of the microemulsion cannot be neglected in the interlayer environment between interacting droplets of the macro-emulsion.

X. POPULATION BALANCE EQUATION

A. General

The Smoluchowski equation described in Sec. III.A. relates to monodisperse colloids, whereas emulsions are polydisperse. Smoluchowski derived another equation describing the time dependence of the evolution of particle size distribution. This equation was later used and analyzed by many authors [161,192–197]. Usually when applying this equation, an assumption of homogeneity in space is made. In emulsions, the flocculation is often followed by creaming, and, hence, the assumption concerning the homogeneity fails.

Experimental studies on emulsion stability involving simultaneous flocculation and creaming are underestimated. Emulsion flocculation has often been studied [198–201] by avoiding or by neglecting creaming. Demulsification of crude oil in water involves simultaneous flocculation and creaming. Wasan et al. [202] studied these two processes separately; coalescence of droplets and the creaming rate of dispersed phase.

Researchers in the field of aerosols studied only an approximate problem involving flocculation and sedimentation of aerosols [203]. Okuyama [204] studied the aerosol behavior in a closed chamber involving coagulation and settling. They assumed that the aerosol concentration and size distribution were spatially uniform and measured the amount of aerosol deposited at the bottom due to sedimentation and measured only the lumped change in aerosol concentration and size distribution only as a function of time.

The general problem of emulsion stability with simultaneous flocculation and creaming has been formulated by Melek and Fogler [15] as a general dynamic equation (GDE) for simultaneous flocculation and creaming derived from a population balance.

In Ref. 15, it was achieved by adding a term to the Smoluchowski equation, describing the change of particle size distribution (PSD) $f(x, v, t)$ due to creaming [last term in Eq. (124)]:

$$\frac{df(x, v, t)}{dt} = \frac{1}{2} \int_0^v \alpha(u, v - u) f(x, u, t) f(x, v - u, t) \, du$$

$$- \int_0^v \alpha(u, v) f(x, v, t) f(x, u, t) \, du - u(x) \frac{\partial f}{\partial x} (x, v, t), \tag{124}$$

where $\alpha(u, v) = \alpha(v, u) > 0$ is the collision probability of particles of volume u.

In this equation, the first integral describes the growth of the number of particles with volume v at the expense of joining particles of volume u and of volume $(v - u)$. The second term describes the decrease in particle number of volume v by coagulation with particles of the kind u.

In the case of flocculation only, the particle concentration remains spatially uniform. In the case of creaming only, the particle concentration always increases in the direction of creaming.

Concentration profiles of smaller particles are qualitatively unpredictable for the case of simultaneous flocculation and creaming. This was predicted by solving the general population balance equation [15].

The changes in the particle size distribution (PSD) at the bottom of a liquid column of height L (i.e., $x_1 = 0$) under conditions of simultaneous flocculation and creaming are shown in Fig. 26 for various times. Initially, the PSD shifts to the right toward larger particle size, owing to flocculation of the smaller particles (see PSD at $t = 3$ h). The larger particles cream out at a much faster rate than the smaller ones, thereby resulting in an observable shift of the PSD back to the left (see PSD at $t = 6$ h). These cyclic changes or shifts in particle size distributions would not have been realized without including the simultaneous effects of flocculation and

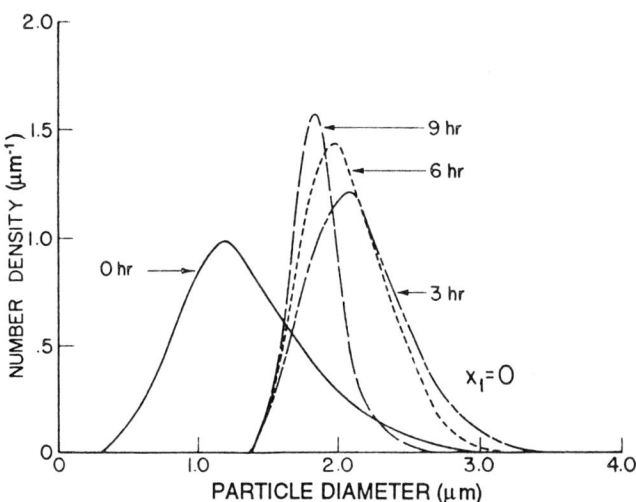

FIG. 26 Changes in the particle size distribution with time at $x_1 = 0$ for the case of simultaneous flocculation and creaming. (After Ref. 15.)

creaming in the population balance equations. In the case of creaming only, the PSD only shifts to the left toward smaller particle sizes; whereas in the case of flocculation only, the PSD only shifts to the right toward larger particle sizes.

B. Experiment

The efficiency of PBE was confirmed by experiment [15,178]. Emulsions of hexatriacontane, $C_{36}H_{74}$, and octacosane, $C_{28}H_{58}$, in water were prepared ultrasonically. The particle size and size distribution were measured with a scanning electron microscope. The irradiation time was chosen so that the resulting emulsion contained a particle size on the order of 1 μm. Electrostatic stability of the emulsion was adjusted by adding magesium sulfate. Irradiation time and concentration of $MgSO_4$ were chosen such that flocculation and creaming will take place at a significance rate.

The particle size distributions (PSD) at various aging times corresponding to $y = 0$ are plotted in Fig. 27. PSDs computed from the theory of simultaneous flocculation and creaming are also shown in this figure for the electrical and physical properties. The initial PSD and electrostatic properties of the emulsion were used for the theoretical predictions. From this figure, one observes that particle size distribution initially shifts to the right toward the larger particle sizes and then to the left toward the smaller particle sizes. This experimentally observed reverse shift in the

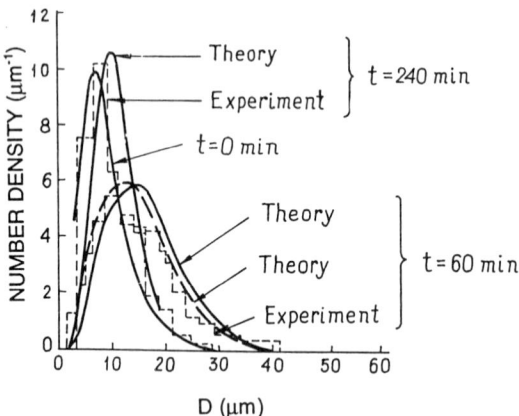

FIG. 27 Changes in the particle size distribution with time at $y = 0$ for the $C_{36}H_{74}$/water emulsion containing 8.3×10^{-3} g mole L^{-1} $MgSO_4$. (After Ref. 15.)

size distribution was theoretically predicted. Flocculation of particles produces the shift in size distribution to the right and creaming produces the shift in size distribution to the left. The cyclic changes in emulsion particle size distribution are due to simultaneous flocculation and creaming. One observes that the agreement between the theory and experiment is good.

In this experiment, the Stern potential was kept low and the stability ratio differed only slightly from 1. In this way, the discussed experiment related more to rapid flocculation.

XI. THE ROLE OF DEFORMATION OF INTERACTING DROPLETS IN KINETICS OF GRAVITATIONAL COAGULATION

Specificity of the droplet coagulation is caused by the mobility of their surfaces and the possibility of the droplet deformation during the interaction. The influence of the droplet surface mobility on the short-range hydrodynamic interaction and the coagulation rate is described in Sec. IV.J. The mechanism of the droplet deformation is very complicated and deserves a separate consideration. Only recently, essential progress is achieved in the understanding and quantifying of the role of deformation [205–209].

A. Close Approach and Deformation of Two Viscous Drops Due to Gravity and van der Waals Forces. Qualitative Approach

To clarify the mechanism of droplet deformation and to take into account its role in the coagulation kinetics (see Sec. IV.J), Yiantsios and Davis [205] considered two drops of the same fluid but of different parameters, which are suspended in a second unbounded fluid and which translate along their line of centers due to buoyancy. For definiteness, they assume that the drops are lighter than the suspending phase and that the smaller drop is above the larger one. Then both of them rise against gravity while their separation decreases with time.

When the gap between the drops is much smaller than either radius, the translational velocities of the two drops are approximately equal, whereas the drag force on each drop due to translation is not exactly equal to the buoyancy force. The difference is balanced by a pressure force the drops exert on each other, which is concentrated in a small area in the vicinity of the near-contact region and which squeezes out the fluid in the gap separating the drops.

Under the action of this pressure, the section of the droplet surface in the gap between droplets flattens (Fig. 28). Provided that the deformation is still unimportant, the small relative velocity can be found from the analysis of Davis et al. [76].

However, after some time, the drop separation will become of the same order of smallness as the deformation of the drops. The assumption that they remain spherical becomes inadequate in the vicinity of the thin gap, and one expect that the deformation would affect the thinning rates of the deformed gap quantitatively or even qualitatively.

Yiantsios and Davis [205] analyze this later stage of evolution prior to coalescence, starting from a position where the minimum separation between the drops is much smaller than either radius but also much larger than their deformations.

The two drops, denoted as phases I and II, have identical viscosity and density ρ', but different radii, a_1 and a_2, respectively in Fig. 28. The

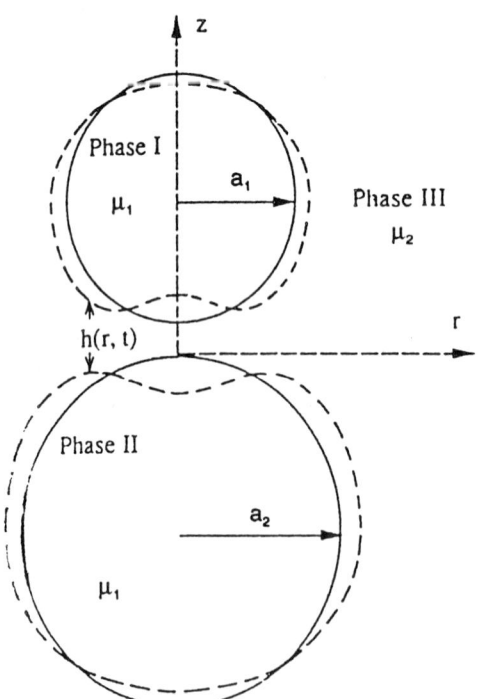

FIG. 28 Definition sketch for two drops in near contact. (After Ref. 205.)

unbounded suspending fluid, denoted as phase III, has viscosity μ and density ρ, where it is assumed that $\Delta\rho \equiv \rho - \rho' > 0$. At time zero, the minimum separation, h_0, between the drops is such that $h_0 \ll \lambda a_2$, and it is assumed that the constant interfacial tension, σ, is sufficiently large that the deformation of the drops is small compared with h_0. Under the action of the constant buoyancy forces, both drops move along their line of centers in the positive z-direction.

For the calculation of the coagulation rate it is important to find the shapes of the two interfaces as the deformation becomes important, and to determine the thinning rates of the fluid film separating the drops. As the gap decreases, molecular forces (i.e., of the London/van der Waals type) become important and cause a relatively fast rupture of the fluid film between the drops.

Before proceeding to next section with the analysis given by Yiantsios and Davis, it is of interest to consider the main results.

B. Weak Influence of Droplet Deformation on Grazing Trajectory and Collision Efficiency

The condition of a weak deformation is usually based on small values of the so-called capillary number Ca based on the relative magnitude of viscous and capillary pressure. Ca represents the ratio of magnitudes of dynamic and capillary pressures. A small value of Ca means that the dynamic pressure is small in comparison with capillary pressure and can cause a small deformation only. For an isolated rising (creaming) drop,

$$Ca \equiv \frac{\mu U}{\gamma},$$
(125)

which follows from Eq. (89) and the equation for capillary pressure $p_\sigma = 2\gamma/a$. At the droplet interaction, analogs of the capillary number are introduced for the estimation of the surface deformation inside the thin gap between droplets in Ref. 205. One of them concerns the deformation of the larger droplet; a second one concerns the deformation of smaller one.

However, for the first estimation, the modified capillary number for the near-contact region

$$C_m = \frac{\mu U a}{\gamma h_0}$$
(126)

is introduced in Ref. 205. With decreasing h_0, the viscous resistance characterizing the lubrication flow field due to the drop relative motion increases rapidly, which is reflected in Eq. (126). For the C_m estimation,

$$h_0 = 0.05a_2 \tag{127}$$

is chosen as an upper limit for the gap flow to be described by lubrication theory. As seen by comparing Eqs. (127) and (126) at the restriction (127), C_m decreases with decreasing a because the translating velocity, the viscous resistance F_L, and the dynamic pressure inside the gap all decrease. An estimation made in Ref. 205 for the values $\mu = 10^{-2}$ cm s^{-1}, $\Delta\rho = 0.1$ g cm^{-3}, and $\gamma = 10$ dynes cm^{-2}, C_m diminishes in the range 5×10^{-3} to 1.2×10^{-5} with $2a$ decreasing in the range 100–5 μm.

In addition, the restriction of weak attractive forces relative to the lubrication force requires that the Hamaker parameter

$$\mathrm{Ha} = \frac{A}{6\pi\mu h_0^2 U} \ll 1 \tag{128}$$

be small compared with unity. Ha changes in the range 10^{-8}–10^{-2} with $2a$ decreasing in the range 5–100 μm for the above-mentioned values of parameters and $A = 10^{-14}$ erg. Thus, in the broad range of parameters, both the lubrication force in the gap and the van der Waals attraction do not cause the essential deformation of the interacting drops at the rather large value of the closest separation according to Eq. (127). Naturally, a further decrease of the interdroplet distance leads to an increase of the lubrication and van der Waals forces that cause the deformation of the droplet. This was analyzed by Yiantsios and Davis [205] (see Sec. XI.C). They used numerical calculations to track the evolution of the shapes of the drops from a relatively underformed state corresponding to the aforesaid values C_m and δ_1 until the gap takes the form of a dimple.

The important result of the calculations is that the essential droplet deformation and its influence on the coagulation rate is possible at very small separations only, that is, approximately at

$$h_{cr}^* \cong 0.002h_0 = 10^{-4}a_2, \tag{129}$$

where condition (127) is used.

In the next section, the validity of this estimation for a broad range of a large set of parameters will be shown until the calculation is accomplished by a numerical method for selected values of parameters.

The important conclusion follows from the comparison of the values of h_{cr}^* according Eq. (129) and h_m according Eq. (82).
If

$$h_{cr}^* < h_m, \tag{130}$$

the calculation of the grazing trajectory can be accomplished by neglecting the weak droplet deformation at the distances $h \geq h_{cr}^*$. In other words, for the validity of the theory of the collision efficiency in the gravitational coagulation (Secs. IV.G–IV.J), condition (130) is an additional necessary condition.

Substitution in condition (130) of h_{cr}^* according to Eq. (129) and h_m according to Eq. (82) yields

$$a_{1cr}^* < \left(\frac{a_1}{a_2} \frac{A}{13.2\pi\Delta\rho g}\right)^{1/4} \sim 30 \ \mu m. \tag{131}$$

The numerical value is obtained by the substitution of the aforementioned values of the parameters and $a_1/a_2 = 0.5$, which was used in Ref. 205 in the numerical calculation of the evolution of the emulsion film.

Thus, at the condition

$$a_1 \leq \sim a_{1cr}^*, \tag{132}$$

the grazing trajectory is not sensitive to the droplet deformation and the theory of collision efficiency for undeformed droplets can be used (Secs. IV.G–IV.J).

The smaller the droplet, the weaker the deformation due to higher static capillary pressure which stabilizes the spherical shape. The smaller the droplet, the larger the distance of the grazing trajectory h_m to the surfaces of the bigger droplet, according to Eq. (84) and, consequently, the weaker the van der Waals interaction along the grazing trajectory and the droplet deformation caused by it. Thus, the physical reasons for the applicability of the collision efficiency of undeformed droplet at sufficiently small dimensions are clear.

However, the important condition (XI.8) establishes a quantitative restriction the exactness of which is determined by the reliability of Yiantsios and Davis theory and numerical calculations (Section XI.C).

C. Different Regimes of Deformation of Interacting Droplets and Their Manifestation in Coagulation

The difference between the drag and buoyancy force on each drop is balanced by equal and opposite forces the drops exert on each other through a thin lubrication layer between them.

Representative calculations are shown in Fig. 29, where F_L is nondimensionalized with $(4\pi/3)\Delta\rho g a_1^3 \lambda^3 (\bar{\mu}\lambda + 1)/(\bar{\mu}\lambda + \lambda)$.

According to Ref. 76, when $h_0 \ll \alpha$, a thin lubrication layer of radial extent $O(\alpha h_0)^{1/2}$ is established between the drops, through which they exert a force, F_L, on each other. Hence, a pressure builds up in that layer which is $P_d^{*III} = O(F_L/\alpha h_0)$ relative to the ambient.

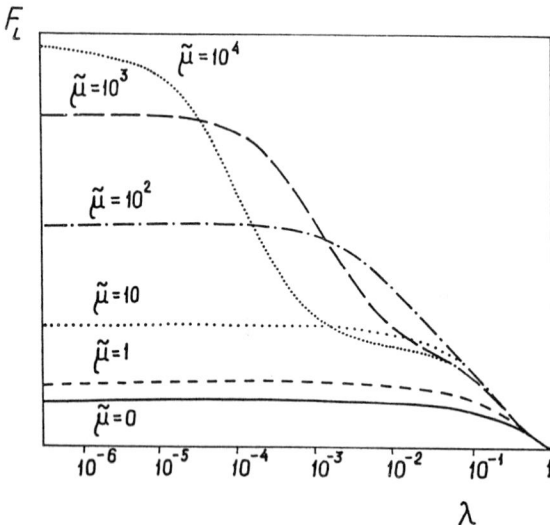

FIG. 29 The lubrication force, F_L, as a function of radii ratio, κ, and viscosity ratio, λ, nondimensionalized with $(4\pi/3)\Delta\rho g a^3 \lambda^3 (\tilde{\mu}\lambda + 1)/(\tilde{\mu}\lambda + \lambda)$. (After Ref. 205.)

Taking into account that F_L does not depend on h_0, one concludes that pressure and droplet deformation increase with decreasing h_0. Applying the theory of droplet deformation to calculations of collision efficiency, the prediction of the evolution of film shape and thickness according to the function $h(r,t)$ (see Fig. 28) is important. For this calculation, a set of equations is derived [205]:

$$-P = \frac{1}{2\delta}\left[\frac{1}{r}\frac{\partial}{\partial r}\left(r\frac{\partial h}{\partial r}\right) - 2\right] - \frac{\delta_1}{h^3}, \tag{133}$$

$$2\pi \int_0^\infty \left(P - \frac{\delta_1}{h^3}\right) r\, dr = 1, \tag{134}$$

$$\frac{\partial h}{\partial t} = \frac{1}{12r}\frac{\partial}{\partial r}\left(rh^3 \frac{\partial P}{\partial r}\right), \tag{135}$$

where dimensionless variables are used: $r = r^*/(\alpha h_0)^{1/2}$, $h = h^*/h_0$, and $P = P_d^*/(F_L/\alpha h_0)$.

Apart from numerical factors which depend on the size ratio κ and the viscosity ratio $\bar{\mu}$, δ and δ_1 are equivalent to the dimensionless parameters, C_m and Ha as defined earlier.

Equation (133) expresses the balance of pressure inside the gap (i.e., capillary pressure and the van der Waals force of attraction). In term of capillary pressure, the surface curvature is expressed through the derivative dh/dr in the usual way [98]. Equation (134) represents the integral force balance and relates the lubrication force and the pressure.

The equation of liquid conservation during film drainage is the basis for the derivation of Eq. (135). The local change of film thickness is caused by the divergence of tangential liquid fluxes, which is expressed by the pressure gradient on the right-hand side of Eq. (135) in the numerical calculations; δ was fixed at 0.025 and δ_1 was varied in the range 10^{-8}–10^{-2}. In Figs. 30a and 30b, successive shapes of the gap thickness for characteristic times are shown for $\delta_1 = 1.3 \times 10^{-6}$ and $\delta_1 = 1.2 \times$

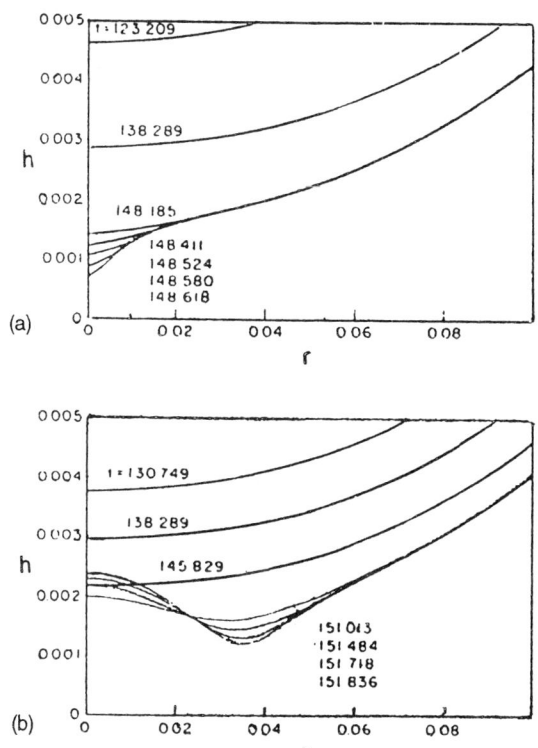

FIG. 30 (a) Shapes of the film thickness for characteristic times when gravity and molecular forces are acting, for $\delta_1 = 1.3 \times 10^{-6}$. (b) Shapes of the film thickness for characteristic times when gravity and molecular forces are acting, for $\delta_1 = 1.2 \times 10^{-6}$. (After Ref. 205.)

10^{-6}. In the first case, rupture occurs at the axis of symmetry, whereas in the second, we have rim rupture. This rupture mode is referred to as "nose rupture." On the other hand, when δ_1 is relatively small, dimpling will occur before rupture, and the rupture will occur along a circle at the rim of the dimple. This rupture mode is referred to as "rim rupture" and leads to the entrainment of a very small drop of the suspending fluid when the two drops coalesce. The effects of van der Waals forces suddenly become dominant and introduce relatively fast changes. This is also shown in Fig. 31, where the minimum thickness, which is at the axis of symmetry, is plotted as a function of time for various values of δ_1 together with the results for $\delta_1 = 0$. The above supports the view taken by other authors that rupture can be described as an instability.

The critical value of δ_{1cr} which demarcates the two different regimes would be of the order of 10^{-6} according Fig. 31.

Thus, at $\delta_1 > \delta_{1cr}$, the collision has to be efficient due to a rapid decrease of the film thickness and its rupture. As to smaller δ_1 values, the question concerning the collision efficiency is open because the rupture time increases with decreasing δ_1 as seen in Fig. 32.

Film rupture bears the features of an instability because once molecular forces become important, changes are fast and occur on a much smaller time scale than the approach due to gravity. The problem of the stability of emulsion and foam films has attracted much attention [210–218] in colloid science.

Initially, the gravitational coagulation and thin-film stability were separate areas in colloid science. However, Yiantsios and Davis [205] make a very valuable effort to combine these scientific directions.

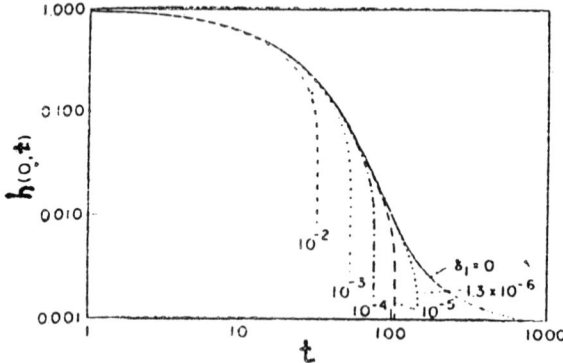

FIG. 31 Minimum thickness versus time for characteristic values of δ_1. The minimum is at the axis of symmetry. (After Ref. 205.)

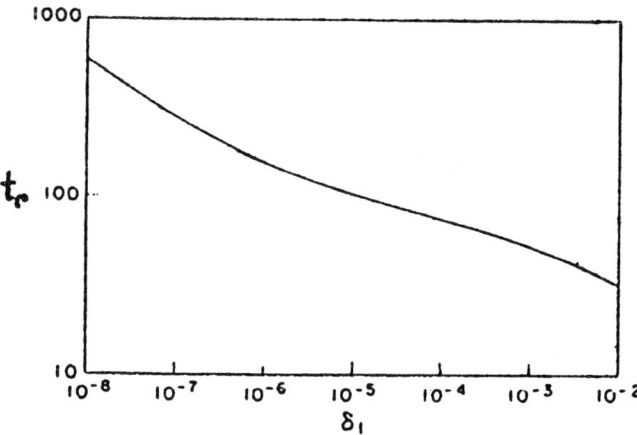

FIG. 32 Rupture time, t_r, as a function of δ_1, when gravity and molecular forces are acting. (After Ref. 205.)

D. Unsolved Problems of Gravitational Coagulation in Emulsions

Yiantsios and Davis were successful due to the efficient application of the lubrication theory. One important component of their investigation omitted in our brief outline concerns the proof of the applicability of this theory. In the following, attention is paid to one restriction.

The authors chose the initial distance separating the two drops [Eq. (127)] as an upper limit for the gap flow to be described by lubrication theory. The restriction of a small initial deformation requires that the modified capillary number for the near-contact region [Eq. (126)] to be small compared with unity. Taking into account Eq. (128) and that the characteristic translational velocity U of the pair may be approximated by $\Delta \rho g a^2 / \mu$, this restriction transforms to

$$\frac{\Delta \rho g a^2}{\gamma} \ll 1. \tag{136}$$

Recently, much attention is paid to emulsion systems with low interfacial tension [219]. Restriction (136) can be operative in such systems.

Yiantsios and Davis considered thin-film stability and rising droplet deformation in droplet interaction in the interconnection. However, they draw only minor attention to next step (i.e., to the calculation of the collision efficiency). Because of this, we tried in Sec. XI.B to pay more attention to the link between the droplet deformation and their collision efficiency.

However, previous authors usually described film rupture as an instability of a steady or slowly varying base state. Many articles consider infinite films of uniform thickness [210–214]. A notable exception is the work of Gumerman and Homsy [215], who analyzed the stability of a nearly flat film between two drops which slowly approach each other. Prior to Ref. 205, this is the only work which correctly identifies the appropriate length, time, velocity, and pressure scales inside the film. However, drawbacks in their work are that the analysis is restricted only to the region inside the film and that the artifical boundary conditions are imposed at the nose and not at the periphery. A work that bypasses this difficulty is by Chen and Slattery [216], who consider rupture as a natural evolution rather than an instability. However, ad hoc boundary conditions are used away from the near-contact region, and a dimpled shape is used as an initial condition, which is also found in an ad hoc way.

As in the work of Chen and Slattery [216] Yiantsios and Davis [205] consider rupture to be a natural evolution rather than an instability. However, the drop shapes are found from first principles, without any a priori assumptions. More specifically, they consider two drops approaching each other due to gravity and allow van der Waals forces, which are initially weak but increase as the drops become close together. Therefore, the previous analysis describes the initial stages of evolution.

A more rigorous approach would account for what is now generally accepted; that is, the pressure inside thin films is not isotropic but rather tensorial in nature [220,221]. However, the "exact" form of this pressure tensor is known only for equilibrium films and in rather special cases. Another important aspect that is not taken into account in the literature and not here either is that the density is not uniform inside these thin films, and the interfaces which are invariably treated as mathematical surface actually are regions of finite thickness across which dramatic variations in density occur. This thickness may be of the same order of magnitude as the films themselves.

XII. DYNAMICS OF BROWNIAN COAGULATION OF DEFORMABLE EMULSION DROPLETS

The stochastic nature of Brownian coagulation makes the description of droplet deformation extremely difficult. There is a single work (Ref. 207) devoted to this important problem. For the characterization of the method developed in Ref. 207, a discrimination between macroemulsions and miniemulsions is important.

Most of the works on liquid film stability in emulsions deal with relatively large drops (millimeter and submillimeter size), where the effect of the Brownian motion is negligible. Such systems are usually called *macroemulsions*. The case of emulsions consisting of droplets of submicrometer to at most a few micrometers in size has been much less investigated (in Ref. 207, these systems are called *miniemulsions*). In particular, it is not known in advance whether or not the small droplets (possessing considerably greater capillary pressure) will deform upon mutual approach.

The method developed in Ref. 207 is based on two main assumptions. Similar to the case of macroemulsions, the approach of the droplets in miniemulsions is supposed to follow the scheme shown in Fig. 33a (see Sec. XII.A). Brownian force, which causes droplet deformation, is expressed through the mean droplet velocity of the directional motion toward the central droplet (Sec. XII.B).

A. Mutual Diffusion of Deformable Interacting Emulsion Droplets

In accordance with Smoluchowski and Fuchs, Danov et al. [207] consider the diffusion of *deformable miniemulsion* droplets toward a given "central" droplet in the presence of potential interaction energy $U_{1,2}$ (see Sec. III.C.1).

The first stage (region A_a in Fig. 33) presents the approach of two spherical or slightly deformed droplets of radius a. The distance z in this region adopts values between $z_d = 2a + h_i$ and infinity (h_i is the minimum distance between the droplet surfaces prior to the film formation). This region can be described by the theory of barrier diffusion of spherical particles (Sec. VIII).

The second stage, below the distance z_d (see region A_d in Fig. 33) represents rapid deformation of the droplet caps and formation of a flat film. This occurs at given value, h_i, of the minimum distance between the droplet surfaces; see Fig. 33 (bottom). h_i is called the *thickness of film formation* (the thickness at which the curvature at the droplet caps inverts its sign). At this stage, the radius, r, of the formed plane-parallel film gradually (but rapidly [218,222]) changes from zero to a certain maximum value (denoted hereafter by r_f). Although the thickness of the film can be assumed to remain constant during this process, the distance between the *mass centers* changes (due to the deformation) from z_d to z_f [223]. It was shown analytically [218,222] and numerically [205] that the rate of film expansion is very high. For small (micrometer size) particles, the plane-

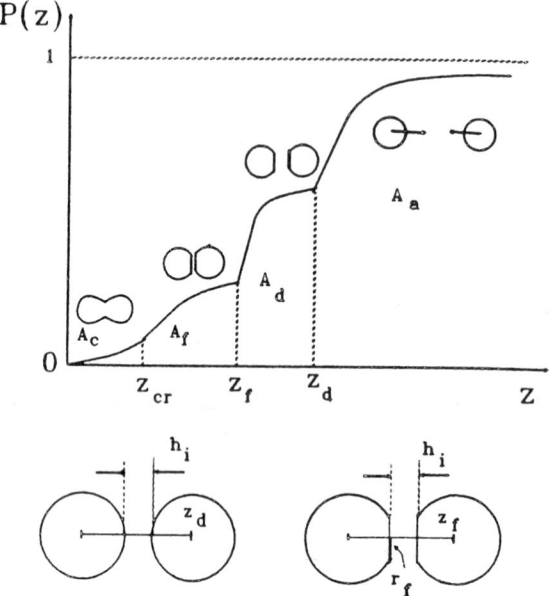

FIG. 33 Top: Schematical presentation of the pair probability distribution function $P(z)$ for deformable coalescing droplets. z is the distance between the mass centers of the droplets. A_a denotes the region of nondeformed droplets approach, A_d denotes the region of droplet deformation, A_f denotes the region of film thinning, and A_c denotes the region of droplet fusion. z_d is the distance at which the deformation occurs, z_f is the distance at which the planar film thinning starts, and z_{cr} is the critical distance of film rupture. Bottom: Droplet configurations at the points z_d and z_f. h_i is the separation between the drop surfaces at which the deformation occurs and r_f is the film radius in the region A_f.

parallel film may appear without being preceded by dimple formation [218].

During the third stage (region A_f in Fig. 33), the film thickness is reduced from h_i to the critical thickness of rupture h_{cr} at (almost) constant radius r_f. The distance between the mass centers of the two droplets in the moment of rupture is z_{cr}. The rate of film thinning depends on different hydrodynamic and thermodynamic factors [205,218,224–226].

In the last stage, after the film rupture (region A_c in Fig. 33), the mass centers of the droplets approach from h_{cr} to $h = 0$ and rapid fusion of the droplets is accomplished.

If the film can be thermodynamically stable, the thinning stops at the equilibrium thickness h_e where the disjoining pressure in the film equilibrates the capillary pressure in the drops. In Ref. [207], attention is paid only to the case of thermodynamically unstable films which always rupture. In this case, the film lifetime depends mainly on the rate of thinning and the critical thickness h_{cr}.

B. Droplet Deformation and Mean Droplet Velocity of Directional Motion

Equation (23) determines droplet velocity V_{12} of directional motion, which corresponds to drop collision rate [Eq. (31)], and expresses the result of the averaging over all drop pairs of the statistical ensemble. In Ref. 207, the droplet deformation is characterized as caused by the droplet approach with the velocity V_{12}. Thus, the droplet deformation is introduced as a result of the averaging over all drop pairs.

From Eq. (23), it is clear that the mean droplet velocity is caused by the force

$$F_T(z) = kT \frac{d}{dz} [\ln P_{12}(z)] + \frac{dU_{12}}{dz} \tag{137}$$

Hence, as a general result we obtain that in the case of steady diffusion, the total force acting on the droplet, F_T, is a superposition of the Brownian (diffusion) force (i.e., $kT\nabla[\ln P_{12}(z)]$) and the potential force which is due to direct interactions between the particles, $\nabla U_{12}(z)$.

From Eqs. (23), (31), and (137), one obtains

$$F_T(z) = \frac{J\gamma(z)}{4\pi z^2 n_\infty P(z)}, \tag{138}$$

where

$$\gamma(z) = \left[\frac{D_{12}(0)}{kT} G(z) \right]^{-1}$$

is the hydrodynamic resistance to the mutual droplet motion.

The calculation of the stability ratio or collision efficiency and correspondingly of the integral in Eq. (36) requires knowledge (i) of the distances z_f and z_{cr} at which the film formation and the film rupture take place (see Fig. 33) and (ii) of the explicit expressions for the interaction energy $U_{12}(z)$ and the mutual diffusivity $D(z)$. The film formation thickness h_i depends on the total force F_T [218,222]:

$$h_i = \frac{F_T}{2\pi\sigma}. \tag{139}$$

From Eqs. (138) and (139), one obtains

$$h_i = \frac{J\gamma(z_d)}{8\pi^2 \sigma z_d^2 n_\infty P(z_d)}.$$ (140)

$P_{12}(z_d)$ was determined by integrating Eq. (32) along with boundary condition (33) and was substituted in Eq. (140). It yielded

$$h_i = \frac{kT}{4\pi\sigma a f_0},$$

$$f_0 = \int_{z_{cr}}^{z^d} \left(\frac{2a}{z}\right)^2 \frac{\gamma(z)}{\gamma(z_d)} \exp\left(\frac{U_{12}(z) - U_{12}(z_d)}{kT}\right) \frac{dz}{2a}.$$ (141)

In Eq. (141), the lower limit in integral 0 is replaced by z_{cr}. It corresponds to the assumption that the contribution from region A_c is much smaller than from the other two regions, A_f and A_d. This assumption is justified when the rupture is rapid in comparison with the film thinning.

C. Evaluation of Droplet Deformation and Stability Ratio

Equation (141) allows a calculation of h_i if $\gamma(z)$, $U_{12}(z)$, and z_{cr} are known. Equation (141) should be solved numerically because of the complex relation between $h_{cr} = z_{cr} - 2a$ and h_i [218,224,226].

The latter is evaluated using the approximate formula of Vrij [227], which can be transformed into

$$\bar{h}_{cr} = \frac{h_{cr}}{h_i} = 0.243 \left(\frac{A_H^2 a^2}{2\pi\sigma^2 h_i^6}\right)^{1/7}.$$ (142)

After the determination of h_{cr} and h_i from Eqs. (141) and (142), one can obtain the stability ratio or collision efficiency according to Eq. (36).

The function $U_{12}(z)$ and $\gamma(z)$ for potential and hydrodynamic interactions between two approaching deformable droplets are specified in Refs. 206, 208, and 209.

Numerically calculated values of the thickness h_i, at which the droplet deformation takes place and of Fuchs factor W_f, are presented. It is shown that the attraction between the droplets increases the values of h_i, whereas the repulsion leads to the opposite effect.

The theory predicts that with a significantly long-range repulsion (e.g., electrostatic and with sufficiently low electrolyte concentration) below some critical ionic strength, the physically relevant root of Eq. (141) (which determines the value of h_i) disappears. This means that the kinetic energy of the droplets is not large enough to overcome the energy barrier accompanying the droplet deformation and the particles behave as nondeformable charged spheres. This energy barrier relates to the region A_d in Fig. 33, corresponding to an increase in the film radius r_f. This is caused by the increase of the electrostatic repulsion and the electrostatic barrier that prevents surface from flattening.

For predominating attractive forces, the growth of the film radius is accompanied by an increase in attraction. The attraction causes the surface flattening, which enhances the attraction force. In other words, this type of coupling between the attractive surface forces and the surface deformation is favorable for the droplet deformation.

As the electrolyte concentration increases and electrostatic barrier decreases, the root h_i of Eq. (141) appears and increases. This growth in h_i is understandable because the lower the electrostatic repulsion, the higher the attractive force that causes the droplet deformation at larger distance h_i.

Most of the numerical results are devoted to calculation of the particle separation, h_i, at which the deformation occurs in the absence of electrostatic repulsion. The latter depends on the Hamaker constant, droplet radius, surface tension, and droplet surface retardation by surfactants.

For micrometer-size droplets, h_i is between 5 and 25 nm and depends only slightly on the droplet radius. For particles smaller than 1 μm, one observes a significant decrease in h_i. The stronger h_i depends on the Hamaker constant. Values for h_i show that the Brownian force may lead to a deformation of the emulsion droplets and formation of planar films during the collision between them. h_i increases when γ decreases. At γ below 1 dyne cm^{-1}, a sharp increase in h_i is observed.

The qualitative explaination of the results is easy. The surface deformation is caused by van der Waals attraction and occurs at a larger distance h_i with decreasing Hamaker constant. The smaller the capillary pressure, the easier the droplet deformation and the greater the distance h_i. Consequently, h_i expands as γ diminishes and a increases.

The authors also investigated the role of the surface mobility by introducing the dimensionless parameter

$$\epsilon_s = \frac{\epsilon^e + \epsilon^f}{1 + \epsilon^e + \epsilon^f}, \tag{143}$$

where the quantities ϵ^e and ϵ^f account for the surfactant diffusivity in the disperse phase and/or media, as well as for the droplet surface properties, and vary within a wide range of values [218,228]. ϵ^f can acquire values between 1 (for tangentially immobile surfaces) and 0.001 (for pure liquids or when the surfactant is soluble only in the droplet phase). Typical values of ϵ_s with surfactants soluble only in the continuous phase are between

0.1 and 1; for systems with surfactants soluble in the dispersed phase, ϵ_s is between 0.001 and 0.01 [229,230].

The enhanced tangential mobility of the droplet interfaces, which strongly affects the rate of film thinning (and hence the film lifetime [218,225]) does not practically change h_i. This means that the increased tangential mobility of the interfaces shortens the time of film drainage but does not change the distance at which the droplet deformation starts. This result is in accordance with the theoretical prediction by Ivanov et al. [222]. They assumed that the shape of deformable approaching droplets depends mainly on thermodynamics factors rather than on hydrodynamic ones.

The numerical calculations show that the stability ratio for deformable and nondeformable spheres of *micrometer size* practically coincide. This result means that for miniemulsions, the mutual approach of the droplets prior to their deformation is the rate-determining stage. The authors explain the result pointing out that the distance h_i is sufficiently large.

D. Current State of Slow Coagulation Modeling

The modeling [207] of thin-film formation, its thinning, and rupture enables one to overcome immense difficulties in the theory of slow coagulation of deformable droplets. On the other hand, Danov [207] write, "We are aware that for complete estimation of the rate of coalescence in a particular system one should provide an additional analysis of the possibility for rupture of the layer between the droplets prior to the formation of planar film, $h_{cr} > h_i$. Since a theory for h_{cr} in the case of spherical droplets is still not available, we cannot say whether this possibility can be realized in miniemulsions." This suggestion can be extended in respect to the case $h_{cr} < h_i$ too. If a theory for h_{cr} in the case of spherical droplets is still not available, "the use of Vrij equation [227] in [207] is questionable. In particular, the conclusion concerning the weak influence of the droplet deformation on the stability ratio is also questionable. The theory in Ref. 207 obviously needs a systematic experimental examination.

According to Yiantsios and Davis [205], it is possible to investigate the droplet deformation without any assumption about the different stages of this process (see Fig. 33) and to investigate the film rupture without the analysis of its instability and without the introduction of h_{cr}.

Thus, at least three models of slow coagulation of the deformable droplets deserve further investigations.

1.   The model elaborated in Ref. 207, with the main feature $h_{cr} < h_i$. The assumptions which must be further examined concern the applicabil-

ity of Vrij equation [207] for h_{cr} and the adequacy of the real film formation and its evolution according to Fig. 33.

2. The model with the assumption $h_{cr} > h_i$.
3. The model based on the approaches developed in Refs. 205 and 207. It ignores the instability and avoids the assumption concerning the stages of the droplet deformation (see Sec. XII.E).

E. Combining Different Methods in Slow Coagulation Theory

The droplet deformation is caused by the lubrication force under conditions of gravitational coagulation and by the regular force according to Eq. (137) during Brownian coagulation. After the introduction of the notion of the regular force caused by the mean velocity of directional motion [207], a substantial similarity in conditions for droplet deformation during gravitational or Brownian coagulation appears. However, the sequence of stages illustrated in Fig. 33 does not agree with the features of the film evolution established in Ref. 205, in particular for the dimple formation (Section XI.C, Fig. 31).

The sequence of stages illustrated in Fig. 33 was observed in macro-emulsions and postulated for miniemulsions. Omitting this assumption can improve the theory of slow coagulation. This improvement can be achieved by the combining methods described in Refs. 205 and 207. Instead of postulating the stage sequence according Fig. 33, a set of equations (133)–(135) can be used to investigate the film formation and its evolution. The additional advantage of this approach is the refusal from the introduction of h_{cr} and Eq. (142) for its estimation.

Although the direct application of the set of equations (133)–(135) is absent in Ref. 207, the prerequisite for its application is created. This prerequisite is the notion of the regular force $F_T(z)$ according to Eq. (138). This force controls droplet deformation under conditions of slow coagulation and it has to be inserted in right-hand side of Eq. (134) instead of the lubrication force responsible for droplet deformation during gravitational coagulation.

The right-hand side of Eq. (134) is a time function similar to the left-hand side. The time dependence $z(t)$ can be established on the basis of Eq. (138) because $V \equiv dz/dt$. It enables one to establish the time dependence for P_{12} and afterward for F_T according to Eq. (138).

From Eqs. (137) and (138), one obtains

$$\frac{J}{4\pi z^2 n_i D(t)} \exp\left(\frac{U_{12}}{kT}\right) = \frac{d}{dz}\left[p \exp\left(\frac{U_{12}}{kT}\right)\right] = \frac{d}{dt}\left[p \exp\left(\frac{U_{12}}{kT}\right)\right] \frac{4\pi z^2 n_\infty P}{J}$$

$$(144)$$

that enables one to accomplish the integration

$$
\frac{J}{J_{sm}} = \frac{J}{16\pi D_{12}^{(0)} n_\infty a} = W^{-1} = \left[4D^{(0)}a \left(2 \int_0^\infty \frac{\exp(2U_{12}/kT)dt}{z^4 D(t)} \right)^{1/2} \right]^{-1},
$$

(145)

where W is the stability ratio.

$$
p_{12} = \exp\left(-\frac{U_{12}}{kT} \right) p_{12}(h_0)
$$

(146)

$$
\times \left\{ 1 - \left[\int_0^t \exp\left(2\frac{U_{12}}{kT} \right) \frac{dt}{D(t)z^4} \right] \left[\int_0^\infty \exp\left(2\frac{U_{12}}{kT} \right) \frac{dt}{D(t)z^4} \right]^{-1} \right\}.
$$

The choice of h_0 corresponding to $t = 0$ is similar to that in Ref. 205; that is, it satisfies conditions (127) and (128) that cause the simplification in Eqs. (145) and (146). With condition (127), z can be replaced by a in integrands of Eqs. (145) and (146).

At any moment t of film evolution, $U_{12}(t)$ and $D(t)$ have to be expressed through the film profile $h(r, t)$, where r is the radial coordinate of the cylindrical coordinate system introduced in Ref. 205 (Section XI.3). Derjaguin's approximation [1,12,231] can be used because the film thickness is very small in comparison with droplet radius:

$$
U_{12}(t) = \int_0^\infty f[h(r, t)]2\pi r \, dr.
$$

(147)

By equating the dissipated energy to the work of the outer forces, Charles and Mason [232] obtained the expression for hydrodynamic resistance of two approaching solid particles at small distances (i.e., in lubrication approximation):

$$
\gamma(t) = \frac{kT}{D(t)} = 6\pi\mu \int_0^\infty \left(\frac{r}{h(r, t)} \right)^3 dr.
$$

(148)

The shape of the thin film is not specified in Eqs. (147) and (148). This enables one to deviate from the assumption concerning the sequence of stages illustrated in Fig. 33 and to accomplish a general investigation of the film evolution following the method of Yiantsios and Davis [205].

From Eqs. (145) and (147), one obtains

$$
W = \left(\frac{2D^{(0)}}{a^3} \int_0^\infty F(t)E(t) \, dt \right)^{-1/2},
$$

(149)

where

$$E(t) = \exp \int_0^\infty f[h(r, t)]2\pi r \, dr, \tag{150}$$

$$F(t) = \int_0^\infty \left(\frac{r}{h(r, t)}\right)^3 dr. \tag{151}$$

From Eqs. (146), (150), and (151), one obtains

$$p_{12}(t) = E^{-1}(t) \left(1 - \frac{\int_0^t E(t')F(t') \, dt'}{\int_0^\infty E(t')F(t') \, dt'}\right)(1 - W^{-1}). \tag{152}$$

The last multiplier on the right-hand side of Eq. (152) is the expression for $p_{12}(h_0)$ which follows from the description of the diffusion flux in the zone with $h > h_0$.

If we further restrict ourselves to tangentially immobile interfaces, we use $F_T(h_0)$ instead of F_L, and the relevant dimensionless variables are the same as in Section XI. According to Eq. (138),

$$F_T\big|_{t=0} \equiv F_T(h_0) = \frac{6\pi\mu D}{W(1 - W)}\frac{a}{h_0}, \tag{153}$$

where Taylor's [64] equation is used:

$$\gamma(0) = \frac{6\pi\mu a^2}{h_0}. \tag{154}$$

From Eqs. (138), (148), (150)–(152), and (154), one obtains the modified Eq. (134) for the investigation of thin-film evolution during slow coagulation:

$$2\pi \int_0^\infty \left(p - \frac{\delta_1}{h^3}\right)r \, dr = \frac{F_T(t)}{F_T(0)} \equiv \frac{\gamma(t)}{\gamma(0)}\frac{p_{12}(0)}{p_{12}(t)} \tag{155}$$

$$= \frac{h_0 F(t)E(t)}{a}\left(1 - \frac{\int_0^t E(t')F(t') \, dt'}{\int_0^\infty E(t')F(t') \, dt'}\right),$$

where

$$\delta_1 = \frac{aA}{6\pi h_0^2 F_T(0)}. \tag{156}$$

In Eq. (133), δ has to be changed to

$$\delta = \frac{F_T(0)}{\gamma h_0}. \tag{157}$$

Thus, a set of equations (133), (135), and (155) together with Eqs. (147), (150), (151), and (156) is the basis for the investigation of thin-film evolution during slow coagulation.

XIII. REVERSIBLE COAGULATION IN EMULSIONS

A. Qualitative Approach

The depth of the secondary potential pit is rather small for small droplets or for a decrease in electrolyte concentration (Sec. V.A). The small (some kT) interaction energy of droplets in a doublet enables a disintegration under action of the thermal movement of the molecules in the media. Thus, at small depths of the potential pit, two processes have to be taken into account. In parallel to doublet formation, singlet collisions and doublet disaggregation occur. The second process is not accounted for in Smoluchowski or Fuchs theories, describing the irreversible coagulation.

During the initial state of coagulation, singlets and doublets exist. The concentration of multiplets is negligible. The quantity of singlets (n_1) decreases while the doublet quantity (n_2) increases. Hence, the aggregation process is retarded while disaggregation process is intensified. After some time, a dynamic equilibrium arises; that is, the quantities of aggregation and disaggregation are equal.

The irreversible coagulation leads to a creaming of the emulsion. The reversible coagulation does not lead to a breakdown of emulsions.

B. Dynamic Equilibrium Singlets–Doublets

In the theory of irreversible coagulation (Secs. III and VIII), its regularity is based on the choice of one particle as central and considering the flux of other particles within the vicinity of the central particle. The same approach is undertaken in theory of dynamic equilibrium. The absence of the flux means that the distribution of other droplets with respect to the central one obeys a Boltzmann distribution:

$$n(r) = n_0 \exp\left(-\frac{U_{1,2}(r)}{kT}\right), \tag{158}$$

where n_0 is the initial particle concentration and r is the distance between the centers of droplets.

The summation of particles in the space where $U_{1,2}(r) < 0$ yields the averaged value of particles in an aggregate bound to a central one:

$$N = 1 + 4\pi n_0 \Gamma, \tag{159}$$

where

$$\Gamma = \int_{v_{1,2}<0} r^2 \left[\exp\left(-\frac{U_{1,2}(r)}{kT} \right) - 1 \right] dr.$$

If the minimum of the second potential pit is sharp, the Muller [233] approximate equation

$$N = 1 + 16\pi a^3 n_0 \sqrt{\frac{2\pi kT}{U''(\kappa h_{min})}} \exp[-U_{1,2}(\kappa h_{min})] \tag{160}$$

is valid. The second derivative can be expressed through $U_{1,2}(\kappa h_{min})$ which transforms Eq (159) into the function $U_{1,2}(\kappa h_{min})$. At small doublet concentration n_2 and in the absence of multiplets,

$$N = \frac{2n_2}{n_1 + n_2} + \frac{n_1}{n_1 + n_2} \cong 1 + \frac{n_2}{n_1} \cong 1 + \frac{n_2}{n_0}, \tag{161}$$

which corresponds to a small N value.

In Ref. 234, the concentration of equilibrium doublets in w/o (water–oil) emulsions was measured, which enabled one to determine U_{min} using Eq. (160) and to estimate the Hamaker constant. Similar investigations for suspension were undertaken in Ref. 235. This is a valuable approach to estimate the Hamaker constant. However, the retardation of the van der Waals forces has to be taken into account in order to make the estimations more reliable.

C. The Retardation and Screening of van der Waals Forces and Their Influence on Reversible Coagulation

As seen in Fig. 14, the coordinates of the secondary minimum corresponds to $\kappa h_m = 5{-}12$. Due to this rather large distance, the frequency dependence of the Hamaker constant may be of importance [236]. Hence, the Hamaker function $A(h)$ characterizing molecular interaction should be introduced.

In Ref. 237, the distance-independent interaction at zero frequency and at nonzero frequency is considered separately:

$$A(h) = A_0 + [A(h) - A_0]. \tag{162}$$

The results of calculations for 36 systems accomplished in Ref. 237 are in rather good accordance with the calculations in other papers. According to Churaev, the system polystyrol–water–polystyrol can be used to estimate the Hamaker function for oil–water systems. In Ref. 238, the influence on κh_{min} and U_{min} are calculated (see Figs. 14 and 15). The corresponding influence on doublet concentration is given in Fig. 34.

With an increase in the droplet separation, the importance of the component A_0 is increasing because of $A(h) - A_0$. The component A_0 is screened in electrolyte concentrations because of dielectric dispersion [239–241]. In Ref. 239, it is proved that an excess electrolyte concentration in DL does not influence the screening of molecular attraction. This screening is incorporated in the Lifshitz theory [240–241]. At a distance

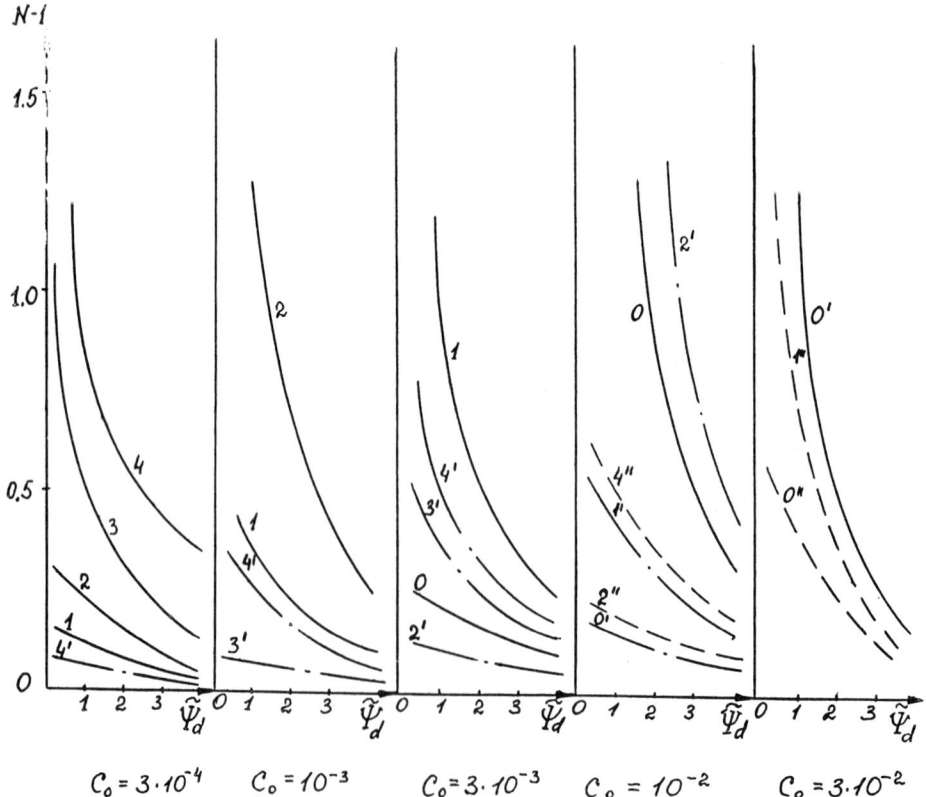

FIG. 34 Dependence on the Stern potential and electrolyte concentration for aggregation number N ($a = 1$ μm). Curves are numerated similar to Fig. 14. Volume part $p = 4/3(\pi a^3 n_0) = 10^{-2}$.

of $\kappa h_{min} \sim 3\text{–}5$ nm, the authors found that the molecular interaction disappeared at zero frequency. In this case, A_0 can be omitted when evaluating the coagulation in the secondary minimum. Figs. 14–16 illustrate this.

D. Influence of Disaggregation on Coagulation Kinetics

The theory by Smoluchowski and Fuchs disregards the disintegration of aggregates which can occur at the sufficiently deep potential pits. This can be generalized by introducing a term accounting for the function of singlets due to the disrupture of doublets [233]:

$$\frac{dn_1}{dt} = -an_1^2 + bn_2, \tag{163}$$

where a and b are coefficients characterizing the probability of the formation and disrupture of aggregates per unit time. In Ref. 233, a simple expression for b depending on the droplet radius, kT, and Γ is obtained. The averaged lifetime of doublet equals b^{-1}. After this time, the singlet quantity preserves (i.e., $dn_1/dt = 0$), and from Eq. (163),

$$n_2 = \frac{a}{b} n_1^2. \tag{164}$$

The substitution of the expressions for a and b into Eq. (164) yields

$$n_2 = 16\pi n_1^2 a^3 \Gamma, \tag{165}$$

where

$$\Gamma' = \int \left[\exp\left(-\frac{U_{1,2}(s)}{kT} \right) - 1 \right] ds,$$

$$s = \frac{r - 2a}{2a}, \tag{166}$$

which is in agreement with Eqs. (160) and (161) if

$$n_2 \ll n_1 \approx n_0. \tag{167}$$

This is a necessary condition for a simple description of the dynamic equilibrium, pointing out that the collisions between doublets and between doublets and singlets cannot be neglected. In other words, the equation for doublets and other multiplets has to be added to Eq. (163) which complicates the expressions for the kinetics [203,204]. As it is seen from Eq. (163), condition (164) is satisfied if the potential pit is not deep and the volume part is not high. In the opposite case, the simple theory of the dynamic equilibrium fails and a more general theory [203,204] must be used.

In multiplets, the interaction energy for a droplet is higher than in doublets. Thus, a slow accumulation of triplets, quadruplets and so on is possible.

If so, the singlet–doublet quasi-equilibrium occurs at the initial stage of coagulation. When multiplets with longer lifetime and higher concentration appears, the reversible coagulation transforms into an irreversible one. In other words, the coagulation reversibility leads to some delay in the irreversible coagulation.

E. Doublet Lifetime

The equation for the averaged doublet lifetime τ_d was obtained [233] in connection with the description of dynamic equilibrium:

$$\tau_d = 2T_d\Gamma', \tag{168}$$

where

$$T_d = \frac{a^2}{D} = \frac{6\pi\mu a^3}{kT}, \tag{169}$$

the averaged time of Brownian collision in the absence of surface forces. For a droplet with $a \sim 1$ μm, T_d is of second order. Doublet lifetime does not depend on the presence of other droplets. In polydisperse emulsions, a doublet consisting of droplets of different radii, a_i and a_j, depends on them; that is,

$$\tau_d(a_i, a_j) = 2T_d(a_i, a_j)\Gamma'(a_i, a_j), \tag{170}$$

where

$$T_d(a_i, a_j) = \frac{(a_i + a_j)^3}{D_{ij}}, \tag{171}$$

where D_{ij} is given by Eq. (24). Equation (170) is generalized by using $U_{1,2}(a_i, a_j)$ and $s_{ij} = [r - (a_i + a_j)]/(a_i + a_j)$. Forthcoming papers will show the applicability of, for instance, enhanced video microscopy in determining doublet lifetimes as well as doublet–singlet equilibria.

XIV. EVALUATION OF THE INFLUENCE OF SURFACTANT ADSORPTION ON FLOCCULATION

A. General

As shown in the preceding sections, the electrostatic repulsion cannot prevent flocculation at millimolar concentration (at least not for small droplets). This electrolyte concentration range is important in practical applications. The stability toward collision and flocculation is provided by the surfactant adsorption. It does not mean that DLVO completely loses its significance. The equations of DLVO theory cannot be directly applied. However, the DLVO concept can be extended to systems stabilized by surfactants, and its part concerning molecular attraction preserves its significance almost completely.

B. Superposition Approximation in DLVO. Extension to Surfactant-Stabilized Emulsions

According to this approximation, the total interaction potential energy for two drops is given by

$$U_{1,2} = U_A + U_R + U_{st} + \Delta U. \tag{172}$$

Here, U_A is the contribution due to the van der Waals interaction, U_R is the electrostatic term, U_{st} accounts for the steric interaction if polymers are adsorbed at the drop surface, and ΔU stands for other possible interactions, as mentioned earlier (structural, depletion, etc.). The superposition approximations is usually applied [1,12,208,209]. In Secs. XIV.E–XIV.F the necessity of the refusal from the assumption concerning the force additivity for some conditions will be shown.

C. Influence of Surfactant-Adsorption Layer on Dynamic Equilibrium Between Singlets and Doublets

In Sec. XIII.C, the dynamic equilibrium between doublets and singlets was described. In Sec. XIV.B, the manifestation of the surfactant-adsorption layer in surface forces was described in the framework of the superposition approximation. In this approximation, the motion of the dynamic equilibrium singlet–doublet can be extended to emulsions stabilized by surfactants. The easiest case is characterized by a definite surfactant-adsorption layer, d. More complicated adsorption layers including loops or tails are omitted here. Also, for this case, an effective layer may be introduced; for example, see equations explained in Refs. 242–244. Israelachvili [138] states that the details in the description of the dependence of surface forces on distance is difficult to justify and proposes the exponential dependence

$$f(h) = B \exp\left(-\frac{h}{d}\right) \tag{173}$$

for electrostatic, steric, hydrophobic, or hydration interaction. The parameter d is positive and determines the range of interaction. For electrostatic interaction,

$$d = \kappa^{-1} \tag{174}$$

and B follows from a comparison of Eqs. (173) and (174) with Eq. (11).

Due to the similarity in the distance dependence for electrostatic and sterical forces, all results obtained in Sec. XIII for the coagulation in the secondary minimum can be used for the case. When steric repulsion predominates and electrostatic repulsion can be neglected, the solution of Eq. (99) for the determination of the location of the secondary minimum can be used. Instead of κh_m, h_m/d must be used. Also, B depends on the adsorption. We restrict ourselves to the case of sufficiently high adsorption corresponding to sufficiently high Stern potential. Thus, Eqs. (158)–(160) can be used.

The main conclusions from Fig. 34 to the estimation of the steric repulsion influence on the doublet–singlet dynamic equilibrium are as follows:

1. The retardation of van der Waals forces and their screening by electrolyte causes the strong decrease of the flocculation similar to the case of electrostatic repulsion.
2. The larger the droplet dimension, the stronger the flocculation.

For $c = 3 \times 10^{-4}$ M $\to d = 17$ nm, $c = 10^{-3}$ M $\to d = 10$ nm, $c = 3 \times 10^{-3}$ M $\to d = 6$ nm, $c = 10^{-2}$ M $\to d = 3$ nm, and $c = 3 \times 10^{-2}$ M $\to d = 1.7$ nm. From model accounting for the retardation and neglection of the screening one can conclude that at an adsorption layer thickness of $d \sim 1.7$ nm, a flocculation of droplets with a radius larger than 0.5 μm is possible. At $d \sim 3$ nm and $d \sim 6$ nm, the flocculation of the droplets with a radius less than 2 μm and 4 μm is prevented. An adsorption layer thickness of $d \geq 10$ nm is sufficient to prevent flocculation in miniemulsions ($a < 10$ μm).

These conclusions drastically change if the retardation is not taken into account. For instance, even at $d \geq 10$ nm, a flocculation of droplets with $a > 1$ μm is possible. A submicron flocculation is possible at $d < 6$ nm and especially for $d < 3$ nm. Polymer adsorption prevents flocculation due to the retardation of the van der Waals forces. This statement is further justified if a screening is taken into account.

It was assumed that the flocculation occurs if less than 10% of droplets are included in doublets. However, for instance, 1% droplets in doublets can also be considered as a measure of flocculation. Taking into account the adsorption layer thickness for different surfactants, a conclusion concerning their influence on stability can be made. Low-molecular-weight surfactants cannot prevent flocculation in the secondary minimum due to the small thickness of their adsorption layer. Above critical concentration of micellization (CMC), the micellization can change the surfactant influence on emulsion stability. This special topic is beyond the scope of this chapter.

The thickness of a globular protein adsorption layer usually does not exceed 2 nm. It means that the flocculation of smaller droplets with dimension less than 2 μm can be prevented. Protein adsorption cannot be efficient in preventing flocculation of the larger droplets.

A polymer adsorption leads to an increase in the adsorption layer thickness and hence is favorable in preventing flocculation. As it is pointed out by Dalgleish [Chapter 5], multiple layers seem to be more easily formed in emulsions containing whey proteins. In his chapter, the thickness of β-casein adsorption is estimated to be 10 nm which is sufficient to prevent flocculation.

D. Some Conclusions Concerning Flocculation in Concentrated Emulsions

The higher the volume fraction, the more favorable are the conditions for flocculation. Hence, conclusions established for dilute emulsions cannot be generally applied to concentrated emulsions. In dilute emulsions, pair collisions predominate. The binding energy is determined by the pair interaction energy. In concentrated emulsions, the ternary and more complicated collisions occur and cause an increase in the potential pit depth (i.e., it enhances flocculation). However, there is an essential link between flocculation regularities in dilute and concentrated emulsions. In the latter case, pairwise interaction is the foundation for the quantifying flocculation theory because the thickness of the zone of the surface force action (0.1 μm) is less than the droplet radius in miniemulsions.

The potential pit depth for droplets in concentrated emulsions can be estimated as the product of the U_{min} for a pairwise interaction multiplied with the quantity of a droplet neighbor. One concludes that the smaller the droplet, the weaker the flocculation in concentrated emulsions. Thus, the decrease in droplet dimensions of an emulsion increases its stability both toward sedimentation and flocculation.

A higher binding energy of droplets which can be achieved in multiple collisions leads to more strict demands to the surfactants used for the flocculation. Contrary to the case for dilute emulsions that a 2-nm protein adsorption layer can prevent flocculation of droplets with a dimension of 1 μm, the situation in concentrated emulsions can be quite different.

An efficient method for providing high aggregative stability of concentrated emulsions is to combine small droplets and thick surfactant adsorption layer. The molecular attraction decreases with the separation determining the conditions of the emulsion stabilization toward flocculation. Correspondingly, such phenomena as the retardation and screening of van der Waals forces are favorable for the stabilization and deserve attention.

E. Effect of Surfactant Adsorption on van der Waals Forces

The surfactant adsorption is usually considered as the source of the steric stabilization. Its possible effect on van der Waals forces important for emulsion flocculation is usually not even mentioned.

The effect of adsorbed layers on van der Waals interaction between particles has been analyzed by Vold [245] and several other authors [246–248] largely based on the Hamaker theory [33]. Ninham and Parsegian [249] and Mahanty and Ninham [250] discussed this effect using the Lifshitz theory. They have shown that the effect is significant when the separation is comparable to the thickness of the adsorbed layers. The next results relate to the modeling of surfaces forces in flotation [251]. The effect of adsorbed layers on van der Waals interaction between alumina particles and bubbles in aqueous media was investigated in Ref. 252. In the absence of adsorbed layers, there is a repulsive interaction. When alumina particles and bubbles are covered with adsorbed layers, the interaction energy changes from repulsive to attractive at small distances. This example demonstrates that the effect can be of interest for the theory of emulsion stability as well.

F. Modeling of Double-Layer Structure at Surfactant Adsorption

At adsorption of nonionic surfactants, the charge determining ions continue to be distributed along the o/w (oil–water) interface. At adsorption of an ionic surfactant, the additional charge determining ions appear to be distributed outside the interface. At adsorption of a polyelectrolyte, the additional charge occupies a layer corresponding to the adsorption-layer thickness. The counterions are distributed outside and inside the adsorption layer. Thus, the electrical double layer consist of two parts. One coincides with the adsorption layer and the second with penetration of the electrolyte at a distance κ^{-1}. The classical Poisson–Boltzmann (P-B) equation describes the external diffuse part of the DL. With respect to the part of DL which coincides with the adsorption layer, the P-B equation has to be generalized to account for additional charge caused by macroion dissociation. Thus, the potential across the adsorption layer and outside of it is described by two different functions. The unknown coefficients of these solutions are determined by the two conditions at the boundary between two parts of the DL. A real potential drop across the external diffuse part of DL can exceed the measured ζ-potential due to protrusion of polymer chain beyond the boundary that retards the electroosmotic slip. The total potential drop across DL as a whole is an analog of the Stern potential. If the thickness of the polyelectrolyte adsorption layer exceeds the Debye length, the potential drop across the adsorption layer can essentially exceed the ζ-potential. The energy of electrostatic interaction is caused by the structure of DL as a whole. It cannot be expressed by classical DLVO theory. The potential distribution inside and outside adsorption layer has to be incorporated in the general equations of DLVO theory for the disjoining pressure and the free electrostatic energy. A simple model is made in Ref. 253. At decreasing surface separation and at increasing overlap of the diffuse parts of DL, the potential at the boundary between adsorption layer and electrolyte increases. Its maximum value is achieved at zero separation and equals the Stern potential. The electrostatic barrier corresponds to a nonzero separation and is determined by the potential value intermediate between the Stern potential and the electrokinetic potential [253].

One concludes that the calculation of the electrostatic interaction based on the ζ-potential value leads to the strong underestimation of the role of the electrostatic component of the surface forces.

XV. SUMMARY

A. General

1. During the last decades, profound success has been achieved in both physico-chemical and physico-mathematical approaches to emulsion stability. This creates the premises for mathematical modeling of the kinetics of simultaneous flocculation and creaming in dilute emulsified systems. The systematic investigations by Melik and Fogler [15] relate to the kinetics of dilute emulsions, whereas in literature, greater attention is paid to more concentrated systems. This situation can be characterized as an underestimation of the importance of the kinetics of dilute emulsions. Identical processes control the kinetics of concentrated emulsion. However, to repeat the approach by Fogler et al. to more concentrated emulsions would be extremely difficult.

2. Creaming, Brownian, gravitational, and orthokinetic (shear) flocculations control the evolution of the droplet size distribution (DSD). The orthokinetic flocculation is omitted in this chapter for the sake of brevity.

3. Four components are necessary for a quantifying of the emulsion kinetics: (i) forces dictating droplet movement and interaction, (ii) droplet loss mechanisms and their rate, (iii) population balance equation (PBE), and (iv) floc structure theory, for instance, the fractal theory.

At faster coalescence times, an emulsion consists mainly of the singlets and temporary doublets. At faster flocculation times, the system consists of flocs.

4. The traditional description of particle–particle interaction caused by van der Waals forces is presented for o/w emulsions only. This approach can be extended to emulsions with other types of interactions and to w/o emulsions as well.

5. The most general and exact approach in emulsion kinetics is achieved by using droplet size distribution (DSD). The time and space dependence of DSD can be established by the solution of the population balance equation. An effective analytical solution of the population balance equation has been elaborated in aerosol systems. A similar approach can be adopted in emulsion science. The efficiency of numerical methods in the solution of PBE has been shown by Melek and Fogler.

6. The high efficiency of the PBE approach can be realized from the condition describing the droplet loss rates.

Thus, the main task in the elaboration of the mathematical modeling of the dilute emulsion kinetics is the derivation of the kernels of PBE on the basis of the experimental–theoretical investigation of the mechanisms of droplet loss and loss rates. Experimental investigations of rapid and slow flocculation are analyzed. It is concluded that the reliable kernels for PBE are available in the case of rapid flocculation but not for slow flocculation.

PBE can be used to predict changes in the droplet size distributions caused by rapid coagulation.

With respect to slow coagulation, PBE has to be used to extract of information concerning the different mechanisms of the droplet loss and the determination of the kernel for PBE. After the determination of the kernel, PBE will be concretized and will become suitable for the prediction of changes in DSD caused by slow flocculation.

B. Rapid Flocculation

1. Experimental investigations of the rate of droplet loss in emulsions are, to a high degree, lacking. Due to this, we are forced to evaluate the reliability of the rapid coagulation theory to be used as the experimental data of the rapid Brownian flocculation in monodisperse suspensions and

in flotation systems. The foundation for this is a substantial similarity of collision process in these systems and in emulsions.

2.   A satisfactory agreement between rapid flocculation theory and experiments carried out in monodisperse suspensions and applied in flotation process makes it possible to recommend PBE for a description of simultaneous creaming and flocculation in emulsions with uncharged or weakly charged droplets. The accuracy of the modeled time evolution of the spectrum of droplets of intermediate sizes and doublets of large droplets subjected to gravitational break may be low.

3.   The rate of loss of droplets of intermediate dimensions is determined by simultaneous action of Brownian movement and gravitation. This regime corresponds to a Pecklet number of the order of 1. A sufficiently comprehensive theoretical description of the interaction and of the rate of coagulation in this regime is lacking.

4.   A quantification of the flocculation rate in the secondary minimum is possible only at a sufficiently accurate description of the attenuation of van der Waals' forces at large distances. An attempt is made to define more exactly in this respect the flocculation in secondary minimum. Similarity of collision efficiencies of flocculation in the primary and in the secondary minimum is substantiated. Using this as the basis, a simple method of calculation of collision efficiency in the secondary minimum is proposed.

C.   Slow Flocculation

1.   Systematic experimental investigations of slow flocculation in emulsions are not available in literature and all investigators of slow Brownian flocculation in monodisperse suspensions reveal a complete disagreement between theory and experiment performed. The gravitational flocculation in the primary minimum has been insufficiently studied even in solid suspensions. With respect to slow coagulation, the reliable kernel for PBE is thereby absent and PBE cannot be used, contrary to the recommendations in Ref. 15.

2.   When applied to concentrated emulsions, it is necessary to take into consideration the experimental and theoretical investigations of slow coagulation in solid suspensions. This means that consideration of the effect of electrostatic component of disjoining pressure in works devoted to concentrated emulsions is nonrealistic.

3.   As to slow flocculation in emulsions, systematic theoretical and experimental investigations are necessary to provide the kernels of the PBE. Experimental investigations of slow flocculation under conditions

slightly complicated by a simultaneous gravitational coagulation have been of current interest.

To suppress gravitational coagulation in the primary minimum, emulsions with a small $\Delta\rho$ should be used. The depth of the secondary minimum can be decreased by lowering the electrolyte concentration.

An experimental investigation of slow coagulation in dilute emulsions is extremely important for colloidal stability in general because the droplet surface is homogeneous. As has been stated by Matijevic and others, a discrepancy between the theory and experiment due to surface heterogeneity and roughness in suspensions exists. It is not unlikely that "one of the major unsolved problems of colloid science" [2] can be solved by systematic investigations of slow coagulation in emulsions.

D. Role of Droplet Deformation and Surface Mobility in Flocculation Rate

1. Yiantsios and Davis (YD) [205] elaborated a method for the investigation of thin emulsion film evolution under the action of a lubrication force arising in gravitational coagulation and derived an equation for distance dependence at early droplet deformation. They consider the film rupture to be more of a natural evolution than an instability. According to the YD theory, the droplet deformation does not influence the grazing trajectory that justifies an introduction of a the simple method to calculate the collision efficiency in miniemulsions. However, with decreasing surface tension or increasing droplet dimension (transition to macroemulsions), the droplet deformation can influence the collision efficiency.

2. In distinction from the YD theory in the DDPIB theory [207], a sequence of different stages of droplet deformation and thin-film instability is postulated. The important notion of a total drag force acting between coagulating droplets due to Brownian movements is introduced in this theory by considering the mean droplet velocity of directional motion toward the central droplet. According to the DDPIB theory, the stability ratio for deformable and nondeformable spheres of micrometer-size practically coincide. For miniemulsions, the mutual approach of the droplets prior to their deformation is the rate-determining stage.

3. A set of equations describing thin-film evolution during slow Brownian coagulation (without the aforesaid dividing the process in separate stages and without assumption concerning h_{cr}) is derived by transforming a set of equations based on YD theory of droplet deformation within gravitational coagulation. In the latter, the lubrication force which causes the droplet deformation is replaced by the total drag force acting between coagulating Brownian droplets on the basis of DDPIB theory.

The proposed modified set of equations describing the simultaneous deformation and Brownian coagulation of emulsion droplets creates reliable conditions providing the negligible droplet deformation during slow Brownian coagulation.

ACKNOWLEDGMENTS

S. S. Dukhin acknowledges financial support from the Norwegian Research Council (NFR), University of Bergen, and the Strategic Technology Programme financed by NFR, Elf Petroleum a/s, Saga Petroleum a/s and Statoil a/s. Cand Scient Harald Førdedal is acknowledged for taking care of technical aspects of the manuscript.

SYMBOLS

Latin

a	Radius (m)
A	Hamaker constant
A_{ij}	Hamaker constant for interaction
b	targent distance
c_i	bulk concentration of surfactant
C	concentration (mol m^{-3})
D	diffusion coefficient of a droplet (m^2 s^{-1})
D	diffusion coefficient of an ion or a surfactant molecule
e	elementary charge (C)
f	friction coefficient (kg s^{-1})
F	Faraday constant (C mol^{-1})
F	force (N)
g	standard acceleration of free fall (m s^{-2})
h	shortest distance between colloidal particles, drops (m)
J	flux (m^2 s^{-1})
k_f	flocculation rate constant
K	Boltzmann constant (J K^{-1})
n_i	number concentration of particles (aggregates) of size i
N_A	Avogadro's constant
p_{12}	pair distribution function, probability that drop 1 is at position **r** relative to drop 2
Pe	Peclet number
$\mathbf{r}(r)$	distance (m)
R	gas constant (J K^{-1} mol^{-1})
R_{ij}	collision radius

Re   Reynolds number
$R_{ij} - a_i + a_j$ collision radius
t time
t_F characteristic flocculation time
T temperature (K)
$u_0(\hat{\mu})$ instantaneous creaming velocity for a droplet of radius
                       a_i given by Eq. (42)
u_0 Stokes creaming rate for a particle of radius a_i given by
                       Eq. (120)
U energy (J)
U, v velocity (m s^{-1})
U_{12} potential function of interaction between drops 1 and 2
V volume

Greek

γ interfacial surface tension (N m^{-1})
Γ surface (excess) concentration (mol m^{-2})
δ diffusion layer thickness (m)
ϵ relative dielectric permittivity (dielectric constant)
ϵ_0 dielectric permittivity of vacuum
χ_r retardation coefficient Eq. (45)
ζ electrokinetic potential (V)
θ angle of rotation
κ reciprocal Debye length (m^{-1})
λ size ratio of two drop, $\lambda = a_1/a_2 < 1$
μ kinematic viscosity (J mol^{-1})
$\hat{\mu} = \mu_1/\mu_2$ viscosity ratio
$\Pi(h)$ disjoining pressure (N m^2)
ρ density (kg m^{-3})
σ surface charge (C m^{-2})
τ characteristic time (s)
Ψ electric potential (V)

REFERENCES

1. B. V. Derjaguin, *Theory of Stability of Colloids and Thin Films*, Nauka,
    Moscow, 1986 [in Russian]; translation: *Theory of Stability of Colloids and
    Thin Films*, Plenum, New York, 1989.
2. J. Gregory, Crit. Rev. Environ. Control *19*:185 (1989).
3. M. von Smoluchowski, Phys. Z. *17*:557, 585 (1916).
4. M. von Smoluchowski, Phys. Chem. *92*:129 (1917).

5. Th.F. Tadros and B. Vincent, in *Encyclopedia of Emulsion Technology*, Vol. (P. Becher, ed.), Marcel Dekker, Inc., New York, 1983, p. 57.
6. E. E. Kumacheva, E. A. Amelina, A. V. Prtsev, and E. D. Shchukin, Kolloidn. Zh. *40*:1214 (1989).
7. K. Larsen, in *Emulsions—A Fundamental and Practical Approach* (J. Sjöblom, ed.), NATO ASI Series Vol. 363, 1991, p. 41.
8. H. J. Junginger, *Emulsions—A Fundamental and Practical Approach* (J. Sjöblom, ed.), NATO ANSI Series Vol. 363, 1991, p. 207.
9. M. van den Tempel, Rec. Trav. Chim. *72*:419, 433 (1953).
10. R. P. Borwankar, L. A. Lobo, and D. T. Wasan, Colloids Surf. *69*:135 (1992).
11. Ph.T. Jacger, J. J. M. Janssen, F. G. Groeneweg, and W. G. M. Agterof, Colloids Surf. *85*:255 (1994).
12. B. V. Derjaguin, N. V. Churaev, and V. M. Muller, *Surface Forces*, Nauka, Moscow, 1985 [in Russian]; translation: *Surface Forces*, Plenum, New York, 1987.
13. R. J. Hunter, *Foundations of Colloid Science*, Vol. 1, Oxford Science Publication, Oxford, 1987.
14. H. Sonntag and K. Strenge, *Coagulation Kinetics and Structure Formation*, VEB Dentscher Verlag der Wissenschaften, Berlin, 1987.
15. D. H. Melik and H. S. Fogler, in *Encyclopedia of Emulsion Technology*, Vol. 3, Marcel Dekker, Inc., New York, 1988, p. 3–78.
16. N. A. Fuchs, Mechanika aerosoley, Izd-vo AN SSSR, 1955; *The Mechanics of Aerosols*, Pergamon, London, 1964.
17. V. M. Voloshchuk and Yu. S. Sedunov, *Coagulation Processes in Disperse Systems*, Hydrometeoizdat, Leningrad, 1975 [in Russian].
18. S. S. Dukhin, N. N. Rulyov, and D. S. Dimitrov, *Coagulation and Dynamics of Thin Films*, Naukova Dumka, Kiev, 1986 [in Russian].
19. B. V. Derjaguin, S. S. Dukhin, and N. N. Rulyov, in *Surface and Colloid Science* Vol. 13 (E. Matijevich, ed.), Wiley Interscience, New York, 1983, p. 71.
20. H. J. Schulze, *Physicalisch-Chemische Elementarvorgange des Flotations Processes*, VEB Verlag der Wissenschaft, Berlin, 1981.
21. J. Sjöblom, (ed.) J. Disper. Sci. & Technol. *15* (1994).
22. W. Stumm & J. J. Morgan, Aquatic Chemistry, Wiley, New York, 1981
23. B. W. Derjaguin and V. M. Muller, Dokl. Akad. Nauk SSSR *176*:869 (1967).
24. E. Dickenson, in *Emulsions—A Fundamental and Practical Approach* (J. Sjöblom ed.), NATO ASI Series Vol. 363, 1991, p. 25.
25. M. P. Aronson, in Emulsions—A Fundamental and Practical Approach (J. Sjöblom, ed.), NATO ASI Series Vol. 363, 1991, p. 75.
26. P. Walstra, in *Gums and Stabilizers for the Food Industry*, Vol. 4, (G. O. Phillips, P. A. Williams, and D. J. Wedlock, eds.) IRL Press, Oxford, 1988, p. 233.
27. S. R. Reddy, D. H. Melik, and H. S. Fogler, J. Colloid Interface Sci. *82*: 116 (1981).
28. D. H. Melik and H. S. Fogler, J. Colloid Interface Sci. *101*:72 (1984).

29. X. Zhang and R. Davis, J. Fluid. Mech. *230*:479 (1991).
30. S. E. Friberg, in *Emulsions—A Fundamental and Practical Approach* (J. Sjöblom, ed.), NATO ASI Series Vol. 363, 1991, p. 1.
31. J. A. Kitcherer and P. R. Musselwhite, *The Theory of Stability in Emulsions in Emulsion Science* (P. Sherman, ed.), Academic Press, London, 1968.
32. B. Bergenståhl and P. M. Claesson, in *Food Emulsions* (K. Larsson and S. Friberg, eds.), Marcel Dekker, Inc., 1989, pp. 41–96.
33. H. C. Hamaker, Physica, *4*:1058 (1937).
34. S. S. Dukhin and B. V. Derjaguin, *Electrokinetic Phenomena in Surface and Colloid Science*, Vol. 7 (E. Matijevic, ed.) Wiley, New York, 1974.
35. B. A. Pailtorpe and W. B. Russel, J. Colloid Interface Sci. *89*:56 (1982).
36. N. F. H. Ho and W. I. Hiquichi, J. Pharm. Sci., *57*:436 (1968).
37. J. Lyklema, *Fundamentals of Interface and Colloid Science*, Academic Press, London, 1993.
38. Ya. I. Rabinovich and N. V. Churaev, Kolloidn. Zh. *41*:468 (1979).
39. Ya. I. Rabinovich and N. V. Churaev, Kolloidn. Zh. *46*:69 (1984).
40. Ya. I. Rabinovich and N. V. Churaev, Kolloidn. Zh. *52*:309 (1990).
41. R. J. Hunter, *Zeta Potential in Colloid Science*, Academic Press, London, 1981.
42. S. S. Dukhin, Adv. Colloid Interface Sci. *44*:1 (1993).
43. S. S. Dukhin, B. V. Derjaguin, and N. M. Semenikhin, Dokl. Akad. Nauk SSSR *192*:357 (1970).
44. B. V. Derjaguin, *Theory of Stability of Colloids and Thin Films*, Plenum, New York, 1989, Chap. 5, Sec. 5.
45. S. S. Dukhin and V. N. Shilov, *Dielectric Phenomena and the Double Layer in Disperse Systems and Polyelectrolytes*, Wiley, Toronto, 1974.
46. J. Kijlstra, H. P. van Leuven, and J. Lyklema, J. Chem. Soc. Faraday Trans. *88*:3441 (1992).
47. B. R. Midmore and R. I. Hunter, J. Colloid Interface Sci. *122*:521 (1988).
48. J. N. Israelashvili and R. M. Pashley, J. Colloid Interface Sci. *98*:500 (1984).
49. P. M. Claesson and H. K. Christenson, J. Phys. Chem. *92*:1650 (1988).
50. H. K. Christenson, P. M. Claesson, J. Berg, and P. C. Herder, J. Phys. Chem. *93*:1472 (1989).
51. R. M. Pashley and J. N. Israelachvili, J. Colloid Interface Sci. *97*:446 (1984).
52. B. V. Derjaguin and Z. M. Zorin, Zh. Fiz. Khim. *29*:1010 (1955).
53. G. Frens and J. Th. G. Overbeek, J. Colloid Interface Sci. *38*:376 (1972).
54. T. W. Healy, A. Homola, and R. O. James, Faraday Discuss. Chem. Soc. *65*:156 (1978).
55. P. Meakin, Phys. Rev. Lett. *51*:1119 (1983).
56. M. Kilb, Phys. Rev. Lett. *53*:1653 (1984).
57. T. Vicsek and F. Family, Phys. Rev. Lett. *52*:1669 (1984).
58. P. Meakin, T. Vicsek, and F. Family, Phys. Rev. *B31*:564 (1985).
59. A. Einstein, Ann. Phys. *17*:549 (1905); *19*:371 (1906).
60. A. Einstein, *The Theory of the Brownian Movement*, Dover, New York, 1956.

61. N. A. Fuchs, Z. Phys. *89*:739 (1934).
62. B. V. Derjaguin, Dokl. AN SSSR *109*:967 (1956).
63. B. V. Derjaguin and N. A. Krotova, *Physical Chemistry of Adhesion*, Izd-vo AN SSSR, Moscow, 1949
64. G. I. Taylor, Proc. Roy. Soc. London *A108*:12 (1924).
65. L. Spielman, J. Colloid Interface Sci. *33*:562 (1970).
66. E. P. Honig, G. J. Roberson, and P. H. Wiersena, J. Colloid Interface Sci. *36*:97 (1971).
67. G. K. Batchelor, J. Fluid Mech. *119*:379 (1982).
68. A. Z. Zinchenko, Prikl. Mat. Mech. *42*:955 (1978).
69. A. Z. Zinchenko, Prikl. Mat. Mech. *44*:30 (1980).
70. A. Z. Zinchenko, Prikl. Mat. Mech. *46*:58 (1982).
71. G. Hetsroni and S. Haber, Int. J. Multiphase Flow *4*:1 (1978).
72. Y. O. Fuetnes, S. Kim, and D. J. Jeffrey, Phys. Fluids *31*:2445 (1988).
73. Y. O. Fuetnes, S. Kim, and D. J. Jeffrey, Phys. Fluids *A1*:61 (1989).
74. S. Haber, G. Hetsroni, and A. Solan, Int. J. Multiphase Flow *1*:57 (1973).
75. E. Rushton and G. A. Davies, Appl. Sci. Res. *28*:37 (1973).
76. R. H. Davis, J. A. Schonberg, and J. M. Rallison, Phys. Fluid *A1*:77 (1989).
77. X. Zhang and R. H. Davis, J. Fluid Mech. *230*:479 (1991).
78. X. Zhang, Ph.D. thesis, University of Colorado, 1992.
79. H. Sonntag, in *Coagulation and Flocculation* (Dobias, ed.), Marcel Dekker, Inc., New York, 1993.
80. E. B. Vadas, H. L. Goldsmith, and S. G. Mason, J. Colloid Interface Sci. *43*:630 (1973).
81. E. B. Vadas, R. G. Cox, H. L. Goldsmith, and S. G. Mason, J. Colloid Interface Sci. *57*:308 (1976).
82. H. Siedentopf and R. Zsigmondy, Ann. Phys. (Leipzig) *10*:1 (1903).
83. R. Zsigmondy, Z. Phys. Chem. *93*:600 (1918).
84. A. Tuorilla, Kolloidchem. Beihefte *22*:193 (1926).
85. B. V. Derjaguin and N. M. Kudravtseva, Kolloidn. Zh. *26*:61 (1964).
86. A. Watillon, M. Romerowski, and F. van Grunderbeek, Bull. Soc. Chim. Belg. *68*:450 (1959).
87. R. H. Ottewill and M. C. Rastogi, J. Chem. Soc. Trans. Faraday Soc. *56*: 866 (1960).
88. N. H. G. Penners and L. K. Koopal, Colloids Surf. *28*:67 (1987).
89. J. Cahill, P. G. Cummins, E. J. Staples, and L. Thompson, J. Colloid Interface Sci. *117*:406 (1987).
90. H. Gedan, H. Lichtenfeld, H. Sonntag, and H.-J. Krug, Colloids Surf. *11*: 199 (1984).
91. N. A. Fuchs, *Uspechi Mechaniki Aerozoley*, Izd-vo AN SSSR, Moscow, 1961 [in Russian].
92. L. A. Spielman and J. A. Fitzpatrick, J. Colloid Interface Sci. *42*:607 (1973).
93. L. A. Spielman and J. A. Fitzpatrick, J. Colloid Interface Sci. *43*:350 (1973).
94. J. Langmuir and K. Blodgett, Gen. Elec. Comp. Rep., July 1945, pp. 45–58.
95. Rybczynski, Bull. Cracovie (A) 40 (1911).

96. Hadamard, Comp. Rend. *152*:1735 (1911).
97. P. F. Saffman and J. S. Turner, J. Fluid Mech. *1*:16 (1956).
98. V. G. Levich, *Physicochemical Hydrodynamics*, Prentice-Hall, Englewood Cliffs, NJ, 1962.
99. A. N. Frumkin and V. G. Levich, Zh. Fiz. Chim. *21*:1183 (1947).
100. S. S. Dukhin, in *Modern Theory of Capillarity* (A. I. Rusanov and F. Ch. Goodrich, eds.), Springer-Verlag, Berlin, 1981.
101. R. D. Vold and R. C. Groot, J. Colloid Sci. *19*:384 (1964).
102. S. J. Rehfild, J. Phys. Chem. *66*:1969 (1962).
103. E. R. Garret, J. Am. Pharm. Assoc. *51*:35 (1962).
104. H. Brauer, *Grandlagen der Ein-und Mehrphasenstronumgen* Verlag Sanerlander, Aarau, 1971.
105. R. G. Boothroyd, *Flowing Gas-Solid Suspensions*, Chapman & Hall, London, 1971, Sec. 2.2.
106. H. S. Muralidhara, R. B. Beard, and N. Senapti, Filt. Sep. *24*:409 (1987).
107. K. L. Sutherland, J. Phys. Chem. *58*:394 (1948).
108. B. V. Derjaguin and S. S. Dukhin, *Trans. Inst. Mining Metall.* *70*:221, 231 (1960).
109. L. I. Levin, *Research into the Physics of Coarsly Dispersed Aerosols*, Izdvo AN SSSR, Moscow, 1961 [in Russian].
110. G. L. Natanson, Dokl. AN SSSR *116*:109 (1957).
111. A. Fonda and H. Herne, in *Aerosol Science* (C. N. Davies, ed.), Academic Press, New York, 1969.
112. W. R. Russel, D. A. Saville, and W. R. Schowalter, *Colloidal Dispersion*, Cambridge University Press, Cambridge, 1989, Chap. 11.
113. S. S. Dukhin, Kolloidn. Zh. *44*:431 (1982); *45*:207 (1983).
114. S. S. Dukhin and B. V. Derjaguin, Kolloidn. Zh. *20*:326 (1958).
115. M. E. Weber, J. Separ. Technol. *2*:29 (1981).
116. A. Nguen Van and S. Kmet, Int. J. Miner. Process. *35*:205 (1992).
117. R. C. Clift, J. R. Grace, and M. E. Weber, *Bubbles, Drops and Particles*, Academic Press, New York, 1978, p. 27.
118. R. H. Yoon and G. H. Luttrel, Miner. Process. Extractive Metal Rev. *5*: 101 (1989).
119. G. H. Luttrel and R. H. Yoon, J. Colloid Interface Sci. *159*:129 (1992).
120. I. O. Protodiaconov and S. V. Uljanov, *Hydrodynamics and Mass Transport in Disperse Systems Liquid–Liquid*, Nauka, Leningrad, 1986, pp. 28–48 [in Russian].
121. V. Ya. Rivkind and G. M. Riskin, Izv. AN SSSR MZhG *N1*:8 (1976).
122. M. E. Weber & D. Paddock, J. Colloid Interf. Sci., *94*:328 (1983).
123. J. H. Masliyah, Ph. D. dissertation, University of British Columbia, Vancouver, Canada, 1970.
124. S. W. Woo, Ph. D. dissertation, McMaster University, Hamilton, Canada, 1971.
125. N. N. Rulyov and E. S. Leshchov, Kolloidn. Zh. *42*:1123 (1980).
126. A. E. Hamielec and A. I. Jonson, Can. J. Chem. Eng. *40*:41 (1962).

127. J. P. Anfruns and J. A. Kitchner, *Flotation* (M. C. Fuerstenau, ed.), A. M. Gaudin Memorial Volume, SME-AIME, New York, pp. 626–637.
128. E. S. R. Gopal, in *Emulsion Science* (P. Sherman, ed.), Academic Press, London, 1969, pp. 69–70.
129. S. S. Dukhin and M. V. Buikov, Zh. Fiz. Chim. *39*:913 (1965).
130. G. A. Martinov and V. M. Muller, Dokl. AN USSR *207*:1161 (1972).
131. L. A. Spielman, Annu. Rev. Fluid Mech. *9*:297 (1977).
132. S. L. Goren and M. E. O'Neil, Chem. Eng. Sci. *26*:325 (1971).
133. S. L. Goren, J. Fluid Mech. *41*:613 (1971).
134. A. J. Goldman, R. G. Cox, and H. Brenner, Chem. Eng. Sci. *22*:637 (1967).
135. N. N. Rulyov, Kolloidn. Zh. *40*:898 (1978).
136. N. N. Rulyov, Kolloidn. Zh. *40*:1202 (1978).
137. V. Shubin and P. Kekicheff, J. Colloid Interface Sci. *155*:108 (1993).
138. J. N. Israelachvili, *Intermolecular and Surface Forces*, 2nd ed., Academic, London, 1991.
139. D. Reay and G. A. Ratelif, Can. J. Chem. Eng. *51*:178 (1973).
140. D. Reay and G. A. Ratelif, Can. J. Chem. Eng. *53*:481 (1975).
141. G. L. Collins and G. J. Jameson, Chem. Eng. Sci. *31*:985 (1976).
142. G. L. Collins and G. L. Jameson, Chem. Eng. Sci. *31*:239 (1977).
143. K. Okada, Y. Akagi, M. Kogure, and N. Yoshioka, Can. J. Chem. Eng. *68*:393 (1990).
144. K. Okada, Y. Akagi, M. Kogure, and N. Yoshioka, Can. J. Chem. Eng. *68*:614 (1990).
145. Wen Jongsong and Batchelor G. K., *Scintia Sinica 28*:172 (1985).
146. G. K. Batchelor and C. S. Wen, J. Fluid Mech. *124*:495 (1982).
147. D. J. Jeffrey and Y. Onoshi, J. Fluid Mech. *139*:261 (1984).
148. R. H. Davis, J. Fluid Mech. *145*:179 (1984).
149. L. D. Reed and F. A. Morrison, Int. J. Multiphase Flow *1*:573 (1974).
150. Ya. I. Rabinovich and A. A. Baran, Colloids Surf. *59*:47 (1991).
151. V. A. Parsegian, N. Fuller, and R. P. Rand, Proc. Nat. Acad. Sci. USA 2750 (1979).
152. R. P. Rand, Annu. Rev. Biophys. Bioeng. *10*:277 (1981).
153. R. P. Rand, V. A. Parsegian, J. A. Henry, L. J. Lis, and M. McAlister, Can. J. Biochem. *58*:959 (1980).
154. H. R. Kruyt, *Colloid Science*, Vol. 1, Elsevier, Amsterdam, 1952.
155. A. S. Dukhin, Kolloidn. Zh. *50*:441 (1988).
156. A. V. Bochko, A. S. Dukhin, E. I. Moshkovski, and A. A. Baran, Kolloidn. Zh. *49*:543 (1987).
157. M. Stimson and G. B. Jeffrey, Proc. Roy. Soc. London A*110*:1117 (1926).
158. N. N. Rulyov, S. S. Dukhin, and V. P. Semenov, Kolloidn. Zh. *41*:263 (1979).
159. P. S. Kisliy, Yu. I. Nikitin, V. M. Melnik, and S. M. Uman, Poroshkovaja Metallurgija *6*:92 (1982).
160. L. A. Spielman and P. M. Cukor, J. Colloid Interface Sci. *43*:51 (1973).
161. U. W. Lee, J. Colloid Interface Sci. *92*:315, (1983).

162. S. S. Dukhin, Yu. V. Shulepov, and J. Lyklema, Kolloidn. Zh. 56:641 (1994).
163. H. Reerink and J. Th. G. Overbeek, Discuss. Faraday Soc. 18:74 (1954).
164. D. C. Prieve and E. Ruckenstein, J. Colloid Interface Sci. 73:539 (1980).
165. R. H. Ottewill and J. N. Shaw Discuss. Faraday Soc. 42:154 (1966).
166. A. Watillon and A. M. Joseph-Petit, Discuss. Faraday Soc. 42 (1966).
167. G. Frens and J. J. F. G. Heuts, Colloids Surf. 30:295 (1988).
168. A. Lips and W. E. Willis, J. Chem. Soc. Faraday Trans., I 69:1226 (1973).
169. G. R. Zeichner and W. R. Schowalter, J. Colloid Interface Sci. 71:237 (1979).
170. G. R. Wiese and T. W. Healy, J. Chem. Soc. Faraday Trans. 66:490 (1970).
171. H. Kihira, N. Ryde, and E. Matijevic, J. Chem. Soc. Faraday Trans. 88: 2379 (1992).
172. H. Kihira, and E. Matijevic, Adv. Colloid Interface Sci. 42:1 (1992).
173. H. Kihira, N. Ryde, and E. Matijevic, Colloids Surf. 64:317 (1992).
174. R. Jullien, Usp. Fiz. Nauk 157:339 (1989).
175. J. T. G. Overbeck, J. Chem. Soc. Faraday Trans. 1 84(9):3079 (1988).
176. L. Koopal and S. S. Dukhin, Colloids Surf. A 73:201 (1993).
177. S. Yu. Shulepov, S. S. Dukhin, and J. Lyklema, J. Colloid Interface Sci. (in press).
178. S. R. Reddy and H. S. Fogler, J. Colloid Interface Sci. 82:128 (1981).
179. S. R. Reddy and H. S. Fogler, J. Phys. Chem. 84:1570 (1980).
180. S. R. Reddy and H. S. Fogler, J. Colloid Interface Sci. 79:105 (1981).
181. A. Kotera, K. Furusava, and K. Kudo, Kolloid Z. Z. Polym. 240:837 (1970).
182. I. B. Ivanov and D. S. Dimitrov, in Thin Films (I. B. Ivanov, ed.), Marcel Dekker, Inc., New York, 1988.
183. J. Czarnecki, Adv. Colloid Interface Sci. 24:283 (1986).
184. T. Tadros and T. G. M. van de Ven, Colloid Polimer Sci. 261:694 (1983).
185. M. V. Ostrovsky and R. J. Good, J. Disper. Sci Technol. 7:95 (1986).
186. F. S. Milops and D. T. Wasan, Colloids Surf. 4:91 (1981).
187. M. Boenrel, A. Graciaa, R. S. Shechtel, and W. H. Waade, J. Colloid Interface Sci. 72:161 (1979).
188. J. Vinatieri, J. Petrol. Technol. (1977).
189. D. O. Shah and R. S. Shechter, Improved Oil Recovery by Surfactant and Polymer Flooding, Academic Press, New York, 1977.
190. H. K. van Bollen and Associates, Inc., Fundamentals of Enhanced Oil Recovery, Pen Well Publishing Co, Tulsa, OK, 1980.
191. G. A. Martynov and S. P. Bakanov, Investigation of Surface Forces, Isd. Nauk, Moscow, 1961, pp. 220–229.
192. S. K. Friedlander and C. S. Wang, J. Colloid Interface Sci. 22:126 (1966).
193. C. S. Wang and S. K. Friedlander, J. Colloid Interface Sci. 24:170 (1967).
194. E. R. Cohen and E. U. Vaughan, J. Colloid Interface Sci. 35:612 (1971).
195. T. R. Waite, Phys. Rev. 107:461 (1957).
196. G. Wilemaki and M. Fixman, J. Chem. Phys., 58:4009 (1973).
197. V. M. Titulaer, Physica A 100:251 (1980).

198. R. M. Fitch and R. C. Watson, J. Colloid Interface Sci. 68:14 (1979).
199. T. Suwa, T. Watanabe, J. Okamoto, and S. Machi, Lobunshi Renbunshu 35:236 (1978).
200. A. L. Smith (ed.), *Theory and Practice of Emulsion Technology*, Academic Press, New York, 1976, Sec. 4.
201. P. Shermann (ed.), *Emulsion Science*, Academic Press, New York, 1968.
202. D. T. Wasan, S. M. Shah, N. Aderangi, M. S. Chan, and J. J. McNamara, Soc. Petrol. Eng. J. 18:409 (1978).
203. S. I. Rassool (ed.), *Chemistry of the Lower Atmosphere*, Plenum, New York, 1973, Chap. 3.
204. K. Okuyama, Y. Kousaka, and T. Yoshida, J. Chem. Eng. Japan 9:140 (1976).
205. S. Yiantsios and R. H. Davis, J. Colloid Interface Sci. 144:412 (1991).
206. J. K. Klahn, W. G. M. Agterof, F. van Voorst Vader, R. D. Groot, and F. Grocneweg, Colloids Surf. 65:151 (1992).
207. K. D. Danov, N. D. Denkov, D. N. Petsev, I. B. Ivanov, and R. Borwankar, Langmuir 9:1731 (1993).
208. K. D. Danov, D. N. Petsev, and N. D. Denkov, J. Chem. Phys. 99:7179 (1993).
209. N. D. Denkov, N. Petsev, and K. D. Danov, Phys. Rev. Lett. 71:3226 (1993).
210. A. Vrij, Discuss. Faraday Soc. 42:23 (1966).
211. E. Ruckenstein and R. K. Jain, J. Chem. Soc. Faraday Trans. II 70:132 (1974).
212. M. B. Williams and S. H. Davis, J. Colloid Interface Sci. 90:220 (1982).
213. A. Sharma and E. Ruckenstein, Langmuir 2:480 (1986).
214. J. P. Burelbach, S. G. Bankoff, and S. H. Davis, J. Fluid Mech. 195:463 (1988).
215. R. J. Gumerman and G. M. Homsy, Chem. Eng. Commun. 2:27 (1975).
216. J.-D. Chen and J. C. Slattery, AIChE J. 28:955 (1982).
217. I. B. Ivanov (ed.), *Thin Liquid Films*, Marcel Dekker, Inc., New York, 1988.
218. F. A. M. Leermakers, Y. S. Sdranis, and J. Lyklema, Colloids Surf. 85: 135 (1994).
219. M. V. Ostrovsky and R. J. Good, J. Colloid Interface Sci. 102:206 (1984).
220. H. T. Davis and L. E. Scriven, J. Statist. Phys. 24:245 (1981).
221. H. T. Davis and L. E. Scriven, Adv. Chem. Phys. 49:357 (1982).
222. I. B. Ivanov, D. S. Dimitrov, P. Somasundaran, and R. K. Jain, Chem. Eng. Sci. 44:137 (1985).
223. N. D. Denkov, P. A. Kralchevsky, I. B. Ivanov, and C. S. Vassilieff, J. Colloid Interface Sci. 143:157 (1991).
224. S. K. Chakarova, M. Dupeyrat, E. Nakache, C. D. Dushkin and I. B. Ivanov, J. Surf. Sci. Technol. 6:17 (1990).
225. I. B. Ivanov, R. K. Jain, P. Somasundaran, and T. T. Traykov, in *Solution*

Behavior of Surfactants, Vol. 2 (K. L. Mittal, ed.), Plenum Press, New York, 1979, p. 817.

226. C. Maldarelli, R. K. Jain, in *Thin Liquid Films* (I. B. Ivanov, ed.), Marcel Dekker, Inc., New York, 1988, p. 497.
227. A. Vrij F. Hesselink, J. Lucassen, and M. van den Tempel, Proc. K. Ned. Acad. Wet. *B73*:124 (1970).
228. I. B. Ivanov, Pure Appl. Chem. *52*:1241 (1980).
229. T. T. Traykov and I. B. Ivanov, Int. J. Multiphase Flow *3*:471 (1977).
230. T. T. Traykov, E. D. Manev and I. B. Ivanov, Int. J. Multiphase Flow *3*: 485 (1977).
231. B. V. Derjaguin, Kolloid Z. *69*:155 (1934).
232. G. E. Charles and S. G. Mason, J. Colloid Sci. *15*:236, (1960).
233. V. M. Muller, Kolloidn Zh. *40*:885 (1978).
234. N. N. Makagorova, O. G. Usiarev, and A. A. Abramson, Kolloidn Zh. *40*: 252 (1978).
235. Yu. M. Chernoberegski and E. B. Golikova, Kolloidn. Zh. *36*:115 (1974).
236. H. Casimir and D. Polder, Phys. Rev. *78*:360 (1948).
237. Ya. I. Rabinovich and N. V. Churaev, Kolloidn. Zh. *52*:309 (1990).
238. N. A. Mishchuk, J. Sjöblom, and S. S. Dukhin, Kolloidn. Zh. *57*:829 (1995).
239. D. J. Mitchell and P. Richmond. J. Colloid Interface Sci. *46*:128 (1974).
240. V. N. Gorelkin and V. P. Smilga, Kolloidn. Zh. *34*:685 (1972).
241. V. N. Gorelkin and V. P. Smilga, in *Surface Forces in Thin Films and Disperse Systems* (B. V. Derjaguin, ed.), Nauka, Moscow, 1972.
242. D. H. Napper, *Polymeric Stabilization of Colloidal Dispersions*, Academic Press, London 1983.
243. P. G. de Gennes, C. R. Acad. Sci. (Paris) *300*:839 (1985); Adv. Colloid Interface Sci. *27*:189 (1987).
244. H. J. Ploehn and W. B. Russell, Adv. Chem. Eng. *15*:137 (1990).
245. M. J. Vold, J. Colloid Sci. *16*:1 (1961).
246. D. W. Osmond, B. Vincent, and F. A. Waite, J. Colloid Interface Sci. *42*: 262 (1973).
247. B. Vincent, J. Colloid Interface Sci. *42*:270 (1973).
248. S. H. Ekaadi, Colloids Surf. *2*:155 (1981).
249. B. W. Ninham and V. A. Parsegian, J. Chem. Phys. *52*:4578 (1970).
250. J. Mahanty and B. W. Ninham, *Dispersion Forces*, Academic Press, New York, 1976.
251. N. V. Churaev, Colloid Polym. Sci. *253*:120 (1975).
252. S. Usui and E. Barough, J. Colloid Interface Sci. *137*:281 (1990).
253. N. I. Zharkih, and S. S. Dukhin, Kolloidn. Zh. *56*(n5) (1994).
254. K. L. Lin and I. Osseo Asare, Solvent Extract. Ion Exchange *2*:365 (1984).
255. J. Bibette, D. Roux, and F. Nallet, Phys. Rev. Lett. *65*:2470 (1990); J. Bibette, T. G. Mason, Hu Gang, and D. A. Weitz, Phys. Rev. Lett. *69*:981 (1992); J. Bibette, D. Roux, and B. Poligny, J. Phys. II (France) *2*:401 (1992).

3

Flexible Surfactant Films: Phase Behavior, Structure, and Applications

ANNE-MARIE BELLOCQ   Centre de Recherche Paul Pascal, Centre National de la Recherche Scientifique (CNRS), Pessac, France

I. INTRODUCTION

It has been clearly and intensively shown by Ekwall [1] that phase equilibria in multicomponent aqueous mixtures of amphiphilic molecules can be richly diverse and intricate. Due to their considerable potential for aggregation, the surfactant solutions show a multiplicity of structures (bilayers, cylinders, spherical micelles) which can organize and produce a great variety of phases [1–6]. In addition to liquid isotropic micellar phases, either optically anisotropic or optically isotropic mesophases occur. The most commonly observed mesophases are lamellar smectic phases and hexagonal phases of infinite rodlike aggregates. Cubic phases are also detected in numerous binary and ternary mixtures. Addition of oil to aqueous solutions of surfactant molecules influences their phase behavior. In favorable cases, with special surfactants or mixtures of surfactant and cosurfactant, the mixtures of water–oil–amphiphile molecules exist as a fluid, transparent isotropic liquid phase called a micro-emulsion [7]. In contrast to emulsions which are kinetically stable, micro-emulsions are thermodynamically stable. In these special mixtures, one still encounters ordered phases but their extent is reduced because mesomorphic regions are replaced by the isotropic liquid microemulsion phase where no long-range order occurs [8]. This phase often exists over a wide range of water and oil concentrations and can be formed with a few percents of surfactant. The configuration of the oil and water domains varies with composition: for small fractions of oil in water or of water in oil, the structure is that of swollen micelles. When the volume fractions of oil and water are comparable, a random bicontinuous structure is found. The oil regions have the structure of a connected random network and the same property also holds for the water regions, the surfactant molecules making a continuous film between the oil and water domain. (For a review, see Ref. 9.)

Up to the eighties, very restricted portions of the phase diagrams of quaternary mixtures have been explored. Very early, the Swedish school attempted to determine the extent and shape of the region of existence of microemulsions in quaternary systems [10–13]. By examination of sections of the phase diagram at several levels of oil, the Swedish authors have established a direct connection between the microemulsion areas and the inverse micellar solutions described by Ekwall et al. [1,10]. The multiphase regions occurring in these mixtures are little known. In contrast, the multiphase regions of quinary mixtures containing salt have received much interest in connection with their potential use in oil recovery [14,15]. A great number of studies have focused attention on the so-called

Winsor III three-phase region in which a microemulsion is in equilibrium with both an organic and an aqueous phase [16–18].

During the last decade, detailed descriptions of the phase diagrams of several ternary (water–oil–surfactant) (W–O–S), quaternary (W–O–S–alcohol), and even quinary (W–O–S–alcohol–salt) systems have been presented (for reviews, see Refs. 8 and 18–20). These studies have provided evidence for several new phases where the surfactant creates surfaces. Thus, in addition to bicontinuous microemulsions, one finds dilute lamellar phases and liquid isotropic phases of randomly connected bilayers called sponge phases [21,22]. In addition, these works have also established that microemulsion systems can give rise to critical points, critical end points, and even tricritical points [23–27]. During the same period, a great deal of work has been devoted to understanding the structure and stability of such phases and only now can a coherent description of phase behavior be given [9]. From all the work that has been done on the states of surfactants in solution, there emerges a fundamental new concept: These phases are often better described as phases of fluctuating surfaces than as phases of particles. In this new approach, the focus is on the interface between hydrophobic and hydrophilic regions, and a free energy is associated directly with this interface, featuring its bending (curvature) elasticity and its topological entropy separately. The more recent studies of microemulsions highlighted the relevance of flexible, thermally fluctuating surfactants sheets. It is shown that thermal fluctuations have to be large because they are, in fact, responsible for the very existence of the microemulsion structure.

The enormous interest shown in the nature of microemulsions to a considerable extent is inspired by the possibility of applying these systems in various contexts in oil recovery. During the last 20 years or so, the literature on microemulsions has been growing at a very fast rate. A great number of studies have focused attention around phase behavior, stability, interfacial properties, structure, transport and dynamical properties, and, more recently, critical behavior. Several reviews on microemulsions are now available [9,16,28–34].

In this chapter (which is by no means exhaustive in view of the vast literature available), I will review some of the experimental facts about phase behavior, structural properties of microemulsions, and new phases evidenced in systems forming microemulsions, in order to emphasize the role of the interfacial films and to highlight the most recent viewpoints on the stability of these phases. Finally, a few selected applications of these organized media will be discussed.

II. PHASE DIAGRAMS

A. Ternary Systems: Water–Surfactant–Alcohol

Prior to describing the phase diagrams of the quaternary systems, we investigate those of the two ternary systems water–pentanol–SDS (WPS) (SDS: sodium dodecylsulfate) and water–hexanol–SDS (WHS). In a second step, we examine the effect of salt on the stability of the phases.

The phase diagrams of the two ternary systems WPS and WHS at 25°C are shown in Fig. 1. In the pentanol system, four one-phase regions are observed; three of them are mesophases: L_α: lamellar phase; H_α: hexagonal phase; R: rectangular phase [8]. These phases already exist in the binary system water–SDS but in very different ranges of temperature and concentrations. L_α and R are only found at very high temperatures ($T > 50°C$) and very high surfactant concentration (SDS > 70%) [35]. In the ternary system, the lamellar mesophase occurs over a wide range of sur-

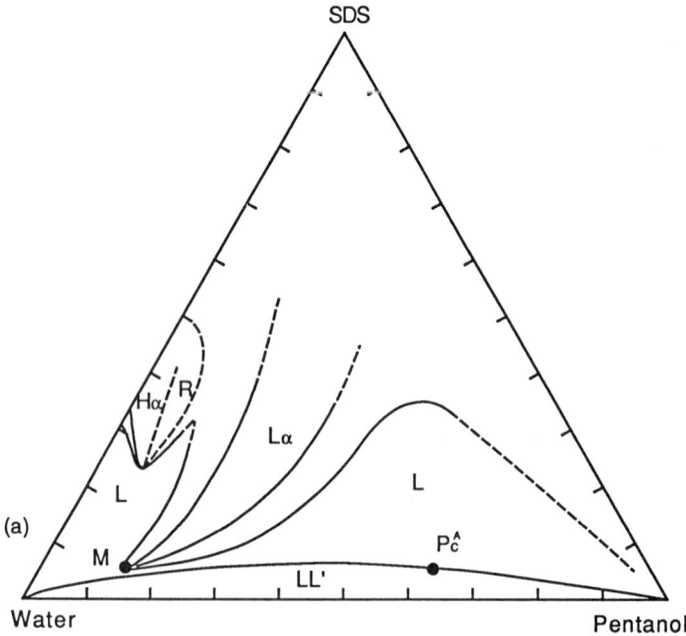

FIG. 1 Phase diagrams at 25°C of the ternary systems (a) water–pentanol–SDS and (b) water–hexanol–SDS. L: isotropic phase; H_α: hexagonal phase; R: rectangular phase; L_α: lamellar phase. The dashed boundaries have not been determined accurately. The point M is an azeotropelike point, P_c^A is a critical point.

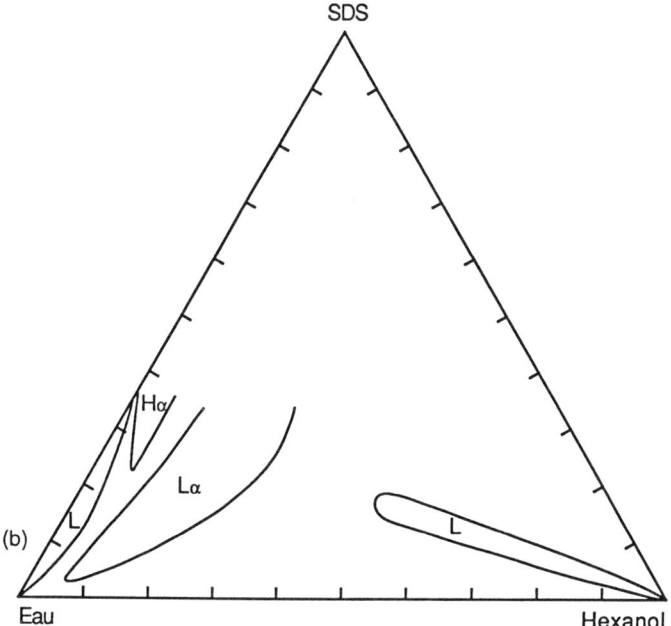

FIG. 1 Continued.

factant concentration. It extends down toward the water corner up to 79% of water, which corresponds to a maximum repeat distance of 100 Å. At 25°C, the pentanol system displays a continuous band of isotropic solution between pure water and pure pentanol. Provided that the surfactant concentration is adjusted, it is possible to go continuously from the water corner to the pentanol corner without any phase separation occurring. The main result of this continuity is the occurrence of a critical point P_C^A in the diagram. The accurate location of this point has been determined by phase composition analysis of several two-phase equilibria [36].

Significant changes in the phase equilibria occur on lowering the temperature. Below 20°C, the isotropic domain L is no longer continuous; it has split into two one-phase regions L_1 and L_2 which expand from the water and pentanol corners, respectively (Fig. 2a). These isotropic regions are separated by a complex multiphase region which involves two three-phase domains, t_1 and t_2. Composition analysis of the coexisting phases in the equilibria t_1 and t_2 at different temperatures shows that these equilibria originate in an indifferent state at 20°C. At 20°C, the points A, B, and C,

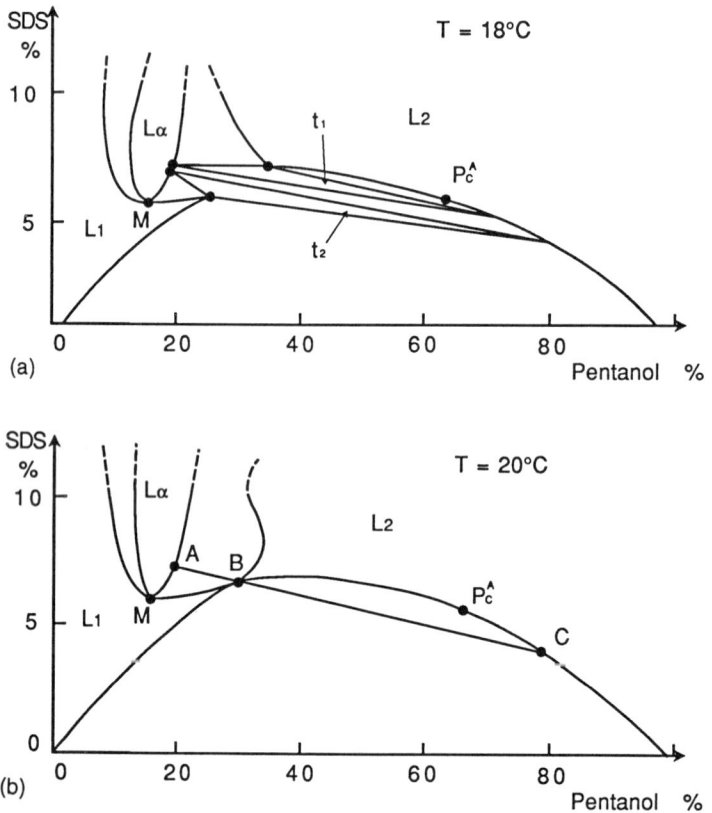

FIG. 2 Phase diagrams of the ternary water–pentanol system at (a) $T = 18°C$ and (b) $T = 20°C$. L_1 and L_2 are isotropic phases; L_α is a lamellar phase. t_1 and t_2 are two three-phase regions. The line ABC is an indifferent three-phase state. (From Ref. 8.)

representative of the three coexisting phases of both equilibria, are located on the same straight line (Fig. 2b). The system is then in an indifferent state. Let us note that the system does not show a particular thermodynamic behavior. At 20°C, regions L_1 and L_2 have one common point; they merge above this temperature.

The ternary system with hexanol (WHS) forms the same mesophases as pentanol (WPS). Replacement of pentanol by hexanol leads to a larger swelling of the lamellar phase which can be prepared with 3.5% SDS instead of 7.5% with pentanol. As a result, the isotropic domain is divided into two regions L_1 and L_2 and the three phases L_1, L_2, and L_α are separated by one three-phase triangle. In addition, this system does not exhibit a plait point at 20°C. This situation resembles that described by Kunieda and Nakamura for other water–ionic surfactant–alcohol systems [37].

1. Effect of Salt

Significant changes in the phase behavior of both systems WPS and WHS occur by the addition of salt (Fig. 3) [38,39]. At the salinity of 20 g L^{-1}, two new isotropic phases L_{3-w} and L_4 corresponding, as we will see later, to a sponge phase (L_{3-w}) and a phase of vesicles (L_4) are observed in the brine rich region of the diagram. Both phases exist in a narrow range of alcohol/SDS ratios. Dilute L_{3-w} samples strongly scatter light and exhibit a flow birefringence. In this phase, the scattered intensity increases with the water content. Dilute L_4 samples are clear and low viscous: as

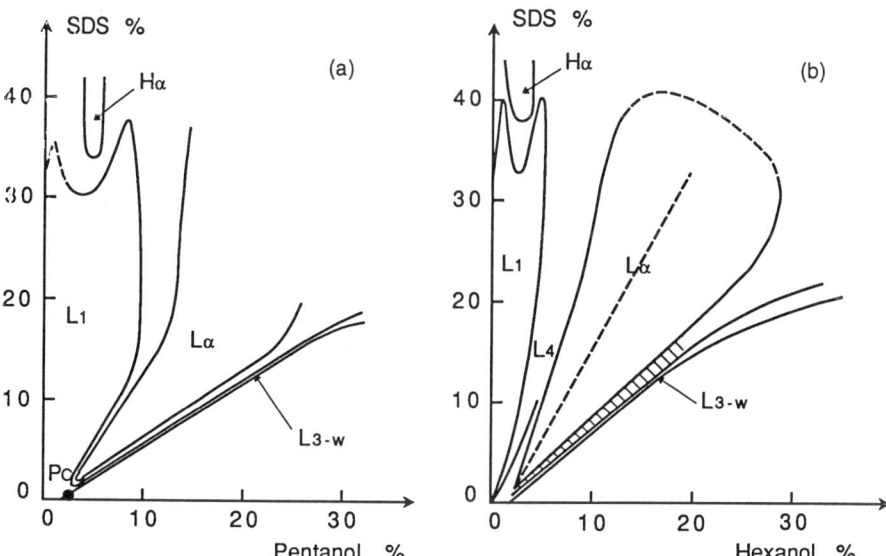

FIG. 3 Partial phase diagrams of (a) pentanol–SDS–water–NaCl system and (b) hexanol–SDS–water–NaCl system at $T = 25$°C, the concentration of NaCl in water is fixed at 20 g L^{-1}. L_1 is a micellar phase; L_α is a lamellar phase; H_α is an hexagonal phase. L_{3-w} and L_4, both with a very narrow range in alcohol concentration, are respectively the sponge phase and the vesicle phase. In the domain located on the left of the dashed line drawn in the hexanol L_α phase, spherulites form spontaneously.

the surfactant concentration increases (above 3%), they become weakly flow birefringent and very viscous. In the domain comprised between L_4 and L_α, an emulsion consisting of a dispersion of lamellar phase in the L_4 phase is found. At low SDS content, typically below 2% of SDS, the emulsion is unstable and a phase separation occurs. In the case of the pentanol system, the L_4 phase only exists in the concentrated domain for SDS contents larger than 2%. In addition, the phase transition from L_4 to L_α seems to be weakly first order. In the three phases L_4, L_α, and L_{3-w}, the surfactant and alcohol molecules build bilayers which fill space with different arrangements. In the L_4 phase, the bilayers form closed aggregates; in the L_α phase, they stack regularly, and finally in the sponge phase, they randomly connect. As previously noted in similar systems [39] in the dilute regime (for SDS below 15% or so), the boundaries of all the phases L_4, L_α, and L_{3-w} are straight lines directed toward the brine corner, therefore corresponding to a constant alcohol/surfactant ratio. Thus, we may consider that any linear path located inside each one-phase region and pointing toward the brine corner corresponds to a dilution line at constant membrane composition. This feature leads to redraw the phase diagram as shown in Fig. 4. This representation, first proposed by Porte et al. [40], clearly shows that the phase transformations are mainly triggered by the variations of the alcohol over surfactant ratio A/S. Thus, for example, concentrated L_{3-w} phases can be obtained from lamellar phases

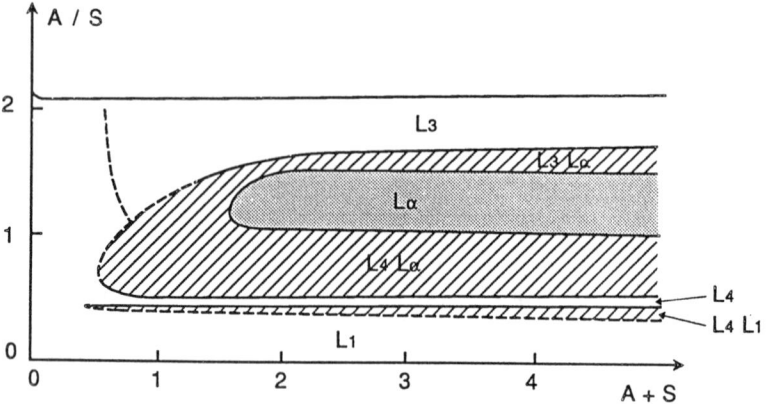

FIG. 4 Brine-rich part of the phase diagram of the system SDS–hexanol–brine (20 g L^{-1} NaCl). The hatched domains correspond to two-phase regions. The phase behavior of very diluted samples ($A + S < 1\%$) is hardly observable: dotted lines (diagrams expressed in ωt).

by increasing the alcohol content in the membrane. Fig. 4 also shows that dilute L_{3-w} phases can also be prepared by dilution of the lamellar phase. A similar pattern of phase behavior has been observed in a few systems containing single-chain ionic surfactant–alcohol and brine [39–47], but generally the L_4 phase is missing [40,44–47].

Regarding the L_α phase, a much larger swelling is obtained in salted systems. In the systems studied here, the repeat distance between the bilayers reaches 1000 Å. Finally, in the case of the hexanol system, the lamellar region can be divided into two subregions. At a low A/S ratio, large multilamellar vesicles referred to as spherulites are dispersed in the lamellar phase; at a high A/S ratio, the lamellar phase is free of these textural defects and exhibits the classical oily streaks defects [39].

Finally, one must note that both phase diagrams present lines of critical points [38,39]. In the case of pentanol, by the addition of salt, the critical point P_c^A shifts toward the brine corner. These points form a line P_c^{1-s} which terminates at the salinity of 3.5 g L^{-1} NaCl at a critical end point P_{ce}^C. A second line of critical points P_c^2 exists in the water-rich region at very low SDS concentrations for salinities above 8.1 g L^{-1} NaCl. These two critical lines are schematically represented in Fig. 5 in a prism with salinity as the ordinate and the ternary system WPS as the base. A line P_c^2 also occurs at very low SDS concentration for the hexanol system.

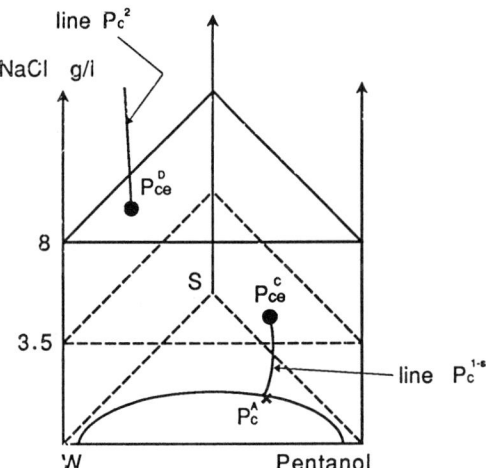

FIG. 5 Representation in a prism of the lines of critical points P_c^{1-s} and P_c^2 for the quaternary system water–NaCl–pentanol–SDS. (From Ref. 27.)

In summary, the main effect of the addition of alcohol and salt to SDS and water mixtures is to produce very swollen lamellar phases, sponge, and vesicle phases. The very existence of these three phases suggest that in these systems, thermal fluctuations are large; so one can expect that they will form microemulsions.

B.   Quaternary Mixtures: Water–Oil–Surfactant–Alcohol

The phase diagrams of two quaternary mixtures made of sodium dodecyl-sulfate (SDS)–water–dodecane and hexanol (system A) or pentanol (system B) have been investigated in details [8,19,48]. In both cases, sections of the three-dimensional diagram which holds constant the water to surfactant ratio have been examined (Fig. 6). These cuts have been choosen because they allow to have a good description of the oil region and also because the ratio water/SDS, termed X in the following, fixes the size of the droplets in the microemulsion phase, or the thickness of the bilayers in the lamellar phase.

In the description of the quaternary mixture, we will emphasize the details of the evolution of the phase equilibria as the X ratio is varied.

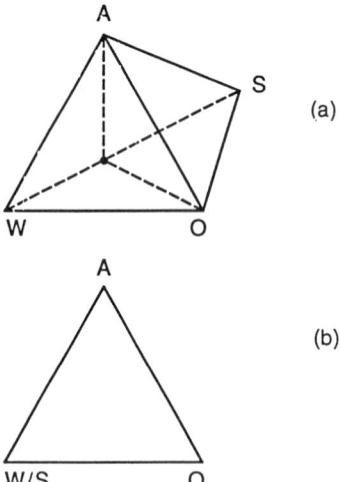

FIG. 6   (a) Three-dimensional representation of the phase diagram of the quaternary water (W)–oil (O)–surfactant (S)–alcohol (A) system in a tetrahedron. (b) Pseudoternary diagram at constant water to surfactant ratio ($X = $ W/S).

We have focused our attention not only on the characterization and the location of the boundaries of the various phases but also on the equilibria between the phases; that is, we have determined the changes in direction of the tie-lines and tie-triangles as X is varied.

1. System A: Water–Dodecane–SDS–Hexanol

Two typical pseudoternary diagrams are shown in Fig. 7. In each cut, three one-phase regions L_{3-0}, L_α, and L_2 are observed. Regions L_2 and L_α correspond respectively to an inverse micellar phase and a lamellar phase. In this latter, the oil content may vary continuously between 0% and 85% (in volume). The third phase L_{3-0} is a sponge phase; while motionless, this phase is isotropic, a flow transient birefringence appears in dilute samples as soon as any disturbance is created. In particular, flow birefringence is easily generated by shaking the sample tube. In addition, this phase scatters light. Minor changes are obtained as the X ratio is increased between 1.55 and 4.3 (in ωt). In both cases, the microemulsion L_2 phase is bounded by a two-phase region in which the L_2 phase coexists with the lamellar phase. The increase of X modifies the extents of the various regions but does not allow to oil-rich and water-rich regions to be connected.

2. System B: Water–Dodecane–SDS–Pentanol

The pseudoternary phase diagrams corresponding to system B are strongly dependent of the X ratio (Fig. 8). In the planes with X below X = 0.76, only the microemulsion phase L_2 is detected in the oil-rich region. For $X = 0.76$, the lamellar phase L_α appears. It can contain very large amounts of oil up to 98% of dodecane and pentanol in volume. Both regions L_2 and L_α exist in all the diagrams for $X > 0.76$, but their extent obviously depends on X. In particular, the maximum swelling of the lamellar phase continuously decreases with X. In the sections corresponding to X above 0.95, two new phases appear, a sponge phase L_{3-0} and an hexagonal phase H_α. The sponge samples display flow birefringence when the oil content is larger than 75%. As X is larger than 3, the L_2 and L_{3-0} domains merge and form a continuous region L which remains continuous up to $X = 5.3$. For this last value, the amount of surfactant with respect to water is no longer sufficient to achieve the continuity of the single-phase domain L. For X above 5.3, L has split into two regions, one rich in oil and alcohol and the other rich in water and surfactant. Moreover, one of the most interesting features occurring in the planes for X above 0.95 is a line of critical points P_c^1 along the coexistence boundary of the microemulsion L_2 region. A detailed analysis of the diagram shows that this critical line P_c^1 develops between P_c^A, located in the face of the tetra-

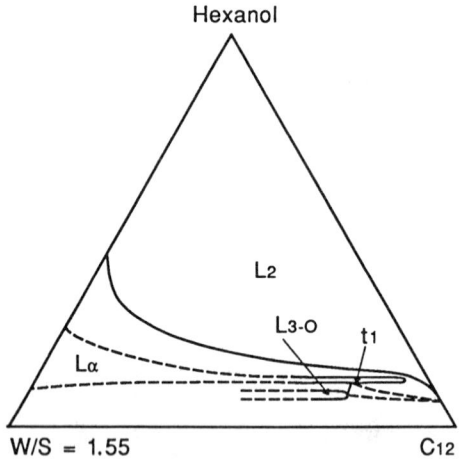

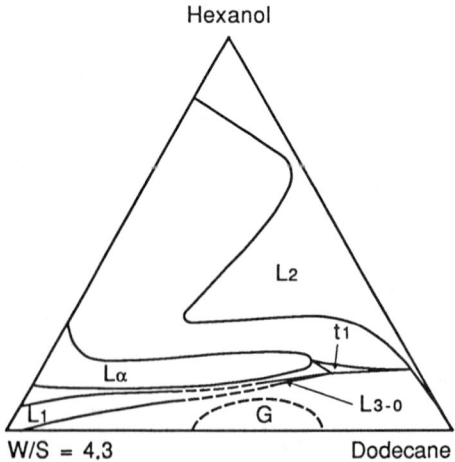

FIG. 7 Sections at constant water to surfactant ratio $X = 1.55$ and $X = 4.3$ of the phase diagram of the water–dodecane–hexanol–SDS system (system A) at 21°C. L_1 and L_2 are isotropic phases. L_α is a lamellar phase and L_{3-0} is an oil-rich sponge phase. t_1 is a three-phase region; G is a gel phase. (From Ref. 34.)

hedron water–pentanol–SDS at $X = 6.6$, and a critical end point P_{ce}^B, in the oil-rich region at $X = 0.95$ (Fig. 9). The direct consequence of this critical line is the occurrence in each X section of a two-phase region (d_1) where two isotropic microemulsions are in equilibrium (Fig. 10). In the sections for $X < 0.95$ and also far from the critical point, the microemulsion separates with a lamellar phase (Fig. 10). The appearance of new phases, on one hand, and that of the critical point, on the other, generate a very complex multiphase region in the sections for X above 1. As an example, Fig. 10 shows the multiphase regions corresponding to $W = 1.55$. It includes five different two-phase regions (d_1-d_5) and two three-phase regions t_1 and t_2; in region t_1, two microemulsion phases (L_2 and L_2) coexist with the sponge phase (L_{3-0}); in region t_2, the middle phase is the lamellar phase, the upper and lower phases are the sponge and the microemulsion phases, respectively. It is worthwhile to note that for the sections for $X < 2$, small variations in the alcohol fractions (a few percents) cause drastic changes in the state of the oil-rich mixtures and lead to phase transformations.

Measurements of phase composition analysis of a large number of two- and three-phase equilibria t_1 and t_2 which involve both isotropic and mesomorphic phases provide evidence that the X ratio has the characteristics of a field variable in the oil-rich mixtures [8,49]; indeed, this ratio takes the same value in all the coexisting phase. As a example, Fig. 11 shows compositions of conjugate phases of 10 samples occurring in region d_1 and two samples in regions t_1 and t_2. These data clearly show that X behaves as a chemical potential. The major consequence of this property is that the phase diagrams experimentally determined, in which X is maintained constant, are true pseudoternary diagrams. The tie-lines and tie-triangles are, indeed, located in the X section considered. Moreover, because X behaves as a chemical potential, we used this variable to approach several critical points of the line P_c^1 at constant temperature [8,49].

3. Comparison of the Phase Diagrams of Systems A and B

From this brief description, it appears that the phase diagrams of systems A and B show several common points and also a few differences. Both systems form in the oil-rich part of the diagram for the same three phases: a microemulsion phase L_2, a lamellar phase L_α, and a sponge phase L_{3-0}. The first two L_2 and L_α, are the direct extension of the micellar and lamellar phases found for the ternary mixtures in the absence of oil. L_α as well as L_{3-0} extend over a large range of water and oil concentrations. Very oily, swollen L_{3-0} and L_α phases can be prepared with only a few percents of surfactant. These high dilutions correspond to distances between the bilayers larger than 1000 Å. In addition, in the oil-rich part of the diagrams, the boundaries of these two phases are straight lines directed toward a point of the alcohol–oil axis. This feature permits the study of these phases along dilution lines which maintain the membrane composi-

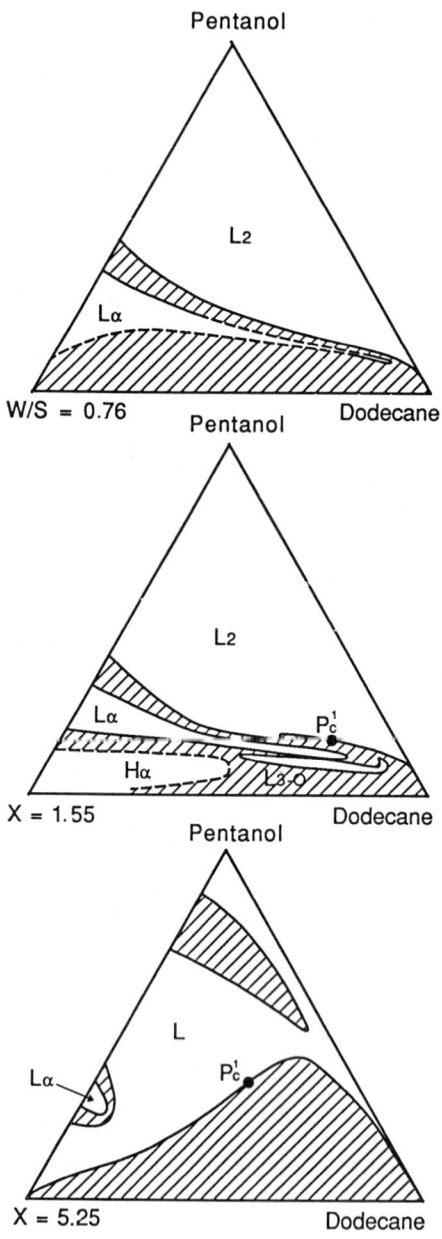

FIG. 8 Sections at constant water to surfactant ratio $X = 0.76$, 1.55, and 5.25 of the phase diagram of the water–dodecane–pentanol–SDS system (system B) at 21°C. The hatched regions are the multiphase regions. P_c^1 is a critical point. L and L_2 are microemulsions, L_α is a lamellar phase, and L_{3-0} is an oil-rich sponge phase. (From Ref. 34.)

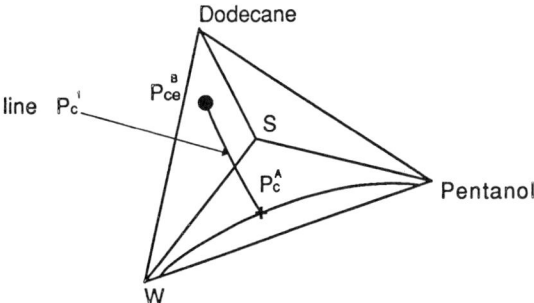

FIG. 9 Line of critical points P_c^1 for the quaternary mixture water–dodecane–pentanol–SDS.

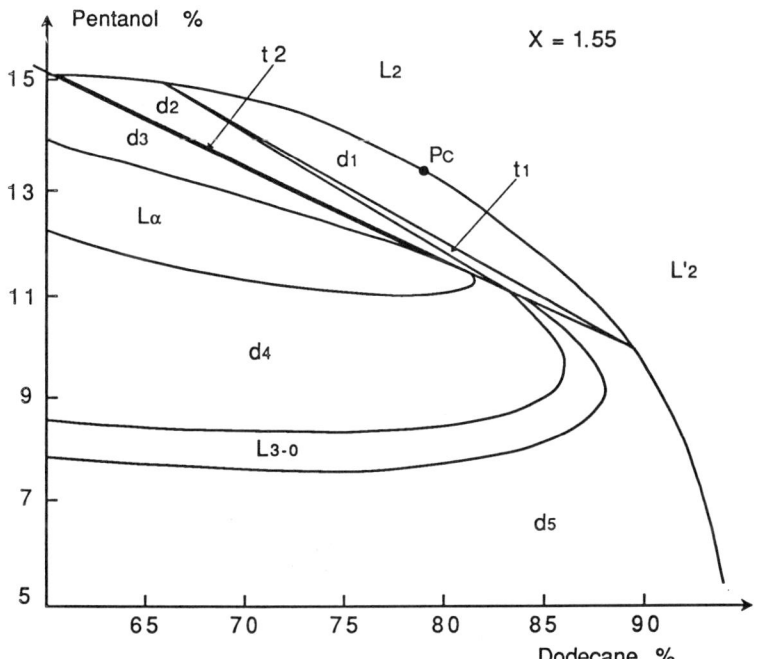

FIG. 10 Magnification of the oil-rich corner of the section for $X = 1.55$ of Fig. 8. d_1 is a two-phase region where two microemulsions L_2 and L_2 are in equilibrium. d_2–d_5 are two-phase regions. t_1 and t_2 are three-phase regions: $t_1 = L_2$–L_{3-0}–L_2. $t_2 = L_2$–L_α–L_{3-0}. (wt %). (From Ref. 34.)

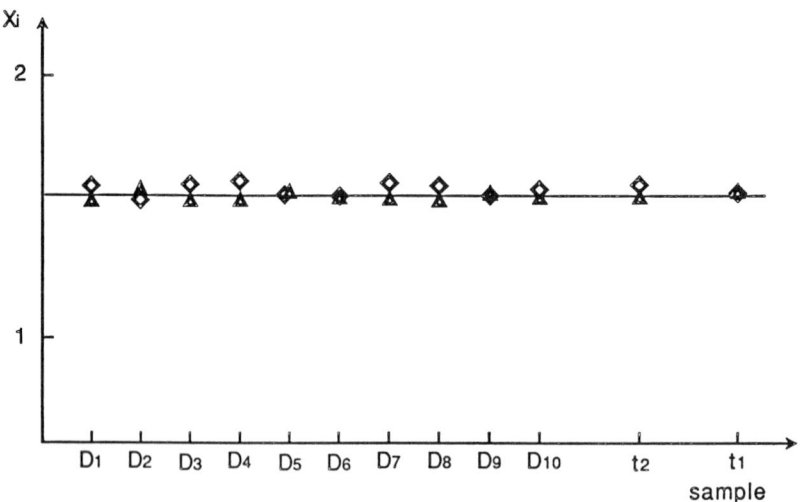

FIG. 11 Values of the water to surfactant ratio (X_i) measured in each phase i for 10 two-phase equilibria (termed D_1–D_{10}) lying in region d_1 and for three-phase equilibria of the regions t_1 and t_2. For all these samples, the global value of X is equal to 1.55 ($T = 21°C$).

tion constant. In both systems, the same sequence of phases L_2–L_α–L_{3-0} is obtained by decreasing the alcohol concentration. This behavior is similar to that previously described for brine-rich–alcohol–SDS mixtures. One must also note that for pentanol and low values of X ($X < 2$), the L_{3-0} phase can be obtained by dilution of the L_α phase.

Two main differences are found as one replaces hexanol by pentanol; the first one is the existence of a critical line in all the sections for $X >$ 0.95 and the second is the possibility of going continuously from the water plus surfactant vertex to the oil vertex when X is larger than 3. Regarding the stability of the microemulsion phase, two characteristic behaviors are evidenced. In the first one, no critical point occurs and the microemulsion region L_2 is bounded by a two-phase region where L_2 phase is in equilibrium with the lamellar phase L_α. The second type is characterized by the occurrence of a critical point and a two-phase region where two micellar phases coexist. As we will see later, this behavior is closely related to the intermicellar interactions.

It is worthwhile to point out that all the equilibria observed in these systems are different from the so-called Winsor equilibria where a microemulsion coexists with an excess oil phase (Winsor I equilibrium, WI) or an excess water phase (Winsor II equilibrium, WII) or both (Winsor III equilibrium, WIII). These equilibria are only obtained by the addition of salt.

C. Quinary Systems: Water–Salt–Oil–Surfactant–Alcohol

1. Oil-Rich Regions

Figure 12 shows the effect of added salt on the stability of the three oil-rich phases L_2, L_α, and L_{3-0} found in the section for $X = 1.55$ of the phase diagram of the water–dodecane–SDS–pentanol system. The three phases still exist at very high salinities, but they are shifted to a lower alcohol concentration. In fact, salt produces the same effect on the structures of the phases and on the topology of the interface as the alcohol does. Indeed, an increase in either salt or alcohol content produces the sequence of phases $L_3 \to L_\alpha \to L_2$.

As one examines the entire diagram, one observes that the phase behavior is largely influenced by the addition of salt. Figure 13 compares three X sections of the phase diagrams obtained in the absence of salt and at the salinity of 20 g L^{-1} NaCl. At low X, salt causes the formation of two-phase equilibria of type Winsor II consisting of a microemulsion phase in

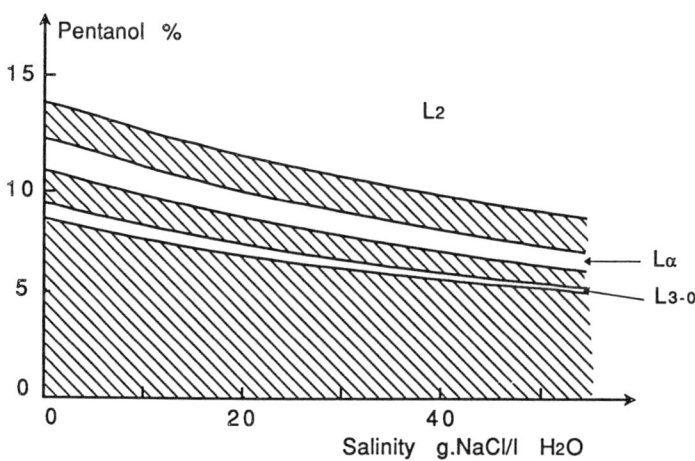

FIG. 12 The system water–NaCl–dodecane–pentanol–SDS. Effect of salinity on the phases microemulsion L_2, lamellar L_α, and sponge L_{3-0}. In the section considered, $X = 1.55$ and the dodecane content $= 78\%$ in wt.

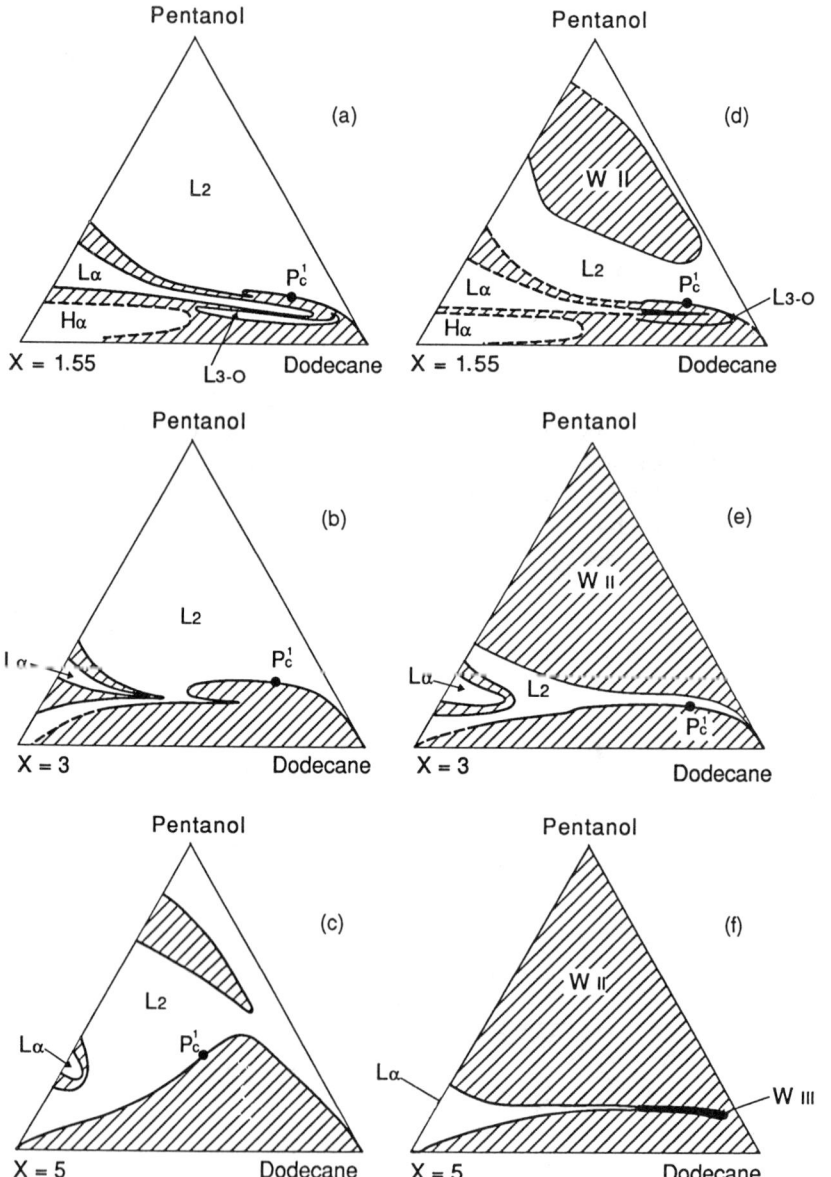

FIG. 13 Effect of salinity on the phase diagram of the water–dodecane–penta-nol–SDS system. (a), (b), and (c) correspond to three X sections obtained without salt. (d), (e), and (f) correspond to the same three sections at a salinity of NaCl 20 g L^{-1}. WII-Winsor II equilibria; WIII-Winsor III equilibria.

equilibrium with an excess of brine and, consequently a drastic reduction of the extent of the microemulsion domain. Upon increasing X, the Winsor II region expands. This trend leads to the disappearance of oil-rich micro-emulsion phases, giving rise to the three-phase body, Winsor III.

2. Brine-Rich Region

The diagrams shown in Fig. 14 correspond to cuts of the water–salt–do-decane–SDS–pentanol quinary system obtained at fixed salinity (20 g L^{-1} NaCl) and fixed oil fraction. When small fractions of dodecane are added to brine–pentanol–SDS mixtures, the L_1 and L_{3-w} phases do not continue to form a single domain but split into two distinct regions. However, for the oil fractions larger than 8%, the L_1 and L_{3-w} phases merge and form again a single domain. Upon increasing the oil fraction in the system, the L_1, L_{3-w}, and L_α phases are shifted to increasingly higher surfactant and alcohol concentrations. This trend is illustrated in Fig. 15 for the lamellar phase; the minimum surfactant concentration required to form a L_α phase increases linearly with the oil content. These various phases are separated by complex multiphase regions including four three-phase regions denoted t_3, t_5, t_6, and WIII. The three first regions exist only at low oil content (<9%), whereas the WIII region develops up to high oil concentrations. Measurements of phase composition analysis of several three-phase equilibria t_3 and t_5 show that the ratio denoted Y between dodecane and SDS concentrations takes the same values in the upper and middle phases of those equilibria. This property is illustrated in Fig. 16, where we plot the ratio Y_i measured in each phase i versus the value Y in the overall mixture. All the points fall along the bisecting line of the graph. The ratio Y has the characteristics of a chemical potential. This behavior is totally identical to that of the ratio X (water/SDS ratio) found in the oil-rich region (Fig. 11).

3. Winsor III Equilibria—Modes of Appearance

For anionic surfactants, the three-phase equilibria Winsor III where a microemulsion (m) coexist with both excess oil (u) and excess aqueous (l) phases are only obtained in presence of salt. The origin of these equilibria have been discussed [23,27,50]. Several modes of appearance may be considered to account for the formation of a three-phase region in multicomponent systems. At constant temperature and pressure, the three-phase region of a quaternary mixture is a volume made by a stack of tie-triangles whose the vertices are located along three curves Γu, Γm, and Γl corresponding respectively to the upper, middle, and lower phases. Depending on the relative positions of these curves, the three-phase region may be generated in several different ways illustrated in Fig. 17. One of these possibilities is the existence of two critical end points P_{ce}^1, and

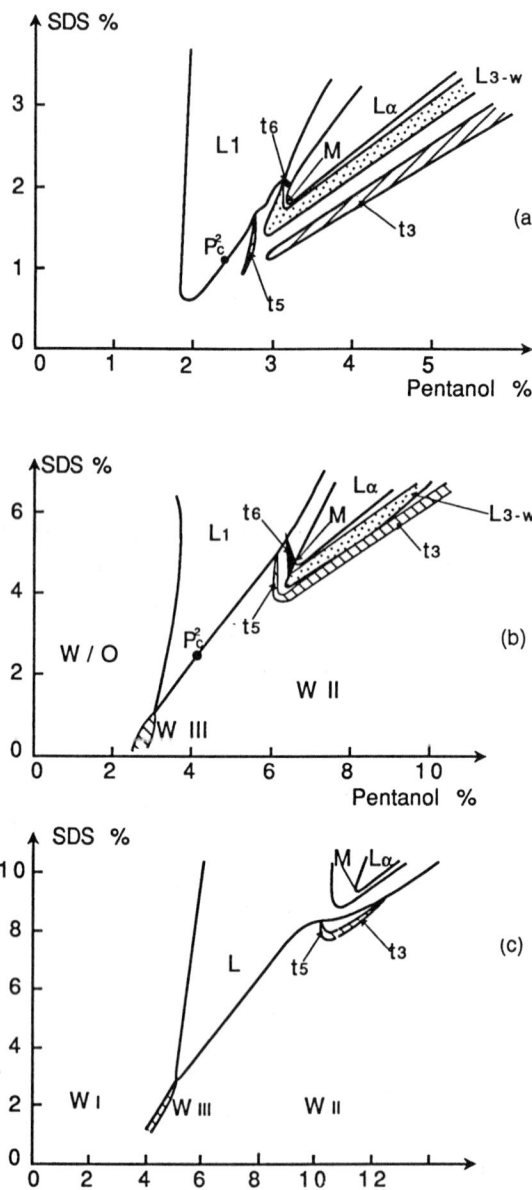

FIG. 14 Three sections of the phase diagram of the quinary mixture water–NaCl–dodecane–SDS–pentanol ($T = 25°C$). In the three sections, the concentration of salt in water is fixed at 20 g L^{-1}. In each section, the dodecane weight fraction is constant: (a) 1%; (b) 4%; (c) 8%. L or L_1: microemulsion; L_α: lamellar phase; L_{3-w}: sponge phase; regions t_3, t_5, t_6, WIII: three-phase equilibria; WI, WII, and WIII-Winsor equilibria.

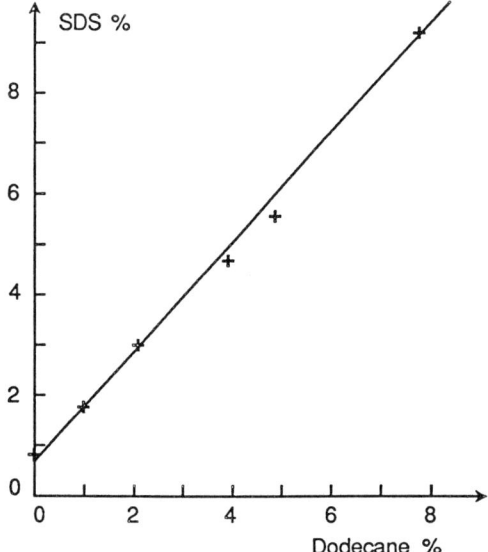

FIG. 15 SDS concentration at the point M (the more dilute lamellar phase, see Fig. 14) as a function of the dodecane concentration.

P_{ce}^2 where respectively the phases u and m, and m and l become identical. The critical end points P_{ce}^1 and P_{ce}^2 are the ends of two critical lines P_c^1 and P_c^2. As the temperature changes, the three-phase region may shrink to a tricritical temperature. At this point, the three phases become simultaneously identical, the critical end points merge, and the critical lines P_c^1 and P_c^2 form a connected single line (Fig. 17c). Two other origins may be considered; it possible that the three-phase volume terminates at a four-phase region or arises by an indifferent state. In this case, the points representative of the phases are along a straight line ABC (Fig. 17d).

The modes of appearance of a three-phase region for a quinary mixture are the same as those for a quaternary one. But the variance of the corresponding states is increased by 1; consequently, at fixed P and T, a three-phase region may result from one tricritical point and develops between two lines of critical end points L_{ce}^1 and L_{ce}^2. For quaternary mixtures, three-phase regions in five component mixtures may originate in an indif-

RECKITT & COLMAN
Laboratoire européen R&D
B.P 835
28011 CHARTRES Cedex
Tél. 02.37.24.78.00 - Fax 02.37.24.78.47

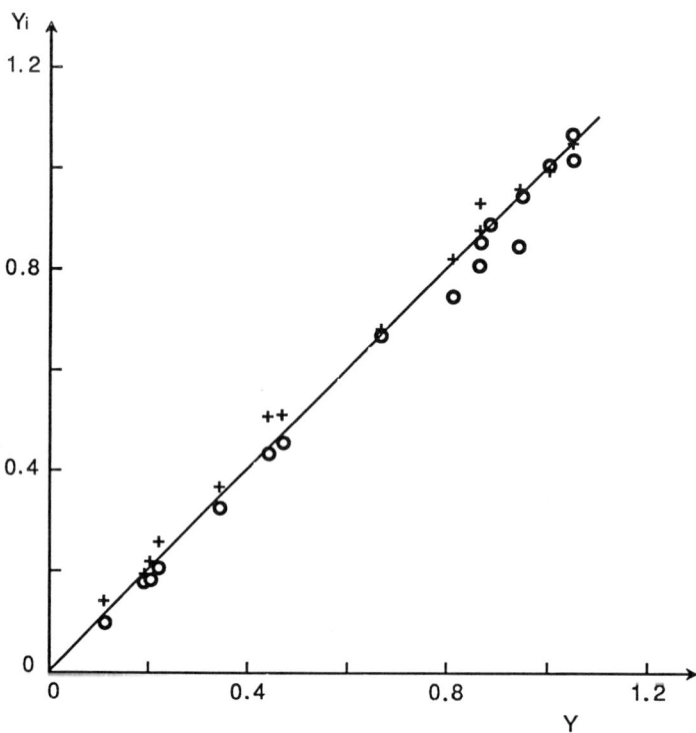

FIG. 16 Values of the dodecane to SDS ratio Y_i measured in the upper and middle phases of the three-phase equilibria t_3 as a function of the ratio Y in the overall mixture.

ferent state. The diagram of the quinary system water–salt–dodecane–pentanol–SDS can be represented by a series of tetrahedrons, each corresponding to a constant brine salinity. In practice, one tetrahedron is itself constructed by examining a series of cuts in which the brine over surfactant ratio is kept constant. This procedure allows one to describe completely the extent of the three-phase region Winsor III (WIII) and to determine its mode of appearance. Our results show that this region, which starts to be observed at 4 g L^{-1} NaCl in water, is generated by an indifferent state (Fig. 18 [27]). In such a state, the compositions of the three coexisting phases are located along a straight line in the four-dimensional space of representation. As salinity goes above 5.5 g L^{-1} NaCl, the volume WIII is bounded in the oil-rich region by a line of critical end points, L^1_{ce}. A second line of such points, L^2_{ce}, occurs in the water-

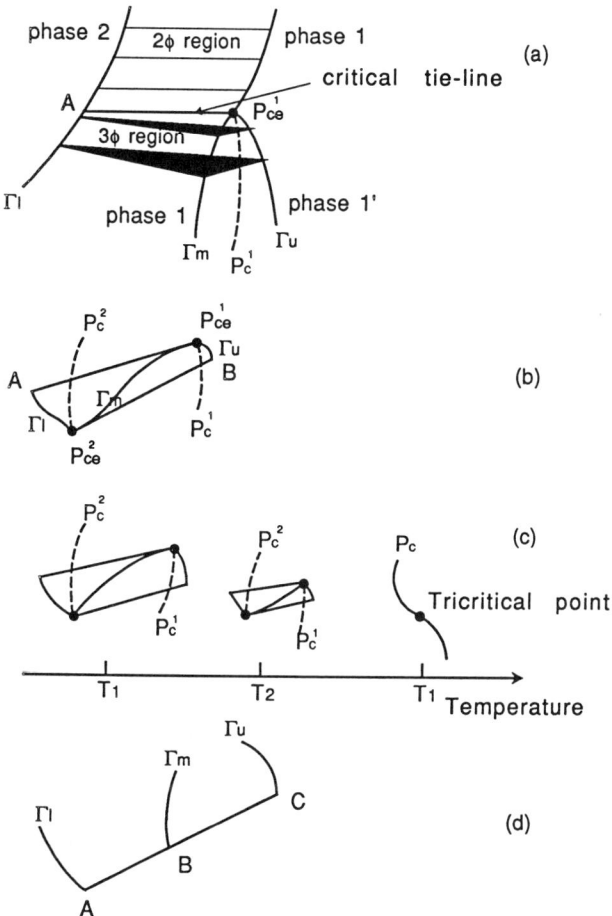

FIG. 17 Modes of appearance of a three-phase region for a quaternary mixture. (a) Critical end point; (b) two critical end points; (c) one tricritical point; (d) one indifferent state. (From Ref. 27.)

rich region at salinities higher than 8.1 g L^{-1} NaCl. Then, the three-phase region WIII appears at two critical end points only when salinity is higher than 8.1 g L^{-1} NaCl. Our experiments show that the lines L_{ce}^1 and L_{ce}^2, originate in the quaternary mixtures H$_2$O–D–P–SDS and H$_2$O–NaCl–P–SDS respectively.

Figure 18 gives the limits of existence of the Winsor III region. At low salinities, the surfaces Su, Sm, and Sl corresponding respectively to the

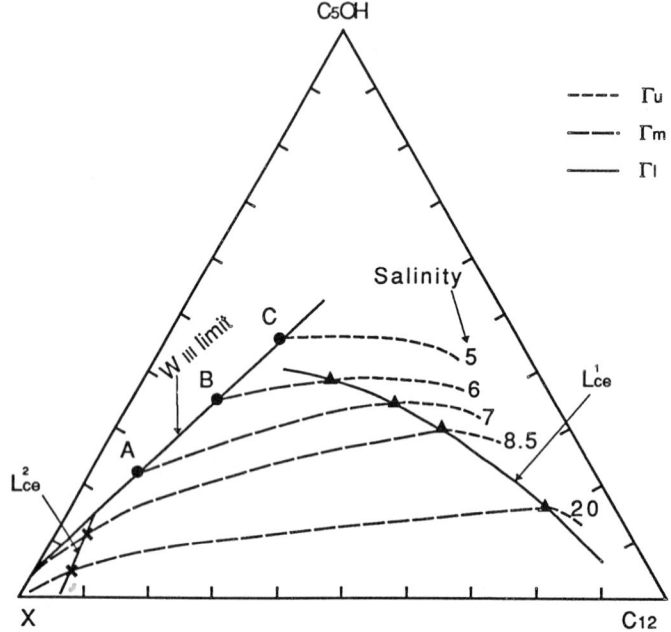

FIG. 18 Projections on the planes X–pentanol–dodecane of the curves Γ_u^g, Γ_m^g and Γ_l^s observed at the salinities 5, 6, 7, 8.5, and 20 g L^{-1} NaCl. The symbols ▲ and + represent respectively the critical end points P_{ce}^1 and P_{ce}^2; the symbol ● correspond to the limit of existence of the three-phase equilibria Winsor III. (From Ref. 27.)

compositions of the oil, microemulsion, and aqueous phases are limited by a straight line ABC. Let us mention that for all the three-phase equilibria close to line ABC, the three phases strongly scatter light. This could indicate that the system is close to a tricritical point. Such a point would result from the connection of the two lines of critical end points L_{ce}^1 and L_{ce}^2. Tricritical points have been discovered in some ionic and nonionic systems [23–26].

III. STRUCTURE AND STABILITY OF MICROEMULSIONS

One common feature to all microemulsions whatever the system, dilute, concentrated or even critical, is the existence of an interfacial film which separates an aqueous microdomain from an oil microdomain. As seen in the preceding section, the microemulsions exist over a wide range of composition, and in some cases, it is even possible to find in the phase diagram a continuous path from oil-rich microemulsions to water-rich microemulsions. Along such a path, the interfacial film adapts itself to the composition of the mixture, and several kinds of microstructures can be obtained. These structures have been reviewed by Auvray [51]. At low oil (or water) concentration, the structure of the microemulsion consists of a dispersion of oil (or water) droplets. As the water (or oil) concentration increases, the bicontinuous structure, first imagined by Friberg [52] and Scriven [53], generally occurs. In this case, oil and water form two intertwined labyrinthic networks separated by the film of surfactant. In systems that we believe to be made with more rigid interfaces, other structures have been identified. In particular, the ternary systems consisting of water–oil–didodecyl dimethyl ammonium bromide form in the oil-rich region connected cylinders [54]. All these structures have been characterized by using different techniques, particularly scattering techniques (light, neutrons, x-rays, electronic microscopy) but also conductivity and nuclear magnetic resonance. These latter methods yield very useful information on the degree of connectivity of the surfactant film and, hence, on the type of structure, discrete or bicontinuous.

A. Water in Oil Microemulsions

A great deal of work has been devoted to the study of quaternary oil-rich microemulsions. Light- and neutron-scattering techniques have been extensively applied to these systems. This has been possible because a dilution procedure was worked out by Graciaa et al. [55] which allows one to vary the concentration of the droplets without modification of their composition. The first studies developed by Dvolaitzky et al. [56] have proved that these microemulsions are dispersions of monodisperse spherical water droplets surrounded by a mixed dense surfactant monolayer in a continuous oil phase.

Scattering measurements provide quantitative information on the micellar size and the second virial coefficient B of the osmotic pressure which is directly related to the intermicellar potential. Results obtained with a large number of systems show that the water core radius is mainly determined by the amphiphilic surface area occupied per the surfactant molecule. This surface Σ, which includes the alcohol molecules, is found nearly constant (about 60 Å^2 in the case of SDS). As the consequence, the radius of the water core is fixed by the water over surfactant ratio [56–58]; it may vary from about 25 to few hundreds of angströms. In oil microemulsions, the second virial coefficient B varies from positive to largely negative values, indicating that interactions may change from hard spheres to largely attractive. The strength of the attractive potential is very sensitive to the size of the droplets and also to the chemical nature of the components. Attractions are found to increase when the micellar size increases, the alcohol chain length decreases, the polar head area of the surface increases, and the molecular volume of oil increases [58,59]. As an example, the effects of the alcohol chain length and the micellar size on the second virial coefficient are shown in Table 1, where the data for three series of microemulsions containing water, SDS, dodecane, and three alcohols are reported.

Vrij first pointed out that the strength of the interactions measured by the virial coefficient is, in general, too large to be due only to the pure van der Waals interactions between the cores of the droplets [60]. In order to explain the origin of attractive interactions, Lemaire et al. [61] assumed a mechanism of interpenetration. The proposed potential takes into account dispersion forces and the quality of the oil continuous medium for the alkyl chains of the interfacial film. A good solvent (short alcanes) leads to hard-sphere interactions, whereas a bad solvent (long alcanes) produces attractions between micelles. In this case, the micelles prefer to interpenetrate each other rather than to be in contact with oil. The effects of the alcohol chain length and of the micellar size are then interpreted as a pure geometrical effect because the interaction is proportional to the overlapping volume of the films. Later, Auvray proposed a mechanism of fusion induced by curvature to interpret the observations made for inverse droplets [62]. A simplified calculation of the virial coefficient B leads to an expression which depends on the shape of the dimer and on the elastic properties of the film. This mechanism also accounts for the variation of B with the alcohol chain length.

As the potential between droplets become strongly attractive a critical behavior is evidenced [63]. This is illustrated in Fig. 19, where the curves of reduced compressibility versus the volume micellar concentration ϕ for three of the microemulsions presented in Table 1 are reported. The heptanol microemulsion is close to a hard-sphere system ($B = +3$). In the two other microemulsions, interactions are more attractive ($B = -3$ for hexanol, $B = -16$ for pentanol). In the case of pentanol, both $\partial\pi/\partial\phi$ and $\partial^2\pi/\partial\phi^2$ are close to zero, which corresponds to a critical point. Then, for this system, it appears that attractions between micelles are strong enough to induce a phase separation between a rich micellar phase and a poor micellar phase. Therefore, the line of critical points P_c^1 and the two-phase region d_1 where two microemulsions coexist observed in all the sections for $X > 0.95$ of the phase diagram of the pentanol system (Figs. 9 and 10) can be interpreted as a liquid–gas transition due to intermi-

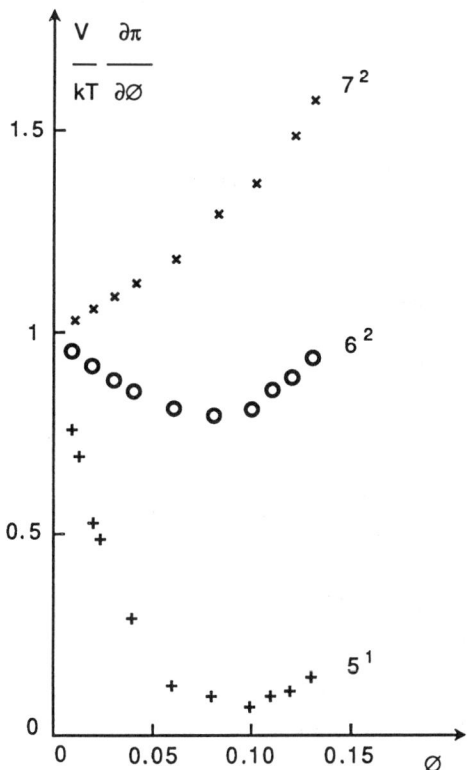

FIG. 19 Normalized osmotic compressibility versus the volumic fraction ϕ for the microemulsions 5^1, 6^2, and 7^2 presented in Table 1. V is the volume of a micelle. (From Ref. 34.)

TABLE 1 Effect of Alcohol Chain Length and W/S Ratio on the Micellar Radius (R) and the Virial Coefficient (B) of Microemulsions Formed with SDS, Dodecane, Water, and Pentanol, or Hexanol, or Heptanol

	Microemulsion				
	Pentanol	Hexanol		Heptanol	
	5^1	6^1	6^2	7^1	7^2
W/S (wt %)	0.95	1.55	2.55	2.25	3.05
R (Å)	45 ± 3	47.6 ± 1	64 ± 4	54 ± 2	66 ± 3
B	−16 ± 2	−0.5 ± 0.5	−3	+6 ± 1	+3 ± 3

cellar interactions. A quantitative interpretation of the diagram of the quaternary system water–dodecane–pentanol–SDS has been made possible [64] by using the intermicellar potential developed by Lemaire et al. [61] and the Baxter state equation for adhesive hard spheres [65]. Interactions may explain the differences observed between the phase diagrams of the systems containing hexanol and pentanol. In the hexanol system, interactions remain moderately attractive even for large micellar sizes, and the calculated demixing line is largely below the experimental curve. For this system, the phase separation is not driven by interactions but results from another mechanism involving probably the curvature energy of the oil–water interface.

B. Concentrated and Bicontinuous Microemulsions

Although a model of interacting particles satisfactorily accounts for light-scattering data obtained over a wide range of droplet volumic fractions ϕ (up to $\phi \sim 0.30$), results of electrical conductivity, transient electrical birefringence [66,67], and ultrasonic adsorption [68] provide evidence that the domain of validity of the droplet model in quaternary microemulsions strongly depends on the micellar potential; that is, on the length of interpenetrability during collisions. When the droplets behave as hard spheres, they may be concentrated without distortion up to random sphere close-packing concentration [69]. The electrical conductivity of these microemulsions remains low; it increases smoothly as the volumic fraction ϕ is increased. As the attractive potential increases, a steep increase of conductivity is found around $\phi_p = 0.10$ [66,67]. This phenomenon has been attributed to percolation [70,71]. In the case of strongly attractive interactions, ϕ_p is very close to the concentration of the critical point ϕ_c. Because water in oil droplets are nonconducting particles, the observation of the electrical percolation phenomenon suggests that the water cores of the droplets become interconnected to ensure the electric charge transport. At the percolation threshold, it is proposed that in quaternary microemulsion, large-size open structures are formed by merging of droplets during collisions, and the droplet structure is replaced by a bicontinuous structure [66–68]. The merging process is the consequence of the large interpenetrability of the droplets and of the fluid character of the interfacial layer. Aggregation of the droplets and exchange of droplet contents were qualitatively evidenced by electrical birefringence [66,67,72] and ultrasonic adsorption.

The bicontinuous structure is also present in the middle phase of the so-called Winsor III equilibria. We have seen previously that by varying salt or alcohol concentration, one gets the sequence of equilibria Winsor

I (WI) → Winsor III (WIII) → Winsor II (WII). In the WI equilibria, an oil-in-water microemulsion coexists with an oil excess; in the WII equilibria, a water-in-oil microemulsion coexists with an excess of water; and in the intermediate region WIII, a microemulsion phase coexists with both a water phase and an oil phase. This sequence of phases realizes a progressive inversion of the curvature. The middle-phase microemulsion would then correspond to the interesting case where the spontaneous curvature of the film tends to vanish. To test this prediction, different techniques have been used: conductivity measurements [70], self-diffusion coefficients measurements [74], electron microscopy [75], and scattering techniques [51,76]. Nuclear magnetic resonance (NMR) measurements give clear evidence of the progressive evolution from a droplet structure to a bicontinuous structure in the Winsor equilibria [77]. Results presented in Fig. 20 show that in the intermediate salinity range, the self-diffusion coefficients of oil and water are of the same order and both oil and water move almost freely, which is characteristic of a bicontinuous structure. More direct information on bicontinuous microemulsions have been obtained using scattering techniques (x-rays and neutron) and the method of contrast variation with deuterated molecules [51,76]. The spectra of middle phases show the following features:

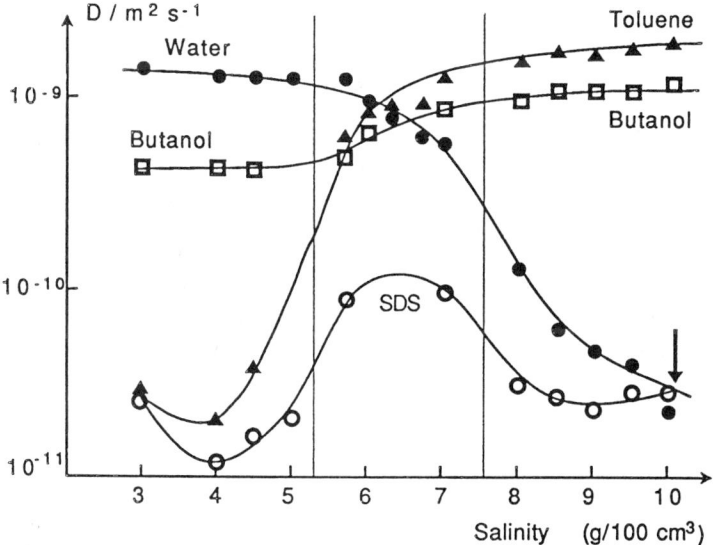

FIG. 20 Self-diffusion coefficients of the constituents of Winsor microemulsions as a function of the brine salinity. (From Ref. 77.)

(a) Scattering at large wave vectors q obeys a Porod law. The Porod law breaks down at a characteristic length scale related to the curvature of the interface.

(b) A pronounced peak at a smaller value of $q = q_{max}$. This gives a measure of the characteristic size ξ of the random microemulsion. The existence of a well-defined length scale is theoretically related to the rigidity of the film. In the inversion zone $(0.3 < \phi_0 < 0.7)$, the variation of ξ with the composition $\xi \sim \phi_o \phi_w / c_s \Sigma$ is in good agreement with the predictions of the models of random bicontinuous structure (c_s is the surfactant concentration and Σ is the polar head area of the surfactant). Finally, the results show that the average curvature of the film varies continuously with the volume fractions of oil ϕ_o and water ϕ_w and vanishes at the inversion point where both ϕ_o and ϕ_w are equal.

C. Thermodynamic Model

The existence of several kinds of microemulsion has been interpreted by considering the curvature energy of the surfactant film and the entropy of mixing of the oil and water domains. Because the relevant structures have curvatures C_1 and C_2 which are much smaller than $1/l$, where l is a typical molecular length, the bending energy can be written phenomenologically as [78]

$$F_b = \frac{1}{2} k \int (C_1 + C_2 - 2C_0)^2 \, dS + \bar{k} \int C_1 C_2 \, dS.$$

The constants k and $\bar{k}$ are respectively the mean and Gaussian bending moduli. The spontaneous curvature C_0 describes the tendancy of the interface to bend toward either the water ($C_0 > 0$) or the oil ($C_0 < 0$). It arises from the competition in the packing of the polar heads and hydrocarbon tails of the surfactant. The theoretical treatment show that the transitions between different microemulsions shapes result from a competition between the bending energy of the surfactant film and the conservation constraints. Steric constraints leading to a large asymmetry of the film and strong spontaneous curvature favor the formation of droplets. When the film has a small spontaneous curvature, it tends to adapt either a flat or saddle-shaped local configuration. This should lead to ordered structures. De Gennes and Taupin first proposed that the reason why the structure is not lamellar but bicontinuous is the existence of thermal fluctuations which distorts the film [79]. They introduced the valuable notion of persistence length, ξ_K, which is exponentially related to the bending constant k by

$$\xi_K = l \exp\left(\frac{4\pi k}{k_B T}\right).$$

l is a molecular length. The interfaces are flat over distances $\xi < \xi_K$ and have independent orientations at larger scales. Thus, the value of k plays an essential role. If $k \sim 10k_B T$–$100k_B T$, as for usual compact films of surfactant, ξ_K is macroscopic and the system is lamellar. If $k \sim k_B T$, as for the films forming microemulsions, then the structure is bicontinuous.

Jouffroy et al. attempted to derive the phase diagram of microemulsions using this concept [80]. They include the effects of the bending energy and the thermal undulations in a model based on a simplified version of that of Talmon and Prager who first proposed a phenomenological model of disordered microemulsions [81]. They divided space into a lattice of cubes filled at random with oil and water and they chose the lattice size to be always equal to ξ_K. Only two-phase equilibria and not the characteristic three-phase coexistence of microemulsion, water and oil, were predicted. However, they found one very important result for applications: The minimum amount de surfactant ϕ_s^m needed to form a microemulsion with the same volume fractions of oil and water is inversely proportional to ξ_K.

In a series of articles by Andelman, Cates, Roux, and Safran (ACRS model) [82], a thermodynamic model was proposed which incorporates the physics of the fluctuations and also predicts the correct phase behavior. By allowing the length scale ξ of the random microemulsion to be determined by the composition of the microemulsion, an idea first introduced by Talmon and Prager [81] and extended by Widom [83] (who did not include the effects of thermal fluctuations on the free energy),

$$\frac{\xi}{a} = \frac{6\phi_o \phi_w}{\phi_s},$$

it is possible to derive phase diagrams which showed the regions of stability of the lamellar and bicontinuous phases as well as the correct multiphase equilibria (a is the thickness of the surfactant monolayer). The free-energy density of the microemulsion phase F_M can be written as the sum of the entropy of mixing per unit volume, S, and the bending energy density, F_b:

$$F_M = F_b - TS$$

with

$$S = -\frac{k_B}{\xi^3}[\phi \ln \phi + (1 - \phi) \ln(1 - \phi)]$$

and

$$F_b = \frac{8\pi\phi}{\xi^3}(1 - \phi)k(\xi).$$

In the ACRS model, the main departure from the previous formulations lies in the treatment of the bending energy. They incorporate in their treatment the renormalization of the bending constant k by the thermal fluctuations, calculated by Helfrich [84] and Peliti and Leibler [85]. These authors found a size-dependent effective bending constant $k(\xi)$ which obeys

$$k(\xi) = k_0 \left[\left(1 - \tau \log \frac{\xi}{a} \right) \right],$$

where $k_0 \equiv k(a)$ is the bare constant denoted previously k and α is a numerical constant. The downward renormalization of k indicates that it becomes very easy to bend a monolayer of size $\xi > \xi_K$ because such a film is already spontaneously crumpled. In the ACRS model, thermal fluctuations are treated on two levels. First, the fluctuations on a scale larger than the size ξ of the domains as for other models taken into account in the entropy of mixing of water and oil. At the same time, the authors have explicitly introduced the effect of thermal fluctuations on a smaller scale by incorporating the renormalization of k. This point appears to be crucial to the stabilization of the middle-phase microemulsions. At constant ϕ, it follows from the above equations that the free energy of the microemulsion is of the form

$$F_M = (A + B \ln \phi_s)\phi_s^3.$$

The free-energy density of the lamellar phase F_{L_α} due to steric repulsion of surfactant sheets is estimated as

$$F_{L_\alpha} = \chi \frac{(k_B T)^2}{4\Pi k_0 a^3} [\phi^{-2} + (1 - \phi)^{-2}]\phi_s^3.$$

(χ is a numerical parameter). The phase diagram calculated by double tangent construction is shown in Fig. 21. The model successfully explains the characteristic features of the phase behavior; in particular, the competition between a bicontinuous microemulsions phase and a lamellar phase. Indeed the model exhibits, at low surfactant volume fraction, a stable middle-phase microemulsion coexisting with extremely dilute phases of surfactant in water and in oil, and at the high surfactant volume fraction a lamellar phase. In addition it also predicts that the transition between these two phases is determined by the value of k_0 (Fig. 22). Indeed, for small k (very flexible surfactant films), the lamellar phase is stable only

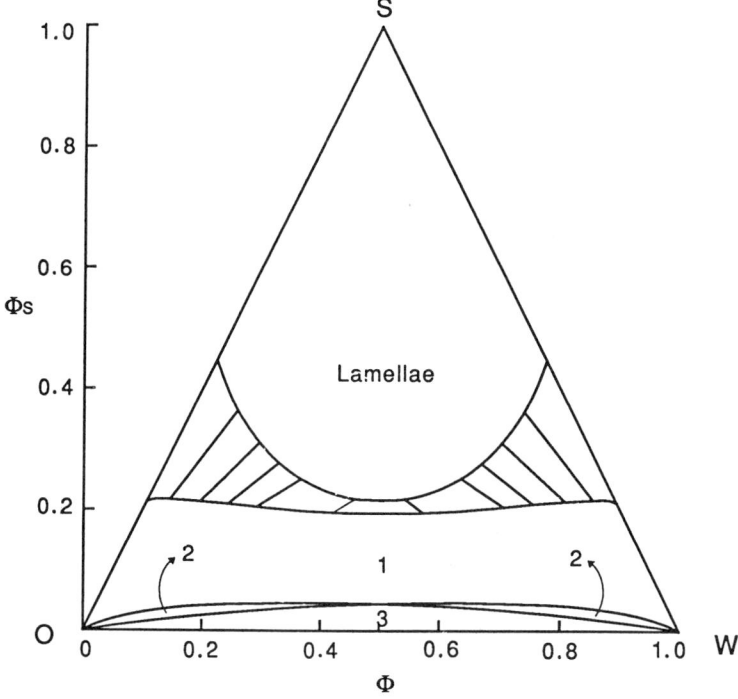

FIG. 21 Typical calculated phase diagram for the case of no spontaneous curvature, showing the transition to the lamellar phase and the associated two-phase region. 1: microemulsion; 2 and 3: Winsor equilibria. This phase diagram has been obtained for a value of $4\pi k_0 = 5k_BT$. (From Ref. 82.)

for very high ϕ_s and the microemulsion is stable for a large range of concentration, although ϕ_s^m is large. When k increases, ϕ_s^m decreases as $\exp(-k_0)$. However, in the meantime, the L_α phase is more stable and the transition microemulsion $-L_\alpha$ occurs at smaller ϕ_s. Following these results, the recipe for making a microemulsion that is stable over a wide range of ϕ_s is to make k small in order to push the lamellar phase at higher ϕ_s. However ϕ_s^m will then be quite large and the microemulsion will use a lot of surfactant. Consequently, the price to pay for having a microemulsion phase made with a very small amount of surfactant is to have a narrow region of stability. The link between theory and experiments in the case of nonionic surfactant systems has been established by assuming that k_0 increases with surfactant chain length [82]. The theory also qualitatively correctly describes the competition between the micro-emulsion and the lamellar phases in an ionic surfactant system [86].

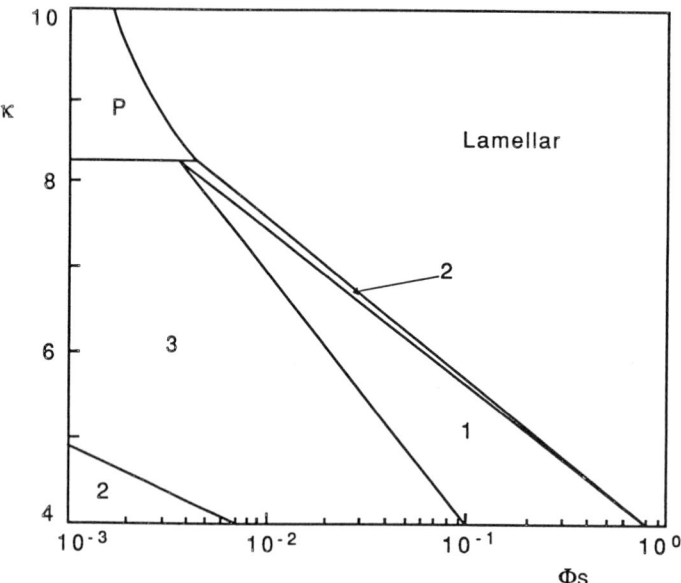

FIG. 22   Cut of the phase diagram in the plane $K = 4\pi k_0/k_B T - \phi_s$. The oil–water ratio is fixed at unity and the spontaneous curvature is zero. The Gaussian bending constant $k = 0$. (From Ref. 82.)

On the basis of the ACRS model, the structure factor of a random microemulsion has been calculated [87]. The model predicts a peak in the intensity scattered by the oil and water at a position $q^* = \pi/\xi$. This is in perfect agreement with the experiments. As expected, the physical origin of the peak resides in the curvature elasticity of the surfactant layer.

In conclusion, the most recent models of microemulsions highlight the relevance of flexible thermally fluctuating films. Thermal fluctuations have to be large, as they are, in fact, responsible for the very existence of the microemulsion structure.

IV.   LAMELLAR L_α PHASES

The most striking property of the lamellar phases encountered in the mixtures forming microemulsions is that the repeat distance d between the layers can be varied continuously upon the addition of an appropriate solvent from molecular sizes to extremely large values (100–1000 Å). Experimental phase diagrams show that the lamellar phase can be swollen with either oillike solvents or with water or brine. The ability to swell the smectic structure with water is associated with strong electrostatic repulsions. The even larger swellings obtained by dilution with oil or brine come from repulsive interactions between bilayers which have a nonelectrostatic origin. This interaction, known as the undulation interaction, arises in the high flexibility of the surfactant bilayers through a mechanism of entropy reduction proposed by Helfrich [88]. Under the influence of thermal fluctuations, a free flexible membrane undulates. When the membrane is constrained in a lamellar phase, it cannot explore the whole space around it. The restriction in its movement leads to an effective repulsive interaction, whose the amplitude is inversely proportional to the bending constant k. An estimate of this undulation interaction using the Landau–De Gennes elastic energy for a smectic leads to a contribution to the free energy V_{und} given by [88]

$$V_{und} = \frac{3\pi^2 (k_B T)^2}{128k} \frac{1}{d^2}.$$

This is a long-range repulsive interaction which can overcome the van der Waals attraction at large distances d. This interaction is responsible for the very large dilution obtained in microemulsion-forming systems.

The description of the lamellar phase can be done at two levels: For lengths larger than the repeating distance d, the system behaves as a regular smectic phase, but for smaller lengths, it can be described in terms of interacting membranes. Consequently, it is possible to relate the macroscopic properties of the phase such as the elastic constants to the microscopic properties of the membranes. Of particular significance are the layer compressibility modulus at constant chemical potential $\bar{B}$ and the smectic splay constant K. $\bar{B}$ is directly related to the microscopic interactions between the membranes, whereas K is related to the membrane-bending modulus k by $K = k/d$.

In order to measure these elastic constants, several technics have been used. There are experiments involving direct measurements of the compressibility; for example, it was possible to access $\bar{B}$ directly by using the force machine of Israelachvili [89]. Other methods are based on the analysis of the order parameter (EPR or NMR) or the analysis of fluctuations. Due to the one-dimensional nature of the order in the smectic state, the long-wavelength fluctuations exhibit the so-called Landau–Peierls instability. The x-ray structure factor can be analyzed by using the Caillé theory and elastic constants can be extracted quite easily. Another way of measuring the elastic constant is to use dynamic light scattering. It can be shown that the control of both the wave vector and the polarization allows one to obtain all the elastic constants independently.

Three of these methods have been applied to systems discussed in this chapter: The first one is based on a accurate determination of the shape of the x-ray structure factor [90–92] the two others are dynamic light scattering [93] and deuterium solid-state NMR [94,95]. All the results show that in the pentanol–SDS and hexanol–SDS systems, the constant k is of the order of $k_B T$. Recent NMR studies indicate that the choice of the cosurfactant is crucial for obtaining small values. In the case of SDS–alcohol bilayers, it is found that k is sensitive to both the amount of alcohol in the membrane and to the alcohol chain length, with typical values increasing between $1.3 k_B T$ and $13 k_B T$ from hexanol to decanol systems [94]. These measurements have allowed one to elucidate the role of the alcohol length. The reduction in k obtained by replacing surfactant by alcohol molecules is due to the thinning of the membrane and also to the increase of the area occupied by the surfactant at the interface [92,94].

Results obtained by scattering methods (x-ray and light) have clearly demonstrated the existence of undulation forces in the absence of long-range electrostatic interactions (water–salt dilution or oil dilution). Indeed, a universal behavior, characteristic of this entropically driven interaction, has been found experimentally for different systems [90,92]. The dependence of the Caillé coefficient η, which is related to the elastic constant $\bar{B}$ with the repeat distance is the same for the SDS pentanol system diluted with either oil or brine. When electrostatic forces exist, corresponding to a pure water dilution of charged membranes, the behavior of η shows that electrostatic interactions dominate.

Here again, as well as for microemulsions, it is essential to emphasize the role of the membrane flexibility characterized by the bending constant k and its consequence for interactions between membranes, namely, the existence of an undulation interaction. This interaction due to thermal fluctuations is responsible for the very existence of extremely diluted lamellar phases. It is also important to point out the role of the Gaussian bending constant $\bar{k}$, which is known to control the topology of the bilayers [78,96]. As mentioned in Sec. II.A, the spherulite defects, corresponding to negative values of $\bar{k}$ are found at a low alcohol surfactant ratio; these multilamellar vesicles disappear as the alcohol content increases.

V. SPONGE PHASES

Sponge phases as well as bicontinuous microemulsions are two examples of the experimental realization of phases of randomly connected fluctuating membranes dominated by thermal fluctuations. A simplified representation of this phase is shown in Fig. 23. This structure has been proposed

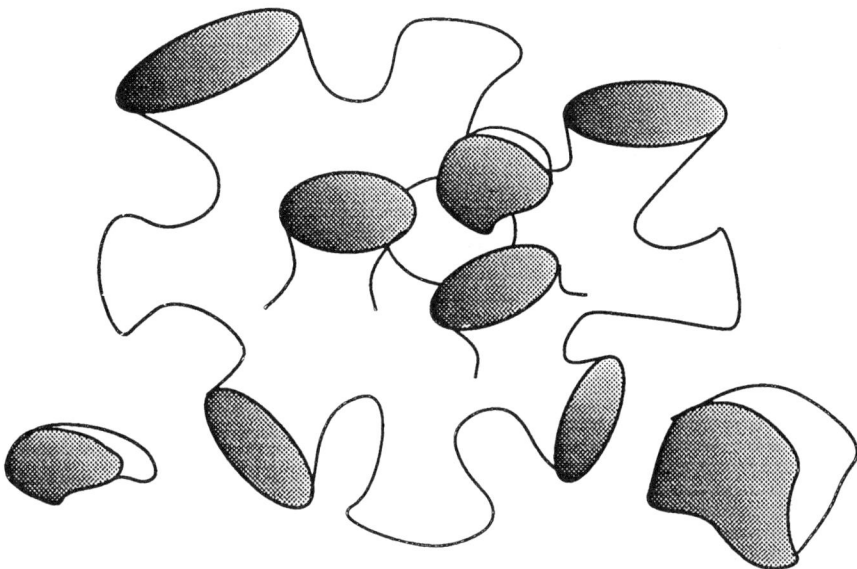

FIG. 23 Schematic representation of a sponge phase.

independently on the basis of theoretical arguments [97] and experimental results [21] in 1988.

Both oil- and water-rich sponge phases have been observed in the phase diagrams of many systems. They are found in the vicinity of swollen lamellar phases stabilized by undulation forces. The first reported phases were quaternary systems [21,22]. They are obtained from the lamellar phase either by dilution or by addition of alcohol. Sponge phases are isotropic phases of randomly connected bilayers. In this phase, the surfactant forms a continuous surface which divides the solvent into two identical domains I and II. In bicontinuous microemulsions, domains I and II are different; I corresponds to water and II to oil.

The continuity of the membrane in sponge phases has been demonstrated in using three experimental techniques: electrical conductivity, neutron scattering and electron microscopy [21,22,98]. The most spectacular conductivity measurements are found in an oil-rich sponge phase L_{3-0}. The solvent is a mixture of dodecane and pentanol and is an insulator. The membrane is an inverted bilayer, slightly swollen with water, which can be considered as a conductor. Measurements of the reduced conductivity (compared to the expected conductivity for the same amount of pure water + charges coming from the surfactant counterions) shows a high conductivity (around 0.65) all along the dilution path, even for the most dilute samples containing a few percents of membrane. A comparison with the conductivity of a nearby droplet microemulsion phase for the same surfactant concentration shows several orders of magnitude differences in the conductivity. This result clearly indicates that the bilayers are connected and that continuous water paths through the membranes exist. The values of 0.65 instead of 1 is easily explained in terms of the tortuosity of the membrane path.

In a water-rich sponge phase L_{3-w}, the conductor is the solvent and the insulator is the membrane; in this case, the reduced conductivity is found to be two-thirds of the conductivity of the pure solvent [21,22]. Again the ratio 2/3 can be interpreted as an obstruction effect produced by the bilayers.

Other evidence for the sponge structure has also been obtained by neutron scattering [21,22]. The scattering intensity can be divided into two regimes: the large-q regime where all the data fall on a universal curve, corresponding to the form factor of a flat bilayer, and the small-q regime where correlations between bilayers lead to a broad peak. This peak shifts by dilution to low q, according to a law expected for the swelling of a sponge phase, namely, $d = \gamma\delta/\phi$; in this relation, d is the size of the sponge, δ is the thickness of the bilayer, and γ is a number, estimated to be 1.5, for a lattice model. Experimentally, depending on the system, γ varies between 1.4 and 1.6.

One of the most striking properties of sponge phases is to have a particular symmetry. Indeed, due to the equivalence of both domains I and II, the sponge phase may exist under two different states: a symmetric state S corresponding to a perfect symmetry between both domains ($\phi_I = \phi_{II}$) and an asymmetric state A where this symmetry is broken ($\phi_I < \phi_{II}$). The transition from the symmetric sponge phase to the asymmetric phase can be either first or second order [96,97,99]. From a theoretical point of view, the description of this transition requires two order parameters, one, η, related to the asymmetry of the membrane and the second, ρ, related to the concentration. The existence of these two order parameters lead to interesting properties; in particular, the correlation function $S(q)$ possesses, when the fluctuations of η dominate, a very original q dependence, different in the A and S phases [99–101]. The topology of the bilayers evolves in the asymmetric phase toward a phase of vesicles far from the S/A transition. The measure of the intensity of light scattering enables one to determine unambiguously the S or A nature of the phase. In addition, in the system brine–pentanol–SDS, a line of maxima of turbidity has been identified as a line of second-order transition between the phases S and A [100].

The structure, stability, and properties of the sponge phase have been described theoretically by Cates et al. in using a modified version of ACRS model developed for microemulsion [96]. The structure consists of locally sheetlike sections of semiflexible surfactant bilayer, connected up at larger distances into a multiply-connected random surface having a preferred structure length scale of order of the persistence length. The basic idea is that, upon dilution, the lamellar phase melts when the repeat distance is of the order ξ_K into the disordered sponge phase stabilized by entropy. In this approach, the bending modulus k_c is the main control parameter. An alternative explanation for the stabilization of the sponge phase has been proposed by Porte et al. [102]. It is based on the phase diagrams of several brine–surfactant–alcohol systems where the transition from lamellar (L_α) to sponge (L_3) occurs by the addition of alcohol. Porte et al. argue that this structural transformation is mainly triggered by variations of the Gaussian constant $\bar{k}$, $\bar{k}$ being determined by the spontaneous curvature of each constituting monolayer.

The theoretical and experimental studies of sponge phases are still an open problem. One of the objectives is to understand the role of defects which can affect the stability of the phase. From a theoretical point of view, Huse and Liebler suggest the existence of a new phase, the sponge with free edges when the energy to create defects is low [103]. Experimentally, it has been shown by conductivity that defects exist in dilute symmetric sponge, before the S/A transition [104]. In addition, a new topology of the diagrams consisting of several lines of maxima of turbidity is observed in the SDS–pentanol–brine at low salinity, suggesting the existence of a teared sponge [105]. Also, recent light-scattering and conductimetry results suggest a spontaneous tearing of the membrane in the very dilute regime [106].

VI. VESICLE PHASES

It has been shown theoretically that phases of unilamellar and multilamellar vesicles are likely to arise instead of the bicontinuous sponge phase when the elastic constants k and $\bar{k}$ associated respectively with the mean curvature and the Gaussian curvature of the bilayer are chosen so that the curvature energy to create a sphere is small [107]. Using different experimental techniques such as conductimetry, light and neutron scattering, and freeze-fracture electron microscopy, it has been shown that a phase of vesicles at thermal equilibrium exists in all the pseudoternary systems made of SDS–brine–alcohol with a chain length varying from C_5

(pentanol) to C_{10} (decanol) [41,108,109]. In all these systems, the vesicle phase, designated L_4, is located between the micellar phase and the lamellar phase. The most interesting feature of this phase is that two regimes can be described. The first one corresponds to a dilute phase of vesicles with the properties of a colloïdal suspension of polydisperse particles; in this first regime, the vesicles are unilamellar and polydisperse with typical sizes increasing between 150 Å for hexanol and 1250 Å for decanol. As the surfactant concentration increases, a second regime with small-sized multilayered vesicles containing less than four layers on average is obtained. In this second regime, the phase exhibits high viscosity and viscoelasticity. The relative stability of unilamellar and multilamellar vesicles depends on the bilayer elasticity. The pentanol system exhibits a behavior different from the other mixtures. Indeed, for this alcohol, a dilute vesicle phase does not exist, only the small onion phase is stable. Besides, the transition from small onions ($N = 4$) to the lamellar phase is more difficult to observe because one does not see a two-phase coexistence region between the concentrated L_4 phase and the lamellar phase. This finding as well as most of the effects of bilayer elasticity are consistent with the theoretical predictions of Simons and Cates [107].

In all systems investigated, the L_4 phase is very narrow in terms of the alcohol/surfactant ratio, which means that a fine-tuning of $\bar{k}$ is required to stabilize the L_4 phase. Likewise, for the other discussed phases, the stability and properties of the vesicle phase can be understood as the result of a competition between entropic and elastic contributions to the free energy [41,107]. Taking into account the renormalization of elastic constants by thermal fluctuations, the radius R of unilamellar vesicles has been computed to vary exponentially with the bending energy [41]. Therefore, the experimental measurement of R by light at very low ϕ_m allows one to obtain the elastic coefficient $E_0 = 2k_0 + \bar{k}_0$ for each alcohol studied. As already mentioned, the bending modulus k has been measured by ^{2}H-NMR in the lamellar phase of samples containing water–alcohol–SDS [94]. So, using these k results and the E_0 values, we obtain a rough estimate of $\bar{k}$. As expected, $\bar{k}$ is negative and decreases as the alcohol chain length increases from almost zero for hexanol to $-22k_BT$ for decanol. Thus, in the bilayers considered here, although both constants k and $\bar{k}$ vary in a large range, their elastic energy remains small and all of them form vesicles [41,108,109].

Multilamellar vesicles are found in both the lamellar phase for the low values of R and the L_4 phase. Those dispersed in the lamellar phase with a typical size of several microns are easily detected by optical microscopy; in contrast, the multilayered vesicles found in the concentrated L_4 phase are much more smaller (0.2–0.5 μm) and consist of only a few layers

[108,109]. In some conditions, stable viscous dispersions where small and large multilamellar vesicles coexist can be prepared. These dispersion correspond to two-phase equilibria involving the L_4 and L_α phases. Due to the high viscosity of concentrated L_4 phases, these mixtures do not separate, even after a long time.

In summary, the brine–SDS–alcohol systems form three different phases of membranes with different topologies: a phase of vesicles L_4, a lamellar phase L_α, and a sponge phase L_3. At fixed surfactant concentration, the sequence of phases L_4–L_α–L_{3-w} is obtained by increasing the alcohol content. Thus, the composition parameter R, defined as the molar alcohol/SDS ratio, appears to be the parameter which controls the bilayer structural transformations from closed vesicles to a stack of membranes,

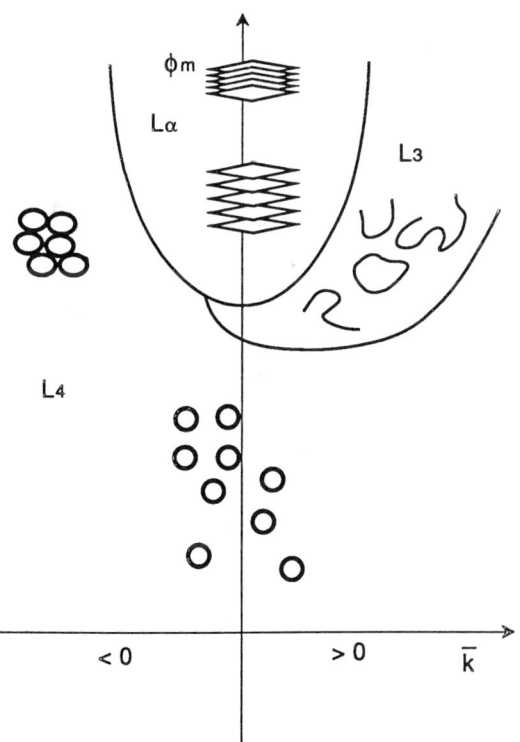

FIG. 24 Schematic phase diagram of membranes. ϕ_m: membrane concentration; $\bar{k}$ Gaussian elastic constant; L_α: lamellar phase; L_3: sponge phase; L_4: vesicle phase. (From Ref. 110.)

and finally to a continuous random surface. Let us emphasize that a similar sequence of phases L_2 (inverse micelles)–L_α–L_{3-0} is found in the oil-rich part of the phase diagrams by decreasing the alcohol content.

These phases of membranes may be considered as phases of surfaces at the thermodynamic equilibrium. The recent theoretical and experimental developments on surfactant solutions allow one to establish a general phase diagram, describing the behavior of phases of membranes as the result of the competition between entropy and elasticity [41,96,97,99,102,107]. As we have seen previously, the latter is expressed as a function of only two parameters, k and $\bar{k}$; k controls the amplitude of the fluctuations and the changes of layer topology (Gauss–Bonnet theorem). The diagram shown in Fig. 24, which corresponds to a fixed value of k, describes the phase behavior in the plane ϕ_m (membrane volume fraction)-$\bar{k}$ [110]. The first consequence of thermal fluctuations is the possibility to dilute a lamellar phase. Indeed, the undulation interaction is responsible for the existence of very swollen L_α phases. Further additions of solvent lead to either sponge phases or vesicle phases, depending on the sign of $\bar{k}$. Then, the concept of topological elastic constant has direct consequences on the phase diagram of interacting membranes. We reach now a point where it is possible to satisfactorily interpret the properties of the phases of surfactant with only two elastic parameters, k and $\bar{k}$. In the SDS–alcohol systems, k is controlled by the alkyl chain length of the alcohol and $\bar{k}$ by its concentration in the interface.

VII. TECHNOLOGICAL RELEVANCE OF MICROEMULSIONS AND AMPHIPHILIC PHASES

Microemulsion formulations were widely used by Australian housewives more than a century before they were described by Schulman en 1940. It was a transparent dispersion of eucalyptus oil in water made by adding soap flakes and white spirit. This formulation was very efficient for the washing of wool. Several commercial products such as liquid waxes, cutting oils and detergent, formulations were patented around 1930. In the seventies, the importance of microemulsions for tertiary oil recovery was the origin of the development of a great number of theoretical and experimental studies of these systems.

The fields of applications of micelles and microemulsions often correspond to those of emulsions and vesicles. The advantage of the first systems is their thermodynamic stability and the very small scale of the microstructure: 10–200 Å. The radii of the emulsion droplets are typically 1 μm which make the system very turbid and unsuitable for some particular applications.

It would be impossible to make a complete catalog of all the industrial and technological applications of microemulsions and amphiphilic phases. Rather than giving a long list of these applications, I have selected a few important examples, emphasizing the particular role played by the disperse medium in each case. Indeed, if most of the applications of reverse micelles are based upon their ability to solubilize substances, they can also be used as a "reservoir" of monomers or to play special roles in absorption processes. Examples of the relevance of these properties are oil recovery, lubrication, detergents, polymerization, preparation of small solid particles, drug delivery etc..

A.   Petroleum Industry—Tertiary Oil Recovery

Due to its enormous economic potential, tertiary oil recovery was of interest well before the energy crisis. Indeed, primary tapping and ordinary water flushing of oil wells allow one to recover only about one-half of the total oil content. Several processes have been considered to improve the recovery. The mechanisms by which microemulsions improve oil recovery have been studied extensively in the past 20 years [15,111]. In order to recover the oil by a flooding process, the interfacial tension between the oil and the injected fluid has to be lower than about 10^{-2} dyne cm^{-1}. In this way, the oil ganglia previously trapped by capillary forces will be displaced, with reasonable values of the injection pressure, and all the residual oil will be recovered.

The crude oil–water interfacial tensions are of the order of a few tens of dynes cm^{-1}. When an appropriate surfactant system is used, the interfacial tension can be reduced to 10^{-3}–10^{-4} dyne cm^{-1}. Although the first studies were made with low concentrations of surfactants, several workers suggested that still lower interfacial tensions could be obtained with high surfactant concentration (microemulsion) systems. The optimum conditions were shown to be satisfied by microemulsions coexisting with both oil and water: middle-phase microemulsions in the so-called Winsor microemulsions [112–116].

The mechanisms of lowering of oil–water interfacial tensions were recently reviewed by Langevin and Meunier [117]. The interfacial tension at a liquid–air interface can never be very small. On the contrary, the interfacial tension at an oil–water interface γ_{ow} can be ultralow, possibly zero, in the thin surfactant layer of high surface pressure (II). In this case, the polar parts of the surfactant molecules are in contact with water, and the hydrophobic chains are in contact with oil: They are able to balance completely the interactions between oil and water. This happens at a particular value of the area per surfactant molecule, a^*, when the interface is said to be "saturated." Then, the system adjusts its surface concentration so that $\Pi(a^*) = \gamma_{ow}$, the interfacial tension between oil in water without surfactant [79].

The zero interfacial tension state is never obtained in practice because the surfactants are always slightly soluble in water and/or in oil. When the saturation state is reached at the interface, micelles (or other aggregates) begin to form in the bulk phases. The interfacial tension γ is then related to the energy difference per unit area between surfactant molecules in the micelles and in the interfacial layer [118]. In the case of microemulsions, two contributions γ_c and γ_e have to be taken into account; γ_c comes from the surfactant bending energy and γ_e from the entropy of mixing. The contribution γ_c has been explicitly calculated by many authors [119–122] De Gennes and Taupin [79] have shown that

$$\gamma_c = \frac{2}{R}\frac{k}{R_0},$$

where k is the mean bending modulus, R is the droplet radius, and R_0 is the monolayer spontaneous radius of curvature. The physical interpretation of this equation is simple: When the plane interfacial area is increased, the minimum cost in energy is obtained if the area per polar head is kept constant. The new plane area must be covered by a surfactant monolayer which is taken from the bulk, that is, from the monolayers around the droplets. The energy needed to unfold the monolayers is $2k/RR_0$ per unit area. This is accompanied by a change in the droplet number which is accounted for by the entropic term γ_e. This change is also accompanied by a change in the topology which introduces a term $\bar{k}/R^2$ so that [79]

$$\gamma_c = \frac{2k + \bar{k}}{RR_0}.$$

The entropic term is less easy to calculate because the exact form of the entropy of mixing of spheres is not known. Several approximations can be found in the literature. They lead to

$$\gamma_e = \frac{k_B T}{R^2}\ln(\alpha\phi),$$

α being a numerical constant depending on the specific approximation employed and ϕ the volume fraction of droplets. It could be noted that the order of magnitude of γ_e is kT/R^2, to be compared to the tension between simple liquids which are of the order of kT/a^2, a being a molecular length. This means that if R is 100 times larger than a, γ will be 10^4 times smaller than usual surface tensions.

This brief description shows the strong importance of the bending elasticity which appears as a second-order term in the surface energy but becomes essential in systems of vanishing surface tension. In the following, we will present experiments performed in a number of Winsor multiphase microemulsions systems, in connection with tertiary oil recovery. The surface tension curves have similar shape in all cases. Consider, for example, the much-studied brine–toluene–SDS–butanol system [114,123,124]. Fig. 25 shows the values of the interfacial tension measured at the oil–microemulsion interface (which is denoted O/M) and at the water–microemulsion interface (W/M) versus the brine salinity S. γ_{OM} decreases with the brine salinity, whereas γ_{WM} increases. At "optimal salinity," S^*, the two surface tensions are equal: $\gamma_{OM} = \gamma_{WM} = \gamma^*$. The interfacial tension was also measured at the oil–water interfaces obtained from the Winsor systems. In the Winsor I and II domains, the surface tension was found to be independent of the degree of dilution of the microemulsion, that is, independent of the presence of droplets in bulk. This proves that the ultralow values of the surface tension at these interfaces arise from the large surface pressure Π in the interfacial monolayer.

In the Winsor III domain, there are two interfaces for each salinity S: an O/M interface and a W/M interface. The larger of the two surface tensions measured at each salinity (γ_{OM} for $S < S^*$ and γ_{WM} for $S > S^*$) is equal to the surface tension γ_{OW} at the O/W interface, obtained by removing the microemulsion phase. As before, this surface tension is independent of the presence or absence of bulk structures in the microemulsion: It results from the pressure Π in the monolayer. The origin of the ultralow surface tension of the second interface (that of lower surface tension) is different. The behavior found is characteristic of the vicinity of a critical consolute point [123,124].

In conclusion, the origin of the lowest interfacial tensions ($\gamma < \gamma^*$) is the vicinity of critical points, whereas that of the highest interfacial tensions ($\gamma > \gamma^*$) is the pressure in the interfacial monolayer. The transition between these two different behaviors, in this particular system, is close to γ^* as evidenced by Fig. 25 [115,116,123,124].

The preceding equations allow to calculate the interfacial tension γ between droplet microemulsions and excess phases and to compare it with the experimental value γ_{exp}. This comparison has been made for two systems (Table 2) [125,126]. Although the calculated values for $\bar{k} = 0$ have the correct order of magnitude and the correct variation with droplet size (salinity), the agreement between measured and calculated tensions is only qualitative. The origin of the discrepancy is not clear.

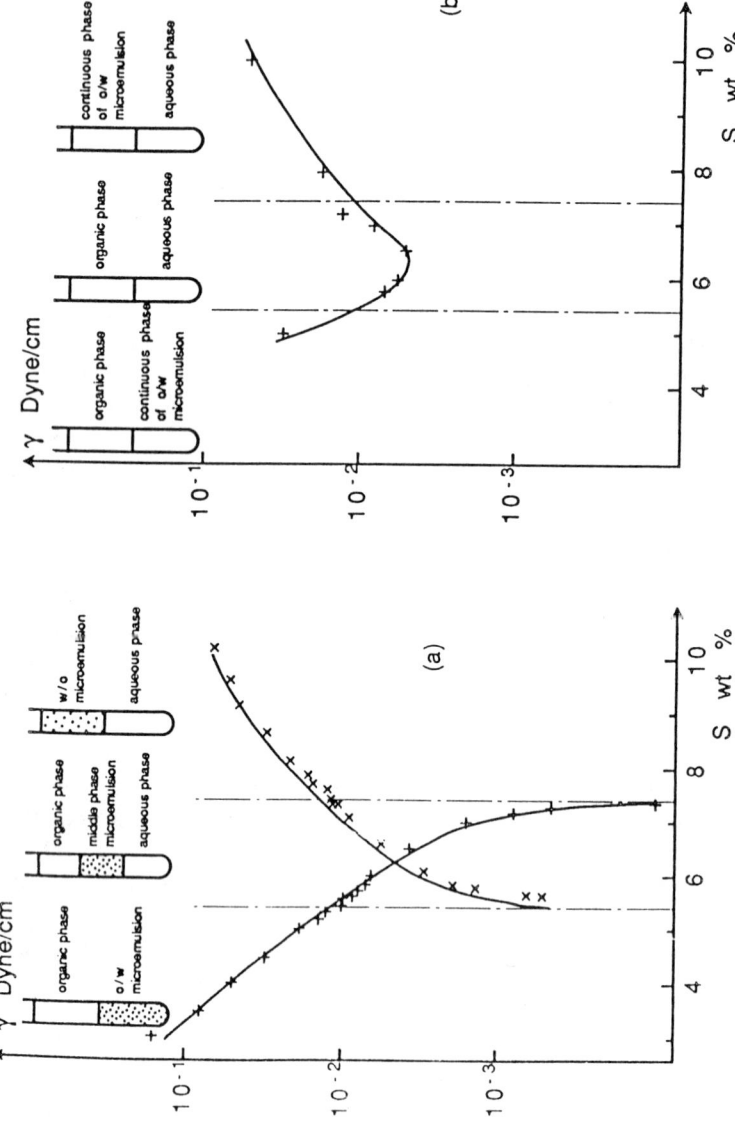

FIG. 25 Toluene–brine–SDS–butanol–system. (a) The interfacial tension γ at the O/M and W/M interfaces versus the salinity S in Winsor equilibria. (b) The interfacial tension γ at the O/W interfaces versus the salinity S. (From Ref. 110.)

TABLE 2

System	S (wt %)	$k/k_B T$	R(Å)	γ_{exp}	γ
SDS	5 (γ_{OM})	0.65	130	2.0×10^{-2}	3.4×10^{-2}
	8 (γ_{WM})	0.65	180	2.3×10^{-2}	1.8×10^{-2}
AOT	0.175	1	69	5.4×10^{-2}	2.0×10^{-1}
	0.474	1	114	5.6×10^{-2}	7.4×10^{-2}

Source: Ref. 117.

As far as oil recovery is concerned, the optimal recovery is obtained when all the tensions are below 10^{-2} dyne cm^{-1}, that is, in practice at the optimal salinity. In principle, the microemulsion can be removed and low surfactant concentrations could be used without lowering the efficiency. But in practice, the surfactants adsorb on the porous rocks and the interfacial tensions become rapidly large. It is then very interesting to use microemulsions in between the oil and water (the microemulsions playing the role of surfactant reservoirs).

It must also be recalled that microemulsions might play a role in other phenomena involved in oil recovery, specially in the wetting of the porous rock (solid-liquid surface tensions) and in the coalescence of oil ganglia (oil bank formation).

B. Polymerization in Organized Media

One of the roles of organized assemblies is to concentrate the reactants in the microenvironments provided by the aggregates. The large diversity of structures produced by amphiphilic systems make these media attractive systems for polymerization. Although much less investigated than emulsion polymerization, a large interest has developed for polymerization in microemulsions [127–135] and more recently in liquid-crystalline phases [136–144]. Preparation of latex in microemulsions allows one to eliminate some disadvantages found in inverse emulsions; namely, the instability of the produced latexes and the high polydispersity of the polymer particles. A process of polymerization of acrylamide within the water core of inverse AOT microemulsions has been described by F. Candau et al.; it leads to latices which are stable transparent, with a small size (*d* < 50 nm) and a narrow size distribution [132,133]. This process was extended to the polymerization of others compounds used in technical applications. The optimization of the systems was by a selection procedure based on the hydrophile–lipophile balance of the surfactant and solubility parameters of the various compounds. Replacement of anionic surfactant by nonionic improved the efficiency. One of the characteristics of the process is the very fast reaction rate a few minutes instead of hours in emulsions. In addition, whatever the initial structure of the organized medium is—dispersed or bicon-tinuous—the final state consists in a uniform dispersion of spherical latex particles. This has been accounted for by the very high flexibility of the surfactant interface and by the conformation of polymer coils which favor the formation of small spheres, produced from the local curvature of the interface. In all cases, the size of the latices is larger than the size of the initial droplets. The latex particles, in fact, result from the merging of 50–100 micelles. This has been attributed to merging of the water pool droplets, inducing the formation of larger aggregates which grow progressively during the polymerization. At the end of the polymerization process, latex particles containing, on average, one high-molecular-weight macromolecule in a collapsed state are formed. This number is of several hundreds in emulsion.

In order to produce structures which are ordered in space (e.g., periodic or anisotropic), polymerization has been performed inside a host structure which acts as a template. Several attempts have been made to obtain new materials using lyotropic phases of different symmetries [136–146]. Potential applications of polymerized lyotropic phases are numerous. Of special interest are the cubic phases, which are bicontinuous systems with a sharply monodisperse pore diameter adjustable in a wide range. Polymerization of this phase could provide microporous materials with potential applications in ultrafiltration, catalysis, or enzyme immobilization, for instance. Laversanne succeeded in polymerizing acrylamide inside the water layers of lamellar, hexagonal, and cubic phases, conserving the liquid-crystalline nature of the lyotropic arrangement [141]. Also, layered materials made of silica polymers separated by surfactant bilayers have been produced through the condensation of silicic acid in the water layers of a collapsed lamellar mesophase [143]. This structure is stable as long as the surfactant template is kept in place. If the surfactant template is destroyed, then the silica polymers collapse or aggregate together into three-dimensional lumps. Attempts to obtain anisotropic silica surfactant composites in a swollen lamellar phase were unsuccessful. Indeed, in this case, the silica polymers had a tendency to segregate out of the lamellar phase, in which case they lost the anisotropy and periodicity of the original lamellar phase [142]. Polymerization was also carried out in other anisotropic structures. Mesoporous silica was synthetized within the water lattice of an hexagonal mesophase [144].

C. Preparation of Ultrafine Particles

Inverse microemulsions have been widely used in material sciences to produce ultrafine particles [145]. These microreactors have served to prepare numerous well-defined nanosized crystallites such as catalysts, semiconductor particles, magnetic colloids, silver chloride particles for photography, and so forth. A first method consists of preparing the particles directly in the micellar system with a suitable chemical reaction. A second method of synthesis takes advantage of the dynamic character of reverse micelles. Upon collision, reverse micelles exchange their water content, which makes possible coprecipitation or chemical reactions from reactants dissolved in separate droplets. Although the size of the crystallites increases with the radius of the water droplet, the relationship between the final size of the particle and the micellar radius is not always clear. Indeed, the final size depends on the experimental conditions in particular on the interactions between the droplets which control the exchange rates of the water content. Generally, reverse micelles are able to limit the size of growing of particles. For instance, when colloidal dispersions of copper metal are prepared in a AOT–water–isooctane microemulsion containing copper ions and hydrazine as a reducing agent, their size increases form 2 to 10 nm upon increasing the water content from 1 to 10 water molecules per AOT molecule. At water contents above 10, the size of the particles remains unchanged, but the polydispersity increases [146]. In this system, the growth of the particles seems to be controlled by the local structure of the micelles at low water content, by the number of copper ions, and by the kinetics of the micellar exchange. By varying the conditions of synthesis, it is possible to produce particles with different shapes. When the reaction occurs in cylindrical reverse micelles, 10% of crystallites formed retains the initial shape of the micelles [146]. Many other metallic particles such as iron, nickel, platinium, rhodium, palladium, iridium, and so on were also prepared in inverse microemulsions [147,148]. Microemulsions also offers interesting possibilities to prepare alloy particles [149].

In photography, as in catalysis, very small and monodisperse particles such as silver chloride are required. An attempt to achieve a better understanding of particle growth has been made on oil–external AOT and nonionic microemulsion by combining x-ray and light-scattering techniques [150]. Two processes were clearly distinguished: crystal formation with a small final diameter (60 Å) and flocculation of these crystals. But the reason for the constant size of the first crystal remains unknown.

In the past few years, there has been much interest in the experimental study of small semiconductor particles in order to understand the photochemistry of surface reactions and the dependence of electronic properties on particle size. The property of exchange between the water cores was used to prepare semiconductors such as CdS, Ag_2S, PbS, $Cd_xZn_{1-x}S$, and so forth. The mixing of two microemulsions of the same size (identical water/surfactant ratio) but containing different reactants in the aqueous cores, for instance $(NO_3)_2Cd$ and Na_2S, lead to the formation of small crystallites with a size ranging from 15 to 50 Å. The polydispersity of these particles strongly depends on the preparation [151–157] and also on the conterion of the surfactant used in the microemulsion. The formation of monodisperse nanoclusters can be very important in the future because nanosized crystallites could present nonlinear physical properties. In the case of CdS particles, nonlinear optical properties have already been observed. This could be extended to other semiconductors differing by their electronic properties.

VIII. CONCLUSION

The multicomponent systems consisting of water–oil–surfactant and alcohol exhibit a very complex phase behavior. In addition to bicontinuous microemulsion phases, these mixtures form sponge phases, vesicle phases, and dilute lamellar phases. All these phases can be considered as phases of fluctuating surfaces for which the bending elastic constant k is of the order of $k_B T$. The considerable theoretical and experimental efforts devoted to the study of these phases have led to an unified interpretation of the structure and stability of these systems. Indeed, one can understand the properties of these phases in terms of a competition between the entropy and the elastic energy expressed as a function of only three parameters: the mean (k) and Gaussian $\bar{k}$ elastic constants and the spontaneous curvature C_0. k controls the amplitude of the thermal undulation modes, whereas $\bar{k}$ governs the topology of the interfaces. The relationships between the molecular parameters (alcohol chain length, area per polar head of the surfactant, alcohol over surfactant ratio, alcane chain length, salinity, etc.) and the elastic constants k and $\bar{k}$ are now clear enough to know how to prepare bicontinuous microemulsions, dilute lamellar phases, sponge phases, or vesicle phases.

ACKNOWLEDGMENTS

A part of the work described in this chapter has been performed at the Centre de Recherche Paul Pascal in collaboration with several scientists; among them I would like to specially thank D. Roux, F. Nallet, and C. Coulon, D. Gazeau, G. Guerin.

REFERENCES

1. P. Ekwall, in *Advances in Liquid Crystals*, Vol. 1 (G. H. Brown, ed.), Academic Press, New York, 1975, p. 1.

2. P. A. Winsor, Chem. Rev. *68*:1 (1968).
3. K. Fontell, in *Liquid Crystals and Plastic Crystals*, Vol. 2 (G. W. Gray and P. A. Winsor, eds., Ellis Howood, Chichester, 1974, p. 80.
4. G. J. T. Tiddy, Phys. Rep. *57*:1 (1980).
5. G. J. T. Tiddy and M. F. Walsh, in *Aggregation Process in Solution*, Vol. 28 (E. Wyn Jones and J. Gornally, ed.), Elsevier, Amsterdam, 1983, p. 151.
6. J. Charvolin, J. Chim. Phys. *80*:15 (1983).
7. T. P. Hoar and Schulman, Nature *152*:102 (1943).
8. A. M. Bellocq and D. Roux, in *Microemulsions: Structure and Dynamics* (S. E. Friberg and P. Bothorel, eds.), CRC Press, Boca Raton, FL, 1987, p. 33.
9. W. M. Gelbart, A. Ben-Shaul, and D. Roux (eds.), *Micelles, Membranes, Microemulsions and Monolayers*, Springer-Verlag, New York, 1994.
10. P. Ekwall, L. Mandell, and K. Fontell, in *Molecular Crystals and Liquid Crystals*, Vol. 8 (P. Ekwall, L. Mandell and K. Fontell, eds.), Gordon & Breach Science Publ., New York, 1969, p. 157.
11. G. Gillberg, H. Lehtinen, and S. Friberg, J. Colloid Interface Sci. *33*:40 (1970).
12. S. E. Ahmad, K. Shinoda, and S. Friberg, J. Colloid Interface Sci. *47*:32 (1974).
13. S. Friberg and I. Burasczenska, in *Micellization, Solubilization and Micro-emulsions*, Vol. 2 (K. L. Mittal, ed.), Plenum Press, New York, 1977, p. 791.
14. R. N. Healy; R. L. Reed, and D. G. Stenmark, Soc. Petrol. Eng. J. *16*:147 (1976).
15. D. O. Shah and R. S. Schecter (eds.), *Improved Oil Recovery by Surfactant and Polymer Flooding* Academic Press, New York, 1977.
16. A. M. Bellocq, J. Biais, P. Bothorel, B. Clin, G. Fourche, P. Lalanne, B. Lemaire, B. Lemanceau, and D. Roux, Adv. Colloid Interface Sci. *20*:167 (1984).
17. A. M. Bellocq, J. Biais, B. Clin, A. Gelot, P. Lalanne, and B. Lemanceau, J. Colloid Interface Sci. *74*:311 (1980).
18. M. Kahlweit, R. Strey, and G. Busse, J. Phys. Chem. *94*:3881 (1990).
19. A. M. Bellocq and D. Roux, in *Progress In Microemulsions* (S. Martellucci and A. N. Chester, eds.), Plenum Press, New York 1985, p. 159.
20. H. Kunieda, K. Nakamura, and A. Uemoto, J. Colloid Interface Sci. *150*: 235 (1992).
21. G. Porte, J. Marignan, P. Bassereau, and R. May, J. Phys. (Paris) *49*:511 (1988).
22. D. Gazeau, A. M. Bellocq, D. Roux, and T. Zemb, Europhys. Lett. *9*:447 (1989).
23. M. Kahlweit, R. Strey, and P. Firman, J. Phys. Chem. *90*:671 (1986).
24. H. Kunieda and K. Shinoda, Bull. Chem. Soc. Japan *56*:980 (1983).
25. H. Kunieda, J. Colloid Interface Sci. *122*:138 (1988).
26. M. Yoshida and H. Kunieda, J. Colloid Interface Sci. *B8*:273 (1990).

27. A. M. Bellocq and D. Gazeau, Prog. Colloid Polym. Sci. 76:203 (1988).
28. L. M. Prince (ed.), Microemulsions. Theory and Practice, Academic Press, New York, 1976.
29. I. D. Robb (ed.), Microemulsions, Plenum Press, New York, 1982.
30. K. L. Mittal (ed.), Micellization, Solubilization and Microemulsions, Plenum Press, New York, 1977.
31. S. Friberg and P. Bothorel (eds.), Microemulsions: Structure and Dynamics CRC Press, Boca Raton, FL, 1987.
32. H. L. Rosano and M. Clausse (eds.), Microemulsion Systems, Marcel Dekker, Inc., New York, 1987.
33. M. Corti and V. Degiorgio (eds.), Physics of Amphiphiles: Micelles, Vesicles and Microemulsions North-Holland, Amsterdam, 1987.
34. A. M. Bellocq, in Physics of Complex and Supermolecular Fluids (S. A. Safran and N. Clark, eds.), Wiley, New York, 1987, p. 41.
35. K. Kekicheff and B. Cabane, J. Phys. 48:1571 (1987).
36. D. Roux and A. M. Bellocq, in Micellar Solutions and Microemulsions (S. H. Chen and R. Jajagopalan, eds.), Springer-Verlag, New York, 1990, p. 251.
37. H. Kunieda and K. Nakamura, J. Phys. Chem. 95:1425 (1991).
38. G. Guérin and A. M. Bellocq, J. Phys. Chem. 92:2550 (1988).
39. F. Auguste, A. M. Bellocq, F. Nallet, D. Roux, and T. Gulik, Langmuir (submitted).
40. G. Porte, J. Appell, P. Bassereau, and J. Marignan, J. Phys. (Paris) 50:1335 (1989).
41. P. Hervé, D. Roux, A. M. Bellocq, F. Nallet, and T. Gulik, J. Phys. II (Paris) 3:1255 (1993).
42. H. Hoffmann, C. Thuning, and U. Munkert, Langmuir 8:2629 (1992).
43. H. Hoffmann, U. Munkert, C. Thuning, and A. Valiente, J. Colloid Interface Sci. 163:217 (1994).
44. W. J. Benton, J. Natoli, S. Qutubuddin, C. A. Miller, and T. Fort Jr., SPE J. 53 (1982).
45. W. J. Benton and C. A. Miller, J. Phys. Chem. 87:4981 (1983).
46. R. Gomati, J. Appell, P. Bassereau, J. Marignan, and G. Porte, J. Phys. Chem. 91:6203 (1987).
47. Ph. Boltenhagen, M. Kleman, and O. D. Lavrentovich, C. R. Acad. Sci. Paris II 315:931 (1992).
48. A. M. Bellocq and D. Roux, in Macro and microemulsions. Theory and Applications (D. O. Shah, ed.), ACS Symposium Series 272, American Chemical Society, Washington, DC 1985, p. 105.
49. D. Roux and A. M. Bellocq, Phys. Rev. Lett. 52:1895 (1984).
50. H. Kunieda, K. Nakamura, and A. Uemoto, J. Colloid Interface Sci. 163: 245 (1994).
51. L. Auvray, in Micelles, Membranes, Émulsions and Monolayers (W. M. Gelbart, A. Ben-Shaul, and D. Roux, eds.), Springer-Verlag, New York, 1994, p. 347.

52. S. Friberg, I. Lapeznski, and G. Gillberg, J. Colloid Interface Sci. *56*:19 (1976).
53. L. E. Scriven, Nature *263*:123 (1976).
54. I. Barnes, S. Hyde, B. Ninham, P. J. Derian, and T. Zemb, Prog. Colloid Polym. Sci. *81*:20 (1990).
55. A. Graciaa, J. Lachaise, A. Martinez, M. Bourrel, and C. Chambu, C. R. Acad. Sci. Paris B *282*:547 (1976).
56. M. Dvolaitzky, M. Guyot, M. Lagues, J. P. Le Pesant, R. Ober, C. Sauterey, and C. Taupin, J. Chem. Phys. *69*:3279 (1978).
57. A. M. Cazabat and D. Langevin, J. Chem. Phys. *74*:3148 (1981).
58. S. Brunetti, D. Roux, A. M. Bellocq, G. Fourche, and P. Bothorel, J. Phys. Chem. *87*:1029 (1983); D. Roux, A. M. Bellocq, and P. Bothorel, in *Surfactants in Solution*, Vol. 3 (K. L. Mittal and B. Lindman, eds.), 1984, p. 1843.
59. T. Dichristina, D. Roux, A. M. Bellocq, and P. Bothorel, J. Phys. Chem. *89*:1433 (1985).
60. A. A. Calje, W. G. M. Agterof, and A. Vrij, in *Micellization, Solubilization, and Microemulsions*, Vol. 2 (K. L. Mittal, ed.), Plenum Press, New York, 1976, p. 779.
61. B. Lemaire, P. Bothorel, and D. Roux, J. Phys. Chem. *87*:1023 (1983).
62. L. Auvray, J. Phys. Lett. *46*:L163 (1985).
63. G. Fourche, A. M. Bellocq, and S. Brunetti, J. Colloid Interface Sci. *89*: 427 (1982).
64. D. Roux and A. M. Bellocq, Chem. Phys. Lett. *94*:156 (1983).
65. R. J. Baxter, J. Chem. Phys. *49*:2770 (1968).
66. A. M. Cazabat, D. Langevin, J. Meunier, O. Abillon, and D. Chatenay, *Macro and Microemulsions Theory and Applications*. (D. O. Smith, ed.) ACS Symposium Series, 272, American Chemical Society, Washington, DC, 1985, p. 75.
67. P. Guering and A. M. Cazabat, J. Phys. Lett. *44*:L601 (1983).
68. R. Zana, J. Lang, O. Sorba, A. M. Cazabat, and D. Langevin, J. Phys. Lett. *43*:L829 (1982).
69. M. Dvolaitzy, M. Lagues, J. P. Lepesant, C. Sauterey, R. Ober, and C. Taupin, J. Phys. Chem. *84*:1532 (1980).
70. M. Lagues, R. Ober, and C. Taupin, J. Phys. Lett. *39*:L-487 (1978).
71. B. Lagourette, J. Peyrelasse, C. Boned, and M. Clausse, Nature *281*:60 (1979).
72. D. Chatenay, W. Urbach, A. M. Cazabat, and D. Langevin, Phys. Rev. Lett. *54*:2253 (1985).
73. A. M. Cazabat, D. Chatenay, P. Guering, and W. Urbach in *Physics of Complex and Supramolecular Fluids* (S. A. Safran and N. Clark, eds.), Wiley, New York, 1987, p. 585.
74. For a review, see B. Lindman and P. Stilbs in Ref. 31, p. 119 and Ref. 32, p. 129.
75. W. Jahn and R. Strey, J. Chem. Phys. *92*:2294 (1988)

76. For a review, see L. Auvray, P. Cotton, R. Ober, and C. Taupin, in *Physics of Complex and Supramolecular Fluids* (S. A. Safran and N. Clark, eds.), Wiley, New York, 1987, p. 449, and Ref. 51.
77. P. Guerin and B. Lindman, Langmuir *1*:464 (1985).
78. W. Helfrich, Z. Naturforsch C *28*:693 (1973).
79. P. G. De Gennes and C. Taupin, J. Phys. Chem. *86*:2294 (1982).
80. J. Jouffroy, P. Levinson, and P. G. De Gennes, J. Phys. (France) *43*:1241 (1982).
81. Y. Talmon and S. Prager, J. Chem. Phys. *69*:2984 (1978).
82. S. A. Safran, D. Roux, M. Cates, and D. Andelman, Phys. Rev. Lett. *57*: 491 (1986); J. Chem. Phys. *87*:7229 (1987); D. Andelman, S. A. Safran, D. Roux, and M. Cates, Langmuir *4*:802 (1988); see also Ref. 9 p. 427.
83. B. Widom, J. Chem. Phys. *81*:1030 (1984).
84. W. Helfrich, J. Phys. (Paris) *46*:1263 (1985).
85. L. Peliti and S. Leibler, Phys. Rev. Lett. *54*:1690 (1985).
86. W. K. Kegel and H. W. Lekkerkerker, J. Phys. Chem. *97*:11124 (1993).
87. S. T. Milner, S. A. Safran, D. Andelman, M. E. Cates, and D. Roux, J. Phys. *49*:1065 (1988).
88. W. Helfrich, Z. Naturforsch. A *33*, 305 (1978).
89. D. Roux, C. R. Safinya, and F. Nallet in Ref. 9, p. 303.
90. C. R. Safinya, D. Roux, G. S. Smith, S. K. Sinha, P. Dimon, N. A. Clark, and A. M. Bellocq, Phys. Rev. Lett. *57*:2718 (1986).
91. D. Roux and C. R. Safinya, J. Phys. (Paris) *49*.307 (1988).
92. C. R. Safinya, E. Sirota, D. Roux, and G. S. Smith, Phys. Rev. Lett. *62*: 1134 (1989).
93. F. Nallet, D. Roux, and J. Prost, J. Phys. (Paris) *50*:3147 (1989); Phys. Rev. Lett. *62*:276 (1989).
94. F. Auguste, P. Barois, L. Fredon, B. Clin, E. J. Dufourc, and A. M. Bellocq, J. Phys. II (Paris) *4*:2197 (1994).
95. B. Halle and P. O. Quist, J. Phys. II (Paris) *4*:1823 (1994).
96. D. A. Huse and S. Leibler, J. Phys. *49*:605 (1988).
97. M. E. Cates, D. Roux, D. Andelman, S. Milner, and S. Safran, Europhys. Lett. *5*:733 (1988).
98. R. Strey, W. Jahn, G. Porte, and P. Bassereau, Langmuir *6*:1635 (1990).
99. D. Roux, C. Coulon, and M. E. Cates, J. Phys. Chem. *96*:4174 (1992).
100. D. Roux, M. E. Cates, U. Olsson, R. G. Ball, F. Nallet, and A. M. Bellocq, Europhys. Lett. *11*:229 (1980).
101. C. Coulon, D. Roux, and A. M. Bellocq, Phys. Rev. Lett. *66*:1709 (1991).
102. G. Porte, J. Appell, P. Bassereau, and J. Marignan, J. Phys. (Paris) *50*:1335 (1989).
103. D. Huse and S. Leibler, Phys. Rev. Lett. *66*:437 (1991).
104. C. Vinches, C. Coulon, and D. Roux, J. Phys. II (Paris) *4*:1165 (1994).
105. C. Coulon, A. M. Bellocq, F. Nallet, D. Roux, C. Vinches, and I. Alibert, J. Phys.
106. M. Filali, G. Porte, J. Appell, and P. P. Feuty, J. Phys. II (Paris) *4*:349 (1994).

107. B. D. Simons and M. E. Cates, J. Phys. II (Paris) *2*:1439 (1992).

108. F. Auguste, A. M. Bellocq, D. Roux, F. Nallet, and T. Gulik, J. Phys. II

109. F. Auguste, A. M. Bellocq, D. Roux, F. Nallet, and T. Gulik, Prog. Colloïd Polym. Sci. *98*:276 (1995).

110. D. Roux and S. Candau, *Images de la Recherche: les systèmes moléculaires organisés*, ed. CNRS, 1994, p. 29.

111. D. O. Shah (ed.), *Surface Phenomena in Enhanced Oil Recovery*, Plenum Press New York, 1979.

112. R. N. Healy, R. L. Reed, and D. G. Stenmark, SPE J. *16*:147 (1976).

113. See also Refs. 15–20 and 177–187 in Ref. 16.

114. A. Pouchelon, J. Meunier, D. Langevin, D. Chatenay, and A. M. Cazabat, Chem. Phys. Lett. *76*:277 (1980); J. Colloid Interface Sci. *82*:418 (1981).

115. A. M. Bellocq, D. Bourbon, G. Fourche, and B. Lemanceau, J. Colloid Interface Sci. *89*:427 (1982).

116. P. D. Fleming, J. E. Vinatieri, and G. R. Glensman, J. Phys. Chem. *84*: 1525 (1980).

117. D. Langevin and J. Meunier in Ref. 9, p. 485.

118. J. Israelachvili, in *Surfactants in Solution*, Vol. 4. (K. Mittal and P. Bothorel, eds.), Plenum Press, New York, 1986.

119. M. Robbins, in *Micellization, Solubilization and Microemulsions* (K. Mittal, ed.), Plenum Press, New York, 1976.

120. C. A. Miller, R. Hwan, W. J. Benton, and T. Fort, J. Colloid Interface Sci. *61*:554 (1977).

121. C. Hu, J. Colloid Interface Sci. *71*:408 (1979); *97*:201 (1984).

122. B. W. Ninham and D. J. Mitchell, J. Phys. Chem. *87*:2996 (1983).

123. A. Cazabat, D. Langevin, J. Meunier, and A. Pouchelon, J. Phys. Lett. *43*: L-89 (1982).

124. A. M. Cazabat, D. Langevin, J. Meunier, and A. Pouchelon, Adv. Colloid Interface Sci. *16*:175 (1982).

125. B. P. Binks, J. Meunier, O. Abillon, and D. Langevin, Langmuir *5*:415 (1989).

126. D. Langevin, D. Guest, and J. Meunier, Colloids Surf. Sci. *19*:159 (1986).

127. J. O. Stoffer and T. Bone, J. Polym. Sci. *18*:2641 (1980).

128. S. S. Atik, J. Am. Chem. Soc. *104*:5868 (1982).

129. Y. S. Leong and F. Candau, J. Phys. Chem. *86*:2269 (1982).

130. A. Jayakrishnan and D. O. Shah, J. Polym. Sci. Polym. Lett. Ed. *22*:31 (1984).

131. H. I. Tang, P. L. Johnson, and E. Gulari, Polymer *25*:1357 (1984).

132. F. Candau, Y. S. Leong, G. Pouyet, and S. J. Candau, J. Colloid Interface Sci. *101*:167 (1984).

133. F. Candau, in *Encyclopedia of Polymer Science and Engineering*, Vol. 9 (M. Crayon and J. Kroschwitz, eds.), Wiley, New York, 1987.

134. P. L. Kuo, N. J. Turro, C. M. Tseng, M. El Aasser, and J. W. Vanderhoff, Macromolecules *20*:1216 (1987).

135. C. Schauber and G. Riess, Makromol. Chem. *190*:725 (1989).

136. J. Herz, F. Reiss-Husson, P. Remp, and V. Luzzati, J. Polym. Sci. Part C 4:1275 (1966).
137. R. Thumathil, J. O. Stoffer, and S. Fribert, J. Polym. Sci. Polym. Chem. Ed. 18:2629 (1980).
138. S. E. Friberg, C. S. Wohn, and F. E. Lockwood, Macromolecules 20:2057 (1987); Y. Bing and G. A. Campbell, J. Polym. Sci. Polym. Chem. Ed. 28: 3575 (1990).
139. D. M. Anderson and P. Störm, in Polymer Association Structures: Micro-emulsion and Liquid Crystals (M. E. Nokaly, ed.), ACS Symposium Series 384, American Chemical Society, Washington D.C., 1989, p. 204.
140. C. Holtzscherer, J. C. Wittman, D. Guillon, and F. Candau, Polymer 31: 1978 (1990).
141. R. Laversanne, Macromolecules 25:489 (1992).
142. M. Dubois, Th. Gulik, and B. Cabane, Langmuir 9:673 (1993).
143. M. Dubois and B. Cabane, Langmuir 10:1615 (1994).
144. C. Kresge, M. Leonowicz, W. Roth, J. Vartuli, and J. Beck, Nature 359 (1992).
145. M. P. Piléni, J. Phys. Chem. 97:6961 (1993).
146. I. Lisiecki and M. P. Piléni, J. Am. Chem. Soc. 115:3887 (1993).
147. M. Boutonnet, C. Anderson, and R. Carsson, Acta Chem. Scand. A 34: 639 (1980).
148. M. Boutonnet, J. Kisling, P. Stenius, and G. Maire, Colloids Surf. 5:209 (1982).
149. R. Touroude, P. Girard, G. Maire, J. Kizling, M. Boutonnet-Kizling, and P. Stenius, Colloids Surf. 67:9 (1992).
150. M. Dvolaitzky, R. Anthare, X. Auvray, C. Petipas, C. Taupin, and C. Williams, J. Disper Sci. Technol. 4:29 (1983).
151. M. L. Steigewald, A. P. Alirisatos, J. M. Gibson, T. D. Harris, R. Korten, A. J. Muller, A. M. Thoyer, T. M. Duncan, D. C. Douglas, and L. E. Brus, J. Am. Chem. Soc. 110:3046 (1988).
152. C. Petit and M. P. Piléni, J. Phys. Chem. 92:2282 (1989).
153. P. Lianos and J. K. Thomas, Chem. Phys. Lett. 125:299 (1986).
154. P. J. Atkinson, M. J. Grimson, R. K. Heenan, A. M. Howe, and B. H. Robinson, J. Chem. Soc. Chem. Commun. 1807 (1989).
155. C. Petit, P. Lixon, and M. P. Piléni, J. Phys. Chem. 94:1598 (1990).
156. L. Motte, C. Petit, P. Lixon, L. Boulanger, and M. P. Piléni, Langmuir 8: 1049 (1992).
157. K. Kurihara, J. Kizling, P. Stenius, and J. H. Fendler, J. Am. Chem. Soc. 105:2574 (1983).

4

Emulsions and Emulsion Stability

PATRICK J. BREEN   Fluid Separation Research, Baker Performance Chemicals, Inc., Houston, Texas

DARSH T. WASAN, YOUNG-HO KIM, and ALEX D. NIKOLOV   Department of Chemical Engineering, Illinois Institute of Technology, Chicago, Illinois

C. S. SHETTY   Nalco Chemical Company, Naperville, Illinois

I. INTRODUCTION

Methods currently available for demulsification can be broadly classified as chemical, electrical, and mechanical. Chemical demulsification is the most widely applied method of treating crude-oil emulsions. It involves the use of chemical additives to accelerate the emulsion-breaking process.

The breaking of a petroleum emulsion is of considerable importance and a necessity for several reasons. Obviously, the quality of the crude oil is highly dependent on residual contents of water and water-soluble contaminants present. Even small amounts of these components can cause unwanted effects in pipelines and refineries. The formulation of the optimum demulsifier for a specific petroleum emulsion is a complicated undertaking. In petroleum systems, asphaltenes and resinous substances comprise a major portion of the interfacially active components of the oil [1,2]. They are large polyaromatic and polycyclic condensed-ring compounds containing heteroatoms. Chemically, they represent the pentane- or hexane-insoluble portion of the oil.

The addition of polyfunctional demulsifying agents to petroleum emulsions serves the same purpose as polymeric electrolytes in aqueous systems. These compounds are adsorbed at the interfaces of droplets and promote coalescence. It is well to remember, however, that the actual demulsifier polymers responsible for emulsion destabilization are usually not very soluble in aqueous phases or aliphatic oils. As most crude oils are more aliphatic than aromatic, the demulsifier is usually dissolved in an aromatic solvent in order to achieve a rapid dispersion throughout the emulsion and, hence, speed the delivery of the demulsifier to the interface. A high degree of agitation will also speed this process and is desirable.

Demulsifiers are most always preferentially oil-soluble blends of several polymer components which generally fall into one of four classes: (a) flocculants, (b) coalescers, (c) wetting agents, and (d) cosolvents. Typically, demulsifier components include polymerized alkoxylated polyglycols, polyglycol esters, and alkoxylated phenol-formaldehyde resins.

The process of demulsification itself is complex and cannot be thought of as the reverse of emulsification [3]. It involves settling, flocculation, and coalescence of drops and it requires disrupting the stabilized layers of the interface.

According to Bancroft [4], the stability of any emulsion is largely due to the nature of the interfacial film that is formed. The stability of this film is strongly dependent on the surfactant adsorption–desorption kinetics, solubility, and interfacial rheological properties such as elasticity and viscosity.

As dispersed droplets gravity-settle and flocculate in a crude-oil emulsion, a biliquid foam structure eventually forms [5], with thin films of oil separating the polyhydric cells of water. The formation and eventual rupture of these structures is governed by hydrodynamic forces between the water drops which depend on such factors as drop size and polydispersity, the presence of asphaltene–resin particles and their suspension rheology, and the rheology of the oil–water interface. A complete analysis of the factors governing the hydrodynamic film stability is available [6]. It has been observed [7,8], however, that demulsifiers speed the rate of film thinning and shorten the time it takes for the film to rupture, although exactly why this is the case remains unclear. Demulsifiers have also been observed to reduce the interfacial viscosity [9], although the ability to do so does not always correlate with performance data.

During the hydrodynamic droplet–droplet interactions, the approaching droplet interfaces behave more or less as deformable single interfaces. At high film thickness (about 1 μm), at first a dimple forms. When the thinning film ruptures at high film thickness, correlations between film stability and the rheological behavior of the single interface can be found [10]. Besides film rheological factors, however, two other major film-stabilizing mechanisms can be present when asphaltenes, resins, or fine solids are present in significant concentration in the crude oil and at the interface. These involve (a) steric stabilization by adsorbed asphaltenes, resins, and/or solids at the interface and (b) the formation of ordered long-range particle structures within the oil film. The various aspects of these three stabilizing mechanisms are the topic of discussion in the next three sections.

II. EFFECTS OF THIN-FILM RHEOLOGY AND DRAINAGE ON DEMULSIFICATION

A. A Model for Film Drainage

The overall coalescence process in demulsification can be conveniently divided into (1) movement of two single noninteracting droplets, (2) deformation of mutual approaching droplets and formation of a plane-parallel film (Fig. 1), and (3) thinning of that film to a critical thickness at which point the film becomes unstable and ruptures, resulting in the droplets combining into a single larger droplet.

The approach of two droplets under the capillary pressure acting normal to the interface causes liquid to be squeezed out of the film into the bulk. This liquid flow results in the convective flux of surfactant in the sublayer (Fig. 2). Therefore, the surfactant concentration at the interface

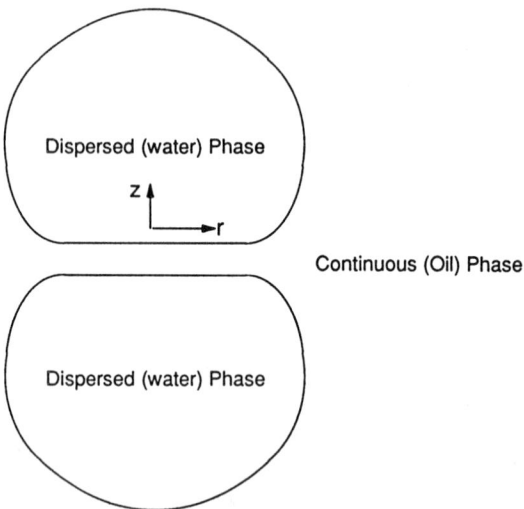

FIG. 1 Mutual approach of two droplets and subsequent formation of plane-parallel film.

is increased in the direction of that flow. The other fluxes associated with the drainage process shown in Fig. 2 include: (1) bulk flux in the droplet, (2) bulk flux in the film phase, and (3) interfacial diffusion flux caused by the concentration gradient at the interface. The bulk fluxes can be conveniently divided into two subsequent steps: (1) diffusional flux up to

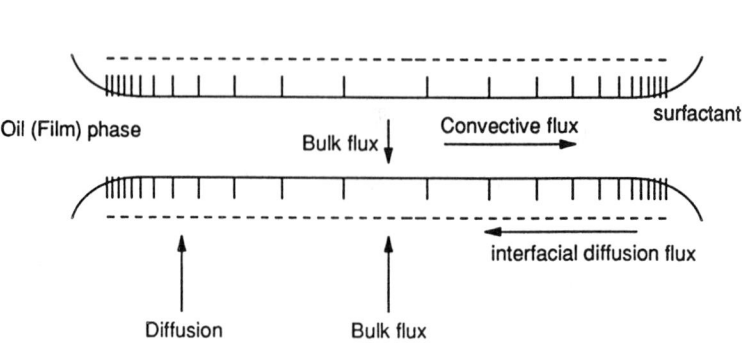

FIG. 2 Surfactant mass balance at the film interface.

a layer adjacent to the film interface and (2) adsorption flux from this layer onto the interface.

The tangential stress balance at the film interface is shown in Fig. 3. The nonuniform surfactant distribution leads to interfacial flow, which, in turn, gives rise to interfacial stress. The difference in concentration along the interface results in differences in the local values of interfacial tension, producing a force (equal per unit length to the gradient in interfacial tension) opposite to liquid flow (Marangoni–Gibbs effect). In addition, the surfactant monolayer may undergo dilating and shearing deformation which also produces interfacial stresses. The sum of the above stresses and the tangential bulk stress from the liquid in the droplet must counterbalance the tangential bulk stress from the film liquid which causes interfacial flow [11].

Reynolds [12] was the first to study the rate of approach between surfaces separated by a draining thin film. His analysis assumed that the two surfaces were both flat and rigid. His expression for the rate of film drainage is presented as Eq. (1):

$$V_{RE} = -\frac{dh}{dt} = \frac{2\pi F h^3}{3\eta A^2}, \tag{1}$$

where F is the force bringing the two forces together, h is the distance between the surfaces, η is the viscosity of the fluid between the surfaces, and A is the area of the surfaces. As has been often pointed out [6], Eq. (1) represents a most conservative prediction. Due to the mobility of the interfaces, the rate of film thinning, and thus the coalescence rate, can be several times greater than predicted by this equation.

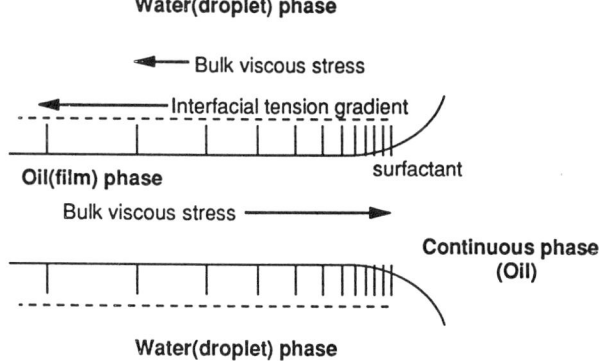

FIG. 3 Tangential stress balance and the film interface.

More recently, a generalized model was developed [11] which accounts for the effect of the mobility of the interfaces on film-thinning phenomena by considering the kinetics of adsorption–desorption of surfactants, partitioning of surfactants, interfacial and bulk diffusion, interfacial rheological properties, and flow in both film and bulk phases. The interfacial mobility was described using Eq. (2):

$$\frac{V}{V_{RE}} = 1 - \frac{3\pi}{h^2 F} \sum_{n-1}^{N} A_n \frac{e^{\lambda_n h}}{1 + \lambda_n h} J_0(\lambda_n), \tag{2}$$

where V is the velocity, V_{RE} is the film drainage velocity from Eq. (1), h is the film thickness, F is the force exerted by the film on the interface, A_n are constants, J_0 is the zero-order Bessel function, and λ_n is the nth root of J_0.

B. Marangoni–Gibbs Effects

Wasan et al. examined the impact of Marangoni–Gibbs effects on coalescence. In Fig. 4, the effect of the interfacial tension gradient on interfacial

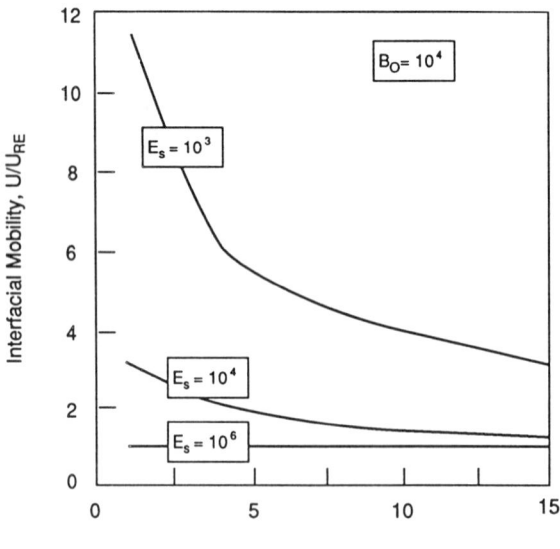

FIG. 4 Interfacial mobility, or dimensionless drainage velocity, versus dimensionless film thickness, at three values of the dimensionless interfacial elasticity.

mobility is shown in terms of the dimensionless elasticity number, E_s. The effect of gradients in interfacial tension on the film drainage time is depicted in Fig. 5. Large gradients in interfacial tension, corresponding to high E_s, cannot be counterbalanced by bulk and interfacial diffusion, and the velocity of film thinning (or the drainage time) is essentially given by Eq. (1). For smaller values of E_s, however, even at a moderate interfacial viscosity (i.e., moderate Boussinesq number), the thinning or approach velocity is several times greater than velocity predictions of Eq. (1). An increase in interfacial viscosity results in decreased interfacial mobility and higher drainage time. Thus, the thin-film drainage model predicts that at low values of interfacial viscosity (i.e., Boussinesq number <10), the Marangoni–Gibbs effect will become the more significant influence on film drainage and the droplet coalescence rate [7,11]. These theoretical findings clearly suggest that in the emulsion coalescence process, the interfacial viscosity must not only be low but gradients in interfacial tension must also be minimized. Such conclusions are reinforced by a number of other works, such as a study published by Hartland and Jeelani [13], who modeled film drainage and surface mobility using calculations of velocity profiles and interfacial tension gradients. Tambe et al. [14] concluded, on the basis of work using decane–water emulsions stabilized with graphite or stearic acid, that effective demulsifiers produce short

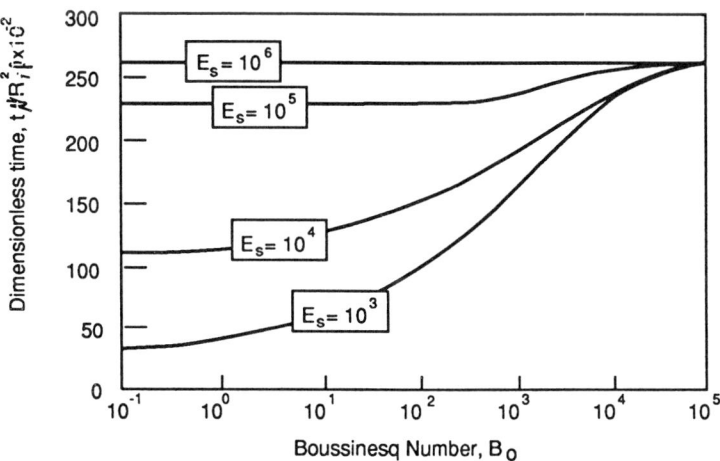

FIG. 5 Dimensionless drainage time for the film to drain from a dimensional thickness, $\bar{h}_i$, to the thickness $\bar{h}_f$, versus Boussinesq number, at various values of the dimensionless interfacial elasticity. (From Ref. 6.)

relaxation times for sudden changes in interfacial tension and significantly lower the dilatational viscoelasticity of the interface. In yet another study, Bhardwaj and Hartland [15] presented data supporting the conclusion that the adsorption of demulsifier to a crude-oil–water interface reverses the interfacial tension gradient and enhances film drainage.

Two factors that significantly influence the magnitude of interfacial tension gradients in the thin film are surface diffusion and surface adsorption. In Fig. 6, drainage time is plotted against the dimensionless interfacial viscosity (Boussinesq number) for various values of the surface diffusivity number. As the number increases, corresponding to increased surface diffusion, the drainage time diminishes, which is evidence that the film surfaces have become more mobile. The cause for this effect may be understood by reference to Fig. 2. If a large surface diffusion occurs in the film surface, the surfactant may diffuse back into the film, diminishing the surface concentration gradient and canceling the Marangoni–Gibbs effect.

A similar effect may be achieved by increasing the rate of surface adsorption to the film surfaces from the dispersed phase. This effect is illustrated in Fig. 7, which reveals that an increased adsorption rate results in a diminished time of drainage, again owing to a cancellation of Marangoni flow

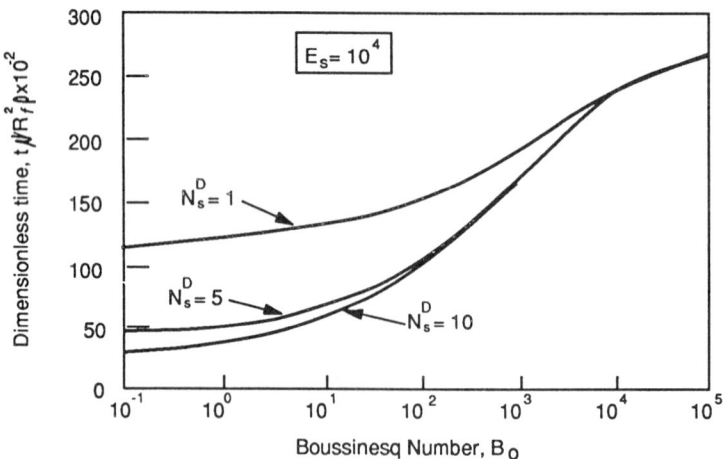

FIG. 6 Dimensionless drainage time for the film to drain from a dimensional thickness, h_i, to the thickness, h_f, versus Boussinesq number, at various values of the surface diffusivity number. (From Ref. 6.)

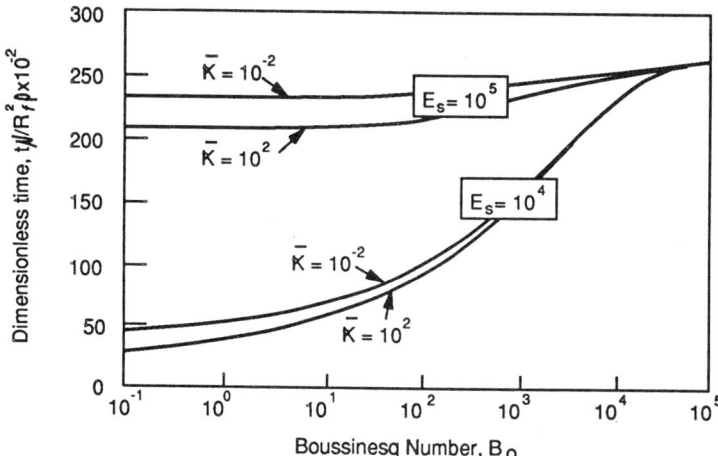

FIG. 7 Dimensionless drainage time for the film to drain from a dimensional thickness, h_i, to the thickness, h_f, versus Boussinesq number, at various values of the rate of surfactant adsorption. (From Ref. 6.)

Figures 8 and 9 compare the effect of selective surfactant solubility on interfacial mobility and drainage time, respectively, for the cases where (1) the surfactant is soluble only in the film phase and (2) the surfactant is soluble only in the drop phase. It is observed that solubilized surfactant in the dispersed phase is most effective in destabilizing droplets with the consequence that emulsions with surfactant soluble in the droplet phase are generally less stable than emulsions with surfactant soluble in the continuous phase, all other things being equal. This conclusion, which is clearly illustrated in Fig. 9, is known as "Bancroft's Rule." The higher interfacial mobility (Fig. 8) and lower drainage times (Fig. 9) are obtained from the case where surfactant is soluble in the droplet phase, whereas lower interfacial mobility and higher drainage times are obtained when the surfactant is soluble only in the film (continuous) phase.

C. Interfacial Studies Using Crude-Oil Emulsions

Kim [16] and Wasan [17] used photomicrograph observations of a water–in–crude-oil emulsion to determine coalescence rates in the presence of a series of different demulsifiers which were ethylene oxide/propylene oxide (EO/PO) copolymers (e.g., RE-1747, RE-1748, RE-1250, and RE-1751) and ethoxylated resins (e.g., RE-1756). Figure 10 is a plot of

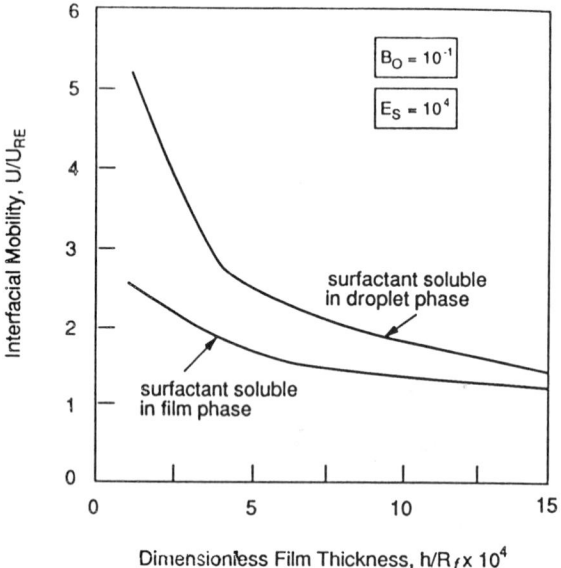

FIG. 8 Interfacial mobility versus dimensionless film thickness, for surfactant soluble in film and droplet phases only.

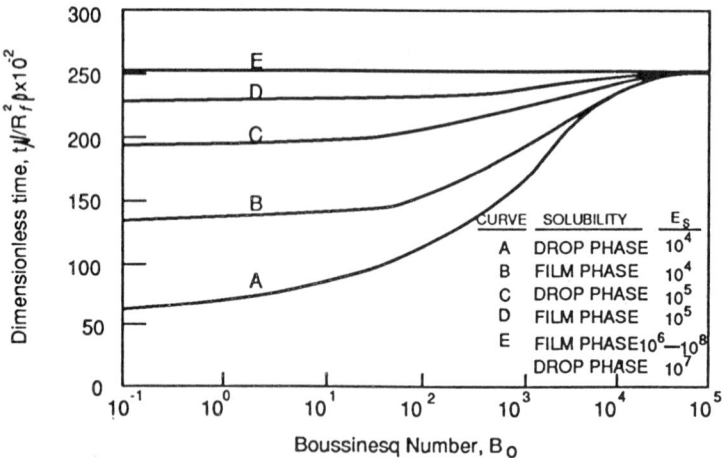

FIG. 9 Dimensionless drainage versus Boussinesq number for surfactant soluble in film and droplet phases.

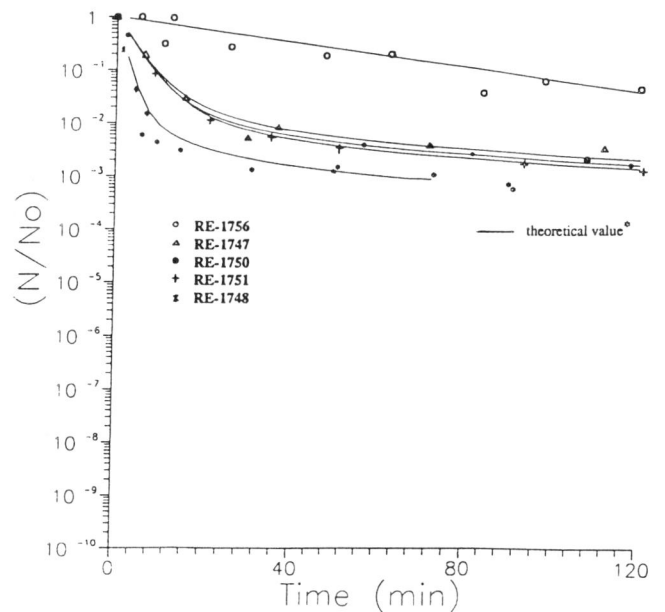

FIG. 10 Dimensionless droplet number versus time for various demulsifiers in the water-in-crude-oil emulsion.

the normalized number of water droplets (N/N_0) as a function of time t for these various emulsion breakers (EB). The concentration of each of the demulsifiers used was 100 ppm. The coalescence rate coefficient (K) and flocculation coefficient (a) were calculated using a theoretical model for emulsion stability developed by Borwankar et al. [18]. Figure 10 shows that, initially, the demulsification process is coalescence rate controlled; however, with the passage of time, the process becomes flocculation rate controlled (see Table 1).

Furthermore, the bottle tests for these systems confirmed the one-to-one correlation between the coalescence rate constants (K) and the demulsifier effectiveness: The best demulsifier, RE-1748, had the highest coalescence rate constant.

Kim [16] and Wasan [17] also measured the interfacial activities of the four emulsion breakers used in the study. The interfacial activity was determined from the slope of the interfacial tension isotherms. In this case, the crude oil containing an emulsion breaker was preequilibrated with water for 3–4 days. The equilibrated water was first separated from the crude oil and then recontacted with crude which did not contain demul-

TABLE 1 Constants of Coalescence and Flocculation Rates in Kinetics of Emulsions Stability

	Demulsifier				
	RE-1748	RE-1750	RE-1747	RE-1751	RE-1756
K	0.746	0.306	0.298	0.320	0.003
$a \times 10^8$	0.765	0.925	1.423	1.974	8.580

Note: K = coalescence rate constant; a = flocculation rate constant.

sifier. The static interfacial tension of this preequilibrated water against the pure crude oil was measured as a function of the water dilution ratio. Table 2 lists the coalescence rate constants and the interfacial activities for the four emulsion breakers. A perfect one-to-one correlation exists between the interfacial activity (δ_i) of the demulsifier, the coalescence rate constant and performance. The dynamic interfacial activity was also measured using the maximum droplet-pressure method [19] for the best and worst demulsifier and it was found that the dynamic interfacial activity was also highest for the best demulsifier, RE-1748, and lowest for the worst emulsion breaker, RE-1756.

In addition to measuring the coalescence rate constants and interfacial activity, partition coefficients were also determined for various demulsification systems. In these experiments, the model oil containing asphaltenes (1 g L^{-1}) in heptol (70% heptane, 30% toluene) was used. Table 3 lists the partition coefficients (K_p) for a series of demulsifiers which are dispersed mixtures of alkylaryl sulfonate, phenolic resins, and polyamines. The partition coefficient is defined as a ratio of the concentration of demulsifier in the water phase to that in the oil phase. In this set of experiments, a

TABLE 2 Coalescence Rate Constants and Interfacial Activities for Various Demulsifiers in Water-in-Crude-Oil Emulsions

	Demulsifier			
	RE-1748	RE-1753	RE-1868	RE-1756
δ_i	1.97	1.05	0.63	0.15
Coalescence rate constant (K)	0.75	0.56	0.32	0.01
Performance	1	2	3	4

TABLE 3 Partition Coefficients and Interfacial Activities for Various Demulsifiers in Water-in-Crude-Oil Emulsions

	Demulsifier				
	RP-4011	R-77	RP-2327	RP-484	No EB
K_p	0.40	0.19	0.19	0.16	0.0
δ_i	2.11	1.03	0.65	0.55	0.05
Performance	1 (0.5 min)	2 (2 min)	3 (10 min)	4 (days)	5

model oil containing demulsifier was contacted with fresh water and the preequilibrated oil was then separated and contacted with fresh water. It can be seen that the best demulsifier, RP-4011, also had the highest partition coefficient. This demulsifier also had the highest interfacial activity, and a strong correlation appears to exist between performance, partition coefficient, and interfacial activity.

Kim et al., developed a novel technique for measuring rheological properties of a water–crude oil–water film [20]. Using this technique, they were able to demonstrate a strong connection between good demulsifier performance (fast coalescence) and dynamic rheological properties, using samples of a Mississippi crude-oil emulsion and the associated brine. The oil had an API gravity of ~21, and extractions with pentane revealed an asphaltene content of 30.5% and a resin content of 35.0%. Demulsifiers used were of the polypropylene oxide type and were obtained from Baker Performance Chemicals, Inc. Table 4 lists the average molecular weights of the demulsifiers used.

Kim et al. [20] carried out dynamic-film-tension measurements by forming an oil film at the tip of a capillary, and then expanding the equilibrated film at a rate of 2.7×10^{-2} mm^2 s^{-1}. This rate corresponds to the rate

TABLE 4 Demulsifier Molecular Weights

Demulsifier	M_w	M_n
RE-2306	40,067	3,985
RE-2307	56,767	7,143
RE-2308	3,900	3,230
RE-2309	79,547	6,227

of film thinning predicted by the Reynolds equation [21]:

$$t = \frac{3\eta r_f^2}{4\Delta p h^2},$$ (3)

where t is the rate of film thinning, Δp is the capillary pressure, h is the film thickness, η is the bulk viscosity of the continuous phase, and r_f is the radius of the circular film between two contacting droplets.

Dynamic-film-tension measurements for several demulsifiers are presented in Fig. 11. The slope of the curves in this figure is the dilatational modulus or dynamic film elasticity, E, as defined by Eq. (4):

$$E = \frac{\delta \gamma}{\delta \ln(A/A_0)},$$ (4)

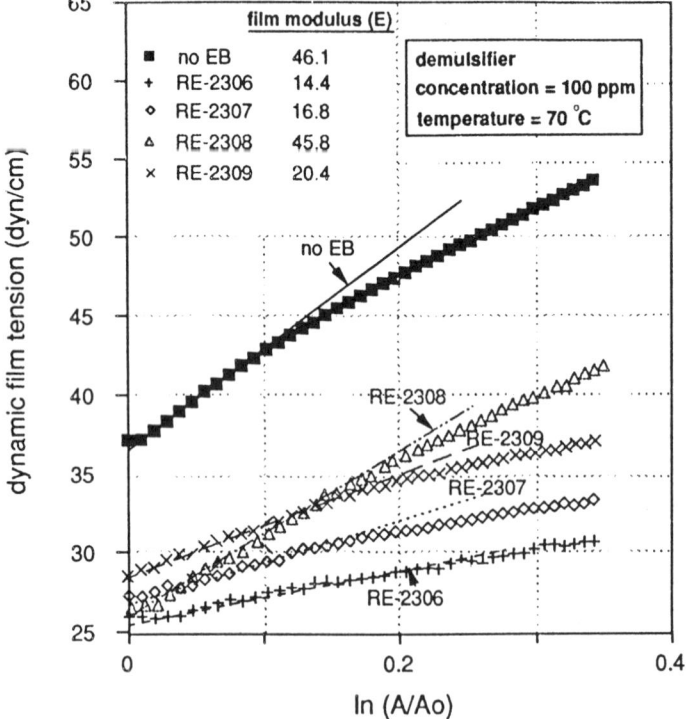

FIG. 11 Film tension as a function of film area at 100 ppm demulsifier, 70°C. The initial slope represents film modulus at 2.7×10^{-2}-mm^2 s^{-1} film expansion rate.

where γ is the dynamic-film tension, A is the expanded film area, and A_0 is the initial film area at zero time. The measured film dilatational modulus, as well as demulsifier performance efficiency, are presented in Table 5. These measurements demonstrate that the efficiency of a demulsifier seems to be inversely related to the dynamic elasticity of the interface.

Kim et al. [20] also conducted film stress–relaxation measurements by rapidly expanding an oil film, with 100 ppm of demulsifier in the oil phase, then recording the film capillary pressure, proportional to the dynamic film tension, as a function of time after the expansion had stopped. Experimental results for the film stress–relaxation experiments are presented in Fig. 12 as plots of film tension versus time. The curves exhibit a region of relatively fast relaxation (<0.2 min.), followed by a much slower period of relaxation. The steepness of the early, fast section of these curves correlates with the performance data in Table 5, where approximated slopes for these early regions are also listed.

The data in Table 6 clearly illustrate the need for a demulsifier to exhibit fast-relaxation kinetics in order to achieve rapid coalescence. This corresponds to a low elasticity of the interface, which should be related to the diffusivity and dynamic interfacial activity of the demulsifier. Fast diffusivity and high dynamic interfacial activity should be especially important for the case of a thinning oil layer between coalescing droplets where demulsifier transport from the bulk will be opposed by the outward drainage [22].

Equilibrium, or static, conditions are achieved for the curve in Fig. 12 after 30–60 mins, and the resultant film tensions are treated as static film tensions. The relative values of these static film tensions do not correlate with performance.

TABLE 5 Data Summary for Demulsifier Performance, Film Modulus and Initial Slope in Film Stress–Relaxation

Demulsifier	Efficiency[a] (%)	Film modulus (dyn cm^1)	Initial slope[b] (dyn cm^{-1})
RE-2306	86	14.4	0.89
RE-2307	42	16.8	0.38
RE-2308	4	45.8	0.15
RE-2309	15	20.4	0.28

[a] Water separation as volume percent after 6 h.
[b] Initial slope from Fig. 12.

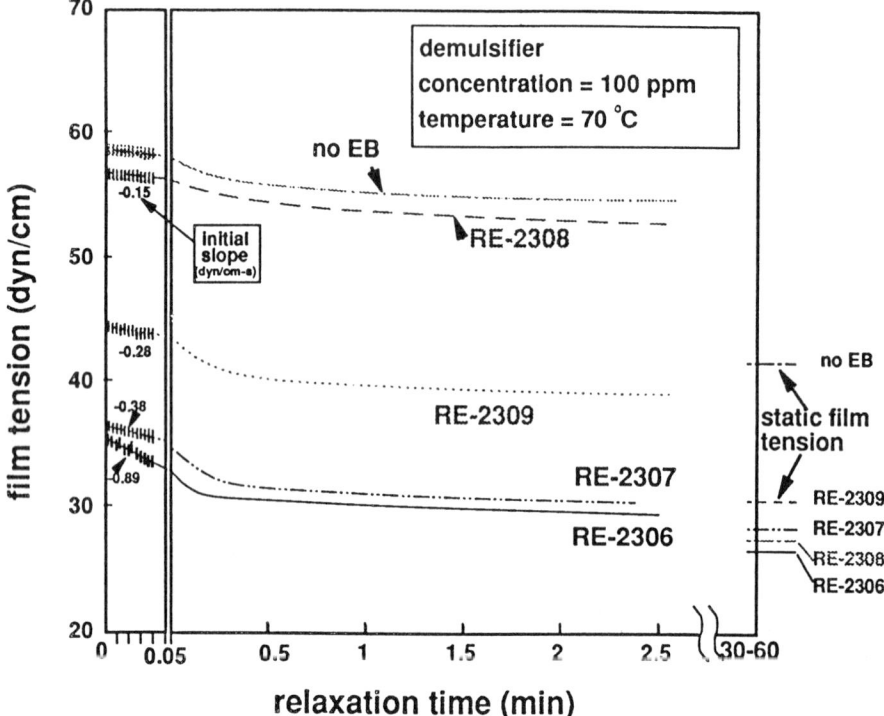

FIG. 12 Film relaxation: film tension as a function of relaxation time at 100 ppm demulsifier concentration, 70°C.

TABLE 6 Static Interfacial Tension and Activity at 100 ppm Demulsifier, 70°C

Demulsifier	Efficiency (%)	Static tension (dyn cm^{-1})	Static activity (dyn cm^{-1})	Dynamic activity (dyn cm^{-1})	Diffusivity, D (cm^2 s^{-1})
RE-2306	86	13.1	1.6	1.6	5.2×10^{-5}
RE-2307	42	13.7	1.9	1.3	5.0×10^{-5}
RE-2308	4	13.4	0.8	0.8	3.2×10^{-9}
RE-2309	15	14.4	1.4	1.2	3.1×10^{-5}

Breen [23] studied the relaxation kinetics of several homologous series of polyoxyalkylates in heptane and toluene and found that increasing levels of polyoxyethylene in the polymers led to slower relaxation kinetics, particularly in heptane as opposed to toluene. This led to the supposition that hydrophilic polyoxyethylene segments in the demulsifier structure should be minimized when attempting to demulsify light, waxy crudes.

In another study by Kim et al. [24], the effects of the individual components of a blended commercial demulsifier product on dynamic film and interfacial tension, elasticity, and diffusivity were examined. As in the previously quoted study, a water–oil–water film on the tip of a capillary was expanded in both dynamic film tension experiments, at an expansion rate of 2.7×10^{-2} mm^2·s^{-1}, and in stress–relaxation experiments. The water-in-crude-oil emulsion studied came from the Louisiana gulf coast. The oil was found to have an API gravity of 28. Three different blended demulsifiers from the Nalco Chemical Co., designated blend-2, blend-15 and blend-19, were used. The compositions of these blends are listed in Table 7. The N-100 and N-253 components were oxyalkylated derivatives of phenol-formaldehyde resins, and the N-499 and N-428 components were diepoxide cross-linked alkoxylated polyglycols.

Film stress–relaxation data were used to calculate the diffusivity of the demulsifiers on the crude-oil surfaces by fitting the film relaxation data to a diffusion-controlled kinetics model [25]:

$$\frac{1}{\Delta \gamma} = \frac{1}{2RT} \left(\frac{1}{2C \sqrt{Dt/\pi}} - a_0 \right), \tag{5}$$

where $\Delta \gamma$ is the difference between the dynamic interfacial tension at time t and at time $t = 0$, C is the bulk concentration of surface active demulsi-

TABLE 7 Blend Demulsifiers and Individual Components

Individual components	Chemical structure	Molecular weight	Blending ratio[a]		
			Blend-2	Blend-15	Blend-19
N-100	} Branched	8,400	7	5	0.5
N-253		135,000	—	—	—
N-428	} Cross-linked	15,400	—	1	0.5
N-499		400,000	3	2	—

[a] Volumetric blending ratio of the individual components.

TABLE 8 Summary of Dynamic Film Properties of Blended Demulsifiers and Their Individual Components

Individual component	Blend-19				Blend-15				Blend-2			
	vol. ratio	ppm	E (mN m^{-1})	$D \times 10^4$ (cm^2 s^{-1})	vol. ratio	ppm	E (mN m^{-1})	$D \times 10^4$ (cm^2 s^{-1})	vol. ratio	ppm	E (mN m^{-1})	$D \times 10^4$ (cm^2 s^{-1})
N-100	5.0	17.7	10.4	1.1	5.0	18.75	10.1	1.4	7.0	21	9.58	1.4
		30	10.1	1.5		30	10.1	1.5		30	10.1	1.5
N-253	3.0	10.6	9.0	42.0								
		30	9.8	40.0								
N-428	0.5	1.7	7.0	4.2	1.0	3.75	6.7	5.0				
		30	10.9	8.0		30	10.9	8.0				
N-499					2.0	7.5	10.0	1.0	3.0	9	9.09	1.2
						30	10.7	1.1		30	10.7	1.1
						30	9.9	6.8		30	10.5	3.6
Properties of the blend		30	7.9	20.0								

fier, D is the apparent diffusivity, and a_0 is the initial interfacial area occupied by a surfactant molecule. From the slope of a plot of $1/\Delta\gamma$, versus $t^{-1/2}$, an apparent diffusivity of the demulsifier can be calculated. The apparent diffusivity should be a weighted average of the diffusivities of the various individual components of the demulsifier. As was observed in Kim's previous study [20], the best-performing demulsifier (blend-19) gave the lowest film elasticity. This demulsifier also had the highest diffusivity and initial interfacial activity.

The calculated dynamic film properties (i.e., the apparent diffusivities and film dilatational moduli for the demulsifier blends and the individual components) are summarized in Table 8. The measurements for blend-19 show the existence of a nonadditive effect between the components: The film modulus, E, obtained for each single component, at 30 ppm, is significantly higher than in the blend. Figure 13 also shows that the interfacial tension and initial interfacial activity of each component at 30 ppm is higher than for the blend.

The interactions between blend components are probably governed by intermolecular interactions in the adsorbed layers. To understand the mechanism of these interactions, further studies are needed. The results

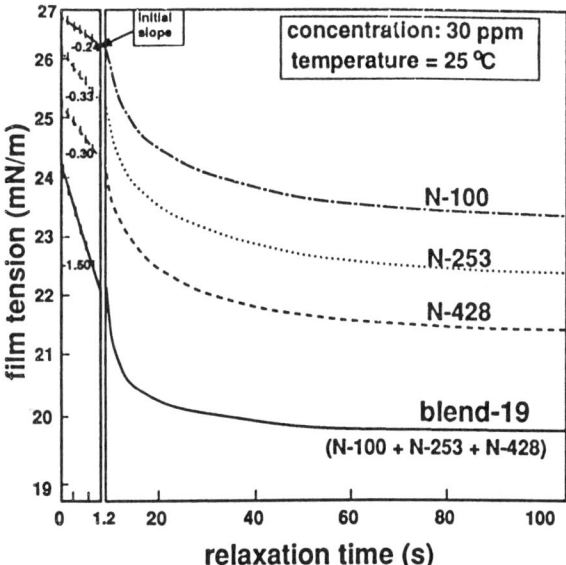

FIG. 13 Film stress–relaxation for blend-19 demulsifier and its individual components at 30 ppm.

of this study clearly demonstrate that the optimum composition of demulsifier blends generally cannot be predicted from the parameters of the individual components because of the possibility of nonlinear interactions between the components, as was found for blend-19.

A study by Mohammed et al. [26] used North Sea Buchan crude oil to demonstrate that demulsifiers serve to reduce the viscoelasticity of the interface. They demonstrated that thick viscoelastic films tended to accumulate in aged films and that some demulsifiers prevented such accumulations. The tendency of asphaltenes to aggregate and form such films at oil–water interfaces has also been studied and documented in several other articles [27–30].

III. ORDERED MICROSTRUCTURES IN THIN FILMS

Johnnott [31] and Perrin [32] long ago observed the phenomenon termed "stratification," the stepwise thinning (through the formation and expansion of thinner spots) of foam films. Later on, stratification of both foam [33–36] and emulsion [37,38] films formed with ionic surfactants was observed and studied. Nikolov et al. [39,40] used a variety of colloidal systems (ionic and nonionic micelles, swollen micelles, and latex suspensions) to demonstrate the universality of stratification phenomena. Using two different methods to obtain stratifying films formed from micellar solutions, they observed with a microscope the following phenomenon: after the film was formed, it immediately began to decrease in thickness. When it became thinner than $\sim$100 nm, the film thickness changed in several steps. The film would then remain for a few seconds in a metastable state with uniform thickness, whereupon one or more dark spots of smaller thickness than the rest of the film would appear and gradually increase in area (see Fig. 14a). The spots would soon cover the entire film which would then remain for several seconds in a new metastable state. Darker spots would then appear and, after they had expanded to cover the entire film, a new metastable state would be reached. This process continued until a stable state was reached. Films formed from latex suspensions (Fig. 14b) stratify in a similar fashion to those formed from micellar solutions with one observable difference: The spots of smaller thickness may be either darker or brighter than the rest of the film. This difference is due to the large diameter of the particles (91 nm in this work), which exceeds the thickness corresponding to the last interfacial maximum.

Using an interferometric technique to measure the film thickness during drainage, Nikolov et al. [41] obtained interferograms similar to that presented in Fig. 15 (film formed from a micellar solution of a nonionic surfac-

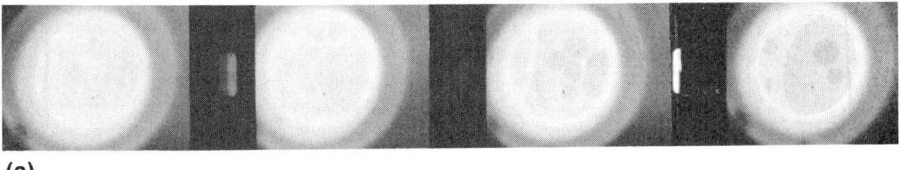

(a)

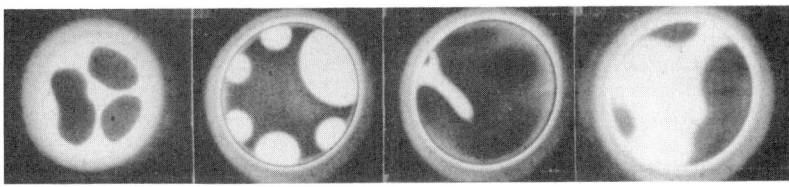

(b)

FIG. 14 Stratification of foam films: (a) 0.1M sodium dodecyl sulfate: (b) formed from 30 wt% latex suspension with particles 91 nm in diameter.

tant). The metastable states of the film appear in the interferogram as steps, where the width of the step is proportional to the lifetime of the respective state. The calculated height of each step is also shown in Fig. 15 and the magnitude is approximately constant for all steps (~10.6 nm). For purposes of comparison, the diameter of a micelle is about 10 nm. In other words, the thickness of a spot was approximately one micellar diameter smaller than the thickness of the surrounding film. The same was true for films formed from latex suspensions.

A. A Model for Film Stratification

Kralchevsky et al. [42] developed a theoretical model to explain the stratification phenomenon and structure formation in thin micellar films based on a diffusive–osmotic mechanism. The driving force for the stepwise thinning of the film is attributed to the existence of a gradient in the chemical potential of the particles. Under the action of the gradient of the chemical potential (i.e., the osmotic pressure gradient), particles or micelles leave the film and vacancies appear in their place. Thickness transitions occur at a given, critical concentration of the vacancies [42], followed by attainment of equilibrium between the film and bulk solution and an arrest of the stepwise thinning [43]. The model permits, for the first time, calculation of the structural contribution to the disjoining pressure of the film that arises from the presence of micellar structures within the film. This structural disjoining pressure, which depends on the effective volume of the particles and the film thickness, is shown to balance the capillary

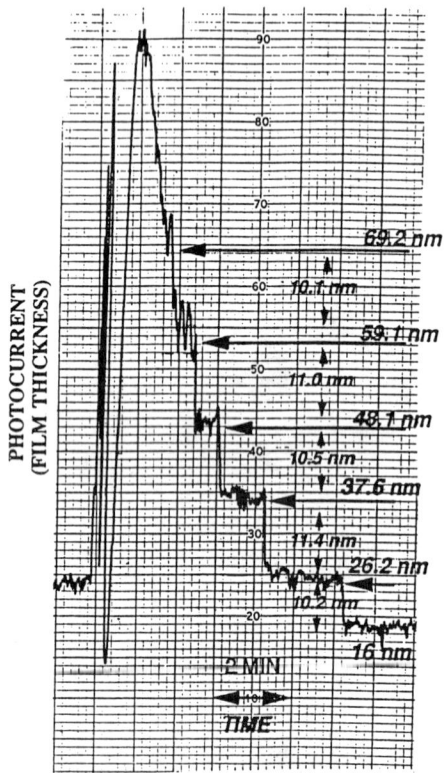

FIG. 15 Interferogram of a film formed from a solution of a nonionic detergent (Enordet AE 1215-30, 0.052M). As the film thins, less light is reflected. The formation of metastable states of uniform thickness is revealed by "steps." The height of a step (arrow) corresponds to the thickness of film. The vertical distance between the steps corresponds to the micelle diameter, about 10 nm. The width of the steps is proportional to the lifetimes of the respective metastable states.

pressure and to stop the film-thinning process. Thus, the film can remain in an equilibrium state at a thickness of the order of several particle diameters. This phenomenon of the formation of long-range ordered microstructures inside the films, over distances on the order of 100 nm, offers a new mechanism for the stabilization of foams, emulsions, and particle dispersions.

Accordingly, any factor decreasing the volume fraction ϕ of the particles should depress the particle ordering inside the film and, hence, the stratification. Nikolov et al. [41] corroborated this statement by using other authors' data for micellar solutions for the same or similar surfactants that they studied to find the trend followed by ϕ when a given system parameter was changed. They defined the particle volume fraction ϕ as $\phi = C_p[4/3(\pi R^3)]$, where C_p is the particle (micellar) number density and R is either the particle radius, R_p, or, for the case of charged particles, the effective particle radius, R_κ, equal to the Debye atmosphere surrounding the particle: $R_\kappa = R_p + 1/\kappa$.

For solutions of sodium dodecyl sulfate (SDS), Nikolov et al. [41] found that increasing the surfactant concentration, C_s, resulted in a decrease in the height of the steps, δ, but enhanced the stratification process which was revealed by both the increased number of transitions and the lifetime of each metastable state. They used data from Sasaki et al. [44] to show that the decrease in δ with increasing C_s is about the same as that of the Debye length $1/\kappa$ (SDS is a strong electrolyte), but the much stronger increase in the micellar (particle) concentration C_p was affecting the decrease in R_κ, to produce an overall increase in ϕ. This explains the enhanced stratification.

Quite different is the effect of a strong electrolyte, NaCl, which was always found to decrease stratification. Indeed, the addition of NaCl in moderate concentrations to micellar solutions of SDS decreases $1/\kappa$ without an appreciable effect on the micellar aggregation number or the critical micelle concentration (CMC). This leads to a decrease in R_κ at virtually constant C_p, which eventually leads to a decrease in ϕ. Above a given electrolyte concentration (specific for each surfactant concentration C_s), the stratification vanished altogether. These results are plotted in Fig. 16 as a phase diagram, with the region on the left of the solid line corresponding to stratifying films and on the right of it to nonstratification. The broken line on the same figure shows the results of Hachisu et al. [45] and Ohtsuki et al. [46] for the order–disorder phase transition in latex suspensions (the ordered phase region is on the left of the broken line). Although the electrolyte concentration in the Nikolov et al. experiments was much higher, owing to the much smaller particle size, the close resemblance of the two plots supports their explanation of the stratification as being a restricted-volume-induced phase transition.

Increasing the neutral electrolyte concentration is known to promote liquid-crystal formation in surfactant systems [47]. Therefore, if stratification was due to the formation of a liquid crystal structure, it should be enhanced by adding electrolyte. However, Nikolov et al.'s results, as well as those of Kenskemp and Lyklema [34], show the opposite trends.

Nikolov et al. [43] found similar parallel trends between micellization behavior and stratification with two nonionic surfactants: Endoret AE 1215-9.4 and AE 1215-30 (both are products of Shell Oil Co. and consist of

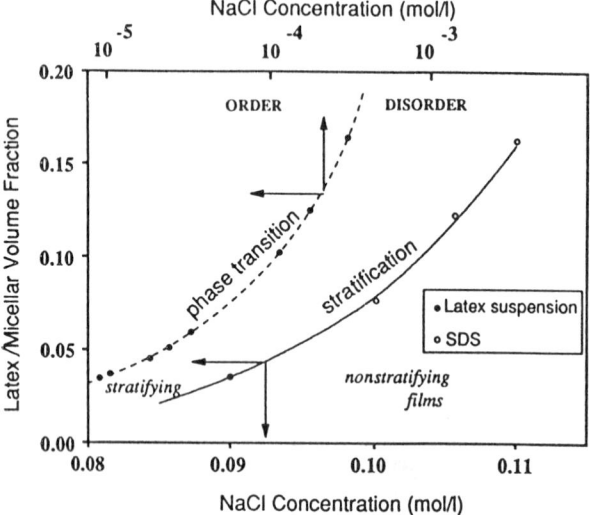

FIG. 16 Volume fraction versus concentration of added NaCl for latex suspension and micelles. The broken line is the experimental curve of Hachtsu et al. [12] for the added electrolyte order-disorder phase diagram for latex dispersion. The results for stratifying–nonstratifying films are plotted also as a phase diagram. The region on the left of the solid line corresponds to stratified and on the right to nonstratified film.

a C_{12-15} primary alcohol ethoxylated with 9.4 or 30 equivalents of ethylene oxide). They found that increasing the surfactant concentration enhanced the stratification. This is obviously a result of the increased particle concentration C_p. The step height δ slightly decreased with increasing C_s but was always approximately equal to the micellar diameter, $2R_p$, determined by dynamic light scattering: for example, $\delta = 15.0 \pm 0.6$ nm versus $2R_p = 15.0 \pm 1.5$ nm for $0.008M$ AE 1215-9.4, and $\delta = 11.2 \pm 0.6$ nm versus $2R_p = 10.6 \pm 2.1$ nm for $0.026M$ AE 1215-30. The stratification with nonionic surfactants is very sensitive to the temperature: whereas the stratification is pronounced at around 25°C, it decreases rapidly with the increase in temperature, and above 35°C it completely disappears for all surfactant concentrations. The effect of the temperature on micellar structuring was studied by Nakagawa et al. [48] for the surfactant $C_{10}H_{21}O(-C_2H_4O)_{12}CH_3$ which shows physico-chemical behavior close to that of Endoret AE 1215-9.5. From their data, Nikolov et al. [41] calculated C_p and R_p (for the Nakagawa solutions), the latter from the second virial coefficient, assuming that the micelles are hard spheres. They found that although R_p did not change with the temperature, C_p decreased, thus leading to a decrease in ϕ, which explains the depressed stratification in the Nikolov et al. experiments.

By preequilibrating a $5.2 \times 10^{-2}M$ solution of Endoret AE 1215-30 with n-decane, Nikolov et al. [41] established that the lifetimes of the metastable states became considerably shorter. The data of Nakagawa et al. [49] on the properties of solutions of swollen micelles of $C_{10}H_{21}O(C_2H_4O)_{12}CH_3$ led to the conclusion that the solubilization of n-decane decreases both C_p and R_p. The ensuing decrease of ϕ explains why solubilization of oil weakens the stratification.

Wasan and Nikolov have published data showing the relevence of film stratification to oil-field emulsions [50]. Figure 17 shows photomicrographs of the various stages of stepwise thinning of a microscopic, horizontal oil film stabilized by asphaltene particles (7 vol%) in a 1 : 1 volume mixture of n-heptane and toluene. At a film thickness higher than about 300 nm, the asphaltene particles inside the film form a random structure which causes the white and dark interference patterns produced in reflected monochromatic light to form a mosaic structure (Fig. 17a). The film is irregular. After a while, a white expanding spot surrounded by a dark rim appears inside the film, with a thickness of about 100 nm (Fig. 17b).

At this point, the film thickness in the region of the spot appears to be much more regular than the surrounding film. Subsequently, the spot expands (Figs. 17c and 17d) until it occupies the entire area of the film. After this, a dark spot appears inside the film, indicating that asphaltene particles remain in the film. The visual evidence of such asphaltene microstructures within a thin film illustrates an additional mechanism of emulsion stabilization not previously associated with asphaltenes.

B. Stratification Due to Silica Particles in Thin Films

Nikolov and Wasan [51] examined the effects of silica particle concentration, size, polydispersity, and surfactant concentration on film structuring and stratification. Figures 18 and 19 show interferograms for silica particles with a diameter of 19 nm at concentrations of 10 and 20 vol% respectively, using a horizontal microscope film technique.

There are three observable thickness transitions for the case of 10 vol% (Fig. 19), the first transition occurring at 160 nm, the second at 120 nm, and the third at 77 nm. The film achieved an equilibrium thickness at about 40 nm and contains one layer of particles inside it. The mean height of the steps is $\sim$38 nm. In this case, the effective diameter of the particles,

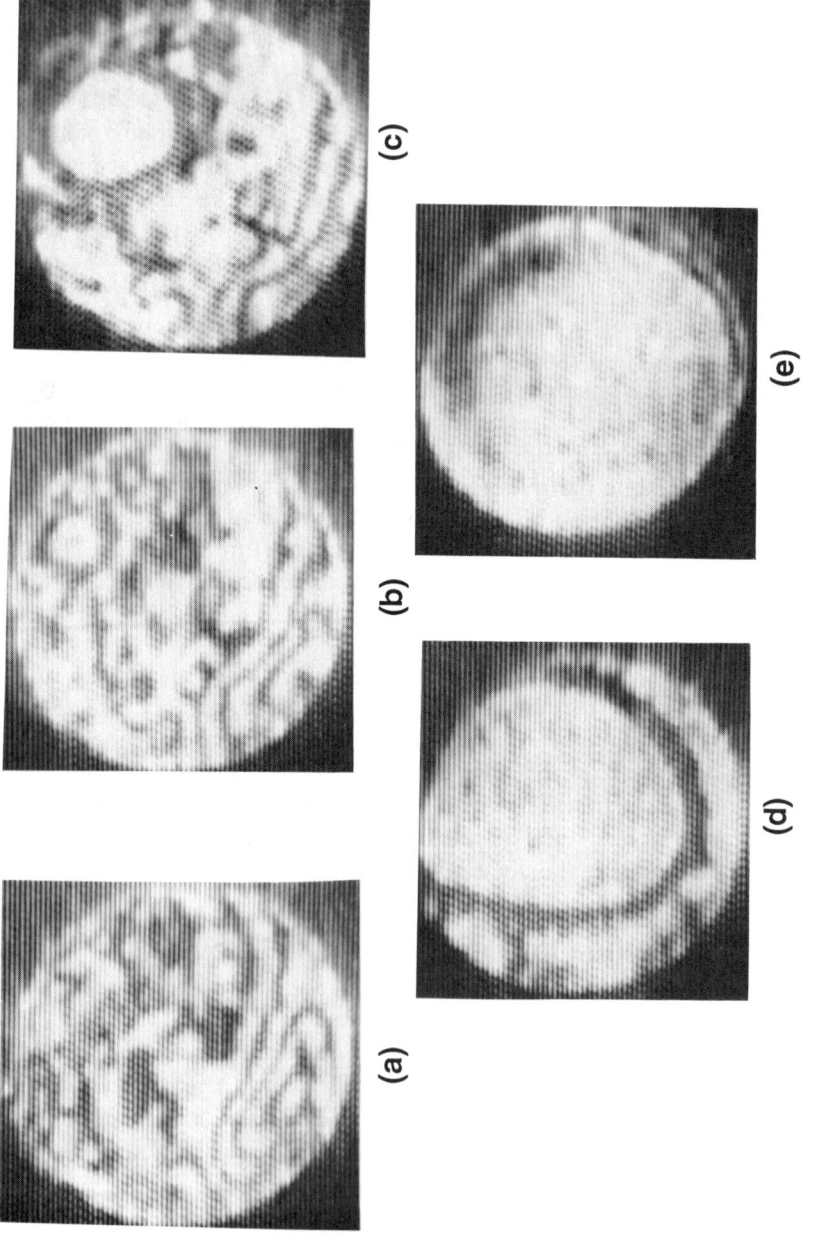

FIG. 17 Sequence of photomicrographs depicting the stages of stepwise thinning of oil emulsion microhorizontal film in the presence of 7 vol% asphaltene. (From Ref. 50.)

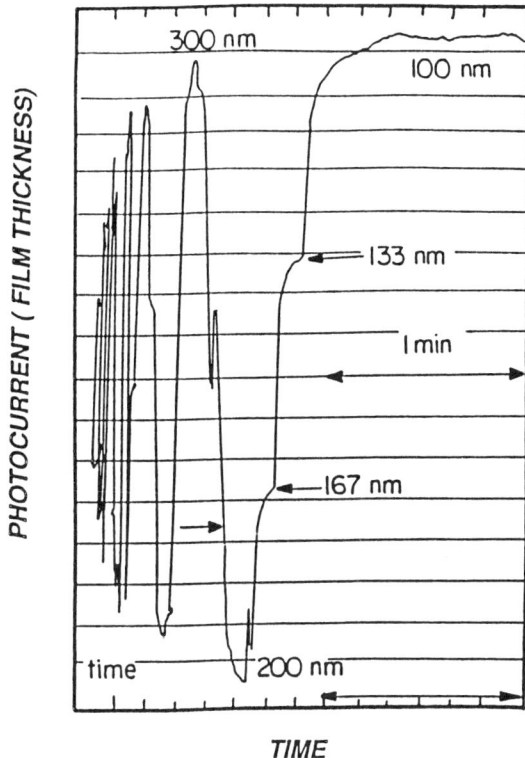

FIG. 18 Photocurrent versus time interferogram of film thinning process of a microscopic horizontal film (film diameter 6×10^{-2} cm) stabilized by silica particles (20 vol%) with a particle diameter of about 19 nm. The thickness at which the transition begins are marked with arrows.

including the Debye length, was found to be ~36 nm, which is close to the mean height of the step. In contrast, there are four thickness transitions in the case of the 20 vol% percent silica hydrosol system, where the effective particle diameter was ~29 nm, which is comparable with the step size of ~30 nm as seen in Fig. 18.

It is important to note that the shape of the steps seen in Figs. 18 and 19 is different. For 20 vol%, the steps are steeper. The authors observed such steep thickness transitions when they examined the stepwise film thinning for nonionic surfactants. The interferogram for 10 vol% shows a more gradual decrease of film thickness during the transitions, indicating

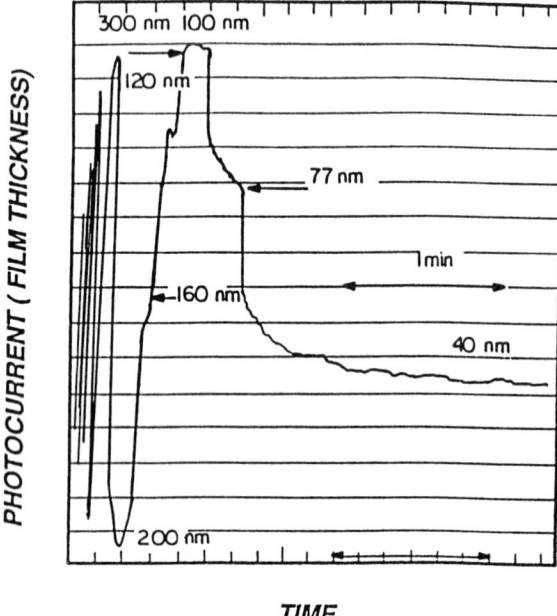

FIG. 19 Photocurrent versus time interferogram of films thinning process of a microscopic horizontal film (with diameter 6×10^{-2} cm) stabilized with silica particles (10 vol%) with diameter of about 19 nm. The thickness at which the transition begins are marked with arrows.

more soft-particle interactions and the electrostatic character of the potential of interaction between silica particles. This shows that for 20 vol% the silica particle interactions inside the film are more like a hard-sphere (HS) interaction. The reason for this is the effect of electrolyte concentration on the Debye length. Ramsay et al. [52], on the basis of a neutron-scattering study of concentrated silica hydrosols, found that the structure factor is similar to that of the HS in which the effective interaction diameter is greater than the diameter of the sol particles. Upon further decrease of the particle concentration to 5 vol%, Nikolov and Wasan [51] saw only one stepwise thickness transition, after which the film ruptured.

   The observation that the number of thickness transitions increases with particle concentration suggests that the particles inside the film form a colloidal, crystallike structure. At high concentrations in the film, ordered microstructures form at large film thicknesses (e.g., film thicknesses of ~200 nm). As particle concentration drops, interactions between particles decrease along with the number of stepwise transitions, resulting in faster thinning of the film.

The experimental observations of Nikolov and Wasan [51], show that in the case of thin liquid films, the colloid crystal structure (CCS) begins to form inside the thinning film at concentrations below 50 vol%, which is the concentration theoretically predicted by Kirkwood and Alder for bulk structure transitions. Therefore, the formation of CCS inside the film at effective particle concentrations less than 54 vol% (which is the concentration corresponding to the simple cubic packing) leads to the conclusion that the particles form a loose structure inside the thinning film; however, this loose structure contains many vacancies (dislocations). This has been theoretically predicted by Chu et al. [53]. Another possible explanation is to assume that domains having CCS are formed locally inside the film, which are at dynamic equilibrium with regions of randomly distributed particles. Such findings have been reported by Yoshiyama and Sogami [54] for deionized, dilute solutions of monodispersed lattices contained between two solid surfaces. The particle concentration in both the domains and regions, where they are randomly distributed, should be the same. Then, the total volume held by the particles and the vacancies related to the dislocations is close to 54 vol%, and this corresponds to a simple cubic structure formation. Thus, the explanation for the experimental observation is that at two or more thickness transitions, the height of the stepwise change is of the order of the effective particle diameter and corresponds to simple cubic packing [43,55].

Nikolov and Wasan [51] investigated the effect of surfactant on the stratification process in the presence of silica particles. Silica hydrosol was preequilibrated with activated charcoal for several days to reduce the concentrations of unwanted surfactants. The charcoal was removed by centrifugation, whereupon the purified hydrosol was used in film-thinning experiments. Figure 20 shows the film-thinning interferograms of a 20 vol%, microscopic horizontal flat film at a film diameter of 6×10^{-2} cm treated with charcoal and in the presence of 10^{-3} mol L^{-1} sodium lauryl sulfate. The number of film-thickness transitions (four) and the final film thicknesses are the same with or without the surfactant. Only the rate of film thinning is different: With surfactant present, the film thins slower and takes longer to reach the equilibrium film thickness. This effect was attributed to the Marangoni–Gibbs effect: The surfactant at the surface creates a surface tension gradient, which reduces the rate of film thinning. When the film thins slowly, the particles inside can pack more efficiently, with resultant fewer vacancies. One can, therefore, expect that the effect of surfactant could also lead to a higher, final film thickness (stable film with more particles inside). According to the authors' vacancy diffusion

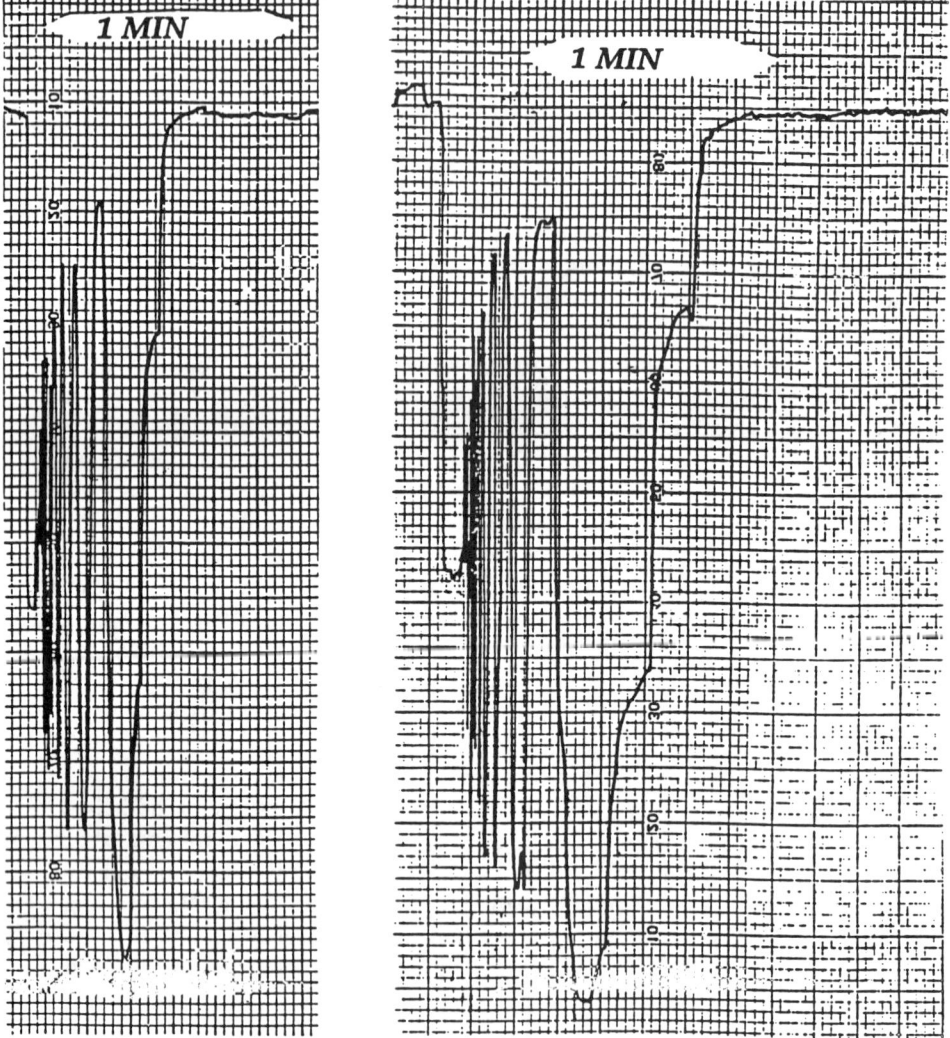

FIG. 20 Photocurrent versus time interferogram of film-thinning process of microscopic horizontal film with a film diameter of about 6×10^{-2} cm stabilized by 20 vol% silica particles with a particle diameter of about 19 nm: left side, after purification by activated charcoal; right side, in the presence of 10^{-3} mol l^{-1} NaLS.

mechanism, if the concentration of vacancies is low, not only will there be an increase in time for the occurrence of each thickness transition but also the probability of forming critical spots will decrease, resulting in a thicker, stable film.

C. Effects of Polydispersity on Stratification

Another factor which plays an important role in effecting interparticle interactions is polydispersity. Nikolov and Wasan [51] studied the stepwise thinning process of a 20 vol% 1:1 mixture of silica hydrosols of sizes 7 and 28 nm. During the film thinning they observed not only the fast formation of many spots at the same time but also the formation of new spots inside the older ones. The stratification began at ~90 nm instead of at 230 nm for the monodispersed hydrosol (of size 19 nm). Furthermore, the spots expanded and coalesced very fast, and then finally the film without particles ruptured. It was, therefore, concluded that even with highly charged particles at 66% effective volume fraction, a stable colloid crystal structure inside the film could not be formed, and because of this, the polydispersity leads to many dislocations inside the colloid crystal structure and less film stability [53].

D. Structural Disjoining Pressure Due to an Ordered Microstructure

At equilibrium and for a plane-parallel film (i.e., at constant film thickness), the capillary pressure is equal to the disjoining pressure. By changing the film radius of the horizontal microscope film, Nikolov and Wasan [51] changed the capillary pressure from 100 to 1000 dyn cm^{-2}. They observed that this change in capillary pressure did not noticeably influence the amplitude of the stepwise thickness transitions. This indicated that the difference between the interparticle interactions creates in the film/meniscus an osmotic pressure which balances the capillary pressure and thus controls the film thickness.

To estimate the osmotic pressure created by silica particle interactions, the authors used the form of a screened Coulombic potential expression derived by Beresford-Smith et al. [56]. Using the methods of statistical mechanics, they have established for the effective pair potential the equation $U_{eff} = U_0 e^{-\kappa\delta}/\delta$ where δ is the distance between the centers of the colloid particles, $1/\kappa$ is the Debye length, and U_0 is the particle surface charge. They have performed numerical calculations of U_0 for colloidal particles (polystyrene latices) of radius $a = 16$ nm and surface charge density of ~4 μC cm^{-2}. For the surface-charge density of silica particles at an ionic strength of 4×10^{-3} mol L^{-1} and pH 9.4, they used the value of 1.5 μC cm^{-2}, calculated from electrokinetic measurements and reported elsewhere [57]. One interesting result of this theory is that, due to the screening effect, the pair interactions become dependent on the volume fraction of particles.

It is known that the silicon dioxide–water system has a low value of the diffuse layer potential and ζ-potential, coupled with extremely high values of titratable charge. This leads to the suggestion that the surface potential is significantly different from Nernst's value and depends of the total electrolyte concentration, and the significant proportion of the titratable surface charge must be balanced off with counterion penetration inside the shear plane. It has now become accepted, for the silica–water interface [58], that the electrokinetic potential or ζ-potential measures the electrostatic potential at, or very near, the beginning of the diffuse double layer (i.e., $\zeta = \psi_d$). To calculate the osmotic pressure, Nikolov and Wasan [51] used the value of the surface charge (1.5 μC cm^{-2}) at the shear plane. Figure 21 shows the osmotic pressure versus volume fraction of the silica hydrosol calculated by the relationship

$$P = \frac{U_0}{\delta}\left(\kappa + \frac{1}{\delta}\right)\frac{e^{-\kappa\delta}}{\delta}, \tag{6}$$

where

$$U_0 = \frac{(z_0 e)^2}{\epsilon}\left(\frac{e^{2a\kappa}}{(1 + \kappa a)^2}\right)(1 + \Phi)^2, \tag{7}$$

where Φ is the volume fraction of silica particles, z_0 is the number of charges per particle (in this case ~ 105), ϵ is the dielectric constant of water, and e is the electron charge.

The steep changing of the osmotic pressure versus particle concentration shows the role of the electrostatic interactions. In such systems dominated by a repulsive interparticle potential, it is known that the disorder–order transition is driven by an entropic effect. The transition is distinguished by the small difference in density between the two coexisting phases and the crystalline order of the denser phase. Nikolov and Wasan [51] observed the formation of stable colloid crystal structures inside the thinning film at a thickness of ~ 0.2 μm. At this film thickness, the interactions between the film surfaces can be neglected and the ordered microstructure inside the film (the denser phase) creates a positive disjoining pressure which balances the capillary pressure [55]. By using the effective potential of the interparticle repulsion proposed by Beresford-Smith et al., they derived the expression for the structural component of the disjoining

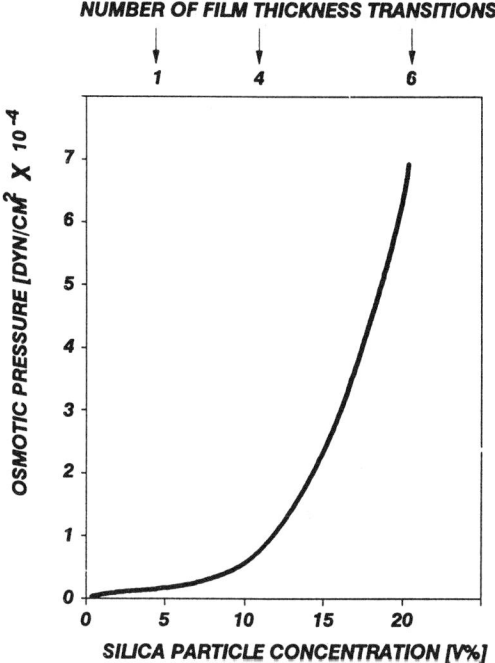

FIG. 21 Osmotic pressure versus volume fraction calculated using the Beresford–Smith potential for the case of silica hydrosol particles.

pressure, Π_{st}, as [45]

$$\Pi_{st} = \frac{U_0}{\delta}\left[\left(\kappa + \frac{1}{\delta_f}\right)\frac{e^{-\kappa\delta}}{\delta_f} - \left(\kappa + \frac{1}{\delta}\right)\frac{e^{-\kappa\delta}}{\delta}\right], \tag{8}$$

where δ_f and δ are the distances between the particles in the film and the meniscus (bulk phase), respectively, assuming that at this concentration particles have hexagonal packing. If one imagines that the film phase and the bulk liquid in the meniscus are brought together, then at equilibrium the excess pressure drop, Π_{st}, will appear at the film–meniscus region and $\Pi_{st} = P_0$. Figure 22 is a plot of the osmotic pressure due to the difference in particle volume concentration between the film and the meniscus. As is known, the capillary pressure depends on the surface tension, σ, and film and tube radius (r_f and R_t, respectively,) and for a thick film is given by the relation $P_0 \approx 2\sigma R_t/(R_t - r_f)$. For this case, capillary pressure was varied from 600 to 1200 dyn-cm^{-2} by changing the film radius. For the pressure of 20 vol% silica particles, if the particle volume

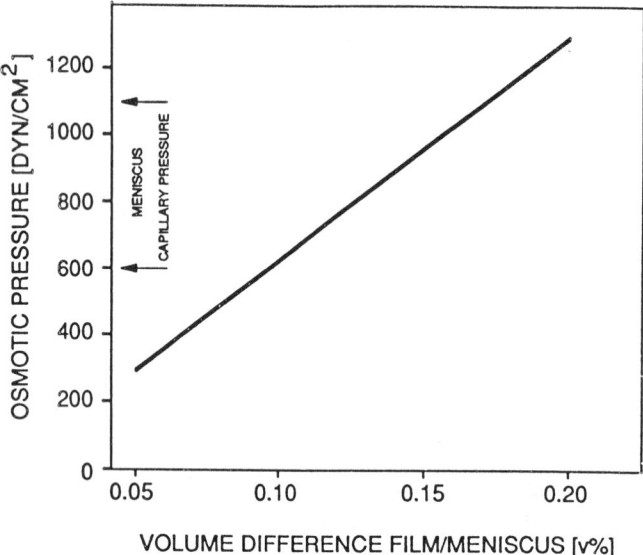

FIG. 22 Osmotic pressure due to the hypothetical difference of particle concentration in the film/meniscus.

density inside the film is higher than for the meniscus, the osmotic pressure will balance the capillary pressure and the film will stop thinning. It is known that such a transition can be distinguished by small differences in the density between the two coexisting phases. This small change in the particle density is due to the formation of a colloid crystal structure inside the film, characterized by Π_{st}. Once the colloid crystal structure is formed, further thinning of the film depends on the equilibrium film/meniscus.

IV. IMPACT OF SOLIDS AT THE INTERFACE ON EMULSION STABILITY

The presence of solid particles at an oil–water interface complicates the measurement of interfacial properties by conventional experimental techniques. When studying the coalescence behavior of solids-stabilized emulsions, however, any interpretation of emulsion stability in terms of interfacial properties must necessarily take into account the effects of solids.

The size and position of a particle at an oil–water interface are important parameters which determine the stability of the emulsion. Nutt [59],

Princen [60], Huh and Mason [61], Winitzer [62,63], and Rapachietta and Neumann [64] have derived the relevant force balance equations to predict the equilibrium position of a particle of known size, density, and contact angle, at a planar liquid–liquid interface of known liquid–liquid interfacial tension. Their calculations reveal that there exists a critical particle size (other parameters remaining the same) above which a particle cannot be supported at the interface. Optimization of demulsification parameters requires an understanding of these and other factors that are responsible for emulsion stability.

Taubman and Koretskii [65] observed that stability of clay-water-in-oil emulsions was due to the formation of surface complexes between the surfactant (which was present in the aqueous phase) and the metal hydroxides of the clay. Such complex formation effectively anchors the solids to the liquid–liquid interface.

In two separate studies, Tambe and Sharma [66,67] investigated the emulsifying effects of finely divided colloidal particles, such as polystyrene microspheres, graphite, or stearic acid, on decane and water. They found, both theoretically [66] and experimentally [67], that such particles caused viscoelastic behavior in the interface which reduced the rate of film thinning.

Tsugita et al. [68] have observed the formation of an interfacial network of clay particles in an oil-in-water emulsion. They attribute the stability of the emulsion to the mechanical resistance offered by this network to drop–drop coalescence.

The mechanical strength of the interface arises due to the particle–particle interactions at the interface which result in an equilibrium packing of the solids. Hydrophobic particles favor an interfacial position because the magnitudes of the surface energies involved are much larger than the thermal or gravitational fields which could displace them [69–71]. The particles form a network structure, the compactness of which depends on the amount of solids present at the interface. When a water drop is placed on such an interface, the latter deforms. A photograph of such a drop is schematically represented in Fig. 23. This observation was utilized by Menon et al. [69,72] to devise a new experimental technique to measure the film tension between a coalescing water drop and a planar oil–water interface covered with hydrophobic particles. The film tension, γ_p, is a measure of the resistance offered by the system to coalescence. For the deformed drop-interface configuration shown in Fig. 23, the pressure balance along the symmetric axis of the drop can be written as [73]

$$2\frac{\gamma_{ow}}{b_t} + Z_t(\rho_h - \rho_l)g + Z_b(\rho_d - \rho_h)g = 2\frac{\gamma_p}{b_b} + Z_m(\rho_h - \rho_l)g, \quad (9)$$

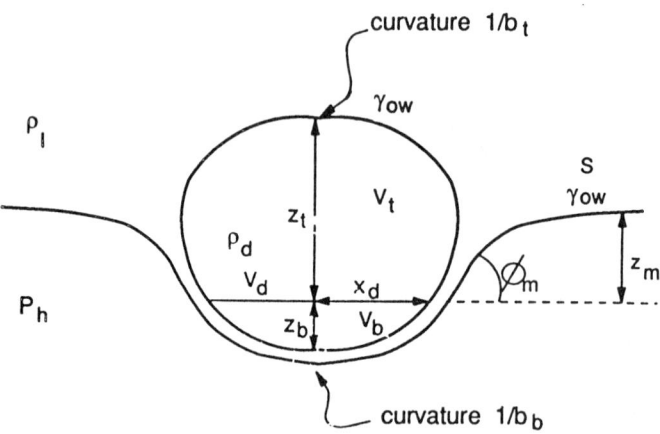

FIG. 23 Profile of a water drop at an oil-water interface containing hydrophobic particles.

where b_t and b_b are the radii of curvature of the top and bottom of the drop, respectively, and Z_t, Z_b, and Z_m are the dimensions of the drop shown in Fig. 23. ρ_d, ρ_h, and ρ_l are the densities of the drop, heavy, and light phases, respectively. In the Menon work, both the drop and heavy phases were identical (water). γ_p is the film tension between the drop and the particle-covered planar oil–water interface.

The force balance along the contact line of the drop can be written as

$$V_t(\rho_d - \rho_l)g + V_c(\rho_d - \rho_h)g = 2\pi\chi_d\gamma_{ow}^s \sin\phi_m + \pi\chi_d^2 Z_m(\rho_h - \rho_l)g, \tag{10}$$

where V_t and V_c are the drop volumes above and below the contact plane. The total drop volume, which can be measured experimentally, is given by

$$V_d = V_t + V_c. \tag{11}$$

V_c can be obtained from the geometry of the spherical cap of the drop:

$$V_c = \frac{\pi\chi_d^3}{3 \sin^3\phi_m} (2 - 3\cos\phi_m + \cos^3\phi_m). \tag{12}$$

The term γ_{ow}^s in the force balance equation is called the "meniscus tension." The meniscus tension is the surface energy of the planar oil–water interface in the presence of particles. Both γ_p and γ_{ow}^s depend on the interfacial concentration of particles. Menon and Wasan [74] found experimentally that a simple correlation exists among the film tension, meniscus tension, and oil–water interfacial tension:

$$\gamma_p = \gamma_{ow} + \gamma_{ow}^s. \tag{13}$$

Such an expression for film tension, independent of disjoining pressure and film thickness, implies that the film of particles and oil that separates the water drop from the bulk water phase is thick. Because the shale dust particles Menon and Wasan [74] looked at had an average diameter of 4 μm, the thickness of the film would indeed be expected to be on the order of a few microns.

The interaction energy between particles at the oil–water interface can be expressed as the difference between the oil–water interfacial tension and the meniscus tension:

$$V_T = \gamma_{ow} - \gamma_{ow}^s. \tag{14}$$

The effect of demulsifier concentration on the film, meniscus, and oil–water interfacial tensions is depicted in Fig. 24 for a concentration of 1.0×10^{-3} g cm^{-2} shale dust particles. The oil phase was a 1 : 1 volume mixture of heptane and toluene and the demulsifier, a blend of Breaxit-126 and Corexit-420 (from Exxon), was added to the oil containing shale dust prior to its contact with the aqueous phase.

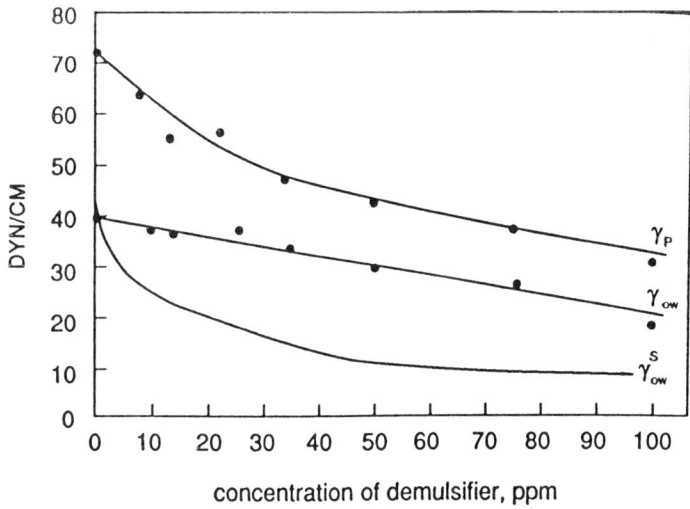

FIG. 24 Effect of demulsifier on film, meniscus, and interfacial tensions.

The film tension dropped from 73 to 30 dyn cm^{-1} as the demulsifier concentration was varied from 0 to 100 ppm. The rapid decrease in the film tension is a measure of the efficiency of the demulsifier in decreasing the resistance of the interfacial layer to coalescence. The meniscus tension and oil–water interfacial tension also decreased but less dramatically (see Fig. 24).

The effect of the demulsifier on particle interaction energy (V_T) is depicted in Fig. 25. V_T goes through a maximum at a concentration of 30 ppm. The demulsifier blend contains an anionic component which adsorbs on the solid particles and charges their surfaces. This increases the repulsive forces between particles, thereby increasing V_T. At a concentration of 30 ppm, there are just enough demulsifier molecules to charge all the particles. Hence, repulsive forces are the highest. Below this concentration, there are not enough demulsifier molecules to charge the particles, whereas above this concentration, there is excess demulsifier left in solution after adsorption. This excess demulsifier acts as an electrolyte and dampens the electrostatic forces of repulsion. Hence, the decrease in V_T at high demulsifier concentrations is expected.

A. Oil Loss in Sludge Layers

In many processes involving the formation of solids-stabilized emulsions, the emulsions prove to be so stable that they do not coalesce in the settlers

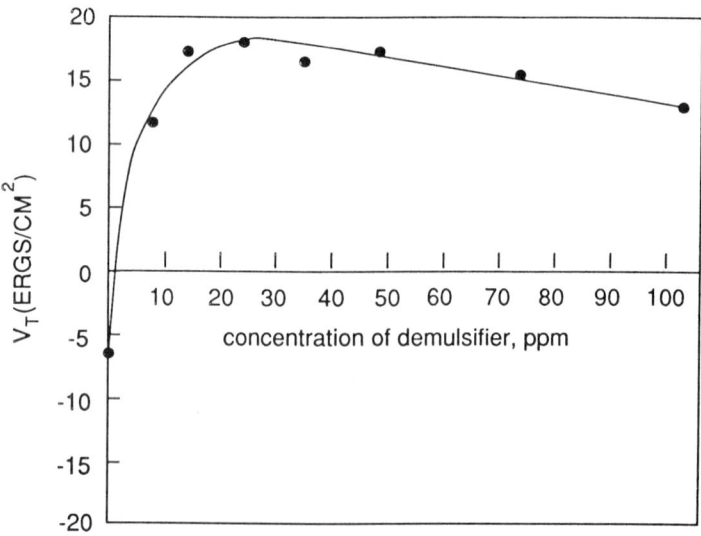

FIG. 25   Effect of demulsifier on particle interaction energy.

or electrocoalescers in spite of the use of chemical or other methods to demulsify them. Such emulsions end up as sludge layers in the coalescing equipment and have to be discarded. Oil trapped in the sludge layer is consequently lost. Economic constraints require that this oil loss be kept to a minimum. Because the sludge layer is composed of solids-stabilized emulsion globules, the factors affecting emulsion formation and stability also affect oil loss.

Figure 26 depicts a typical sludge layer. Oil loss occurs due to two phenomena: (i) oil entrapment in the voids between emulsion droplets (V_{ov} in Fig. 26) and (ii) oil entrapment in the interstices between particles on each water droplet (V_{os} in Fig. 26).

Any estimation of oil loss would have to account for oil lost by both of these modes. Menon et al. [75] developed a semiempirical model to predict oil loss in terms of basic system properties such as oil–water interfacial tension and contact angle. The final expression for oil loss is

$$V = 0.2505 \frac{\rho_o}{\rho_s} \frac{(K_2 \gamma_{ow}^\alpha + 1 - \cos\theta)^3}{(K_2 \gamma_{ow}^\alpha - \cos\theta)^2}, \tag{15}$$

where ρ_o and ρ_s are the densities of the oil and solid, γ_{ow} is the oil–water interfacial tension and θ is the three-phase contact angle measure through the water phase. K_2 and α are constants that need to be obtained experimentally. Using values of $K_2 = 20.2$ and $\alpha = 0.44$, Menon et al. [75] were able to describe the variation of oil loss with surfactant concentration for a water-in-oil emulsion stabilized by shale dust particles. Figure 27 shows that the effect of Aerosol-OT in oil on the oil loss. The solid dots represent experimental values while the solid line represents Eq. (15). The agree-

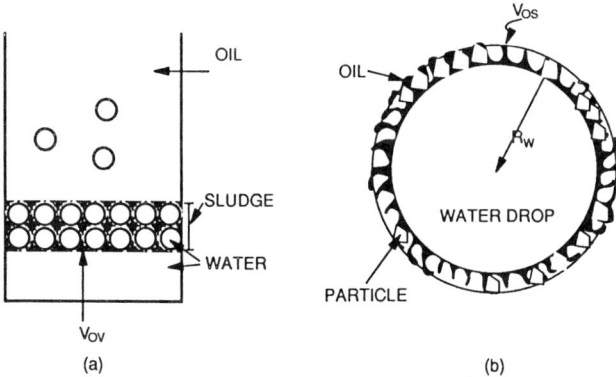

(a)   (b)

FIG. 26 (a) Schematic of the sludge layer; (b) the packing of particles on a water drop.

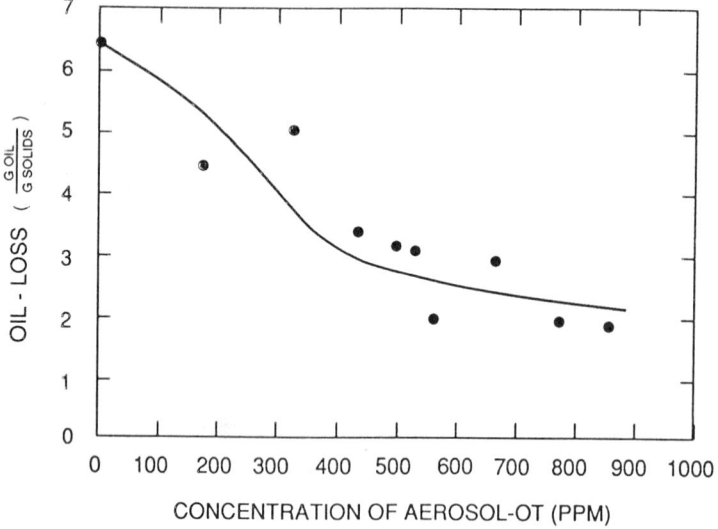

FIG. 27 Variation of oil loss with concentration of surfactant.

ment between the model and experiment is quite good. K_2 and α are independent of surfactant type; hence, such a model can be utilized to evaluate various surfactants for minimizing oil loss.

V. OTHER FACTORS INVOLVED IN EMULSION DESTABILIZATION

A. Effects of pH

Sjoblom et al. [76] conducted a study investigating destabilizing mechanisms in several different crude oil, by varying pH and temperature and by adding a variety of alcohols, amines, and fatty acids. The crude oils investigated contained aromatic compounds between 10% and 50% by weight and saturated hydrocarbons from 40% to 85% by weight [77]. Nonspecified polar compounds determined as precipitates in acetone varied form 2% to 13% by weight, whereas the amount of asphaltenes was strikingly low, between 0% and 1.5%. Microscopic studies revealed the occurrence of wax particles in the crudes. The authors deduced that the interface was most likely built up by nonspecified polar compounds, waxes (as lipids), and asphaltenes. Small wax particles were also incorporated in the film. Inorganic particles such as clays were not detected in the emulsions nor in the crudes [77]. pH was found [76] to influence the stability of the emulsions in such a way that intermediate pHs gave rise to some instability, whereas extremely low or high pHs restored or even enhanced stability. Figure 28 summarizes the results from the pH adjustments in buffer solutions on the stability of the different crude oil emulsions based on four different crude oils. The influence is not uniform. Only one emulsion (based on crude oil E) shows a complete separation in the pH interval 5–7. Two others (i.e., C and D) show some instability around pH 6 and 9. The influence of pH on emulsions based on crude oil G is rather negligible. Although intermediate pH values seem to cause instability, the range and degree of instability is very dependent on the crude-oil emulsion being considered.

Sjöblom et al. [76] interpreted the lack of instability at extreme pHs to be an indication of a low amount of ionizable and polar groups in the interfacial film. Consequently, the interfacial film should have a low surface-charge density at extreme pHs. As extreme pH should alter wettability of emulsion-stabilizing particle, a lack of such particles in the emulsions studied is also indicated.

Tambe and Sharma [78] studied the effects of pH on the stability of a model solids-stabilized emulsion. They found that increasing pH favored

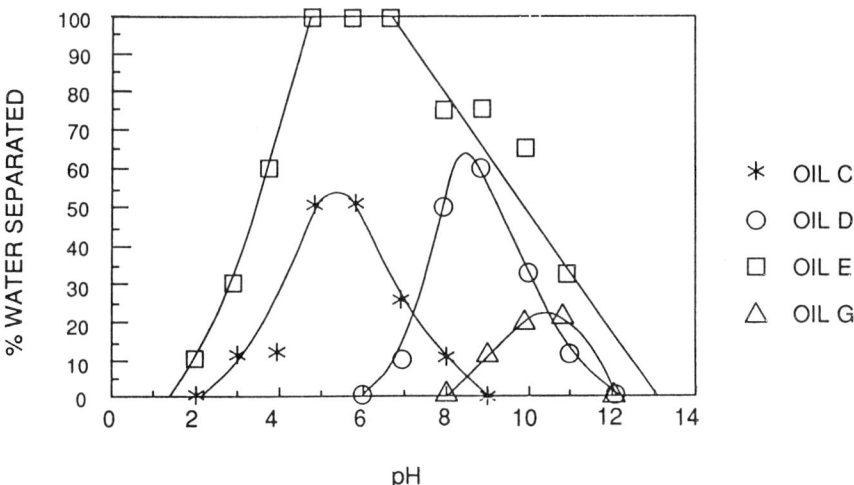

FIG. 28 Separation of water from crude-oil emulsions based on crude oils, C, D, E, and G as a function of pH in buffered systems. Preparation according to procedure A. The aqueous phase was 50% (v/v).

TABLE 9 Effects of Different Chemical Additives on Crude-Oil Emulsion Stability; the Separation of Water from the Emulsion was Measured After 30 min and 24 h, Respectively

	Water separation (%)					
	Separation time, 30 min			Separation time, 24 h		
Chemical additive (1.5%)	C	D	G	C	D	G
Ethanol		0				
Butanol	100	50	80	100	85	95
Pentanol		36			80	
Hexanol		17			17	
Decanol		0			0	
2-Propanol (IPA)	40	15	20	100	20	20
Isobutanol		45			56	
2-Pentanol		30			85	
2-Methyl-1-butanol	30			80		
3-Methyl-2-butanol	30			80		
Cyclohexanol		35			44	
Benzyl alcohol	100	90	90	100	100	90
1,2-Ethanediol	20	10	0	100	10	10
1,2-Butanediol	20	0	0	100	0	0
Propylamine		0			0	
Pentylamine		40			40	
Hexylamine	60	85	80	65	90	80
Octylamine		85			95	
Decylamine		95			100	
1-Hexyl-methylamine	30			30		
Diethylamine		0			4	
Benzylamine		40			40	
Triethanolamine		10			10	
Ethoxylated (8 EO) decylamine	30			30		
Formic acid		10			10	
Butanoic acid		35			35	
Pentanoic (valeric) acid		0			0	
Hexanoic (caproic) acid		0			0	
Heptanoic acid		0			0	
Octanoic acid		0			0	
Nonanoic acid		0			0	
Isobutanoic acid		0			0	

TABLE 9 (Continued)

Chemical additive (1.5%)	Water separation (%)					
	Separation time, 30 min			Separation time, 24 h		
	C	D	G	C	D	G
Isohexanoic acid		0			0	
Oleic acid	0			0		
Methyloleate		0			0	
Acetone	0			0		
2-Butanone		10			10	
Acetaldehyde		0			0	
2-Ethoxyethanol		0			0	
2-Butoxyethanol		50			50	
Butyldiglycol		35			35	
Toluene	0			0		

the formation of water-in-oil emulsion. They also found that the presence of inorganic ions in the aqueous phase destabilized oil-in-water emulsions, and phase inversion resulted when the wettability of the solid particles was varied.

B. Effects of Cosurfactants on Emulsion Stability

Sjöblom et al. [76] examined the relative water separation times for an sizable number of monomeric alcohols, amines and organic acids in emulsions of crude oils C, D and G. The data are presented in Table 9 and give further information about the oil–water interface and active sites there. First, it was shown that fatty acids do not give rise to instability of the emulsions, whereas some amines are efficient up to a certain analytical concentration, after which the efficiency declines. In view of this, the oil–water interface is at least partly built up by carboxylic groups being compatible with added fatty acids, which will be linked with hydrogen bonds to the interface without causing any large defects in the chemical or physical properties of the interfacial film. When an amine is added, a strong interaction between the nitrogen group (a base) and the acid groups present in the interfacial film will take place. As a consequence of this interaction, the properties of the interfacial film will be drastically modified and most likely the film will be too hydrophilic to stabilize the aqueous droplets. The amount of amine spent in the direct interaction with the interfacial groups should be rather low.

The mechanism behind a destabilization when using a medium-chain alcohol seems to be quite different. Wasan et al. [79–81] have observed a similar destabilization effect originating from a medium-chain alcohol. They studied the influence of cosurfactants, such as *n*-hexanol, on crude-oil–aqueous-surfactant systems. They found that the cosurfactant speeded up the destabilization process and attributed the higher coalescence rates to a reduction in interfacial rigidity. Quantitatively similar processes can explain the instability found in the Sjöblom et al. work, when a medium-chain alcohol was added. It is also tempting to draw a parallel to the structure of the so-called microemulsions [82–91], where the interfacial region is made very dynamic and flexible when a medium-chain alcohol is, in substantial quantities, combined with an ionic surfactant. Such an interface will obviously show considerable fluctuations and loss in rigidity upon balancing the surfactant [92,93].

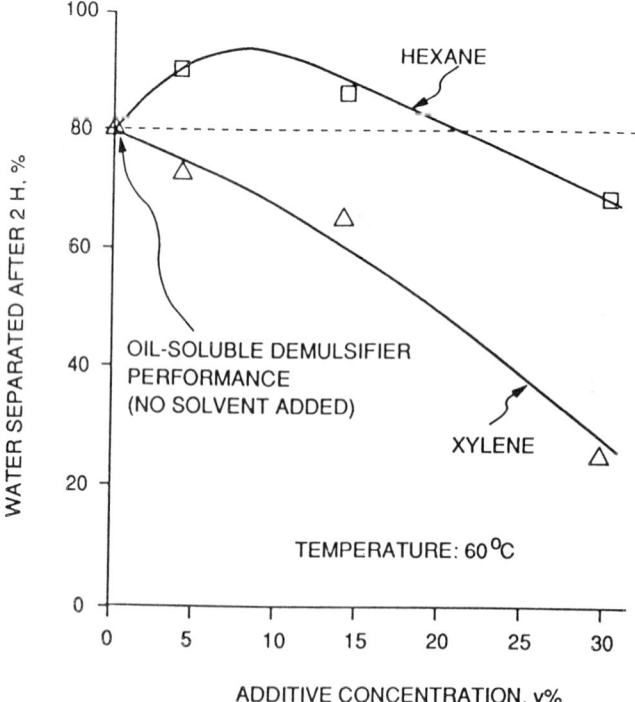

FIG. 29 Effect of the type of solvent on demulsifier efficiency of water-in-crude-oil emulsions. The temperature was 60°C.

Rodriguez [94] conducted an extensive investigation in which the aliphatic/aromatic ratio of a Long Beach, California crude oil was varied by the addition of hexane and xylene in various amounts. At a temperature of 60°C, in the presence of 100 ppm of a conventional demulsifier, an improvement in water separation rate was noted for hexane concentrations up to ~23 vol%, after which water separation became slower. Addition of xylene, however, only resulted in increasingly slower rates of water separation. These effects are illustrated in Fig. 29 and prove that the addition of an organic cosolvent can have widely varied effects. Rodriguez explains the effects in Fig. 29 from the standpoint of asphaltene solubilization/flocculation; the addition of hexane flocculates the asphaltenes, reduces the effective volume they occupy, and thereby removes them from the interface, whereas xylene increases the emulsifying capabilities of the asphaltenes by solubilizing and dispersing them more effectively throughout the oil. For the case of hexane, eventually a point is reached where the solubility of the demulsifier is also affected to the degree that a dropoff in water separation results; most demulsifiers have very low solubilities in aliphatic solvents such as hexane.

Rodriguez also demonstrated that the effects of cosolvents are strongly dependent on temperature. Crude oil can be regarded as a solute (asphaltenes)–solvent (aliphatic and aromatic organic liquids) system with a characteristic Θ-point. The Θ-point represents a transition from a net mutual repulsion among solute segments to a net mutual attraction [95]. A net mutual attraction will lead to asphaltene aggregation and a less stable emulsion. The addition of hexane reduces asphaltene solubility, promotes aggregation, and was shown by Rodriguez to shift the Θ-point to a lower temperature. Thus, the optimal temperature for demulsification will depend heavily on the aliphatic–aromatic composition of the crude oil.

ACKNOWLEDGMENTS

This work was supported in part by the U.S. Department of Energy and by the National Science Foundation.

REFERENCES

1. K. G. Nordli, J. Sjöblom, J. Kizing, and P. Stenius, Colloids Surf. *57*:83 (1991).
2. J. Sjöblom, L. Mingyuan, A. A. Christ, and T. Gu, Colloids Surf. *66*:55 (1992).
3. V. B. Menon and D. T. Wasan, Encyc. Emulsion Technol. *2*:1 (1985).
4. W. D. Bancroft, J. Phys. Chem. *17*:501 (1913).

5. C. S. Shetty, A. D. Nikolov, and D. T. Wasan, J. Disper. Sci. Technol. *13*: 121 (1992).
6. D. A. Edwards, H. Brenner, and D. T. Wasan, in *Interfacial Transport Processes and Rheology*, Butterworth–Heinemann, Boston, 1991.
7. Z. Zapryanov, N. Malhotra, N. Aderangi, and D. T. Wasan, Int. J. Multiphase Flow *9*:105 (1983).
8. I. B. Ivanov, Pure Appl. Chem. *52*:1241 (1980).
9. P. Berger, C. Hsu, and J. P. Arendell, Soc. Petrol. Eng. 457 (1987).
10. M. A. Krawczyk, Ph.D. thesis, Illinois Institute of Technology, Chicago, IL, 1990.
11. A. K. Malhotra and D. T. Wasan, Chem. Eng. Commun, *55*:95 (1987).
12. O. Reynolds, Phil. Trans., Roy. Soc. London *A177*:157 (1886).
13. S. Hartland and S. A. K. Jeelani, Colloids Surf. A: Physicochem. Eng. Aspects *88*:289 (1994).
14. D. Tambe, J. Paulis, and M. M. Sharma, J. Colloid Interface Sci. *171*:463 (1995).
15. A. Bhardwaj and S. Hartland, Ind. Eng. Chem. Res. *33*:1271 (1994).
16. Y. H. Kim, PhD thesis Illinois Institute of Technology, Chicago, IL, 1995.
17. D. T. Wasan, in *Emulsions—A Fundamental and Practical Approach* (J. Sjoblom, ed.), Kluwer Academic Publishers, Amsterdam, 1992, pp. 283–295.
18. R. P. Borwankar, L. A. Lobo, and D. T. Wasan, Colloids Surf. *69*:135 (1992).
19. R. L. Kao, D. A. Edwards, D. T. Wasan, and E. Chen, J. Colloid Interface Sci. *148*:257 (1992).
20. Y. H. Kim, D. T. Wasan, and P. J. Breen, Colloids Surf. *95*:235 (1995).
21. G. E. Charles and S. G. Mason, J. Colloid Sci. *15*:236 (1960).
22. M. A. Krawczyk, D. T. Wasan, and C. S. Shetty, Ind. Eng. Chem. Res. *30*: 367 (1991).
23. P. J. Breen, Langmuir *11*:885 (1995).
24. Y. H. Kim, A. D. Nikolov, D. T. Wasan, H. Diaz-Arauzo, and C. S. Shetty, J. Dispers. Sci. Technol. (in press).
25. Par. J. F. Baret and R. A. Roux, Kolloid-Z. *225*:139 (1967).
26. R. A. Mohammed, A. I. Bailey, P. F. Luckham, and S. E. Taylor, Colloids Surf. A: Physicochem. Eng. Aspects *80*:223 (1993).
27. K. A. Ferworn, W. Y. Svrcek, and A. K. Mehrotra Ind. Eng. Chem. Res. *32*:955 (1993).
28. G. Gonzalez and A. M. Travalloni-Louvisse, SPE Prod. Facilities 91 (1993).
29. R. A. Mohammed, A. I. Bailey, P. F. Luckham, and S. E. Taylor Colloids Surf. A: Physicochem. Eng. Aspects *80*:237 (1993).
30. J. Sjoblom, L. Mingyuan, A. A. Christy, and T. Gu, Colloids Surf. *66*:55 (1992).
31. E. S. Johnnott, Phil. Mag. *70*:1339 (1906).
32. J. Perrin, Ann. Phys. (Paris) *10*:160 (1918).
33. H. G. Bruil and J. Lyklema, Nature Phys. Sci. *233*:19 (1971).
34. J. W. Kenskemp and J. Lyklema, in *Adsorption at Interfaces*, ACS Symposium Series Vol. 8, American Chemical Society, Washington, DC, 1975, p. 191.

35. S. Friberg, St. E. Linden, and H. Saito, Nature *251*:494 (1974).
36. E. Manev, J. E. Proust, and L. Ter-Minassian-Saraga, Colloid Polym. Sci. *255*:1133 (1977).
37. E. Manev, S. V. Sazdanova, and D. T. Wasan, J. Disper. Sci. Technol. *5*: 111 (1984).
38. P. M. Kruglyakov, Kolloid-Z. *36*:160 (1974).
39. A. D. Nikolov, D. T. Wasan, P. A. Kralchevsky, and I. B. Ivanov, in *Ordering and Organization in Ionic Solutions, Proceedings of Yamada Conference XIX*, World Scientific, Singapore, 1988.
40. A. D. Nikolov and D. T. Wasan, J. Colloid Interface Sci. *133*:1 (1989).
41. A. D. Nikolov, D. T. Wasan, P. A. Kralchevsky, and I. B. Ivanov, Colloids Surf. *67*:139 (1992).
42. P. A. Kralchevsky, A. D. Nikolov, D. T. Wasan, and I. B. Ivanov, Langmuir *6*:1180 (1990).
43. A. D. Nikolov, D. T. Wasan, N. D. Denkov, P. A. Kralchevsky, and I. B. Ivanov Prog. Colloid Polym. Sci. *82*:87 (1990).
44. T. Sasaki, M. Hattori, S. Sasaki, and K. Nukina, Bull. Chem. Soc. Jpn. *48*: 1397 (1975).
45. S. Hachisu, Y. Kobayashi, and A. Kose, J. Colloid Interface Sci. *42*:342 (1973).
46. T. Ohtsuki, S. Mitaku, and K. Okano, J. Appl. Phys. *17*:627 (1978).
47. J. W. McBain, G. C. Brock, R. D. Vold, and M. J. Vold, J. Am. Chem. Soc. *60*:187 (1938).
48. T. Nakagawa, K. Kuriyama, and H. Inoue, in *Proc. 12th Symp. of Colloid Chemistry*, Chemical Society of Japan, Tokyo, 1959, pp. 29–37.
49. T. Nakagawa, K. Kuriyama, and H. Inoue, J. Colloid Sci. *15*:268 (1960).
50. D. T. Wasan and A. D. Nikolov, First International Congress on Emulsions, Paris, 1993.
51. A. D. Nikolov and D. T. Wasan, Langmuir *8*:2985 (1992).
52. J. D. F. Ramsay, R. G. Avery, and L. Benest, Faraday Discuss. Chem. Soc. *76*:53 (1983).
53. X. Chu, A. D. Nikolov, and D. T. Wasan, Langmuir *10*:4403 (1994).
54. T. Yoshiyama and I. S. Sogami, Langmuir *3*:851 (1987).
55. A. D. Nikolov and D. T. Wasan, J. Colloid Interface Sci. *133*:13 (1989).
56. B. Beresford-Smith, C. Chan, and D. T. Mitchell, J. Colloid Interface Sci. *105*:216 (1985).
57. D. G. Hall and H. M. Rendall, J. Chem. Soc. Faraday Trans. I *76*:2575 (1980).
58. R. Hunter, in *Foundations of Colloid Science*, Vol. II, Clarendon Press, Oxford, 1989, p. 735.
59. C. W. Nut, Chem. Eng. Sci. *12*:133 (1960).
60. H. M. Princen, in *Surface and Colloid Science*, Vol. 2 (E. Matijevic, ed.), Wiley, New York, 1969, pp. 1–84.
61. C. Huh and S. G. Mason, J. Colloid Interface Sci. *47*:271 (1974).
62. S. Winitzer, Separ. Sci. Technol. *8*:45 (1973).
63. S. Winitzer, Separ. Sci. Technol. *8*:647 (1973).

64. A. V. Rapachietta and Neumann, J. Colloid Interface Sci. *59*:555 (1977).
65. A. B. Taubman and A. F. Koretskii, in *Advances in Colloid Chemistry*, Nauka, Moscow, 1973.
66. D. E. Tambe and M. M. Sharma, J. Colloid Interface Sci. *162*:1 (1994).
67. D. E. Tambe and M. M. Sharma, J. Colloid Interface Sci. (in press).
68. A. Tsugita, S. Takemoto, K. Mori, T. Yoneya, and Y. Otami, J. Colloid Interface Sci. *95*:551 (1983).
69. V. B. Menon, Ph.D. thesis, Illinois Institute of Technology, Chicago, IL, 1986.
70. P. Pieranski, Phys. Rev. Lett. *45*:569 (1980).
71. D. Y. C. Chan, J. D. Henry, Jr., and L. R. White, J. Colloid Interface Sci. *79*:410 (1981).
72. V. B. Menon, A. D. Nikolov, and D. T. Wasan, J. Colloid Interface Sci. *124*: 317 (1988).
73. S. Hartland and R. W. Hartley, in *Axisymmetric Fluid–Fluid Interfaces*, Elsevier, Amsterdam, 1976.
74. V. B. Menon and D. T. Wasan, Colloids Surf. *29*:7 (1988).
75. V. B. Menon, R. Nagarajan, and D. T. Wasan, Separ. Sci. Technol. *22*:2295 (1987).
76. J. Sjöblom, H. Soderlund, S. Lindblad, E. J. Johansen, and I. M. Skjarvo, Colloid Polym. Sci. *268*:389 (1990).
77. E. J. Johansen, I. M. Skjarvo, T. Lund, J. Sjöblom, H. Soderlund, and G. Bustrom, Colloids Surf. *34*:353 (1989).
78. D. E. Tambe and M. M. Sharma, J. Colloid Interface Sci. *157*:244 (1993).
79. D. T. Wasan, J. J. McNamara, S. M. Shah, K. Sampath, and N. Aderangi, J. Rheol. *23*:181 (1979).
80. D. T. Wasan, K. Sampath, and N. Aderangi, AIChE Symp. Ser. *76*:93 (1980).
81. V. B. Menon and D. T. Wasan, in *Encyclopedia of Emulsion Technology. Applications*, Vol. 2 (P. Becker, ed.), Marcel Dekker, Inc., New York, 1985.
82. I. Danielsson and B. Lindman, Colloids Surf. *3*:391 (1981).
83. T. P. Hoar and J. H. Schulman, Nature *152*:102 (1943).
84. M. Zulauf and H. F. Eicke, J. Phys. Chem. *83*:48 (1979).
85. A. M. Cazabat, D. Langevin, and A. Pouchelo, J. Colloid Interface Sci. *73*: 1 (1986).
86. D. J. Cebula, R. H. Ottewill, and J. Ralston, J. Chem. Soc Faraday Trans. I *77*:2585 (1981).
87. R. H. Cole, G. Delbos, P. Winsor IV, T. K. Bosaer, and J. M. Moreau, J. Phys. Chem. *89*:3338 (1985).
88. B. Lindman, P. Stilbs, and E. Moseley, J. Colloid Interface Sci. *83*:569 (1981).
89. B. Lindman and P. Stilbs, in *Microemulsions* (Friberg, S. and Bothorel, P., eds.), CRC Press, Boca Raton, FL, 1987.
90. T. Warnheim, E. Sjöblom, U. Henriksson, and P. Stilbs, J. Phys. Chem. *88*: 5420 (1984).
91. J. Biais, L. Odberg, and P. Stenius, J. Colloid Interface Sci. *86*:350 (1982).

92. P. G. de Gennes and C. Taupin, J. Phys. Chem. *86*:2294 (1982).
93. B. P. Binks, J. Meunier, O. Abillon, and D. Langevin, Langmuir *5*:415 (1989).
94. O. C. Rodriguez, M. S. thesis, Illinois Institute of Technology, Chicago, IL, 1992.
95. R. J. Hunter, in *Foundations of Colloid Science*, Clarendon Press, Oxford, 1987.

5

Food Emulsions

DOUGLAS G. DALGLEISH Department of Food Science, University
of Guelph, Guelph, Ontario, Canada

I. INTRODUCTION

Food emulsions are probably the only types of emulsion with which everybody is acquainted. Unskimmed milk and cream are emulsions, as are, for example, butter, margarine, spreads, mayonnaises and dressings, creamers, some fruit drinks, processed cheeses, ice creams, and whippable toppings. Despite their variety, there is nothing special about them; they are formulated in the same way and are governed by the same principles as are other emulsion systems. Specific requirements for these emulsions are that they must possess long-term stability (i.e., they may need to have a shelf life of several months or possibly years) and that they must be edible (i.e., they should contain only ingredients which are acceptable for human consumption). Clearly, both of these requirements place restrictions on some of the possible ways of formulating the emulsions. Because there is increasing interest in providing foods based on what are perceived to be "natural" ingredients, so the maximum use of naturally available materials such as proteins or phospholipids becomes important in the final product. Therefore, to understand the formulation and behavior of food emulsions, it is necessary to understand as much as possible about these types of emulsifiers, because they may not behave just in the way that "classical" small-molecule emulsifiers are expected to behave. For example, phospholipid molecules may interact with each other to form lamellar phases or vesicles; they may interact with neutral lipids to form a monolayer or multilayer around the lipid droplets, or they may interact with proteins which are either adsorbed or free in solution. Any or all of these interactions may occur in the one emulsion. Depending on which reaction predominates, the properties of the emulsion system will be affected.

Unfortunately for those who have to formulate emulsions, it is rarely possible to consider a food emulsion simply as oil coated with one or a mixture of surfactants. Almost always there are other components whose properties may need to be considered along with those of the emulsion droplets themselves. For example, various metal salts may be included in the formulation (for example, Ca^{2+} is nearly always present in food products derived from milk), and there may also be polysaccharide gums present to increase the viscosity or yield stress of the continuous phase and prevent creaming of the emulsion. In addition, it is almost always the case that some protein is free in solution, having either not adsorbed at all or having been displaced by other surfactants. Any of these materials (especially the metal salts and the proteins) may interact with the adsorbed material which forms the surface layer of the emulsion droplets, and cause instability (flocculation, gelation) of the emulsions. This will be especially true at high temperatures, and it is again a feature of many food products that they have to undergo a heating process so as to achieve or maintain sterility. These heating treatments vary from simple pasteurization (72°C for 15 s) to retort sterilization (120°C for 10 min) to ultrahigh-temperature (UHT) treatments (140°C for 4 s). During heating, proteins present in the product may become denatured and may cause immediate or delayed flocculation or gelation, which are generally deleterious to the food product.

Food oils are almost always triglycerides, and they also play a part in determining the properties of an emulsion, because many of them are partly crystalline at room temperature. There is little evidence that the particular fatty acids in the oil alter adsorption of surfactant, but the crystallinity of the oil affects the homogenization process and is also critical in processes such as partial coalescence, providing texture to various foods. So, in formulation it is important to know whether an oil is saturated or not, so as to give it a melting profile suitable for the product. Although emulsions can be made using fat or oil (basically reflecting the source of triglycerides), no distinction will be made here; the term "oil" will be used throughout to mean liquid or solid triglycerides of animal or vegetable origin.

Oil-in-water (O/W) emulsions (e.g., creams, coffee creamers, cream liqueurs, and mayonnaise) are mainly fluid, although they may have partly crystalline oil phases. Stability of these emulsions may be maintained by adsorption of small-molecule emulsifiers, protein molecules, or aggregates of protein molecules (casein micelles, egg yolk granules), or by mixtures of these. Not all of the emulsions are required to have very high long-term stability because for whipped toppings and ice cream mixes, the emulsion needs to be stable for some time but then must be capable of being destabilized to form the final product. On the other hand, emulsions such as evaporated milks and mayonnaises are required to be stable to flocculation and creaming for long periods of time so as to minimize coalescence and phase separation of the oil. Although stability may be enhanced by the presence of materials giving viscosity or a yield stress to the continuous phase (e.g., gums), it is not good practice to rely solely upon these effects to stabilize the emulsions.

Water-in-oil (W/O) emulsions (butter, margarines, spreads) are not stabilized simply by forming an adsorbed layer of surfactant which minimizes the effects of interparticle collisions. Important factors in these emulsions are the crystallinity of the oil phase, the presence of rigid surfactants on the O/W interface, and the presence of agents which increase the viscosity of the aqueous phase in the droplets. Mechanical stabilization is, therefore, more important in these emulsions. In some cases, it is possible to make water-in-oil-in-water (W/O/W) emulsions by homogenizing a W/O emulsion in presence of suitable surfactants [1].

This chapter will be concerned almost exclusively with liquid O/W emulsions, mainly because they are the most exhaustively studied and the principles for their behavior are the most thoroughly established, not necessarily because they are the most important of the emulsions. For example, no description of emulsions in meat products or in bread mixes and cake batters is given, as these are less understood from a fundamental point of view than are the more simple O/W emulsions. What is attempted here is a description of the structures of emulsion droplets and how these affect the properties of the emulsion.

II. SURFACTANTS

A. Small-Molecule Surfactants

There is a range of surfactants which can be used in the formulation of emulsions. Small-molecule surfactants (monoglycerides and diglycerides, sorbitan esters of fatty acids, polyoxyethylene sorbitan esters of fatty acids, phospholipids, and many others) generally contain long-chain fatty acid residues, which provide the hydrophobic group which binds to the lipid phase of the oil–water interface and causes adsorption. The head groups of these emulsifiers are more varied (Fig. 1), ranging from glycerol (in monoglycerides and diglycerides) and substituted phosphoglyceryl moieties (in phospholipids) to sorbitan highly substituted with polyoxyethylene chains [2]. The differences between these emulsifiers are generally expressed in terms of the hydrophile–lipophile balance (HLB), so that the more oil-soluble surfactants have low HLB numbers and the more water-soluble ones have high numbers, with the value of 7 being treated as "neutral" [3]. The presence of emulsifiers of a low HLB number favors the formation of W/O emulsions, and a high HLB number promotes W/O emulsions, although this is not a completely hard and fast rule [135].

Because these molecules adsorb strongly to the oil–water interface and have few steric constraints to prevent them from packing closely, they produce low interfacial tensions [4] and lower the Gibbs interfacial energy. However, they do not generally give highly cohesive or viscous surface layers, so that the adsorbed layers of these molecules may be relatively easily disrupted, and this is used in certain types of emulsions.

Of these surfactant molecules, the phospholipids (lecithins) are in a class of their own, as they are capable of forming particular structures (e.g., bilayers and vesicles), which, in turn, may interact with the O/W

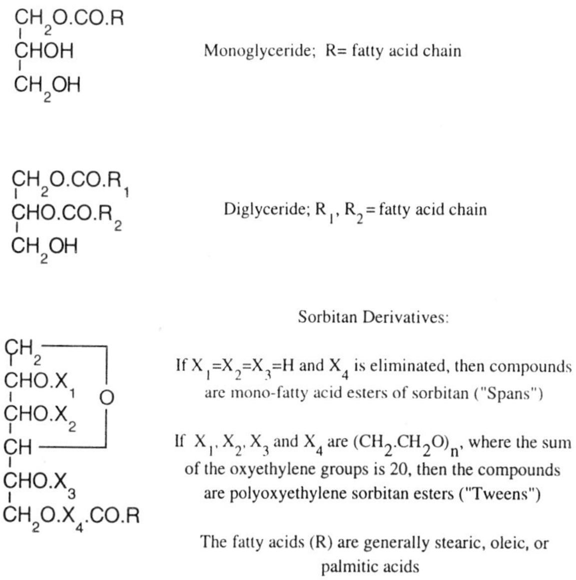

CH$_2$O.CO.R
CHOH   Monoglyceride; R= fatty acid chain
CH$_2$OH

CH$_2$O.CO.R$_1$
CHO.CO.R$_2$ Diglyceride; R$_1$, R$_2$= fatty acid chain
CH$_2$OH

Sorbitan Derivatives:

CH$_2$
CHO.X$_1$ O
CHO.X$_2$
CH
CHO.X$_3$
CH$_2$O.X$_4$.CO.R

If X$_1$=X$_2$=X$_3$=H and X$_4$ is eliminated, then compounds are mono-fatty acid esters of sorbitan ("Spans")

If X$_1$, X$_2$, X$_3$ and X$_4$ are (CH$_2$.CH$_2$O)$_n$, where the sum of the oxyethylene groups is 20, then the compounds are polyoxyethylene sorbitan esters ("Tweens")

The fatty acids (R) are generally stearic, oleic, or palmitic acids

FIG. 1 Structures of some typical emulsifiers.

interface. Experience suggests that these materials do not behave as simply as the small molecules on the interface, and this will be discussed in a later section.

B. Proteins

Proteins, on the other end of the scale of molecular complexity, act as emulsifiers but behave differently from the small molecules, because of their individual molecular structures, and, indeed, it is the particular proteins present which give many food emulsions their characteristic properties. Most, if not all, proteins possess defined three-dimensional structures which are maintained in solution, unless they are subjected to disruptive influences such as heating [5]. It is unlikely that the peptide chains of proteins dissolve significantly in the oil phase, as they are quite hydrophilic; it is more likely that the major entities penetrating the interface are the side chains of the amino acids (Table 1). The structure of the protein itself will ensure that these points of contact are not closely packed, and as a result, the adsorption of a protein reduces the interfacial tension less than does the adsorption of small molecules (Table 2). Although some proteins are excellent emulsifiers, it is evident that not all

TABLE 1 Hydrophobic and Hydrophilic Amino Acids in Side Chains of Protein Molecules

Hydrophobic	Hydrophilic
Tryptophan	Glutamic acid
Isoleucine	Aspartic acid
Tyrosine	Histidine
Phenylalanine	Serine
Proline	Threonine
Leucine	Arginine
Valine	

Note: Both lists are in descending order; amino acids not mentioned are of indeterminate nature.
Source: After Ref. 6.

proteins can adsorb strongly to an O/W interface, either because they are strongly hydrophilic or because they possess rather rigid structures which do not allow the protein to adapt to the interface. Examples of such proteins are gelatin, which forms poor emulsions because of its hydrophilic character and is a large rather rigid molecule [8], and lysozyme, which does adsorb to O/W interfaces but tends (presumably because of its relatively inflexible structure) to be a poor emulsifier [9].

Because adsorption of proteins occurs via hydrophobic amino acids, it seems a reasonable assumption that a measurement of surface hydrophobicity [10] would allow prediction of the emulsifying power of a protein [11]. However, surface hydrophobicity, as determined by the binding of probe molecules [12], may be an unreliable measure because of the capac-

TABLE 2 Reduction of Interfacial Tension by Proteins and Surfactants

Molecule	$\gamma_0 - \gamma$ (mN m^{-1})	Oil phase	Ref.
β-Casein	25	Tetradecane	13
β-Lactoglobulin	21	Tetradecane	96
Gelatin	15	Tetradecane	8
Phosvitin	12	Tetradecane	13
Lysozyme	17	Toluene	27
$C_{12}E_2$ surfactant	35	Tetradecane	7
Glycerol monostearate	28	Sunflower oil	4

ity of the protein to change conformation once it has adsorbed (see below), with the effect that adsorption becomes stronger with time. Equally, the apparent hydrophilicity may also be misleading, as evidenced by the egg protein, phosvitin, which is a fairly good emulsifier [13,14] despite having more than 50% of its residues composed of phosphoserine [15], which is charged and, of course, hydrophilic.

C. Adsorption and Protein Conformation

Much research has recently been aimed at determining the behavior of adsorbing proteins, and it seems likely that most of the proteins which adsorb to interfaces are capable of changing conformation either as they adsorb or shortly afterward. Of course, surface denaturation is not a new concept [16,17], but it underlines the possibility that unfolding of the protein after adsorption will liberate more hydrophobic areas which then interact further with the O/W interface (Fig. 2).

A number of sources confirm that proteins change their conformation when they adsorb to liquid or solid interfaces. Spectroscopic studies of lysozyme, for example, show a decrease in the secondary structure caused by adsorption to polystyrene latex [18]. Proteins adsorbed to a surface and subsequently desorbed by the action of small molecules may also have an altered conformation [19], confirming that the changes caused by adsorption may be irreversible. Adsorbed proteins may also interact chemically by formation of intermolecular disulphide bonds to form oligo-

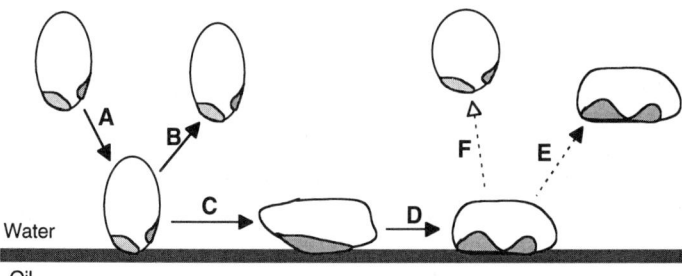

FIG. 2 Adsorption of protein on an oil–water interface. The protein diffuses to the interface (A), and makes contact with it through hydrophobic areas (shaded). It may immediately diffuse away (B) or begin to change conformation (C), until, after some time (D), full surface denaturation has occurred. If the protein is removed, for example, by added surfactant (E), it may retain its denatured conformation or return to a conformation similar to its native state (F).

mers, as has been shown for adsorbed β-lactoglobulin and α-lactalbumin [20], although such reactions are not favored in solution unless the proteins are denatured by heating [21]. Recently, several studies have demonstrated that the heat of denaturation of adsorbed proteins, as measured by differential scanning calorimetry (DSC), is very much diminished, again suggesting that unfolding on the surface has occurred [22–24]. In some cases (e.g., lysozyme and α-lactalbumin), this surface denaturation is at least partially reversible, but in others (e.g., β-lactoglobulin), adsorption causes irreversible changes in the protein molecules [24].

Caseins tend to be a special case. Because these proteins do not contain much rigid secondary structure [25] and because they possess considerable numbers of hydrophobic residues [26], they adsorb well [27,28]. However, because of their flexible nature, it is impossible to determine whether conformational changes occur during adsorption, as neither spectroscopic changes (e.g., as in circular dichroism (CD) spectroscopy) nor DSC are capable of demonstrating conformational changes in these proteins.

It is worth remembering that the behavior of adsorbed protein in emulsions (for instance, intermolecular interactions such as those mentioned above) may simply be a result of the local concentration within the adsorbed interfacial layers. Generally, we know [29,30] that for proteins the interfacial concentration (surface excess, Γ) is of the order of 1–3 mg m^{-2}, and that the adsorbed layers are generally less than 5 nm thick [31], so it is simple to calculate that in the interfacial region a monolayer of protein has a concentration of about 500 mg ml^{-1}, that is, 50% (Table 3). It is impossible to make a protein solution of this concentration because of its extremely high viscosity, so direct comparisons are not possible, but it should be evident that the protein in the adsorbed layer may be in a favorable position for intermolecular interactions, because the molecules are close to one another and adsorption holds them in position so that diffusion is very slow. Therefore, the adsorbed layer of protein may partake more of the nature of a gel than a solution; this is at least partly the reason why many adsorbed proteins form highly viscous interfacial layers. It must be remembered that these will be essentially two-dimensional gels, with each molecule occupying about 11 nm^2 of interface (calculated on the basis of a molecule of 20,000 Da molecular weight and a surface coverage of 3 mg m^{-2}; this fits well with the expected dimensions [34] of a globular protein of this weight and is much larger than the 0.5–2.5 nm^2 per molecule which has been found for modified monoglycerides [35]). It is, therefore, not surprising that adsorption can alter the behavior of proteins. Moreover, the formation of such concentrated layers has relatively little to do with the overall bulk concentration of the protein in solution, which may give stable emulsions at relatively low bulk concentrations (although this de-

TABLE 3 Calculated[a] Interfacial Properties of Proteins (Assumed Molecular Mass of 20,000 Da)

Protein load (mg m^{-2})	Area/ molecule (nm^2)	Equivalent radius[b] (nm)	Interfacial concentration[c] (mg ml^{-1})		
			10 nm	5 nm	2 nm
3	11.1	1.9	300	600	(1500)[d]
2	16.6	2.3	200	400	(1000)[d]
1	33.2	3.3	100	200	500
0.8[e]	41.5	3.6	80	160	400

[a] Ideal calculation, which assumes no specific properties of individual proteins.
[b] Calculated assuming each molecule occupies a circle.
[c] Calculated for three different thicknesses of the interfacial layer, as found for high and low coverage with caseins [32] and for β-lactoglobulin [33,77].
[d] Figures greater than about 1000–1200 are impossible from density considerations.
[e] Selected because this is about the lowest effective Γ to give a stable emulsion with caseinate [32] in the absence of other surfactant material.

pends of course on the amount of oil and the interfacial area to be covered). Table 3 also shows the possible limits for the formation of multilayers. Caseins form extended layers about 10 nm thick, and even at a Γ of 3 mg m^{-2} have a concentration of about 300 mg ml^{-1}. Conversely, whey proteins form much thinner layers (about 2 nm thick) and will have to form multilayers if Γ is more than about 2 mg m^{-2}, as there is no further space available for monolayer adsorption beyond that point.

One factor which may have considerable importance on the emulsifying properties of proteins is their quaternary structure. For example, in milk the caseins exist in aggregates of considerable size (casein micelles) containing between about 500 and 10,000 individual protein molecules [36]. These particles act as the surfactants when milk is homogenized [37]. On the other hand, caseinate prepared from milk can exist in a much less aggregated state [38] and is superior to the micelles in emulsifying properties [39]. Simply, this may be envisaged as a concentration effect; with casein in its natural micellar form, the protein has a much lower concentration (in terms of numbers of particles) than does a solution of caseinate. Therefore, during homogenization, the nonmicellar casein will find its way to the interface more readily than the micelles. Molecules such as β-lactoglobulin also show changes in quaternary structure as a function of pH [40], and these may be related to the protein's surfactant properties [41].

As we consider the details of the proteins which are used in emulsions, we find that, with a few exceptions, all of the detailed research has been performed on relatively few proteins. Of these, the caseins (α_{s1}, α_{s2}, β, and κ) and whey proteins (α-lactalbumin and β-lactoglobulin) predominate. This is principally because these proteins are readily available in pure and mixed forms in relatively large amounts; they are all quite strongly surfactant and are already used in the food industry. Other emulsifying proteins are less amenable to detailed study by being less readily available in pure form (e.g., egg yolk proteins and lipoproteins). Many other available proteins are less surface active than the milk proteins, for example soya isolates [42], presumably because they exist as disulphide-linked oligomeric units rather than as individual molecules [43]. Even more complexity is encountered in the phosphorylated lipoproteins of egg yolk, which exist in the form of granules [44], which themselves can be the surface-active units (e.g., in mayonnaise) [45].

Therefore, in the detailed descriptions of model emulsions given below, we will find that nearly all of them (especially where the surfactant proteins are considered at the molecular level) concern themselves with the milk proteins.

III. EMULSIFYING ACTIVITY OF SURFACTANTS

Ideally, it is necessary to make estimates of the potential of given surfactants for forming emulsions. The problem is to define techniques which are method independent, that is, which give absolute results, or at least give results applicable to specific methods for preparing emulsions. Two methods have been widely used, those of Emulsifying Activity Index (EAI) and emulsifying capacity (EC). In the second of those, a known quantity of protein is dissolved and then oil is added in a blender. This forms a crude emulsion, and further aliquots of oil are added until the emulsion inverts or free oil is seen to remain in the mixture. This ostensibly gives the weight of oil which can be emulsified by the defined weight of protein. However, clearly this is dependent on the particular blender because what is important in emulsion formation is not the weight of oil but its interfacial area. Thus, if the emulsion is made of large droplets, it will consume less protein than if small droplets are present. The conditions of emulsion formation are therefore critical to the method, as it is possible to obtain different results at different blender speeds. Therefore, the method may not be generally transferable between different laboratories, nor is it in any sense an absolute measure.

To measure the EAI, an emulsion is made and the particle size is estimated, usually by turbidimetric methods [46]. The assumption is then made that all of the protein is adsorbed to the interface, and so a measure of emulsifying potential can be measured. Although it provides more information than EC, the method has two major defects: The first is that it is by no means certain that all of the available protein is adsorbed, or that it is adsorbed as a monolayer. Indeed, it is known that at concentrations of protein of more than about 0.5% (with oil concentration of 20%), some of the protein remains unadsorbed, even after powerful homogenization where the concentration of protein is the limiting factor in the determination of the sizes of the droplets [32,47]. If homogenization is less extensive, then the proportion of protein adsorbed decreases. The second major problem in determining EAI is simply the difficulty in determining the particle sizes and their distribution. Perhaps it is no exaggeration to say that there is no totally correct method of measuring the true size distribution of suspended particles; certainly, there is none which is really rapid, convenient, and free from error (see the following section). Generally, the particle sizes in determinations of EAI are measured by turbidity of diluted suspensions of the emulsions. However, this may introduce errors into the calculation, especially if the emulsion is composed of fine droplets or if the distribution of particle sizes is broad.

Ideally, to describe an emulsion, the particle size distribution and the amounts of particular surfactants adsorbed to the oil–water interface need to be defined. In addition, although this is more difficult to determine, it is desirable to know the state of the adsorbed material (e.g., its conformation, which parts of the adsorbed molecules protrude into solution and are available for reaction, etc.). This, of course, represents an ideal which it is rarely possible to achieve, but the explanation of the behavior of emulsions, and perhaps the design of new ones, may depend on this knowledge.

IV. FORMATION OF EMULSIONS

Emulsions in foods are generally formed using colloid mills or high-pressure homogenization. In the former, the oil–water–surfactant mixture is passed through a narrow gap between a rotor and stator; the stresses imposed on the mixture are sufficient to break up the oil into droplets, which are then coated with surfactant [3]. This method is used mainly when it is not important to produce very small droplets of emulsion, as it has a lower size limit of about 2 μm in diameter. The technique is used to manufacture mayonnaises and salad creams, where stability depends less on the presence of very small particles than on the overall composition and viscosity of the preparation. In liquid emulsions, smaller particles are required to prevent creaming.

High-pressure homogenization is used to produce these smaller droplets. First, a coarse emulsion of the ingredients is formed using a blender
or similar device, and this mixture is then passed through the homogenizing valve, at pressures varying from 6.8 to 23.8 MPa (1000 to 3500 psi).
This high-pressure flow through the valve creates turbulence, which pulls
apart the oil droplets, after which the surfactant molecules adsorb to the
newly created interface [48]. If the adsorption is not rapid, or if there is
insufficient surfactant present, then recoalescence of the oil droplets occurs [49]. Apart from the mechanical properties of the system, the sizes
of the emerging droplets depend on the power input; that is the homogenization pressure (Fig. 3), the number of passes [50], and the amount of
surfactant present (Fig. 4) [32]. The particle size may be limited by the
characteristics of the homogenizer if there is a large excess of surfactant;
on the other hand, if only small amounts of surfactant are present, then

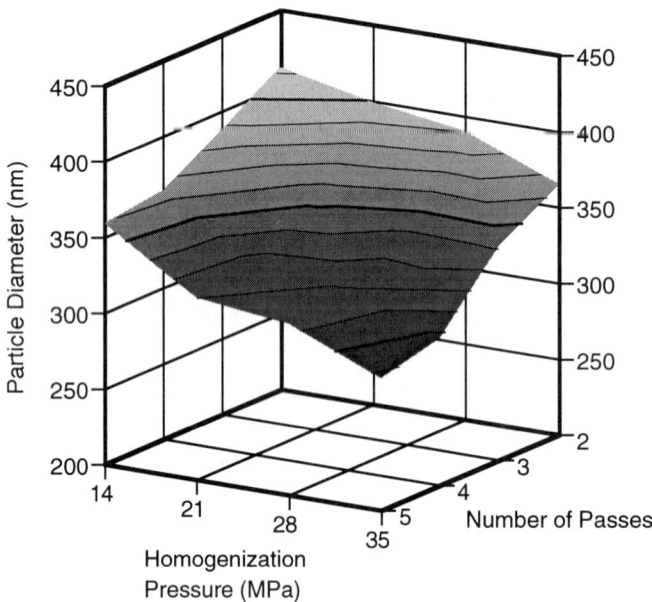

FIG. 3 Diameters of particles in homogenized milk, as a function of homogenization pressure and of the number of times the milk has been passed through the
homogenizer (S. M. Tosh and D. G. Dalgleish, unpublished results). The results
shown are for milk homogenized using a Microfluidizer (Microfluidics, Inc, New-
ton, MA) at a temperature of 40°C. Particle sizes were measured using photon
correlation spectroscopy. Similar results have been obtained by Robin et al. [50].

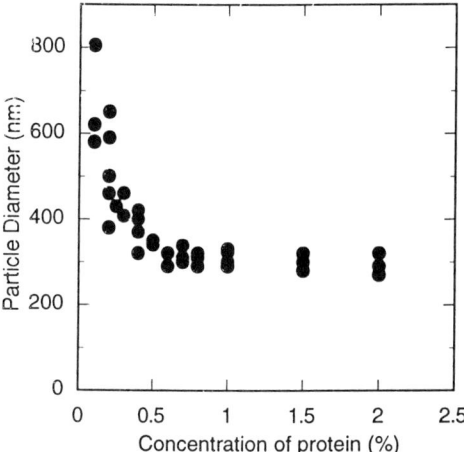

FIG. 4 Limitation of particle size in emulsions by the amount of surfactant present. Diameters of particles formed by homogenization of soya oil (20% w/w) into solutions containing casein, whey protein isolate, or casein + lecithin, as functions of the concentration of protein present. Results shown are a compendium from Refs. 32, 47, and 103. Emulsions were prepared using a microfluidizer and average sizes were measured using photon correlation spectroscopy.

it is this which limits the sizes of the particles. Thus, as the compositions of products are reformulated, the sizes of the emulsion droplets in them may well also change.

In addition to recoalescence and increased droplet size, a further manifestation of insufficient macromolecular surfactant may be seen in the phenomenon of bridging flocculation, that is, the clustering of the emulsion droplets as they emerge from the homogenizer. This process occurs because one surfactant molecule becomes adsorbed onto two separate oil droplets. Such behavior is possible for macromolecules, which may have two parts of the molecule capable of adsorption but not for small-molecule surfactants. Proteins can form bridges in this way [51], and even more commonly, natural aggregates of proteins such as casein micelles can induce clustering of the oil droplets [52]. The solution to bridging flocculation is to incorporate more surfactant (which need not be macromolecular) so as to provide enough material to cover the nascent interface. In the case of clustering by particles, a second-stage homogenization at low pressure (3.4 MPa, 500 psi) is generally sufficient to break down the bridging aggregates and to separate the clustered fat globules. Clearly, however, such treatment will be inapplicable to clusters bridged by single macromolecules, which cannot be broken up in this way.

V. MEASUREMENT OF PARTICLE SIZES AND SIZE DISTRIBUTIONS IN EMULSIONS

Having formed an emulsion by homogenization or other means, it is necessary to characterize it, specifically in terms of its size distribution. This is important in a number of respects: knowledge of the size distribution provides information on the efficiency of the emulsification process, and because monitoring of any changes in the distribution as the emulsion ages gives information on the stability of the system. Also, such size distributions may be important when emulsion systems or processes are patented. However, the measurement of size distributions or even the average sizes of emulsion droplets is not simple, despite the existence of a number of potentially useful methods.

The most direct method, and one which is perhaps least subject to errors, is electron microscopy [53]. This technique can be used to determine the number-average size distribution, providing that (i) a fully representative sample of the emulsion is prepared, fixed, mounted, sectioned, and stained without distortion, (ii) a sufficient number of particles are measured to ensure statistical accuracy of the distribution, and (iii) that proper account is taken of the effects of sectioning. All of this requires considerable time and effort, and so, this is not a technique to be used routinely to determine size distributions. It may be used as a standard against which to compare other methods.

Many of the more rapid methods tend to emphasize one end or other of the distribution and, therefore, to give biased results. Measurement of the particle sizes by conductivity (Coulter counter), for example, is insensitive to the presence of small particles, so it will not register particles below about 0.8 μm [54,55]. For fine emulsions, therefore, this technique will lead to underestimation of the population of small particles. Techniques involving light scattering, which are probably the most widely used, also have some disadvantages. The simplest of these methods depends on the measurement of turbidity at one or a number of wavelengths [46,56]. Turbidity, or apparent absorbance of light, simply measures the total amount of light scattered as it passes through a cuvette containing diluted emulsion. Although the method is rapid and may be performed in any laboratory possessing a spectrophotometer, it cannot be used to give the true distribution of particle sizes, but simply to give an average. If required, it can be assumed that the particles form a distribution of known shape, but this, of course, assumes that the distribution is known beforehand, leading to a circular argument. The turbidimetric method also emphasizes the larger particles because they scatter more light, and so the method tends to underestimate the contribution of smaller particles.

A number of commercial instruments measure the distributions of particle sizes by determining the intensity of light scattered from a diluted sample at specific scattering angles in the range 0°–30° (integrated light scattering, ILS). With knowledge of the scattering properties (i.e., the Mie scattering envelope) of the particles [57], software is used to calculate the most probable distribution, knowing that

$$I(\theta) = \int_0^\infty I(\theta, r)G(r) \, dr, \tag{1}$$

where $G(r)$ is the distribution of radii, $I(\theta)$ is the measured light intensity at angle θ, and $I(\theta, r)$ is the scattered intensity from a particle of radius r at angle θ.

This does not necessarily yield the absolute distribution, for two main reasons. The first of these is that the angular range is generally too restricted to allow measurement of small particles; these scatter isotropically, whereas large particles preferentially scatter in the forward directions, so that to measure the distribution accurately, a span of scattering angles between 0° and 150° is essential [58]. The result of the limitation of the angular range is that the contribution of small particles tends to be underestimated, and the instruments do not attempt to estimate the contribution of particles smaller than 0.1 μm. Unfortunately, many food emulsions contain particles ranging from 50 nm to 1 μm, and this is precisely where many instruments are least accurate. A further problem with any light-scattering method is that the accuracy of the calculated distribution depends on how well the optical properties of the emulsion droplets (i.e., their real and imaginary refractive indices) can be defined. In all cases, the droplets are assumed to be spherical, but it may be necessary to make assumptions about the structures of the interfacial layers. An emulsion droplet is essentially a coated sphere [59], which is characterized by refractive indices of the core and the coat, and these need not be equal; in fact, they may be quite different. Calculations of the scattering behavior of emulsion droplets may, therefore, depend on the presumed structures of the particles.

Dynamic light scattering (DLS) offers an alternative means of measurement [60]. This technique does not measure the total amount of light scattered but the dynamics of the scattered light over short time periods. Usually, the light scattering is measured at a fixed angle of 90°, and a correlation function is measured. This is essentially a weighted sum of exponentials, which depend on the diffusion coefficients of the particles through the aqueous medium:

$$g^{(1)}(\tau) = a \int_0^\infty G(r)I(r, \theta) \exp(-K^2 D(r)\tau) \, dr, \qquad (2)$$

where a is a constant, K is the light-scattering vector for the scattering angle θ, and $D(r)$ is the diffusion coefficient for the particle in the solution.

The weighting factors $I(r, \theta)$ depend on the amounts of light scattered by particles of different sizes; that is, they depend on the Mie scattering of the particles at the angle of measurement. As with ILS, the calculation of the true size distribution depends on the knowledge of the detailed light-scattering properties of the emulsion droplets. In addition, the fit of theory to the true correlation function is ill-conditioned [61], so that the size distribution which is obtained is to a certain extent dependent on the technique used to fit the correlation function. Once again, it is observed that this technique usually overemphasizes the contribution of larger particles to the size distribution. This is partly because they tend to scatter more (i.e., have higher weighting factors), but also because of the nature of the correlation function itself, as the information about the small particles is contained only in the short-time part of the function, whereas information about the large particles is contained at all points.

Figure 5 shows a number of results on the distribution of particle sizes in a particular emulsion, as determined by different methods. It is evident that considerable differences exist. Of the different techniques used, it was evident that a custom-built ILS spectrometer measuring between angles of 4° and 145° was capable of demonstrating that a considerable population of particles with diameters <0.1 μm were present in the mixture, and that none of the other methods could do so; that the small particles existed was confirmed by electron microscopy [62]. The DLS methods were dependent on the configuration of the instrument but provided some information about the small particles. These measurements also illustrated the general principle that the more accurate the answer required, the more laborious and time-consuming is the experimental procedure, which makes it difficult to use the technique routinely.

Although it is not particularly serious in the case of simple emulsions, it is necessary to ensure that no dissociation of particles is caused by the high dilution which is required for accurate light-scattering experiments. This may lead to dissociation of flocculated material, or to the breakdown of complex interfacial layers (e.g., those formed by casein micelles on the oil–water interfaces in homogenized milk). Although a method for measuring light scattering in concentrated solutions exists [63], very few experiments involving emulsions have used the technique, and its full effectiveness remains to be demonstrated.

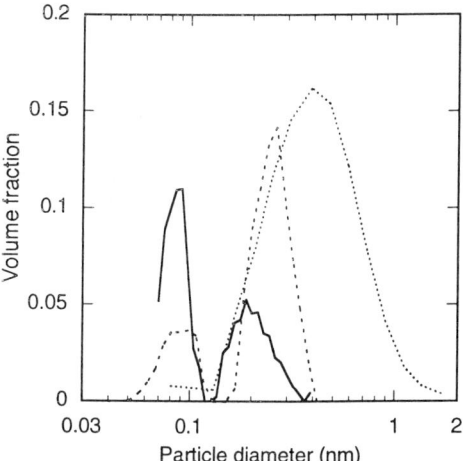

FIG. 5 Particle size distributions in emulsions (20% w/w soya oil) stabilized by caseinate, measured by three different instruments. Solid line—integrated light scattering using a range of scattering angles from 4° to 150°; broken line—measured using photon correlation spectroscopy at a fixed scattering angle of 90°; dotted line—measured by diffraction at scattering angles between 0° and 30°. Note the limited response of the second and third methods to smaller particles. (Data adapted from Ref. 62.)

Despite all of these problems, light-scattering methods are the most effective means of obtaining information about the size distributions of particles in emulsion systems. Especially, these may be used in a comparative mode, to measure changes which occur in the suspensions. All of the methods can detect whether aggregation is occurring, so that they may all be used to detect instability of emulsions. It is under these circumstance that the light-scattering methods come into their own, as they are almost unique in allowing the kinetics of aggregation to be studied on a real-time basis [64]. Simply, the fact that the particle size is increasing can be determined without any particular attributes of the particles needing to be known.

VI. STRUCTURES OF EMULSION DROPLETS

The structures of the interfacial layers in emulsion droplets might be expected to be simple when small-molecule emulsifiers are used, but this is not necessarily the case, especially when not one but a mixture of surfactant molecules is present. Although simple interfacial layers may be formed where the hydrophobic moieties of the surfactants are dissolved in the oil phase and the hydrophilic head groups are dissolved in the aqueous phase, it is also possible to form multilayers and liquid crystals close to the interface [65]. These, of course, depend on the nature and the concentrations of the different surfactants. Interactions between surfactants generally enhance the stability of the emulsion droplets, because more rigid and structured layers tend to inhibit coalescence. Also, mixtures of different surfactants having different HLB numbers appears to provide structured interfacial layers, presumably because of the different affinities of the surfactants for the oil–water interface [66].

Specifically, phospholipids may form multilamellar structures around the oil–water interface, and presumably these layers will have different spacing depending on the amount of hydration [67]. So, although the major adsorption of phospholipids at low concentration is likely to be in the form of monolayers, it may be possible to produce more complex structures when large amounts of phospholipid are present or when other surfactants are coadsorbed to the interface.

Undoubtedly, the most complex structures are produced when the surfactants are proteins, because of the great many conformational states which are (at least potentially) accessible to such molecules. Increasingly, this is becoming an area of interest because of the implications of conformational change on the reactivity and functionality of the proteins. Flexible molecules such as caseins may be considered to adsorb as if they were heteropolymers [68] because of their presumed high conformational mobility [25]. Certainly, adsorbed β-casein exhibits different susceptibility to attack by proteolytic enzymes, compared with the protein in solution [69]. Indeed, adsorption to different materials causes differences in the conformation of the adsorbed molecule; for example, the protein seems to have somewhat different conformations when adsorbed to n-tetradecane–water and soya oil–water interfaces [70]. As a result of these measurements, it has been demonstrated that model hydrocarbon–water systems are not necessarily suitable for describing triglyceride–water systems.

Nevertheless, much is known about the structure of adsorbed β-casein, certainly more than is known for any other food protein. A variety of techniques has been used, and the different methods give good agreement. The first evidence was that β-casein adsorbed to a polystyrene latex caused an increase in the radius of the particle by 10–15 nm [71]. Later studies using small-angle x-ray scattering confirmed this and showed, in addition, that the bulk of the mass of the protein was close to the interface, so the interfacial layer was not of even density throughout [72]. This was confirmed by studies of neutron reflectance, where again the majority of the mass of the protein was shown to be close to the interface [73]. Only a relatively small portion of the mass of the protein protrudes from the tightly packed interface into solution, but it is this part which determines the hydrodynamics of the particle and which is almost certainly the source of the steric stabilization which the β-casein affords to emulsion droplets [71]. It is to be noted that all of the studies just described were performed on latex particles or on planar interfaces; however, it has also been demonstrated that the interfacial structures on emulsion droplets resemble those of the model compounds [31,72]. Although detailed control of emulsion droplets during their formation is impossible, as was described earlier, it is possible to break down the surface layers using proteolytic enzymes [74], and by comparison with the behavior of model systems under similar conditions, it is possible to demonstrate that the proteins seem to have similar conformations in the model systems and emulsions [75].

Using proteolytic enzymes, it is also possible to demonstrate which part of the β-casein is protruding into the solution. There are many sites in the molecule where, in principle, trypsin can attack the adsorbed protein, but in practice, it has been shown that sites close to the N-terminal of the protein are most readily attacked, suggesting that they are the most accessible, presumably because they form part of the adsorbed layer which protrudes into solution [69]. This region of the molecule is the one which would be expected to have this function, as it is the most hydrophilic and highly charged part of the protein; thus, for β-casein, it is possible to predict the conformation of the protein in the adsorbed state from a study of its sequence [69]. It seems that β-casein is perhaps the only protein for which this kind of prediction can be done; most other proteins (even the other caseins) have much less distinctive hydrophilic and hydrophobic regions and, therefore, have conformations which are more difficult to predict [75]. From studies of the structure of adsorbed α_{s1}-casein in model systems and in emulsions, it is established that neither the most accessible sites for trypsinolysis [76] nor the extent of protrusion of the adsorbed protein into solution [75] can be readily predicted.

This difficulty of ascertaining the structure of the adsorbed protein becomes more acute for globular proteins. In these cases, the adsorbed layer is much thinner than it is for the caseins, so that layers of β-lactoglobulin appear to be of the order of 1–2 nm thick instead of about 10 nm for the caseins [31,33,77]. From hydrodynamics and scattering experiments, it is even possible to suggest that the thickness of the adsorbed layers is smaller than would be expected from the protein in its natural conformation, so that these measurements of the size of the adsorbed protein suggest that adsorption causes it to change conformation. This may be confirmed by other techniques. It is known that adsorption of β-lactoglobulin causes the formation of intermolecular disulphide bonds [20], which does not occur when the molecules are in their native conformations in solution. Further, detailed studies of the DSC of emulsions containing β-lactoglobulin [24] have shown (Fig. 6) that (i) the protein when adsorbed to the oil–water interface in emulsions loses its heat of denaturation (i.e., shows no intake of heat which can be associated with denaturation, presumably

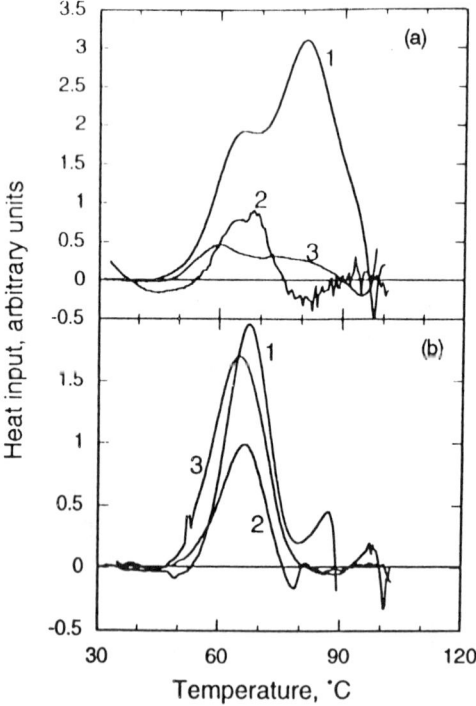

FIG. 6 Conformational changes in adsorbed proteins. Differential scanning calorimetry of proteins in solution and in emulsions. (a) Solution of β-lactoglobulin, with some α-lactalbumin. Curve 1, 1% protein in solution; curve 2, 1.5% protein adsorbed on oil/water interfaces in 20% w/w soya oil emulsion; curve 3, the same emulsion after the protein had been desorbed by addition of Tween-20. Lack of recovery of the original DSC pattern shows irreversible denaturation. (b) Solution of α-lactalbumin. Curve 1, 1% protein in solution; curve 2, 1% protein adsorbed on oil–water interfaces in 20% w/w soya oil emulsion; curve 3, the same emulsion after the protein had been desorbed by addition of Tween-20. Recovery of the DSC pattern on desorption suggests renaturation of the protein. (Data adapted from Ref. 24.)

because the protein is already surface denatured) and (ii) that if the protein is desorbed from the interface by treatment with detergent (Tween-20), it can be seen to be denatured irreversibly (i.e., no recovery of the denaturation endotherm is seen). This may be contrasted with the behavior of α-lactalbumin, which loses its heat of denaturation when adsorbed but recovers its original thermal behavior when the protein is competitively desorbed by Tween [24]. These studies confirm that different proteins show quite different behavior patterns when they are adsorbed to oil–water interfaces. Although it is not possible to study the caseins in this way because they show no heat of denaturation even in solution; they are useful as baseline models for studies of the other proteins.

From the combination of these studies, it is possible to conclude that the adsorption of proteins during emulsion formation leads to at least partial and sometimes complete denaturation of the molecules, although it may be partially or completely irreversible. It is thus potentially possible (although rare in practice) to define the structure of the interfacial layer in terms of the extent of denaturation of the adsorbed proteins and whether they form an extended monolayer. To this may be added two complicating factors: the possibility that multilayers, rather than monolayers, are formed and the possibility that specific proteins may exhibit variable behavior depending on the conditions. Caseins are capable of the latter behavior; for example, it is possible to prepare stable emulsions containing 20% w/w of soya oil, with as little as 0.3% casein, and in these the surface coverage has been measured to be a little less than 1 mg m^{-2}. The hydrodynamic thickness of the adsorbed layer in these emulsions was found to be about 5 nm. In emulsions prepared with larger amounts of casein (1–2%), the surface coverage is increased to 2–3 mg m^{-2} and the thickness of the adsorbed layer is about 10 nm (i.e., about twice that of the layer at lower surface coverage) [32]. It has been suggested that this is a result of the adsorbed casein molecules adopting two different conformations: one at low coverage, where the proteins have to cover a maximum area of surface (about 48 nm^2 mol^{-1}), and one at high coverage, where the molecules are more closely packed (about 13 nm^2 mol^{-1}). It is not thought that the caseins form multilayers in these emulsion particles, particularly because the binding curve is smooth as the concentration of casein is increased and does not show steps such as are typical of the formation of multilayers [47].

However, it is clear that under some conditions, caseins and other proteins can form multilayers; this has been demonstrated for adsorption to planar interfaces, where large surface excesses are easily generated and multilayers are formed [27]. There is less evidence for this in emulsions, although some high surface coverages (up to about 10 mg m^{-2}) have been measured [78], which must demand that there is more than a single layer on the oil–water interface because it would be impossible to pack this amount of protein into a monolayer as explained earlier. It is not clear why in some cases monolayers and in some cases multilayers are formed, although it is likely that the physical conditions of homogenization may be important. Also, differences in the methods of preparing the caseins may be relevant.

Multiple layers seem to be more easily formed in emulsions containing whey proteins. Perhaps because these proteins are originally globular and form thinner layers, which cannot extend far into solution from the interface (Table 3), they are forced to form multilayers. Because the whey proteins project less into solution than do the caseins, they may be less effective at sterically preventing the approach of additional molecules which go to form the multilayers. Finally, because they change conformation when they adsorb, they may offer new possibilities for interaction with incoming whey proteins from solution. There is evidence for multiple layers of whey proteins from both planar interfaces and emulsion droplets [47,79]. However, although these multiple layers exist, there is no definite evidence which links them to changes in the functional properties of the emulsions. Nor is it well determined how stable the multiple layers are, compared with a monolayer. It is very difficult, if possible at all, to simply wash adsorbed proteins from the monolayers formed on the interfaces of oil droplets [80]. However, the outer portion of multilayers may be more readily displaced because it is held in place by protein–protein interactions only, which may be weaker than the forces which lead to adsorption. Generally, therefore, the properties of the multilayers are poorly understood.

As a final degree of complexity, food emulsions may be stabilized by particles. Perhaps the most common are the protein "granules" from egg yolk in mayonnaise [45], and casein micelles in products such as homogenized milk. Both of these are known to be adsorbed to the oil–water interface as complex particles, which do not necessarily dissociate to their individual proteins [81,82]. During the homogenization of milk, casein micelles are disrupted at the oil–water interface so that they adsorb either whole or in fragments. Indeed, once a micelle has adsorbed, it can spread around the interface [53,83]. Thus, the fat droplets in homogenized milk are surrounded by a membrane which must contain some of the original fat globule membrane (phospholipid and protein [84]) but is primarily constituted of semi-intact casein micelles. Likewise, the oil–water interface in mayonnaise is partly coated by the granular particles formed from the phosphoprotein and lipoprotein constituents of egg yolk [85].

Stabilization of emulsions by means of these aggregates of protein is less efficient than by the proteins when they are present in the molecular state, simply because the efficient formation of the emulsion depends on rapid coverage of the newly formed oil surface in the homogenizer. A particle containing many molecules of protein will encounter a fat surface less frequently than an equivalent amount of molecular protein. Thus, although it is possible to prepare homogenized milk with the proportions of casein and fat which occur in natural milk (a ratio of about 1.5 w/w), it is not possible to use 1% w/w of micellar casein to stabilize an emulsion containing 20% oil [39]. On the other hand, 1% casein in a molecular form (sodium caseinate) is quite sufficient to form a finely dispersed, stable emulsion with 20% oil. Therefore, unless the micelles will confer some specific advantage on the functional properties of the emulsion, it is generally more effective to use caseinate. With egg granules, the situation is different, inasmuch as the individual proteins from the egg granules are not readily available in a purified form analogous to caseinate. In this case, the choice between particulate and molecular forms of the protein do not arise.

VII. FORMATION AND CHANGES OF THE INTERFACIAL LAYER

In many, if not all, food emulsions, more than one surfactant is present, so that mixtures of proteins, small-molecule surfactants (oil soluble and water soluble), and lecithins may be present. The result of this is that the interfacial layer may contain more than one type of molecule. The properties of the emulsion (the sizes of the droplets, the functionality) will, in turn, depend on which of the molecules in the formulation is actually on the interface.

It has been shown that, in general, proteins in emulsions formed at neutral pH and moderate temperatures show no selectivity for the interface. Thus, there is no preferential adsorption when a mixture of α-lactalbumin and β-lactoglobulin is homogenized with oil; the amounts of protein which are adsorbed are strictly in proportion to their concentrations [86]. The same is true when a mixture of caseinate and whey protein is used as the surfactant in an emulsion [47]. One case where preferential adsorption has been observed is when β-casein is used to displace adsorbed α_{s1}-casein, and vice versa, so that there is a possibility that these two proteins adsorb according to thermodynamic equilibrium [87]. Even this observation is complicated, however, because it may apply only to mixtures of highly purified caseins; when the experiment is repeated using commercial sodium caseinate (where a similar result is to be expected), the results are much less clear [88,89]. It is sometimes possible to form an emulsion using one protein and then to attempt to displace that protein from the interface with another; however, usually the protein which is first on the interface resists displacement [90]. Because of its high surface activity and flexibility, β-casein appears to be the best displacing agent, of the proteins tested to date, but it is by no means always capable of displacing an already adsorbed protein. It can displace α_{s1}-casein and α-lactalbumin from an interface [87,91], but the process is more complex with adsorbed β-lactoglobulin [92], especially if the emulsion containing the β-lactoglobulin has been allowed to age before the β-casein is added.

This behavior is not surprising because proteins are adsorbed to the interface by many independent points of contact. For all of these to become desorbed at once is extremely unlikely, and so the spontaneous desorption of a protein molecule is very rare; this is why it is very difficult to simply wash proteins from the oil–water interface [80]. Replacement of an adsorbed protein molecule by one from solution must require a concerted movement of the two molecules; as parts of one are displaced, they are replaced by parts of the other, until finally one of the two proteins is liberated into the bulk solution. Even this process, although more likely than the spontaneous desorption, is by no means certain to succeed, especially if the adsorbed protein had been on the interface for some time and had been able to make bonds with neighboring molecules. Moreover, given the very high concentration of protein in the adsorbed layer (see earlier text), it may even be difficult for a second type of protein to penetrate the adsorbed layer to initiate the displacement process. Therefore, although thermodynamic considerations may favor one protein over another, kinetic factors militate against rapid exchange.

Nonetheless, there do seem to be factors which influence the competition between proteins. As has been suggested above, flexibility may be an important criterion (because β-casein is in many cases an effective displacing agent). Thus, whatever increases flexibility may lead to increasing competitiveness. The most obvious example of this is α-lactalbumin; in its native state, this protein has a globular structure which is partly maintained by the presence of one ion of calcium within the molecule [93]. Removal of this Ca^{2+} leads to the protein adopting a "molten globule" state, whose tertiary structure is altered [94], and this leads to increased flexibility and competitiveness at the interface [95]. The removal of the Ca^{2+} can be achieved by chelation with complexing agents or by reducing the pH, and under these conditions α-lactalbumin outcompetes β-lactoglobulin for adsorption to the interface [33,41,95]. To some extent, the competition can be reversed by reneutralizing, so that there is dynamic competition between the proteins for the interface, not simply preferential adsorption during emulsion formation [33].

However, if true competition between species is required, then it appears to be necessary to have small surfactant molecules present as well as proteins [96]. In such a situation, there is competition between the proteins and small molecules as well as between the proteins themselves. But in such a case, instead of the desorption of a protein requiring the inefficient process of simultaneous detachment at all points, or the slow creeping displacement of one protein molecule by another, it is possible for a number of small molecules to displace a protein by separately replacing the individual points of attachment. It is known that small-molecule surfactants are capable of efficiently displacing adsorbed proteins (Fig. 7), although the details of the reactions depend on the type of surfactant and whether it is oil or water soluble [97–100]. Water-soluble surfactants are capable of removing all of the adsorbed protein from the oil–water interface, although they may require a molecular ratio of about 30 : 1

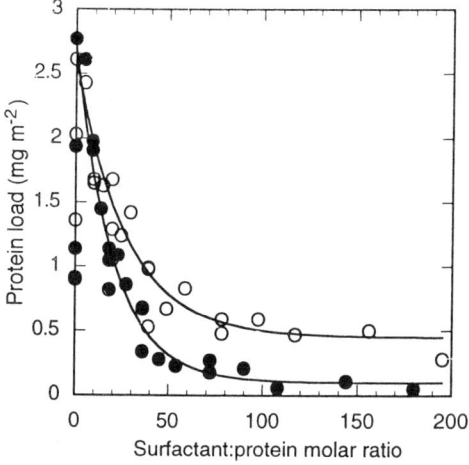

FIG. 7 Displacement of protein from interface by small-molecule surfactants. Protein load of emulsions (20% w/w soya oil) with caseinate at concentrations in the range of 0.25% to 2%, containing Tween-60 (water soluble, filled symbols) or Span-60 (oil soluble, open symbols) at concentrations from 0% to 1%. Protein load decreases with increasing ratio of surfactant to protein, and all points lie on the same curves, with the exception of the points measured in the absence of surfactant. Note that the oil-soluble surfactant displaces less of the casein than the water-soluble one. (Data adapted from Ref. 89.)

surfactant : protein [96]. At lower ratios, some, but not all, of the protein is displaced. Oil-soluble surfactants (low HLB numbers) are, in general, less efficient at completely displacing protein [89,100,101]. For solubility reasons, these surfactants cannot be added to the emulsion once it has been formed, but must be incorporated at the time the emulsion is produced in the homogenizer.

In addition to competing with proteins for adsorption to the oil–water interface, both during formation of the emulsion and its subsequent storage, some small-molecule surfactants also facilitate the exchange reactions of the proteins themselves. For example, although α-lactalbumin and β-lactoglobulin do not compete well with each other under normal circumstances at neutral pH [86], the presence of Tween causes the adsorption of α-lactalbumin to be favored over β-lactoglobulin [96]. Presumably the presence of surfactant enables a more thermodynamic equilibrium to be established, rather than the extremely slow kinetically determined exchange which normally occurs in the presence of the two proteins.

It is, therefore, apparent that "real" food emulsions are likely to behave in a more complex way than are simple model systems studied in the laboratory. This may be especially important when lecithins are present in the formulation. Although these molecules are indeed surfactants, they do not behave like, for example, Tweens. For example, they do not displace proteins efficiently from the interface, even though the lecithins may themselves become adsorbed [102]. They certainly have the capability to alter the conformation of adsorbed layers of caseins, although the way in which they do this is not fully clear; it is possibly because they can "fill in" gaps between adsorbed protein molecules [103]. In actual food emulsions, the lecithins in many cases contain impurities, and the role of these (which may also be surfactants) may confuse the way that lecithin acts [104]. It is possible also for the phospholipids to interact with the protein present to form vesicles composed of protein and lecithin, independently of the oil droplets in the emulsion. The existence of such vesicles has been demonstrated [105], but their functional properties await elucidation.

Therefore, in a real food emulsion, the composition of the interface may be exceedingly complex. Probably all of the types of surfactant present will be adsorbed to some extent, but it is at present impossible to do more than broadly predict what the composition of the interfacial layer will be, especially when the emulsion may be subjected to a variety of environmental changes (e.g., changes in pH and various sterilization procedures). Likewise, the prediction of stability or otherwise, and other functional properties of the emulsion, which depend on the composition and structure of the adsorbed layer, will become extremely complex.

VIII. STABILITY OF FOOD EMULSIONS

Food emulsions need to be stable because many of them are designed to have a long shelf life; for example, mayonnaise and homogenized milks may be required to have shelf lives of several months. The aim of the food technologist is to maintain the structure of the emulsion droplets and to prevent their coagulation, creaming, and especially coalescence, which leads to irreversible phase separation, or, on the other hand, coagulation and gel formation. Creaming, being gravity-driven, can be controlled in a number of physical ways, such as producing smaller droplets and increasing the viscosity of the continuous phase, but whether or not aggregation or coalescence occur depend on the composition and properties of the adsorbed surface layers on the oil droplets.

Different food emulsions are stabilized by a variety of mechanisms, but it has been shown that the structures of the interfacial layers of the emulsion droplets can be very complex, and so it is not in all (indeed in most) cases possible to unequivocally define exactly why an emulsion may be stable or unstable. One need only consider the problem of age-gelation and instability in heated milks [106] to understand the difficulties which attend the determinations of the causes of instability in stored-food emulsions. Of the two major mechanisms for stability (DLVO and steric stabilization), it is not clear whether stabilization by a DLVO type of mechanism [3] is relevant to food emulsions. This may be for the reason given for casein micelles [107], that the surface layers of the particles will begin to overlap before appreciable electrostatic repulsion will be experienced between them. Because of the extended nature of adsorbed layers of protein, the actual thickness of the adsorbed layer may be longer than, or at least of the order of, the Debye length [65], and so the full repulsive potential between the surfaces cannot be developed. Perhaps an alternative way of looking at this is that the primary minimum of energy between the approaching particles is not deep and that temporary aggregates will fall apart rapidly.

Classical DLVO theory shows that an increase in ionic strength should lead to instability, that is, coagulation of emulsion droplets. For emulsions stabilized by proteins, this is not necessarily true. Some evidence is available for casein-stabilized particles, whereby ionic strengths of $>1M$ NaCl do indeed cause the emulsion to coagulate [55], but this applies under only rather specific conditions. The addition of calcium ions in quite low concentrations (<10 mM) can destabilize casein-stabilized emulsions [108–110], but it is probable that this is a specific ion effect. An equivalent amount of ionic strength in the form of NaCl does not cause coagulation

[108], and it is known that Ca^{2+} binds to caseins, which may lead to the formation of calcium bridges between the emulsion droplets rather than to their coagulation as a result of general ionic strength effects. Also, binding of Ca^{2+} to adsorbed caseins should neutralize some of the negative charge on the emulsion droplets and thus permit close approach and flocculation or aggregation; however, measurement of the ζ-potential of latices and emulsions shows a decrease in the absolute magnitude of the ζ-potential as Ca^{2+} is added [110,111]. But this change is smaller than would be expected for the total neutralization of the charge, so it may not be sufficient to destabilize the emulsion by minimizing charge repulsions.

If stabilization is to be achieved by a mechanism other than DLVO, an obvious candidate is steric stabilization (although definitions of this phenomenon are rather variable). For casein-coated emulsion droplets, the extended layer of protein can almost certainly provide this kind of stabilization. As two emulsion droplets approach closely, the adsorbed layers may try to interpenetrate or to distort one another. These interactions lead to increases in enthalpy and decreases in entropy, consequences which reduce the possibility of spontaneous aggregation. In addition, the simple act of close approach causes the formation of a high concentration of macromolecules between the particles, with consequent increase in osmotic pressure, so that water will flow in and tend to push apart the two droplets [3]. Although casein is the most obvious example because it extends far from the oil–water interface, it is possible that all proteins when adsorbed protrude from the interface sufficiently far (2–3 mn) to allow these mechanisms to be active. Even when adsorbed casein and β-lactoglobulin have been digested by proteolytic enzymes, the emulsions are not necessarily destabilized [32], so even some of the peptides remaining on the interface seem to have the capacity to maintain stability, even though they protrude much less into the solution than do the original proteins. It has not been established how thick a layer of protein or peptide is necessary to provide adequate steric stabilization of emulsion droplets.

Both charge-dependent and steric stabilization mechanisms prevent the close approach of particles, so that flocculation is prevented, and if flocculation (or aggregation) cannot occur, then creaming will be slow, provided that the emulsion droplets are small to start with. If aggregation and creaming are slow, then coalescence is unlikely under normal storage circumstances. However, the presence of protein in the adsorbed layers will also help to prevent coalescence. Although the precise properties are still a subject of dispute, it seems self-evident that the viscoelastic properties of an interfacial layer help to define whether the membranes on adjacent oil droplets will rupture to allow the flow of oil from one to another. Some proteins form very viscous interfacial films [91,112], and these may well be important when coalescence is to be controlled; also the lack of mobility of adsorbed proteins, compared with the rapid movement of small surfactants, must lead to greater stability toward coalescence.

An additional aid to stability of food emulsions may be the incorporation of polysaccharide stabilizers. With few exceptions, these act simply to increase the viscosity of the continuous phase of the emulsion, so that they slow down the kinetics of flocculation of the droplets and also slow creaming [113]. They do not themselves participate in any reaction with the emulsion droplets themselves. Thus, they may prolong the life of an unstable emulsion but do not provide absolute stability. An alternative class of polysaccharides form structures in the continuous phase, by forming a weak gel which is not only viscous but also possesses a yield stress. These additives have a more pronounced effect on stability, because the yield stress prevents the movement of droplets which have insufficient energy to overcome the stress [114]. Thus, creaming, for example, may prevented because the gravitational effect is too weak to overcome the yield stress of the continuous phase. Not all polysaccharides form gels; specifically, xanthan and carrageenans form significant gels in the presence of divalent ions [113]. However, these polymers may also destabilize the emulsions by a mechanism of depletion flocculation [115].

A few polysaccharides do appear to interact with emulsion droplets. Of these, the best known example is probably carrageenan. It is established from studies in milk and with isolated components from milk that κ-carrageenan can interact with κ-casein [116]. Presumably, therefore, emulsions which contain carrageenans and caseins should show this interaction, and because the casein is likely to be found on the droplet interface, it is likely that the carrageenan will be found there as well. This is likely to result in a highly stable particle, which will have very strong steric stabilization because the carrageenan molecules may protrude far into the solution from the emulsion interface [117].

Heating may be an important factor in the destabilization of emulsions, especially if the emulsions contain proteins which can be thermally denatured. There is an extensive literature, for example, on the destabilization of homogenized milks (especially when concentrated) by heating [118]. Heat will generally facilitate the interactions between emulsion droplets and promote the aggregation process, but as it alters the conformations of the proteins involved, they may also interact to form new structures, such as gels. Especially when whey proteins are used to form the emulsions, it is possible to form gels [119], which, in effect, are protein gels with filler particles (i.e., the emulsion droplets). It seems that the gels are formed by the interaction of protein in solution with protein on the oil–water interface, because they are not formed when the amount of soluble protein is low [120]. However, it is possible to make gels from fine emulsions containing 20% oil and as little as 2.5% whey protein [121]; normally, gelation of whey protein only occurs at concentrations in the range of 8% and greater. Therefore, the presence of the interactive filler particles of the emulsion droplets considerably enhances the gelling capability of the protein.

The effects of heat on emulsions depend primarily on the type of protein which is present; however, other factors are important; for example, the pH and the presence in solution of specific minerals. Thus, an emulsion which is stable to heating at one pH value may not be stable at another [121]; especially, it is difficult to produce heat-stable emulsions at pH values about 4, using milk proteins as emulsifying agents; other emulsifiers have to be used to offset the tendency of the proteins to aggregate when heated in this pH region.

IX. KINETICS OF THE DESTABILIZATION OF EMULSIONS

Generally, emulsions are considered to be unstable, thermodynamically, because the presence of large areas of the oil-water interface provides positive free energy; surfactants only decrease the interfacial free energy. Thus, in the end, it is kinetics which determines whether emulsions will exist long enough to be useful. Some of the factors governing the kinetics have already been described in the preceding section, but it may be appropriate to consider some more mechanistic descriptions of the process.

If it is established that interaction of emulsion droplets is at least possible, despite the stabilizing action of the adsorbed layers, the process of destabilization will consist of aggregation, followed, possibly, by coalescence. These two processes must be sequential, because coalescence is likely only to result from the prolonged proximity of droplets brought about either by creaming or aggregation followed by creaming [114]. For an unstirred emulsion, it would be expected that the aggregation kinetics should follow Smoluchowski kinetics for perikinetic flocculation, in the presence of an energy barrier [64,122]. However, this does not appear to be generally true. Although Smoluchowski kinetics have been shown [123] to apply to the destabilization of dilute suspension of casein micelles (which we may take as an example of a protein-stabilized colloid similar to an emulsion), there is little evidence to show that the simple Smoluchowski formulation is appropriate for the destabilization of emulsions (Fig. 8). What should be observed if all particles are equally reactive is a linear growth of the volume-average particle size; what tends to be observed is a lag time where growth of molecular weight is slow, followed

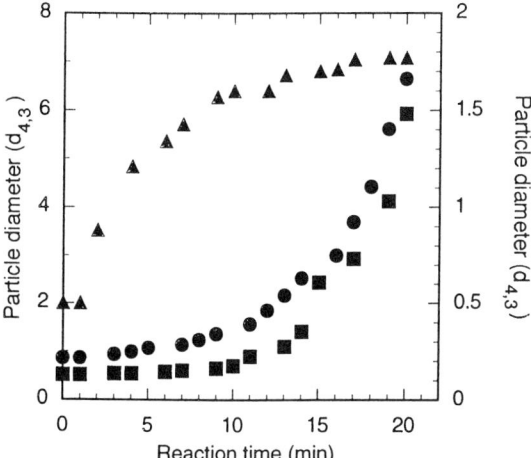

FIG. 8 Kinetics of destabilization of emulsions by calcium addition, acidification, or ethanol. Growth of average particle diameter $d_{4,3}$ with time, showing different mechanisms. The results are for 20% (w/w) soya oil emulsions stabilized by caseinate. ●—0.5% caseinate emulsion diluted into a 17 mM solution of $CaCl_2$ at pH 7.0; ■—1% caseinate emulsion diluted into buffer at pH 5.2; ▲—1% caseinate emulsion diluted into a 40 (v/v) solution of ethanol at pH 7.0.

by a steadily increasing rate of reaction [108,109]. Most of the relevant experiments have been conducted in dilute solution (to allow light scattering to be measured), but even in whole emulsions, the growth of particle size is not that which is predicted by Smoluchowski [47].

To explain this type of behavior, it is necessary to suggest either that there is a first aggregation reaction which is slow and rate determining, followed by one which is rapid, or that the interaction of particles depends on their size, being more efficient as the size increases. The first of these is difficult to explain, as the emulsion droplets are expected to be isotropic, and it is hard to see how some kind of preferred reaction can occur. The second can be explained at least partly by invoking the concept of fractality [124,125]; if the aggregates have a fractal form, because of random aggregation, then the space taken up by an aggregate particle will be dependent on the size of the monomer and the fractal dimension. In effect, the collision diameter of the aggregate will be larger than expected, so that its effective concentration will be increased and its reactions will be faster than if the emulsion droplets in the aggregate coalesced. The apparent lag phase of the aggregation reaction can, therefore, be seen to arise simply from the fractal structures of the aggregate particles. Note that fractality is not in itself a driving force; it is simply the result of the random aggregation of particles; however, it can be seen that fractality can exercise control over the reaction.

Of course, such a mechanism will properly only be calculable for large particles, where the concept of self-similarity of structures can be used. Small aggregates might be expected to form by a more general Smoluchowski mechanism, and only when they reach a critical size can the concept of fractality be applied. However, in some cases, it is not possible even to see the initial aggregation, so there may still be some questions to be raised about the mechanism as a whole. It does seem to be the best available at the moment, if we are to regard the emulsion droplets as possessing isotropic surfaces, which do not have specific "hot spots" through which reaction may take place [126]. If reactive sites can be identified, it is then possible to use a polyfunctional approach to describe the kinetics; essentially, this leads to the same kind of kinetics as the fractal description, with a slow lag phase being followed by explosive growth of the particle size, because large particles react faster than do small ones, having more reactive centers [64,108,124,127].

The kinetics of orthokinetic flocculation of emulsions (i.e., when the unstable emulsion is being stirred or agitated) can perhaps be more easily explained because theory predicts that larger particles will aggregate faster than small ones. Slow early aggregation of the emulsion becomes faster as time goes on because of this [122]. The fractal concept, as described earlier, can also be used to describe the reactions because it describes how the effective particle size of the aggregates is, in fact, larger than would be expected from simple consideration of the degree of polymerization. In a stirred system, the shearing of the solution enhances particle growth, but it can also limit the size of the aggregates which can form, because the effect of shear is also to break up the largest particles [124], unless they rapidly coalesce, in which case only rehomogenization can reduce the particle size. So, precipitation can be avoided (at least while the shearing is applied), even though the emulsion suspension remains unstable.

The preceding discussion applies to an emulsion which is inherently unstable, so the source of the instability is present from the time that the emulsion is formed, or arises from the addition of some other destabilizing material (e.g., Ca^{2+} [108] or small surfactant [128,129] to the emulsion). It is possible, however, to propose another mechanism of destabilization, where the stability of the emulsion is decreased with time by the action of some chemical reaction. The breakdown of a layer of adsorbed protein by proteolytic enzymes present in the food product is a possible cause of instability, especially in milk-based products [130]. Such proteolytic enzymes may arise from microbiological contamination at an early stage in manufacture; often a heat treatment will kill the microorganisms but will leave some of the proteases in an active state. Such a mechanism has been proposed for the age-gelation of homogenized milks. Alternative mechanisms for the destabilization of the surfaces can be chemical reactions between molecules in solution and those on the interface; once again, these may modify the structure or charge of the interfacial layers, or form cross-links between particles so as to lead to instability. Although these reactions appear to be relatively rare, they may become of considerable importance when the emulsions must be shelf-stable for considerable periods.

X. CONTROLLED INSTABILITY OF EMULSIONS—PARTIAL COALESCENCE

Even though emulsions generally need to be as stable as possible, there are several food products where the opposite effect is desired, so a shelf-stable emulsion can be controllably destabilized when required. Such emulsions are the basis for products such as whipped toppings and ice cream and generally depend for their effect on the processes of partial coalescence [131–133] and destabilization by whipping air bubbles into the mixture, at which time the interfacial layer of the emulsions may be mechanically broken and liquid oil spread around the air–solution interface.

Because of the importance of ice cream as a product, much has been written on its structure and formation [134], and the process can only be summarized here. In toppings and ice cream (and indeed simply in whipping cream), it is first necessary to produce a stable emulsion. Ice cream mix is a complex mixture, but the initial emulsion is basically homogenized milk, containing an admixture of small-molecule surfactants as well; in whipped toppings, the emulsion is made with oil and a surfactant mixture, which may or may not contain protein; and in cream, the natural membrane of phospholipid and protein surrounds the milk fat. In all of these, it is necessary to have some small-molecule emulsifiers so as to exchange with, and weaken the rigidity of, the adsorbed layer of protein [98]. The second essential is that the fat or oil in the formulation is partly crystalline; neither completely liquid nor completely solid oil will perform optimally. If the oil is partly crystalline, then the emulsion droplets may not be truly spherical but may have protrusions of crystals on their surfaces.

When such an emulsion is subjected to shearing forces, the emulsion droplets will be subject to partial coalescence, where the protruding fat crystals on different droplets catch together [135]. This is followed by breakage of the interfacial layer, which allows the linking of the emulsion droplets together. Although liquid fat will flow from one droplet to another under these circumstances, complete coalescence cannot be achieved because of the presence of the crystalline oil. The end result is that the emulsion droplets form a network, linked by crystalline oil. This effect is enhanced when the emulsion is not simply sheared but whipped so as to incorporate air into the product. Air interfaces are highly disruptive of the interfaces of emulsion droplets, especially if the interfacial layer is not too strong (which is why small-molecule surfactants are used in the formulations), and the result is that the partial coalescence of the droplets occurs preferentially around the air bubbles. The end product is a foam, in which partially coalesced oil droplets form a framework around the air bubbles to give a stable product. Although this is the basic process, the formulations of the original emulsions may be refined by the addition of stabilizers and flavor compounds, and of course in ice cream, the crystals of ice which are formed during the cooling of the final foam are also important texturizers. It is the emulsion, however, which defines the basic structure of the ice cream, although there must be as many detailed formulations as there are companies making the product.

XI. CONCLUDING REMARKS

The object of making food emulsions is to provide a stable and controllable source of food, whose texture, taste, and nutritional and storage properties are acceptable to the consumer. Although the possible ingredients are limited by the constraints of healthy nutrition, it is nevertheless evident that within the available range there is ample opportunity for variation in the properties of the emulsions, for instance the particle size, and the composition of the stabilizing layer of the interface, which, in turn, influence the stability and functional behavior of the emulsion.

Nonetheless, many emulsions used in foods have their roots in established formulations, and an understanding of why certain emulsions behave as they do is still not established in a number of cases. It has been pointed out that not all of the aspects of the stability of emulsions are known, in terms of the mechanisms which may be operating to maintain the stability of the systems. Also, in many real food emulsions, the path followed during formation is critical, emphasizing once again that emulsions are not equilibrium systems and that the same overall composition may not necessarily lead to the same product. At this point, our knowledge is short and needs to be extended. For example, the heat treatment of ingredient proteins either before or after the formation of an emulsion may critically affect the behavior of the emulsion. As the emulsions contain more ingredients, the level of complexity required to understand the details of their formation and properties is increased. The challenge for the future is to be able to describe and control some of the most complex emulsions so as to enable greater functional stability for these food systems.

REFERENCES

1. K. Larsson, J. Dispers. Sci. Technol. *1*:267 (1980).
2. N. Krog, in *Food Emulsions*, (K. Larsson and S. E. Friberg, eds.), Marcel Dekker, Inc., New York, 1990, pp. 203–247.
3. E. Dickinson and G. Stainsby, *Colloids in Food*, Applied Science Publishers, Barking, U.K. 1982.
4. N. Krog, in *Microemulsions and Emulsions in Foods* (M. El-Nokaly and D. Cornell, eds.), ACS Symposium Series 448, American Chemical Society, Washington, DC, 1991, pp. 139–145.
5. E. Li-Chan and S. Nakai, in *Food Proteins* (J. E. Kinsella and W. G. Soucie, eds.), American Oil Chemists' Society, Champaign, IL, 1989, pp. 232–251.
6. C. C. Bigelow. J. Theor. Biol. *16*:187 (1967).
7. E. Dickinson, S. R. Euston, and C. M. Woskett, Prog. Colloid. Polym. Sci. *82*:65 (1990).
8. E. Dickinson, D. J. Pogson, E. W. Robson, and G. Stainsby, Colloids Surf. *14*:135 (1985).
9. A. Kato, N. Tsutsui, N. Matsudomi, K. Kobayashi, and S. Nakai, Agric. Biol. Chem. *45*:2755 (1981).
10. E. Keshavarz and S. Nakai, Biochim. Biophys. Acta *576*:269 (1979).
11. S. Nakai, J. Agric. Food Chem. *31*:676 (1983).
12. A. Kato and S. Nakai, Biochim. Biophys. Acta *624*:13 (1980).
13. E. Dickinson, J. A. Hunt, and D. G. Dalgleish, Food Hydrocolloids *4*:403 (1991).
14. A. Kato, S. Miyazaki, A. Kawamoto, and K. Kobayashi, Agric. Biol. Chem. *51*:2989 (1987).
15. B. M. Byrne, A. D. van het Schip, J. A. M. van de Klundert, A. C. Arnberg, M. Gruber, and G. Ab, Biochemistry *23*:4275 (1984).
16. F. MacRitchie and N. F. Owens, J. Colloid Interface Sci. *29*:66 (1969).
17. C. A. Haynes and W. Norde, Colloids Surf. B *2*:517 (1994).
18. W. Norde, F. MacRitchie, G. Nowicka, and J. Lyklema, J. Colloid Interface Sci. *112*:447 (1986).
19. W. Norde and J. P. Favier, Colloids Surf. *64*:87 (1992).
20. E. Dickinson and Y. Matsumura, Int. J. Biol. Macromol. *13*:26 (1991).
21. E. A. Foegeding, in *Food Proteins* (J. E. Kinsella and W. G. Soucie, eds.), American Oil Chemists' Society, Champaign, IL, 1989, pp. 185–194.
22. K. D. Caldwell, J. Li, J.-T. Li, and D. G. Dalgleish, J. Chromatogr. *604*: 63 (1992).

23. C. A. Haynes, E. Swilinsky and W. Norde, J. Colloid Interf. Sci. *164*:394 (1994).
24. M. Corredig and D. G. Dalgleish, Colloids Surf. B *4*:411 (1995).
25. C. Holt and L. Sawyer, Protein Eng. *2*:251 (1988).
26. H. E. Swaisgood, in *Advanced Dairy Chemistry: 1. Proteins* (P. F. Fox, ed.), Elsevier Science Publishers, Barking, U.K., 1992, pp. 63–110.
27. D. E. Graham and M. C. Phillips, J. Colloid Interface Sci. *70*:415 (1979).
28. E. Tornberg, Y. Granfeldt, and C. Hakansson, J. Sci. Food Agric. *33*:904 (1982).
29. M. C. Phillips, Chemistry and Industry 170–176 (1977).
30. D. E. Graham and M. C. Phillips, J. Colloid Interf. Sci. *70*:427 (1979).
31. D. G. Dalgleish and J. Leaver, in *Food Polymers. Gels and Colloids* (E. Dickinson, ed.), Royal Society of Chemistry, Cambridge, 1991, pp. 113–122.
32. Y. Fang and D. G. Dalgleish, J. Colloid Interface Sci. *156*:329 (1993).
33. J. Hunt and D. G. Dalgleish, Food Hydrocolloids (in press).
34. S. G. Hambling, A. S. McAlpine, and L. Sawyer, in *Advanced Dairy Chemistry: 1. Proteins* (P. F. Fox, ed.), Elsevier Science Publishers, Barking, U.K., 1992, pp. 141–190.
35. R. D. Bee, J. Hoogland, and R. H. Ottewill, *Food Colloids and Polymers: Stability and Mechanical Properties* (E. Dickinson and P. Walstra, eds.), Royal Society of Chemistry, Cambridge, 1993, pp. 341–353.
36. C. Holt, Adv. Protein Chem. *43*:63 (1992).
37. H. Oortwijn, P. Walstra, and H. Mulder, Neth. Milk Dairy J. *31*:134 (1977).
38. H. S. Rollema, in *Advanced Dairy Chemistry: 1. Proteins* (P. F. Fox, ed.), Elsevier Science Publishers, Barking, U.K., 1992, pp. 111–140.
39. J. A. Hunt and D. G. Dalgleish, unpublished results.
40. H. Pessen, J. M. Purcell, and H. M. Farrell. Biochim. Biophys. Acta *828*:1 (1985).
41. J. Hunt and D. G. Dalgleish, J. Agric. Food Chem. *42*:2131 (1994).
42. Y. Kamata, K. Ochiai, and F. Yamauchi, Agric. Biol. Chem. *48*:1147 (1984).
43. P. Plietz, G. Damaschun, J. J. Müller, and K.-D. Schwenke. Eur. J. Biochem. *130*:315 (1983).
44. D. Causeret, E. Matringe, and D. Lorient, J. Food Sci. *56*:1532 (1991).
45. D. N. Holcomb, L. R. Ford, and R. W. Martin, Jr, in *Food Emulsions* (K. Larsson and S. E. Friberg, eds.), Marcel Dekker, Inc., New York, 1990, pp. 327–367.
46. K. N. Pearce and J. E. Kinsella, J. Agric. Food Chem. *26*:716 (1978).
47. J. Hunt and D. G. Dalgleish, Food Hydrocolloids *8*:175 (1994).
48. P. Walstra, in *Food Structure and Behavior* (J. M. V. Blanshard and P. Lillford, eds.), Academic Press, London, 1987, pp. 87–106.
49. P. Walstra, in *Encyclopedia of Emulsion Technology* (P. Becher, ed.), Marcel Dekker, Inc., New York, 1983, pp. 57–128.
50. O. Robin, N. Remillard, and P. Paquin. Colloids Surf. A *80*:211 (1993).
51. E. Dickinson, F. O. Flint, and J. A. Hunt, Food Hydrocolloids *3*:389 (1989).
52. L. V. Ogden, P. Walstra, and H. A. Morris. J. Dairy Sci. *59*:1727 (1976).

53. W. Buchheim and P. Dejmek, in *Food Emulsions* (K. Larsson and S. E. Friberg, eds.), Marcel Dekker, Inc., New York, 1990, pp. 203–246.
54. E. Dickinson, R. H. Whyman, and D. G. Dalgleish, in *Food Emulsions and Foams* (E. Dickinson, ed.), Royal Society of Chemistry, Cambridge, 1987, pp. 40–51.
55. P. Walstra and H. Oortwijn, J. Colloid Interface Sci. *29*:424 (1969).
56. P. Walstra, J. Colloid Interface Sci. *27*:493 (1968).
57. K. Strawbridge and F. R. Hallet, Macromolecules *27*:2283 (1994).
58. K. Strawbridge and J. Watton, Can. J. Spectrosc. *36*:53 (1991).
59. K. Strawbridge and F. R. Hallet, Can. J. Phys. *70*:401 (1992).
60. D. G. Dalgleish and F. R. Hallett, Food Res. Int. *28*:181 (1995).
61. F. R. Hallett, T. Craig, J. Marsh, and B. Nickel, Can. J. Spectrosc. *34*:63 (1989).
62. K. B. Strawbridge, E. Ray, F. R. Hallett, S. M. Tosh, and D. G. Dalgleish, J. Colloid Interface Sci. *171*:392 (1995).
63. D. S. Horne and C. Davidson, Colloids Surf. A *77*:1 (1993).
64. A. Lips, T. Westbury, P. M. Hart, I. D. Evans, and I. J. Campbell, in *Food Colloids and Polymers: Stability and Mechanical Properties* (E. Dickinson and P. Walstra, eds.), Royal Society of Chemistry, Cambridge, 1993, pp. 31–44.
65. B. Bergenståhl and P. M. Claesson, in *Food Emulsions* (K. Larsson and S. E. Friberg, eds.), Marcel Dekker, Inc., New York, 1990, pp. 41–96.
66. J. V. Boyd, C. Parkinson, and P. Sherman, J. Colloid Interface Sci. *41*:359 (1972).
67. J. M. M. Westerbeek and A. Prins, *Microemulsions and Emulsions in Foods* (M. El-Nokaly and D. Cornell, eds.), ACS Symposium Series 448, American Chemical Society, Washington, DC, 1991, pp. 146–160.
68. F. MacRitchie, Colloids Surf. A *76*:159 (1993).
69. J. Leaver and D. G. Dalgleish, Biochim. Biophys. Acta *1041*:217 (1990).
70. J. Leaver and D. G. Dalgleish, J. Colloid Interface Sci. *149*:49 (1992).
71. D. G. Dalgleish, Colloids Surf. *46*:141 (1990).
72. A. R. Mackie, J. Mingins, and A. N. North, J. Chem. Soc. Faraday Trans. *87*:3043 (1991).
73. E. Dickinson, D. S. Horne, J. S. Phipps, and R. M. Richardson, Langmuir *9*:242 (1993).
74. D. G. Dalgleish and J. Leaver, J. Colloid Interface Sci. *141*:288 (1991).
75. D. G. Dalgleish, Colloids Surf. B *1*:1 (1993).
76. M. Shimizu, A. Ametani, S. Kaminogawa, and K. Yamauchi, Biochim. Biophys. Acta *869*:259 (1986).
77. A. R. Mackie, J. Mingins, and R. Dann, in *Food Polymers, Gels and Colloids* (E. Dickinson, ed.), Royal Society of Chemistry, Cambridge, 1991, pp. 96–111.
78. M. Britten and H. J. Giroux, J. Agric. Food Chem. *41*:1187 (1993).
79. M. Rosenberg and S. L. Lee, Food Struct. *12*:267 (1993).
80. F. MacRitchie, J. Colloid Interface Sci. *105*:119 (1985).
81. D. G. Dalgleish and E. W. Robson, J. Dairy Res. *52*:539 (1985).

82. H. Oortwijn and P. Walstra, Neth. Milk Dairy J. *33*:134 (1979).
83. P. Walstra and H. Oortwijn, Neth. Milk Dairy J. *36*:103 (1982).
84. A. V. McPherson and B. J. Kitchen, J. Dairy Res. *50*:107 (1983).
85. M. A. Tung and L. J. Jones, Scanning Electron Microsc. *3*:523 (1981).
86. E. Dickinson, S. E. Rolfe, and D. G. Dalgleish, Food Hydrocolloids *3*:193 (1989).
87. E. Dickinson, S. E. Rolfe, and D. G. Dalgleish, Food Hydrocolloids *2*:397 (1988).
88. E. W. Robson and D. G. Dalgleish, J. Food Sci. *52*:1694 (1987).
89. H. Singh, S. E. Euston, P. Munro, and D. G. Dalgleish, J. Food Sci. *60*: 1124 (1995).
90. E. Dickinson, J. Food Eng. *22*:59 (1994).
91. E. Dickinson, S. E. Rolfe, and D. G. Dalgleish, Int. J. Biol. Macromol. *12*: 189 (1990).
92. D. G. Dalgleish, S. E. Euston, J. A. Hunt, and E. Dickinson, in *Food Polymers, Gels and Colloids* (E. Dickinson, ed.), Royal Society of Chemistry, Cambridge, 1991, pp. 485–489.
93. M. Ikeguchi, K. Kuwajima, and S. Sugai, J. Biochem. *99*:1191 (1986).
94. O. B. Ptitsyn, in *Protein Folding* (T. E. Creighton, ed.), W. H. Freeman, New York, 1992, p. 243.
95. E. Dickinson and Y. Matsumura. Colloids Surf. B *3*:1 (1994).
96. J.-L. Courthaudon, E. Dickinson, Y. Matsumura, and A. Williams, Food Struct. *10*·109 (1991).
97. J.-L. Courthaudon, E. Dickinson, and D. G. Dalgleish, J. Colloid Interface Sci. *145*:390 (1991).
98. J.-L. Courthaudon, E. Dickinson, Y. Matsumura, and D. C. Clark, Colloids Surf. *56*:293 (1991).
99. J. A. de Feijter, J. Benjamins, and M. Tamboer, Colloids Surf. *27*:243 (1987).
100. E. Dickinson and S. Tanai, J. Agric. Food Chem. *40*:179 (1992).
101. E. Dickinson, R. K. Owusu, S. Tan, and A. Williams, J. Food Sci. *58*:295 (1993).
102. J.-L. Courthaudon, E. Dickinson, and W. W. Christie. J. Agric. Food Chem. *39*:1365 (1991).
103. Y. Fang and D. Dalgleish, Colloids Surf. B *1*:357 (1993).
104. C. H. McCrae and D. D. Muir, J. Dairy Res. *59*:177 (1992).
105. Y. Fang and D. G. Dalgleish, Langmuir *11*:75 (1995).
106. V. R. Harwalkar, in *Advanced Dairy Chemistry: 1. Proteins* (P. F. Fox, ed.), Elsevier Science Publishers, Barking, U.K., 1992, pp. 691–734.
107. T. A. J. Payens, J. Dairy Res. *46*:291 (1979).
108. S. Agboola and D. G. Dalgleish, J. Food. Sci. *60*:399 (1995).
109. J. Chen, E. Dickinson, and G. Iveson, Food Struct. *12*:135 (1993).
110. E. Dickinson, J. A. Hunt, and D. S. Horne, Food Hydrocolloids *6*:359 (1992).
111. D. G. Dalgleish, E. Dickinson, and R. H. Whyman, J. Colloid Interface Sci. *108*:174 (1985).

112. E. Dickinson, B. S. Murray, and G. Stainsby, J. Chem. Soc. Faraday Trans. *84*:871 (1988).
113. E. M. Ozu, I. C. Baianu, and L-S. Wei, in *Physical Chemistry of Food Processes*, Vol. 2 (I. C. Baianu, H. Pessen, and T. F. Kumosinski, eds.), Van Nostrand Reinhold, New York, 1993, pp. 487–517.
114. P. Walstra, in *Food Emulsions and Foams* (E. Dickinson, ed.), Royal Society of Chemistry, Cambridge, 1987, pp. 242–257.
115. H. Luyten, M. Jonkman, W. Kloek, and T. van Vliet, in *Food Colloids and Polymers: Stability and Mechanical Properties* (E. Dickinson and P. Walstra, eds.), Royal Society of Chemistry, Cambridge, 1993, pp. 224–234.
116. K. Osawa, R. Niki, and S. Arima, Agric. Biol. Chem. *48*:627 (1984).
117. D. G. Dalgleish and E. R. Morris, Food Hydrocolloids *2*:311 (1988).
118. A. W. M. Sweetsur and D. D. Muir, J. Soc. Dairy Technol. *35*:126 (1982).
119. G. Masson and R. Jost, Colloid Polym. Sci. *264*:631 (1986).
120. R. Jost, R. Baechler, and G. Masson, J. Food Sci. *51*:440 (1986).
121. J. A. Hunt and D. G. Dalgleish, J. Food Sci. *60*:1120 (1995).
122. M. von Smoluchowski, Z. Phys. Chem. *92*:129 (1917).
123. D. G. Dalgleish, Biophys. Chem. *11*:147 (1980).
124. E. Dickinson and A. Williams, Colloids Surf. A *88*:317 (1994).
125. P. Walstra, T. van Vliet, and L. G. B. Bremer, in *Food Polymers, Gels and Colloids* (E. Dickinson, ed.), Royal Society of Chemistry, Cambridge, 1991, pp. 369–382.
126. P. J. Flory, Faraday Discuss. Chem. Soc. *57*:7 (1974).
127. E. Dickinson, R. K. Owusu, and A. Williams, J. Chem. Soc. Faraday Trans. *89*:865 (1993).
128. H. D. Goff and W. K. Jordan. J. Dairy Sci. *72*:18 (1989).
129. N. M. Barfod, N. Krog, G. Larsen, and W. Buchheim. Fat Sci. Technol. *93*:24 (1991).
130. T. A. J. Payens. Neth. Milk Dairy J. *32*:170 (1978).
131. M. A. J. S. van Boekel and P. Walstra, Colloids Surf. *3*:109 (1981).
132. D. F. Darling. J. Dairy Res. *49*:695 (1982).
133. K. Boode and P. Walstra, in *Food Colloids and Polymers: Stability and Mechanical Properties* (E. Dickinson and P. Walstra, eds.), Royal Society of Chemistry, Cambridge, 1993, pp. 23–30.
134. K. G. Berger, in *Food Emulsions* (K. Larsson and S. E. Friberg, eds.), Marcel Dekker, Inc., New York, 1990, pp. 367–444.
135. D. F. Darling and R. J. Birkett, in *Food Emulsions and Foams* (E. Dickinson, ed.), Royal Society of Chemistry, Cambridge, 1987, pp. 1–29.

6

Alkyd Emulsion: Instability and Drying Properties

BJÖRN BERGENSTÅHL and GUNILLA ÖSTBERG Institute for Surface Chemistry, Stockholm, Sweden

I. WATER-BORNE PAINTS

Generally the use of water-borne paints has increased during recent years. The main reason is the growing awareness of the health and environmental aspects of the organic solvents traditionally used. By far, the most extensively used water-borne paints are polymeric dispersions, latex paints, which are made by emulsion polymerization directly in the water phase. Another type which is receiving increasing attention at the moment is water-borne alkyd paints. In alkyd emulsions, the alkyd oil is emulsified in water. The emulsion is stabilized with external surfactants or by stabilizing groups incorporated in the alkyd (self-emulsifiable alkyds) [1].

One reason for the growing interest in alkyd emulsions is the inadequacy of latex dispersion in some applications; for example, when penetration into the painted substrate is needed to achieve a good water resistance. Latex paints also contain large amounts of organic coalescing agents. Alkyd emulsions are a competitive alternative from an environmental viewpoint because they do not contain any organic solvents [2,3]. They also have the advantage of being made from renewable materials. An alkyd oil is a polymer made of fatty acids, polyols, and a dibasic acid.

One drawback with alkyd emulsions is the longer drying time compared to solvent-borne alkyds and limited colloidal stability. An alkyd emulsion paint is a complex system containing many different components. In addition to the dispersed alkyd, it contains pigments, dispersing agents (for stabilizing the pigments), thickeners (for adjusting rheological properties), wetting agents, biocides, and so forth. All of these components influence the colloidal stability of the paint. A crucial factor, however, is the colloidal stability of the pure alkyd emulsion. The aim of this chapter is to discuss these aspects from a viewpoint which can be generalized even to other types of emulsions.

II. STABILITY AND INSTABILITY

To discuss the stability and instability of alkyd emulsions, one must first identify the major mechanisms of instability. Four main mechanisms can be identified:

1. Creaming or sedimentation: separation caused by the gravitational motion of the particles in the emulsion.
2. Flocculation: aggregation of droplets due to collisions. The identity of the aggregated droplets remains as separated droplets.
3. Coalescence: The droplets fuse together and the identities of separate droplets are lost.
4. Ostwald ripening: diffusion transport of dissolved material from smaller droplets into larger ones.

In all emulsions, all of these instability processes may occur, in principle, simultaneously. However, each of them strongly depend on various key parameters and on particle size (Table 1).

The *creaming process* is, in principle, well described by the Stokes' law. However, one deviation is that Stokes' law assumes that the creaming rate of the individual droplets are independent of their neighbor. In alkyd emulsions with concentrations of 30–60% (v/v) the sedimentation as predicted by Stokes law is strongly reduced. A semiempirical approach to

TABLE 1 Mechanisms of Instability

Key parameter	Size dependency[a]	Relevant to alkyd emulsions
1. Creaming Density difference, concentration	r^2	Yes
2. Flocculation The repulsive barrier	r^{-3} (Brownian) r^0 (Shear)	Yes Yes
3. Coalescence The interfacial properties	r^{-1}	Yes
4. Ostwald ripening The solubility of the dispersed phase in the continuous phase	r^{-3}	No, because the high-molecular-weight alkyd is virtually insoluble in water

[a] Given as sedimentation rate or time scale to double the radius.

this reduction is given by Buscall [4] with

$$v_{\text{Buscall}} = v_{\text{Stokes}} \left(1 - \frac{\phi}{\phi_{\text{close packing}}} \right)^{k\phi_{\text{close packing}}} \tag{1}$$

where k is an empirical constant (approximately 5) and $\phi_{\text{close packing}}$ is the maximum volume fraction.

Stokes' law assumes that the macroscopic viscosity can be used to estimate the sedimentation rate. In many formulations, thickeners are used to control the rheological properties of the paint. The flow properties in the presence of polymeric thickeners is usually strongly non-Newtonian. For estimations of the settling rate, the apparent viscosity at very low shear rates should be used (the shear rate of a creaming droplet is usually assumed to be shear rate $= v_{\text{Sedimentation}}/r$).

Several thickeners (associative thickeners) are also able to create a reversible network through the liquid, totally inhibiting creaming (Bingham plastic flow properties).

The rupture of the film separating the droplets is the crucial part of the *coalescence process* (the process in which two or several droplets form one). Vrij [5] analyzed the rupture of thin films due to self-propagation of interfacial waves at the oil–water interface and found that the amplitude of the waves increases if the dimensions of the film are sufficiently large compared to the magnitude of the derivative of the repulsive pressure. The effect was expressed in the following inequality:

$$\frac{d\Pi}{dh} < \frac{-2\pi^2\gamma}{l^2_{film}} \qquad (2)$$

where l_{film} is a characteristic dimension of the film (for instance, the diameter), γ is the oil–water interfacial tension, Π is the repulsive pressure between the approaching droplets, and h is the thickness of the film.

This approach deals with one possible mechanism for the film rupture, which is a clear oversimplification (for an updated more complete review, see Ref. 6). However, it points to a pronounced size dependency of the process. Alkyd emulsions have to be finely dispersed to be stable (against creaming) and to produce even paint films. The typical size distribution of a commercial alkyd emulsion (Fig. 1) shows that the dominating particle

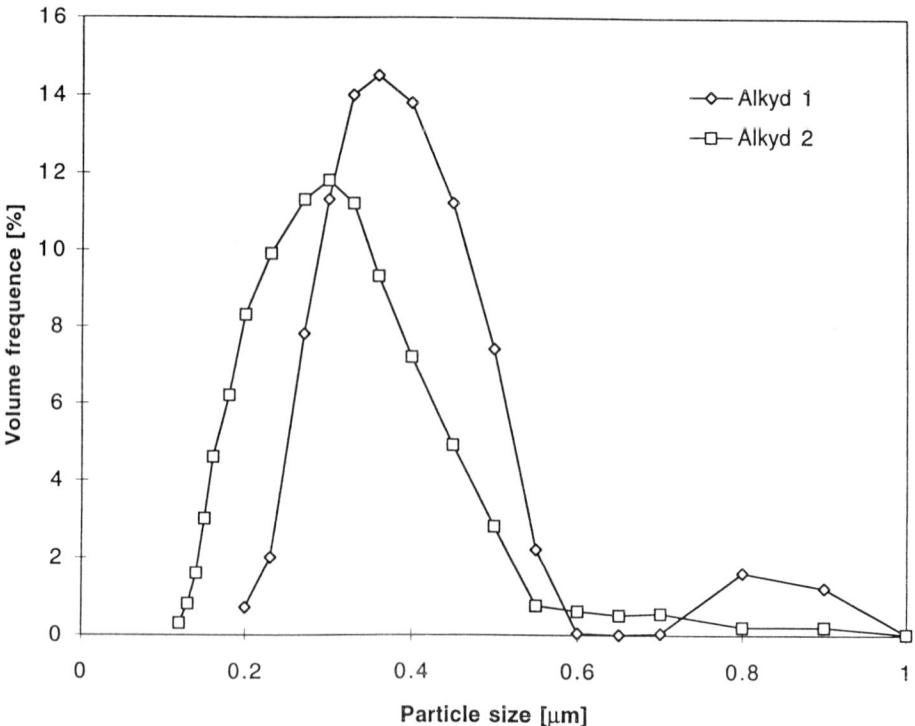

FIG. 1 The particle size distribution for two commercial alkyd emulsions. The particle size has been determined using laser light diffraction.

sizes are in the submicron range, clearly demanding extremely thin films (typically below 100 Å) to obtain instability according to Eq. (2).

Flocculation is collisions causing the formation of permanent flocs. The collision mechanisms may include:

1. Brownian motion (perikinetic flocculation)
2. Sedimentation motion (gravity induced flocculation)
3. Motion in a shear field (orthokinetic flocculation)

The rate of loss of particles can be obtained from the rate of collision (if we assume that there is a certain probability that a collision lead to permanent adhesion). In Table 2, the collision frequency is calculated as a function of the number concentration of droplets in the emulsion and the increase in particle size as a function of time, if all flocculated particles coalesce.

TABLE 2 Comparison Between Various Flocculation Rate Mechanisms in Emulsions

dN/dt	dr/dt	$r(t)$	Ref.
Brownian flocculation			
$\dfrac{dN}{dt} = -\dfrac{4}{3}\dfrac{k_B T}{\eta}N^2\dfrac{1}{w_B}$	$\dfrac{dr}{dt} = \dfrac{2}{3}\dfrac{k_B T}{\eta}\dfrac{\phi}{\pi r^2}\dfrac{1}{w_B}$	$r = \left(r_0 + \dfrac{\phi 2k_B T}{9\pi\eta w_B}\right)^{1/3}$	7
Shear-induced flocculation: laminar flow			
$\dfrac{dN}{dt} = -\dfrac{16}{3}N^2 S r^3\dfrac{1}{w_S}$	$\dfrac{dr}{dt} = \dfrac{4\phi r S}{3\pi}\dfrac{1}{w_S}$	$r = r_0\exp\left(\dfrac{4\phi St}{3\pi}\cdot\dfrac{1}{w_S}\right)$	8
Shear-induced flocculation: turbulent flow			
$\dfrac{dN}{dT} = -6\pi\beta\sqrt{\dfrac{\epsilon_{int}}{\eta}}r^3 N^2\dfrac{1}{w_r}$	$\dfrac{dr}{dt} = \dfrac{3\beta}{2}\sqrt{\dfrac{\epsilon_{int.}}{\eta}}\phi r\dfrac{1}{w_T}$	$r = r_0\exp\left(\dfrac{3\beta}{2}\sqrt{\dfrac{\epsilon_{int.}}{\eta}}\phi t\cdot\dfrac{1}{w_T}\right)$	9
Shear-induced flocculation in turbulent flow expressed as a function of apparent shear rate			
$\dfrac{dN}{dT} = -6\pi\beta\sqrt{\dfrac{c_1\hat{S}^3}{\eta}}r^3 N^2\dfrac{1}{w_T}$	$\dfrac{dr}{dt} = c_2\dfrac{\hat{S}^{3/2}}{\eta^{1/2}}\phi r\dfrac{1}{w_T}$	$r = r_0\exp\left(c_2\dfrac{\hat{S}^{3/2}}{\eta^{1/2}}\phi r\dfrac{1}{w_T}\right)$	

Notes: The instability is expressed in terms of change in number concentration (dN/dt), the increase in particle radius (dr/dt), or as the particle radius as a function of time.

The flocculation rate in terms of particle concentration has been transformed to growths in particle size by using the chain rule $dN/dt = (dN/dr)(dr/dt)$ and by treating the particle concentration as a function of the particle radius: $N = 3\phi/4\pi r^3 \Rightarrow dN/dr = -9\phi/4\pi r^4$. N = particle concentration; t = time; k_B = Boltzmann constant; T = absolute temperature; η = viscosity; ϕ = volume fraction dispersed phase; w_B = stability factor for the Brownian flocculation according to Fuchs [10]; w_S = a stability factor for shear-induced flocculation; w_T = a stability factor for flocculation in a turbulent flow; S = the shear rate; $\hat{S}$ = an apparent shear rate in a turbulent flow field (the apparent shear rate is calculated as if the flow field was laminar); β = a constant describing the relation between the diffusive flux in a turbulent flow field and the liquid velocity; c_1 and c_2 = constants depending on geometry and liquid, respectively.
Source: Ref. 12.

Calculation of the shear-induced flocculation is made both under laminar and turbulent flow.

Some interesting observations from Table 2 are that the Brownian flocculation (in terms of dr/dt) rapidly declines with increasing radius, the sedimentation-induced flocculation has a weak dependence of the particle radius, and the shear-induced flocculation increases with increasing particle radius. This gives the characteristics of a shear destabilization: an extended lag phase with an apparently stable system followed by a rapid destabilization.

The above description of flocculation process includes interparticle interactions, which protect the particles from fusing after a collision event, in the form of a stability factor, w, where $1/w$ is the probability for attachment after a collision event. The stability factor is a function of the ratio between the energy of the collision event and the repulsive barrier (for a modern extensive review of the repulsive barrier between colloidal particles, see Ref. 11).

III. MECHANICAL STABILITY

The mechanical destabilization of alkyd emulsions is an interesting application of the presented models for the destabilization kinetics. Destabilization has been characterized by measurements of the particle size after high shear treatment [12–14]. Emulsions were exposed to shear forces in a colloidal mill, Ultra Turrax. The shear rate was calculated from the velocity of the rotor.

According to the rate equations in Table 2, the particle size after shearing for a constant time under turbulent conditions is expected to be proportional to $e^{\dot{s}^{3/2}}$. In Fig. 2 it can be seen that the logarithm of the diameter of the emulsion droplets increases in proportion to $\dot{s}^{3/2}$ as expected from kinetic theory. However, at low shear rates, the emulsion appears to be stable and the apparent shear rate must overcome a critical value to allow coalescence. The existence of a critical shear rate to overcome the repulsive barrier of a dispersion has also been shown theoretically by Zeichner and Schowalter [15].

Figure 2 shows an example of the reproducibility of the method. At the two lowest shear rates, five similar samples were sheared, and at the two higher rates, four similar samples were sheared.

A. Influence of Initial Droplet Size

The influence of the initial droplet size on the mechanical stability of emulsions stabilized with LA 13 is shown in Fig. 3. With this emulsifier,

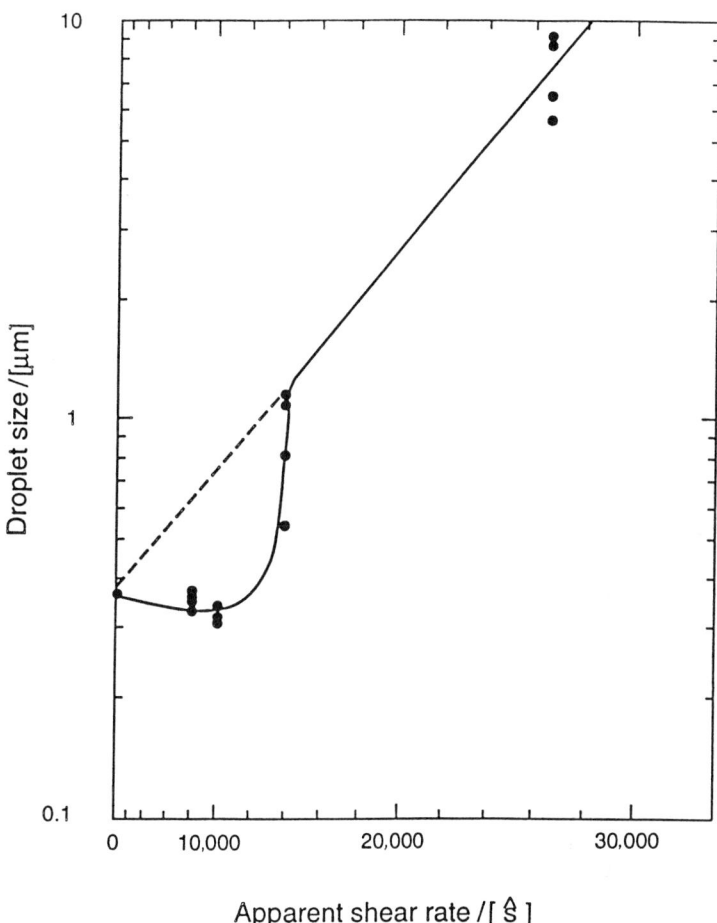

FIG. 2 The droplet size (the volume mean diameter) of an emulsion after 10 min of shear. The logarithm of the droplet size is given as a function of apparent shear rate $\hat{s}^{3/2}$. The model emulsion is stabilized using a nonionic emulsifier with 23 EO groups. (From Ref. 12.)

it was possible to obtain larger droplets by decreasing the concentration of LA 13; see Table 3. An initial droplet size of 7 μm gave an unstable emulsion that increased in droplet size even at low shear rates. The decrease in droplet size at 20,000 s⁻¹ is probably due to rehomogenization producing new smaller droplets. Decreasing the initial droplet size to 2

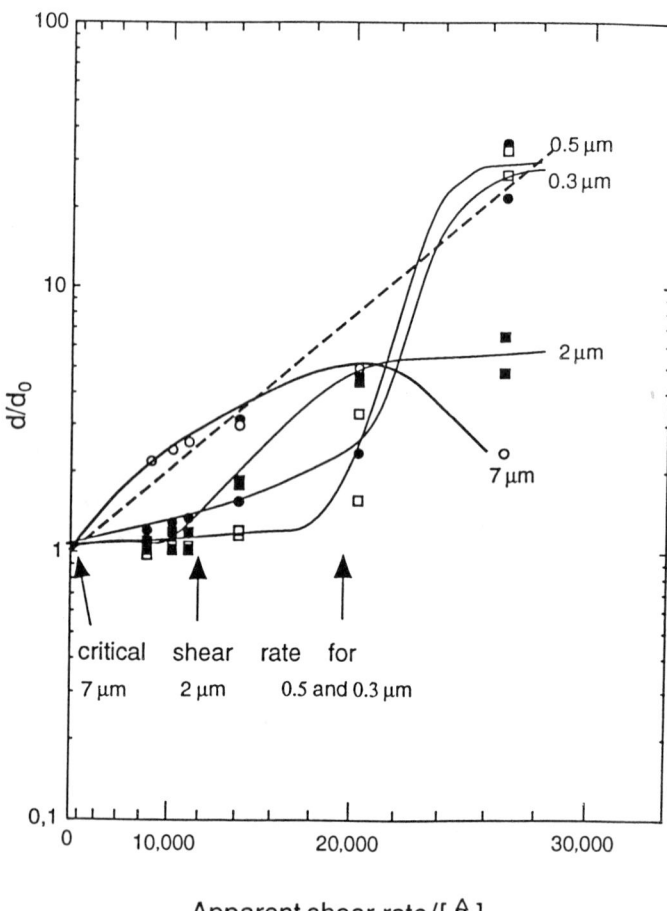

FIG. 3 The relative droplet size (the volume mean) of emulsions with different original droplet size after shear. The logarithm of the relative droplet size is given as a function of apparent shear rate $s^{3/2}$. The dashed line indicates the relative increase according to Table 2. All emulsions are stabilized by a nonionic LA 13 (fatty acid monoethanolamide ethoxylate with 13 EO, from Akzo Nobel). (From Ref. 12.)

μm improved the stability. A further decrease to 0.3–0.6 μm gave even more stable emulsions. Within this range, however, stability is similar.

The hydrodynamic forces bringing the particles together can, as a first approximation, be estimated to be proportional to the surface area of the particle, whereas the stabilizing barrier is proportional to the diameter of

TABLE 3 Droplet Size after Homogenization for Emulsions Produced Using High-Pressure Homogenization (800 bars)

Emulsifier	Emulsifier concentration (% of oil phase)	Droplet size (volume mean diameter, μm)	Estimated apparent thickness of the emulsifier layer (A)[a]
LA 13[b]	5	2.06	150
	5.5	0.49	40
	6	1.95	173
	6.5	0.49	48
	10	0.33	50
	13	0.31	60
DoBS[c]	0.5	0.89	6
	1	0.82	11
	2	0.82	23
	5	0.55	36
	10	0.36	48

Note: The droplet size after the homogenization is measured. The apparent thickness of the emulsifier layer is estimated assuming that all emulsifier is present at the interface.
[a] Estimated using the area weighted average value.
[b] Fatty acid monoethanolamide ethoxylate with 13 EO, from Akzo Nobel
[c] Sodium dodecylbenzenesulfonate
Source: Ref. 14.

the droplets. In agreement with this we observed an increase in the critical shear rate from <4000 s^{-1} for droplets of 7 μm (diameter) to about 15,000 s^{-1} for droplets <0.6 μm. In the calculations of Zeichner and Schowalter [15], a lower critical shear rate for flocculation in the primary minimum with increasing particle size was also obtained, even though the effect was reduced due to hydrodynamic interactions. However, in their calculation, they used electrostatically stabilized particles where the repulsive barrier is relatively long ranged. In this experiment, the repulsive barrier is caused by a short-range steric stabilization.

When the critical apparent shear has been overcome and the shear-induced flocculation has started, the coalescence rate seems to be independent of the initial particle size. Consequently, the various aspects of drainage and film rupture that should depend on the particle size seem not to change the kinetic description of these experiments.

If it is assumed that all of the emulsifiers are located at the surface of the droplets, which is an overestimation because the emulsifier is also soluble in the water and the oil phase, an average apparent thickness of the emulsifier layer can be calculated. A comparison of this calculated apparent thickness with the length of one molecule of the emulsifier gives an estimation of whether there is enough emulsifier to cover the surface. The calculated apparent thicknesses at different droplet sizes are shown in Table 3. If these values are compared to 75 Å, which is the length (calculated from the bond length) of a fully stretched LA13 molecule, it can be seen that the emulsions with larger droplets have an excess of emulsifier compared to emulsions with smaller droplets. This shows that the measured instability in Fig. 3 is actually due to the droplet size and not to a shortage of emulsifier.

The stability of the emulsions with the anionic emulsifier DoBS was not as sensitive to initial droplet size as that of the emulsions with the nonionic emulsifier, see Fig. 4. Larger droplets were obtained by decreasing the pressure and temperature during emulsification because they could not be obtained merely by decreasing the emulsifier concentration. The calculated apparent thicknesses of the emulsifier layer for these emulsions are shown in Table 3. All the emulsions in Fig. 4 had an excess of emulsifier if the thickness is compared with the length of one DoBS, sodium dodecylbenzene sulfate, which is about 15 Å. This value is from a measurement of the layer thickness in a liquid-crystalline phase. The electrostatically stabilized DoBS emulsions are obviously different from the sterically stabilized emulsions. The critical level of the shear rate is high and there is no sensitivity to the initial droplet size.

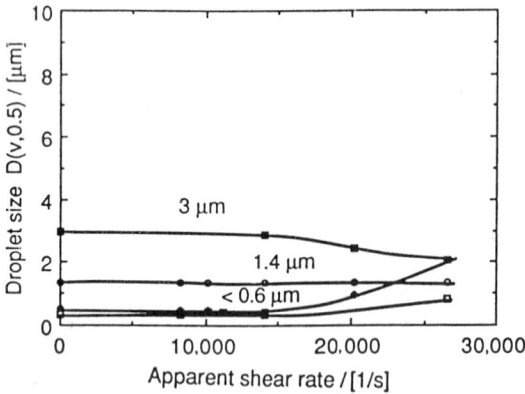

FIG. 4 The mean droplet size after shear for emulsions stabilized by an anionic emulsifier (DoBS). Emulsions with three different original droplet sizes have been investigated. (From Ref. 12.)

B. Influence of the Concentration of the Emulsifier

With the anionic emulsifier DoBS, an excess of emulsifier was used in the experiments shown in Fig. 4. Experiments when the concentration was decreased and the droplet sizes were relatively small that a pronounced instability was obtained when there was a shortage of emulsifier (0.5% of DoBS in Table 3). The instability of the emulsion with the lowest DoBS content clearly shows the need for complete coverage to obtain stability toward shear. It has also been shown by Ivanov and Dimitrov [16] that the drainage, and thereby the resistance toward coalescence, is orders of magnitude faster between two droplets covered by an incomplete emulsifier layer than between droplets covered by a dense layer.

C. Different Types of Emulsifiers

The influence of the length of the EO chain on the shear stability of emulsions stabilized with a homologous series of nonionic emulsifiers is shown in Fig. 5. The critical shear rate necessary to destabilize the emulsion is almost constant, independent of the length of the EO chain. This observation agrees with the general view that the long-range interactions determine the flocculation step, and thereby the critical shear intensity. The probability of coalescence, on the other hand, seems to decrease with increasing length of the hydrophilic part of the emulsifier, particularly because lower concentrations of those with longer ethylene oxide chains were used.

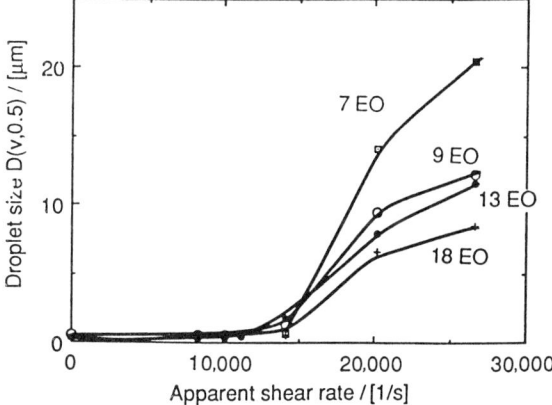

FIG. 5   The mechanical stability as a function of the length of the polar group of the emulsifier. Fatty acid monoethanolamide ethoxylate with 7, 9, 13, and 18 EO groups have been investigated. (From Ref. 14.)

The same stability is, however, also obtained when the anionic emulsifiers (DoBS and $C_{16}/C_{18}EO_2S$) are used. The nonionic steric stabilizers perform well, but the anionic emulsifiers may be just as powerful despite the absence of steric hindrance. These experiments, however, have been performed in the absence of ionic additive normally used in paints (e.g., driers and dispersing agents). If the solubility of ionic additives is significant in the water phase, this reduces the stability of electrostatically stabilized emulsions.

The lower stability of the emulsion with the fatty alcohol sulphate compared to that with the corresponding ethoxylated product is explained by the low solubility of the former product. It has a Krafft point (the lowest temperature at which the emulsifier crystals dissolve and form a micellar solution) of about 52°C. It was possible to make an emulsion with small droplets that was intact at 20°C, but it flocculated at the lowest shear rate. A possible explanation is that most of the emulsifier precipitates, resulting in a shortage of emulsifier at the surface and therefore poor stability. The emulsifier may also be in an undercooled state and the shear forces may induce crystallization of the emulsifier, which results in flocculation.

IV.   DRYING PROPERTIES

A.   Drying of Alkyds

Drying of a paint based on an oxidatively drying alkyd involves both a physical and a chemical part. The physical drying is the evaporation of the solvent (water in the case of an alkyd emulsion) and the formation of a film. The physical drying is followed by the chemical drying when the alkyd polymerizes. The polymerization is an autoxidation of the unsaturated fatty acids in the alkyd.

For practical reasons, a rapid drying is desired. To enhance the drying time of both solvent-borne and water-borne alkyds, driers are added. Driers are metal salts of organic acids. Common dries metals are cobalt, calcium, lead, and zirconium, and the organic acids are often 2-ethyl hexanoic acid, linoleic, or naphtenates [17]. Conventional driers are solved in an organic solvent but emulsifiable types also exist. For alkyd emulsions, both types can be used. To achieve optimal drying, different types of driers are used. Primary driers such as cobalt and manganese work by catalyzing the autoxidation.

B.   Drying of Water-Borne Alkyds

The drying performance of alkyd emulsion paints differ from solvent-borne alkyd paints although the mechanism is the same. Alkyd emulsions show a slow initial drying and bad through drying compared to white spirit-based formulations. The drying time is also rapidly increased when the paint is stored. The dried film is usually softer and more water sensitive than a film from a solvent-borne alkyd. This is at least partly caused by the emulsifier acting as a plasticizer in the film. The reason for the reduced drying properties are not fully understood. Interaction between the drier and other paint components are, however, believed to be the major reasons [18]. The interactions can include:

1. Complex formation with neutralizing agents, emulsifiers, dispersing agents, and the binder (resulting in a less active drier or incompatibility with the resin)
2. Adsorption onto pigments

For example, complex formation between cobalt and amines, used as neutralizing agents, is known to reduce cobalt's catalytic activity [19,20]. Water is also a ligand to cobalt. Adsorption of dryers onto pigments are known, but its effect on drying properties is not clear.

It is also believed that the solubility of the drier in the alkyd and the aqueous phase influence the activity of the drier. The drier can be distributed within the emulsion. A high solubility in the aqueous phase would enhance the complex formation and the adsorption of the drier to pigments. If the drier has to migrate from the water phase to an alkyd of high viscosity in the late stage of drying (after most of the water has evaporated), it may not be evenly distributed in the resin.

The pH of the emulsion has a great influence on the solubility of driers in the alkyd and the aqueous phases [2]. A high pH favors a distribution to the alkyd phase. When pH is lowered, the drier is redistributed toward the aqueous phase. This has been observed both for cobalt and calcium driers, whereas zirconium driers are located in the alkyd phase over a broad pH range. The distribution depend on the solubility product of the drier according to

$$Co\,(RCOO)_{2_{alkyd}} \rightleftarrows Co\,(RCOO)_{2_{aq}} \rightleftarrows Co^{2+}_{aq} + 2RCOO^-_{aq}.$$

A more hydrophobic counterion to the metal favors the solubility in the alkyd phase. Usually, the pH is about 7–8 in a paint formulation but is decreased during storage due to hydrolysis of the alkyd. If pH is too high, the metal will precipitate as metal hydroxides, giving poor drying properties.

Anionic emulsifiers used to stabilize the alkyd emulsion can also form complexes with the driers. It has been shown that the presence of sodium dodecyl sulphate (SDS) and ethoxylated fatty alcohol sulphate (C_{12}/C_{14} EO3-S) reduces the concentration of cobalt in the alkyd phase. SDS forms a liquid-crystalline phase with cobalt that separates from the water and the alkyd phase. Nonionic emulsifiers, on the other hand, do not influence this distribution. A longer drying time has also been observed in emulsions containing SDS as emulsifier compared to emulsions of nonionic emulsifiers. Anionic emulsifiers, on the other hand, have the advantage of giving more stable emulsions at lower concentrations than nonionic emulsifiers.

To overcome these drying problems, attempts have been made with precomplexed and emulsifiable driers. To reduce the problems arising from surfactants in the paint film, the use of types that can copolymerize with the alkyd has been proposed. One such new type recommended for alkyd emulsions is monoethanolamide ethoxylate based on polyunsaturated fatty acids. The hydrophobic part of this type of surfactant contains the same degree of unsaturation as the alkyd resin, which should give them the ability to copolymerize with the binder during the curing stage [3].

REFERENCES

1. R. Hurley, *Handbook of Coatings Additives* (L. J. Calbo, ed.), Marcel Dekker, Inc., New York, 1987.
2. G. Östberg, B. Bergenståhl, and K. Sörenssen, J. Coatings Technol. *64*:33 (1992).
3. K. Holmberg, Prog. Organ. Coatings *20*:325 (1992).
4. R. Buscall, J. W. Goodwin, R. H. Ottewill, and Th. F. Tadros, J. Colloid Interface Sci. *85*:78–86 (1982).
5. A. Vrij, Discuss. Faraday Soc. *42*:23–33 (1966).
6. P. Walstra, in *Encyclopedia of Emulsion Technology. Vol. 4*, P. Becher (ed.), Marcel Dekker, Inc., New York, 1995.
7. M. von Smoluchowskij, Phys. Z. *17*:593 (1916).
8. M. von Smoluchowski, Z. Phys. Chem. *92*:129 (1917).
9. V. G. Levich, *Physicochemical Hydrodynamics*, Prentice-Hall, Englewood Cliffs, NJ, 1962.
10. N. Fuchs, Z. Phys. *89*:736 (1934).
11. J. Israelachvilli, *Intermolecular and Surface Forces*, Academic Press, London, 1992.
12. G. Östberg, B. Bergenståhl, and M. Hulden, J. Coatings Technol. *66*:(832): 37 (1994).
13. G. Östberg and B. Bergenståhl, J. Coatings Technol. *66*(838):37 (1994).
14. G. Östberg, B. Bergenståhl, and M. Hulden, *Colloids Surf.* 94:161 (1995).
15. G. R. Zeichner and W. R. Schowalter, AIChE J. *23*:243 (1977).
16. I. B. Ivanov and D. S. Dimitrov, in *Thin Liquid Films* (I. B. Ivanov, ed.), Marcel Dekker, Inc., New York, 1988.
17. F. J. Buono and M. Feldman, *Encyclopedia of Chemistry and Technology*,

3rd ed., Vol 8, (M. Grayson and D. Eckroth, eds.), John Wiley & Sons, New York, 1979, pp. 34–49.
18. J. H. Bieleman, in *Surface Coatings. Raw Materials and Their Usage, Volume 1*, The New South Wales University Press, Kensington, Australia 1993.
19. M. H. Dean and G. Skirrow, Trans. Faraday Soc. *54*:849 (1958).
20. J. H. Bieleman, *Farbe & Lack 94*:434 (1988).

7

Clinical Development of Perfluorocarbon-Based Emulsions as Red Cell Substitutes

ROBERT J. KAUFMAN   HemaGen/PFC, St. Louis, Missouri

I. INTRODUCTION

The search for solutions to temporarily replace the oxygen transport function of blood has been underway in academia, the military, and industry for close to 50 years. Recent advances in perfluorocarbon emulsion technology indicate that the problems which have impeded development of useful medical products in this field may now be moving toward solution.

The historical background, mechanism of oxygen transport, and fundamental aspects of perfluorocarbons and their emulsions has recently been reviewed [1–4]. This review will focus on new advances in the clinical development of perfluorocarbon emulsions.

II. TEMPORARY OXYGEN TRANSPORT

Since the pioneering experiments of Clark and Gollan [5], Sloviter and Kamimoto [6], and Geyer et al. [7], much research has been devoted to developing medically useful oxygen transport products based on perfluorocarbons (PFCs) to temporarily replace the oxygen transport function of whole blood. Green Cross was the first company to reach clinical trials with an emulsion containing two perfluorocarbons, 70% perfluorodecalin and 30% perfluorotripropylamine, using pluronic as the primary surfactant. This emulsion, called Fluosol DA®, contained 10% PFC by volume and had to be stored in the frozen state [2].

Clinical results with Fluosol DA in oxygen transport were discouraging. In studies of severely anemic surgical patients (Hematocrit [Hb] < 10) who refused transfusions on religious grounds, Fluosol failed to provide more than a transient boost in arterial oxygen tension [8]. Fluosol DA did contribute 28% of the oxygen consumption of these patients. However, this gain in oxygen consumption due to the PFC was offset by dilution of the hemoglobin. After infusion of Fluosol DA, the net oxygen carried by the red cells, the plasma, and the PFC was the same as that carried by the red cells and plasma before infusion (Table 1).

The conclusions from Gould's study were that Fluosol DA was safe but ineffective. To achieve a circulating PFC concentration of 5% required

TABLE 1 Hemodynamics and Oxygen Transport Properties Before and After Fluosol DA Administration in Eight Patients

Property	Before Fluosol DA	After Fluosol DA
Oxygen delivery (ml min^{-1} m^{-2})	235 ± 27	197 ± 32
Oxygen consumption (ml min^{-1} m^{-2})	109 ± 13	88 ± 11
Arterial oxygen tension (torr)	356 ± 24	430 ± 19[3]
Venous oxygen tension (torr)	40 ± 3.9	78.2 ± 3

Note: Data were obtained at peak arterial oxygen content after Fluosol DA administration. Values given as the mean ± SE.

The difference between values before and after Fluosol DA is significant ($p < .05$).
Source: Data from Ref. 8.

an infusion of 2800 ml of Fluosol DA, 90% of which is aqueous. Therefore, concentrated emulsions would be necessary to achieve a PFC concentration at which significant oxygen transport occurs without diluting hemoglobin.

Additional clinical trials revealed that complement activation occurred in 5–10% of the patients treated with Fluosol DA. The surfactant, pluronic F-68, was subsequently implicated as the cause of this effect [9]. The clinical experience with the first-generation product, Fluosol DA, demonstrated that future products needed to be more concentrated, free from pluronic surfactants, and possess better storage stability.

Two new products in development appear to have solved these problems: Oxyfluor™ from HemaGen/PFC and Oxygent™ from Alliance Pharmaceutical. Oxyfluor is a 40% v/v emulsion of perfluorodichlorooctane (PFDCO) stabilized using egg yolk phospholipid and safflower oil. In equilibrium with 100% oxygen at 37°C, Oxyfluor carries 17.2 vol% oxygen. Oxyfluor has a shelf life of greater than 1 year at room temperature.

Perfluorodichlorooctane (PFDCO) has a molecular weight of 471, boils at 155°C, and has a density of 1.76 g ml^{-1}. It is a lipophilic PFC and is soluble in many hydrocarbons and other organic solvents. The ozone depletion potential (ODP) of PFDCO is <0.25.

Perfluorodichlorooctane was discovered in a collaboration between HemaGen and 3M scientists when animal studies using perfluorodecalin (PFD) and perfluorooctyl bromide (PFOB) demonstrated that some PFCs caused pulmonary hyperinflation in rats and dogs. The phenomenon is characterized by an increase in respiratory rate, an increase in functional residual capacity, a decrease in inspiratory capacity, and a reduction in arterial oxygen tension. Gross necropsy in rats or dogs at the time of peak effect showed lungs which remained inflated after removal from the chest. In rats, the lung volume from PFD-treated rats measured by water displacement is two to three times the control. Histologic evaluation of hyperinflated lungs did not reveal any structural defects. Except in extreme cases such as perfluorodimethlycyclohexane, this effect is not lethal in rats. Pulmonary hyperinflation was completely reversible with time, generally peaked at 4–7 days postinfusion and disappeared completely by 30–60 days postinfusion.

A key problem in identification of new candidate molecules was to find PFCs that did not cause pulmonary hyperinflation but retained sufficient vapor pressure to leave the body in a reasonable time frame. To identify compounds with both properties, new PFCs were tested in rats for lung inflation by measurement of lung volume by water displacement and for liver clearance by gas chromatographic analysis of tissue at 2 and 14 days postinfusion. The results from a comparison of the lung volumes induced by PFD, PFOB, PFDCO, and saline in rats is shown in Table 2.

TABLE 2   Effect of Three PFCs on Lung Inflation in the Rat

Treatment	Gross appearance	Lung volume[a,b]
PFD	Marked hyperinflation	2.20
PFOB	Modest hyperinflation	1.10
PFDCO	Normal, collapsed	0.90
Saline	Normal, collapsed	0.87–0.91

[a] ml/per 100 g of body weight.
[b] Values are mean of 10 rats.

The results of the rodent tissue clearance studies for PFDCO are shown in Fig. 1. The liver half-life is about 8 days and the PFDCO drops below the level of detection by 60 days postinfusion.

Candidate PFCs that passed the rodent tests were then tested in baboons for effects on pulmonary function as well as vital signs, hematology, coagulation, and clinical chemistry. Historically, increases in respiratory rate correlate well with a decrease in inspiratory capacity and arterial blood gas tension, and an increase in functional residual capacity. PFDCO was compared to PFOB and PFD because they were the most widely studied PFCs in the field. PFD, which exhibited the largest rodent lung volume, had increased respiratory rate to 200% of baseline by day 1 postin-

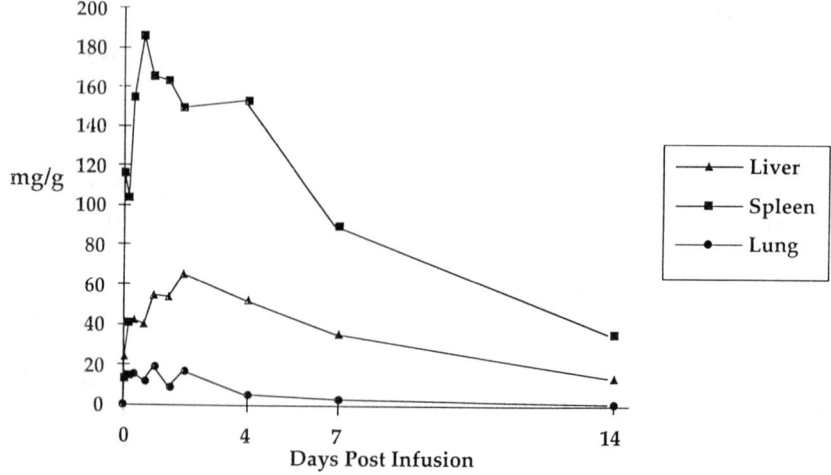

FIG. 1   Clearance rate of PFDCO from the liver, spleen, and lungs.

fusion in the baboon. PFOB, which marginally inflated lungs in rats, surprisingly increased respiratory rate to 400% of baseline by 4 days postinfusion in the baboon and still exhibited elevated respiratory rate 60 days postinfusion (Fig. 2). The respiratory rate in PFDCO-treated baboons rose only to 140% of baseline by 7 days postinfusion and was in the normal range thereafter. PFDCO had only modest ($\leq$20%) changes in inspiratory capacity, functional residual capacity, and arterial blood gases. PFD and PFOB were similar in their marked effect on pulmonary function [10]. Baboons treated with these PFCs had a 35% decrease in inspiratory capacity and a 200% increase in functional residual capacity between days 4 and 7 postinfusion. Blood gas oxygen tension fell below 60 mm Hg at day 4 postinfusion in animals treated with PFD or PFOB, whereas PFDCO-treated animals remained in the normal range throughout the postinfusion period.

Stability studies on Oxyfluor were conducted at 4°C and 30°C for 12 months and for 6 weeks at 40°C. Stability-indicating analyses included particle size by laser light scattering and Coulter counter, and pH. Chemical analyses included phophatidyl choline, free fatty acids, and lysophophatidyl choline content.

In the accelerrated aging studies (40°C), pH dropped from 7.54 to about 7.15 after 6 weeks, whereas free fatty acids (FFA) rose from 0.235 to 0.270 mg KOH g^{-1} (Fig. 3). At 30°C, the pH dropped from 7.54 to 6.70 after 12 months, whereas FFA rose from 0.235 0.510 mg KOH g^{-1} (Fig.

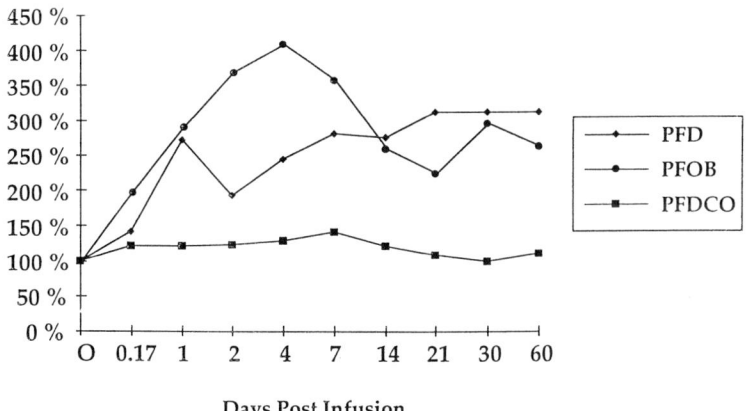

FIG. 2 Comparison of PFD, PFOB, and PFDCO on respiratory rate as a percent of baseline in the baboon at 8cc PFC-kg^{-1}.

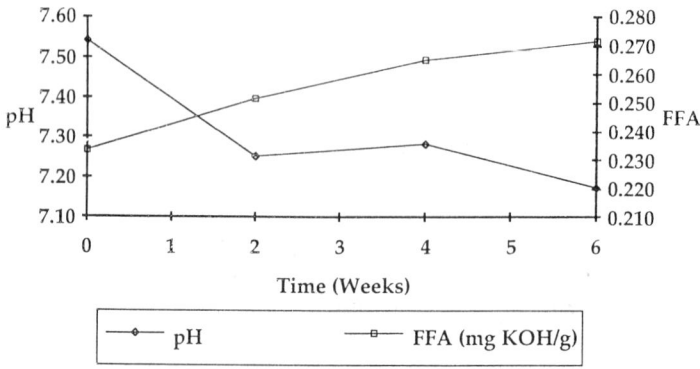

FIG. 3 Free fatty acid generation (mg KOH-g^{-1}) and pH drop in Oxyfluor stored at 40°C for 6 weeks.

4). At 4°C, the pH was virtually unchanged, whereas FFA rose only 0.300 mg KOH g^{-1} (Fig. 5). These changes are similar to changes seen in intravenous fat emulsions and are indicative a stable emulsion.

High-pressure liquid chromatography (HPLC) analyses of the emulsions for phosphatidyl choline (PC) and lyso-phosphatidyl choline (LPC) correlated with the pH and FFA concentrations at all temperatures (Figs. 6–8). The hydrolysis of PC to LPC and FFA with concommitant pH drop are well known in the fat emulsion field and appear to be similar for PFC emulsions.

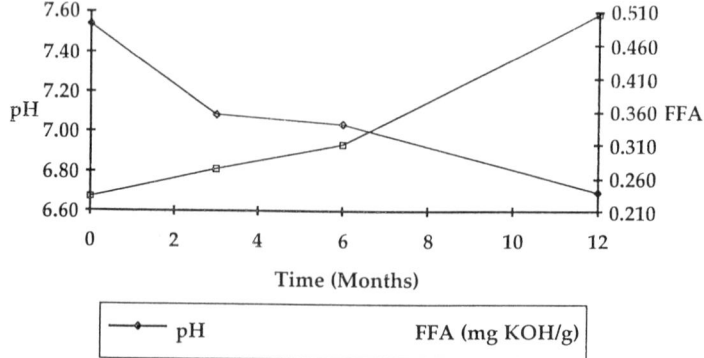

FIG. 4 Free fatty acid generation (mg KOH-g^{-1}) and pH drop in Oxyfluor stored at 30°C for 1 year.

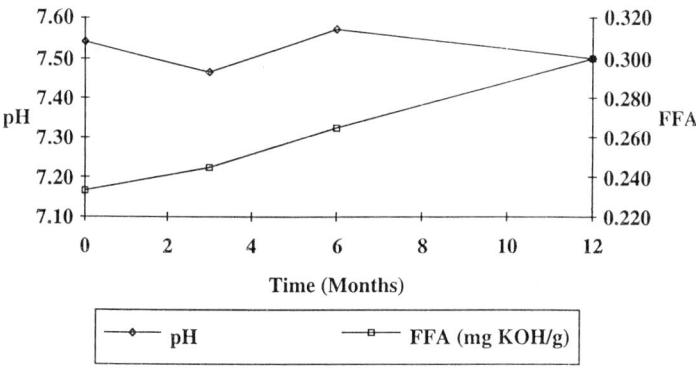

FIG. 5 Free fatty acid generation (mg KOH-g^{-1}) and pH drop in Oxyfluor stored at 4°C for 1 year.

Preclinical animal studies with Oxyfluor in models of shock resuscitation [11] and surgical anemia [10] have shown efficacy for oxygen transport. In contrast to traditional shock models, this model utilized compromised tissue oxygenation as an end point. Groups of dogs were hemorrhaged to a mixed venous oxygen tension of ≤25 mm Hg and held there for 10 min. They were resuscitated with either Oxyfluor (4cc PFC kg^{-1} or 15cc kg^{-1} of emulsion) or lactated Ringer's solution (15 ml kg^{-1}) while breathing 100% oxygen. Dogs resuscitated with Oxyfluor had nor-

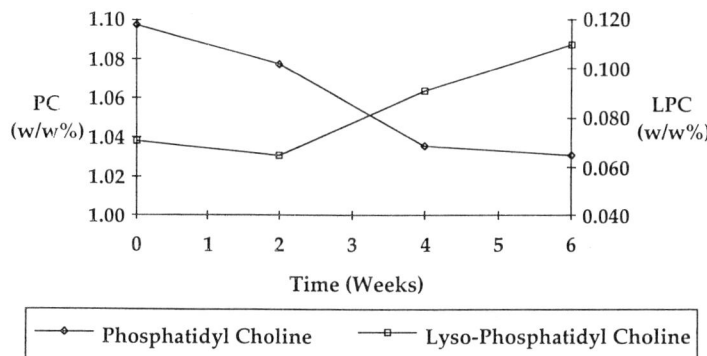

FIG. 6 Phosphatidyl choline (% w/w) and lyso-phosphatidyl (% w/w) in Oxyfluor stored at 40°C for 6 weeks.

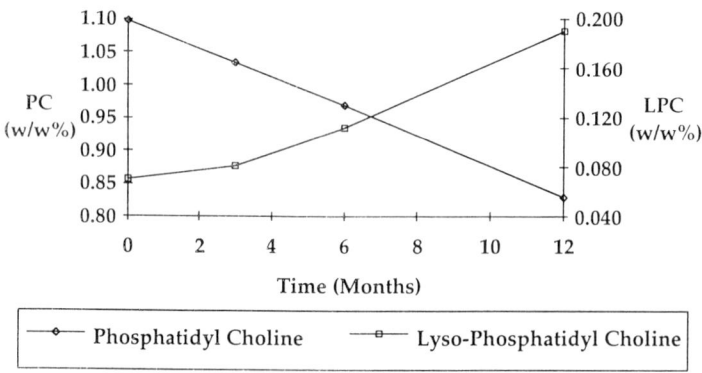

FIG. 7 Phosphatidyl choline (% w/w) and lyso-phosphatidyl (% w/w) in Oxyfluor stored at 30°C for 1 year.

mal mixed venous oxygen tension postresuscitation and all survived. Dogs resuscitated with lactated Ringer's solution only returned 80% of a normal mixed venous oxygen tension and only 62.5% survived (Fig. 9).

In the model of surgical anemia, dogs were hemodiluted to a hematocrit of 20% and then infused with Oxyfluor and allowed to breath 100% oxygen. Following the administration of Oxyfluor, the arterial oxygen tension, venous oxygen tension, and oxygen consumption were all increased significantly for 90 min. Results show that 40% of the consumed oxygen was

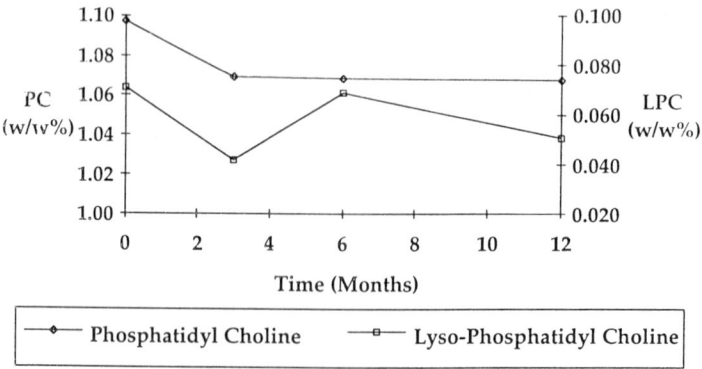

FIG. 8 Phosphatidyl choline (% w/w) and lyso-phosphatidyl (% w/w) in Oxyfluor stored at 4°C for 1 year.

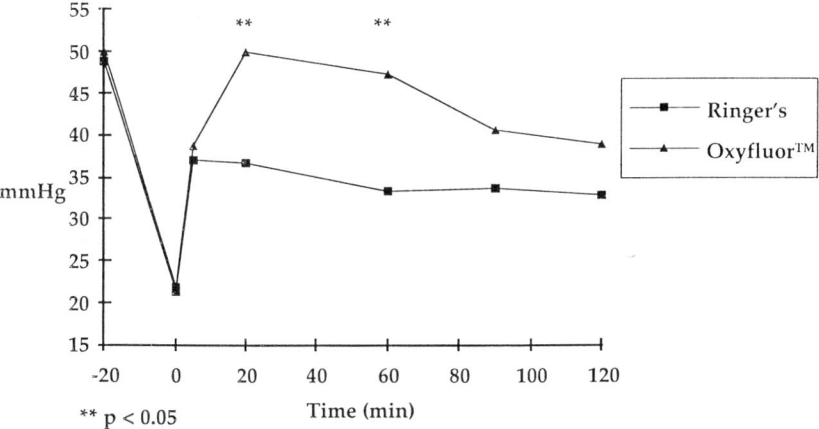

FIG. 9 Effect of 4cc PFC-kg^{-1} of Oxyfluor on mixed venous oxygen tension in shocked dogs.

derived from the circulating PFC for up to 90 min postinfusion. This dramatically illustrates the ability of PFCs to deliver significant amounts of oxygen to tissues (Fig. 10).

A Phase 1 clinical trial in healthy human volunteers has been completed with Oxyfluor. This trial involved three dosing levels with six treatment

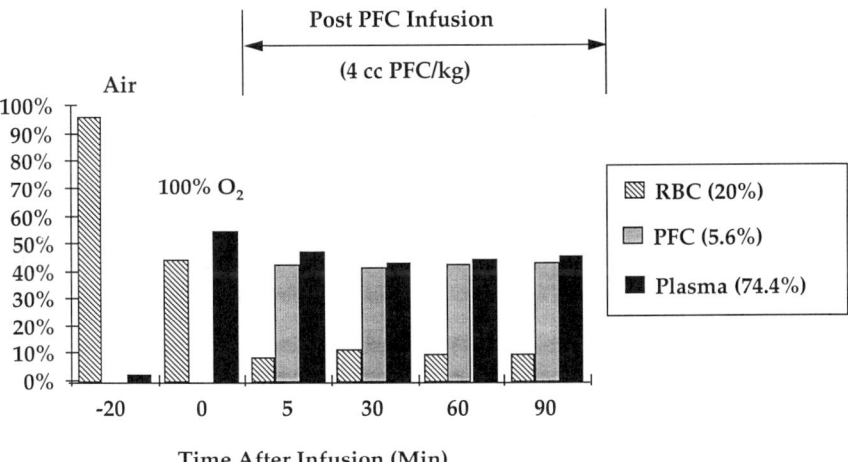

FIG. 10 Fractional contribution to oxygen consumption of 4cc PFC-kg^{-1} of Oxyfluor in anemic dogs.

and six control subjects at each dosage level. The study was randomized within each dosing level and was double blinded. The subjects were monitored for vital signs, hemodynamics, clinical chemistry, hematology, coagulation, pulmonary function, and pharmacokinetics for 30 days postinfusion. There were no reportable adverse events, bleeding abnormalities, respiratory abnormalities, cardiac abnormalities, or complement activation. Dose-related flulike symptoms were observed 4 h postinfusion and included fever, chills, nausea, leukocytosis and polymorphonuclear neutrophils (PMN) shift, increased heart rate, and lowered diastolic blood pressure. Only the high-dose group showed consistent fever, which peaked at 101°F 8 h postinfusion (Fig. 11). The diastolic blood pressure dropped in the high-dose group to a nadir of 55 mm Hg at 4–12 h postinfusion (Fig. 12) while the high-dose group heart rate rose to 105 beats min^{-1} in the same time frame (Fig. 13). The flulike symptoms resolved by 24 h postinfusion without intervention and appeared to be the direct consequence of phagocytosis of the PFC particles.

The leukocytosis that accompanied the flulike symptoms began at 4 h in the high-dose group and peaked at 24 h postinfusion (Fig. 14).

In addition, a dose-responsive, mild thrombocytopenia was observed 2 days postinfusion (Fig. 15) with a nadir of 125,000 μl^{-1}. Platelet numbers returned to normal by 4 days postinfusion. There was no evidence of bleeding associated with the drop in platelet count.

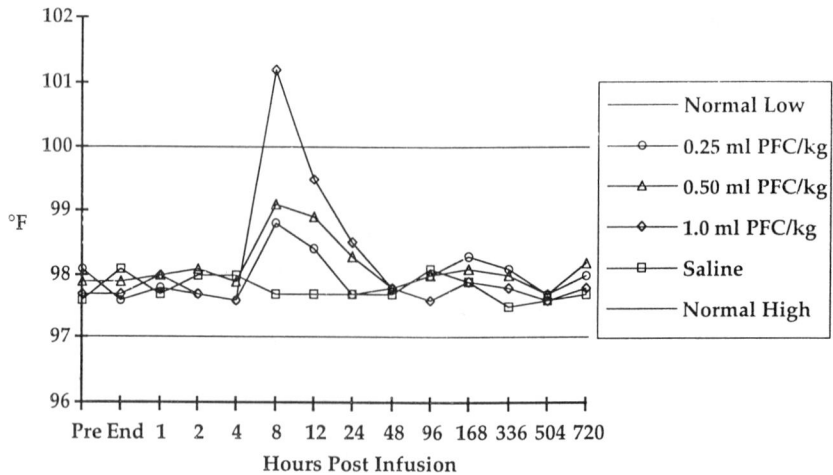

FIG. 11 Effect of Oxyfluor on body temperature (°F) of human subjects after infusion at various doses.

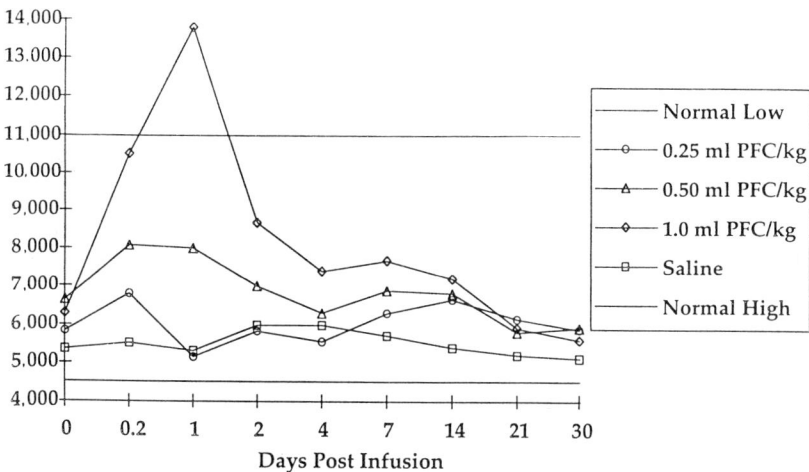

FIG. 12 Effect of Oxyfluor on disastolic pressure (mm Hg) of human subjects after infusion at various doses.

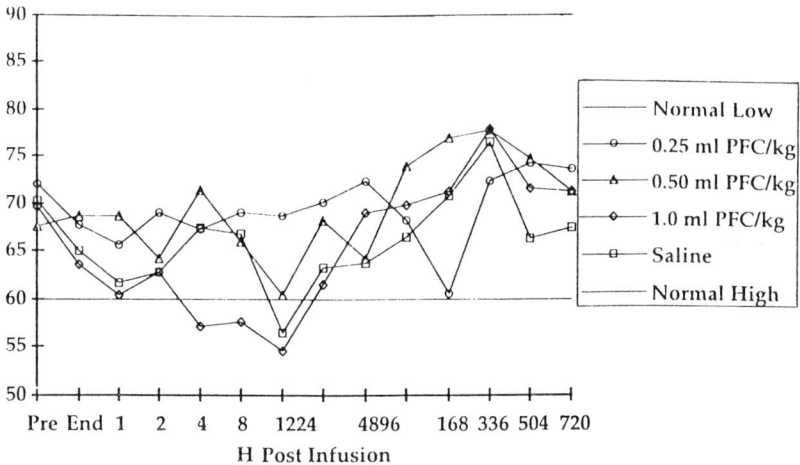

FIG. 13 Effect of Oxyfluor on heart rate (beats-min^{-1}) of human subjects after infusion at various doses.

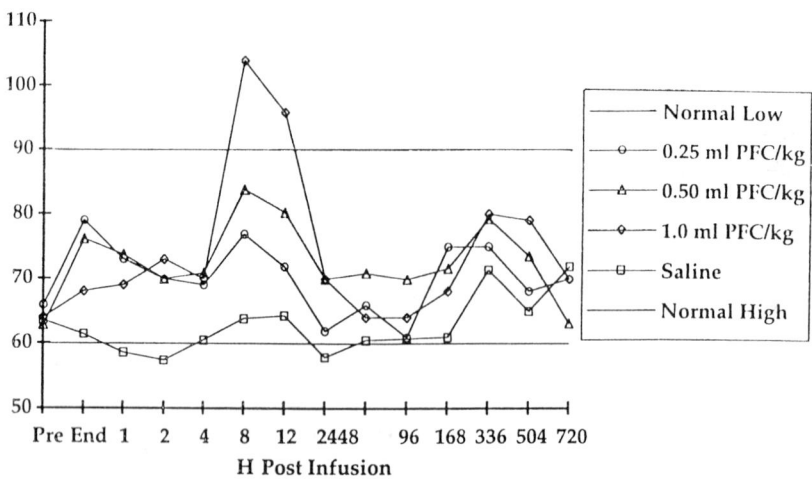

FIG. 14   Effect of Oxyfluor on white cells (WBC-μ^{-1}) of human subjects after infusion at various doses.

The pharmacokinetics of Oxyfluor were dose responsive with a blood half-life of 2 h at 1cc PFC-kg^{-1} as measured by gas chromatographic analysis of subjects blood.

Animal studies have indicated that the flulike symptoms triggered by Oxyfluor (and other PFC emulsions) appear to be due to phagocytosis of

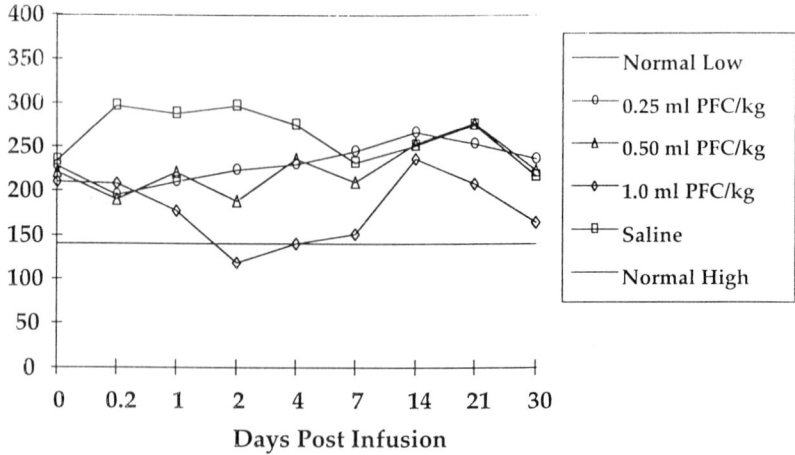

FIG. 15   Effect of three doses of Oxyfluor on platelet number (X1000) over time.

the particles by macrophages and subsequent release of cytokines and arachadonic acid metabolites. Pretreatment of mice with dexamethasone [12] followed by 8cd PFC-kg^{-1} of Oxyfluor resulted in a significant reduction in the quantity of circulating interleukin-6 (Fig. 16).

Subsequent clinical investigations will focus on the safety of Oxyfluor in surgical patients.

The other PFC-based product in clinical development, Oxygent is a 90% w/v (46% v/v) emulsion of PFOB using egg yolk phospholipid. The emulsion is reported to be stable at room temperature for at least 1 year. PFOB is lipophilic, has a molecular weight of 499, boils at 141°C, and has a density of 1.92 g-ml^{-1}. The ODP of PFOB is between 0.25 and 2. PFOB has tissue clearance characteristics similar to PFDCO.

Oxygent has been tested in dog models of surgical anemia with similar results to Oxyfluor [13]. Clinical trials of Oxygent have been completed in both healthy human volunteers and surgical patients [14,15]. Flulike symptoms were also observed at 4–6 h postinfusion which were resolved by 24 h postinfusion. Thrombocytopenia was also observed in these patients at the highest dose with the nadir of 140,000 μl^{-1} occurring 2 days postinfusion.

A study of a similar PFOB-based product, Imagent BP, demonstrated reversal of flulike symptoms in swine using dexamethasone, ibuprofen, or indomethacin [16].

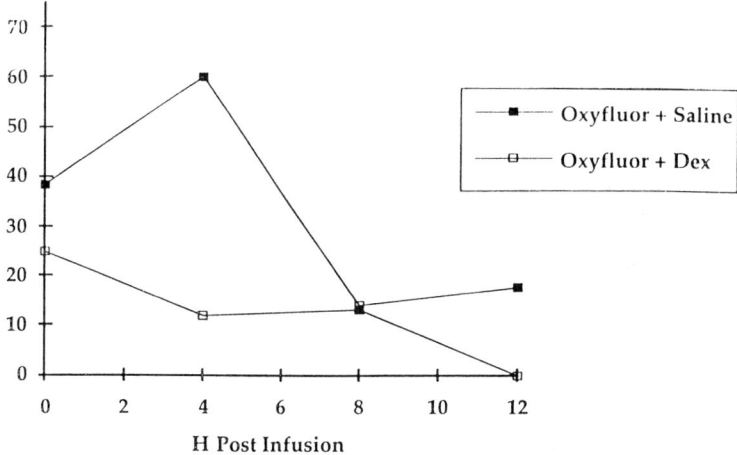

FIG. 16 Effect of 0.1 mg-kg^{-1} of dexamethasone on IL-6 (ng-ml^{-1}) production induced by Oxyfluor in mice.

III. ELIMINATION OF GASEOUS MICROEMBOLI DURING CARDIOPULMONARY BYPASS SURGERY

For years there have been anecdotal reports of neurological and neuropsychological deficits following a cardiopulmonary bypass (CPB). In a landmark study, Shaw et al., using a battery of neurological and neuropsychological tests, quantitatively defined the extent and timing of the problem [17]. The results indicated that 7 days postCPB, 61% of patients have neurological deficits and 79% have neuropsychological deficits, and that 1 year later, the deficits had persisted in about half of these patients (17% and 38%, respectively). The cause of these deficits have not been documented but are believed to be either platelet microaggregates caused by the circulation through the CPB apparatus or gaseous microemboli (GME) generated during cannulation, hypothermia, and oxygenation of the blood. In 1990, two groups developed data that strongly suggested that a significant portion of these deficits were due to GME.

Taylor used retinal fluorescein angiography to compare patients who had bypass with bubble oxygenators to patients who had bypass with membrane oxygenators [18]. There was more than a 50% reduction in lesions in the membrane oxygenator patient group. Moody performed autopsies on patients recently expired post-CPB and found numerous small capillary arteriole dilations in the brain which were thought to be due to GME [19]. In contrast, brains from patients having non-CPB procedures did not have these lesions.

Spiess et al. found that PFC emulsions could protect animals from cerebral emboli in a decompression model in rats [20]. This work sug-

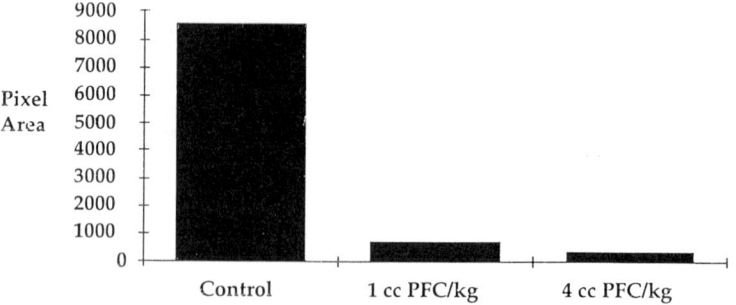

FIG. 17 Effect of 1 and 4cc PFC-kg^{-1} of Oxyfluor on capillary dropout in dogs undergoing CPB.

gested that PFC emulsions might be useful in eliminating GME from CPB patients and result in an improvement in their neurological and neuropsychological outcomes. Using retinal angiography, Taylor found that Oxyfluor significantly reduced the frequency of GME and the area of circulatory dropout in dogs undergoing CPB [21]. The data, shown in Fig. 17, clearly illustrates that low doses of Oxyfluor (1cc PFC kg^{-1}) produce a tenfold reduction in GME.

IV. PERCUTANEOUS TRANSLUMINAL CORONARY ANGIOPLASTY

Perfluorocarbons have been found to eliminate the transient myocardial ischemia induced during balloon inflation during percutaneous transluminal coronary angioplasty (PTCA) in animal models [22] and in human clinical trials [23,24]. In the clinical trial, symptomatic patients with single lesions of 70% or greater stenosis of the coronary artery undergoing PTCA were studied. All patients underwent a preliminary inflation without perfusion. Subsequent inflations were done with either oxygenated lactated Ringer's, oxygenated Fluosol DA, or unoxygenated Fluosol DA. The data showed that oxygenated Fluosol DA maintained an ejection fraction identical with the baseline 45 s post-balloon-inflation while there was a 35% reduction in ejection fraction in the controls. Fluosol DA was approved by the Food and Drug Administration (FDA) in December 1989 for use as an adjunct in high-risk PTCA patients. Unfortunately, improved catheter technology has reduced the market for this product to such an extent that the manufacturer has withdrawn the product from the market.

V. MYOCARDIAL INFARCTION

Numerous studies have suggested that PFCs have beneficial effects in animal models of myocardial infarction (MI). Generally these studies have investigated two approaches to treating MI: as a method of oxygenating the myocardium distal to the coronary artery occlusion [25,26] or as an adjunct to thrombolysis to prevent reperfusion injury [27]. The dog studies of reperfusion injury showed a 60% reduction in the ratio of the area of necrosis to the area at risk (A_N/A_R). Mechanistically, this was attributed to a reduction in circulating neutrophils and reduced neutrophil chemotactic ability in the Fluosol DA-treated dogs. It is still not known if this is just an effect of the pluronic surfactant.

A large clinical trial of Fluosol DA in reduction of reperfusion injury, TAMI-9, has been completed [28]. In this study, 430 patients with acute MI symptoms of less than 6 h duration and ST elevations of at least 0.1

mV in 2 of 6 leads were randomized into 2 groups. The first group received 100 mg tissue plasminogen activator (TPA) and 15 ml kg^{-1} of Fluosol DA; the second group received TPA alone. The clinical end points were global ejection fraction, regional wall motion, infarct size, and clinical outcome. There were no significant differences in global ejection fraction, wall motion, or infarct size between the two groups. There was a significant increase in pulmonary edema (45% versus 31%) and a reduction in the incidence of recurrent ischemia (6% versus 11%) in the Fluosol DA-treated group.

No other products are in clinical trials for this indication.

VI. CANCER THERAPY

Oxygenation of hypoxic tumors during radiation or chemotherapy has been a long-term goal of radiologists and oncologists. PFC emulsions have been investigated in animal models for their ability to sensitize tumors to radiation and chemotherapeutics and, consequently, reduce the rate of tumor growth [29]. Studies with oxygen electrodes have proven that animal tumors show an increased oxygen tissue tension when infused with PFCs while breathing carbogen [30]. Tumors appear to preferentially accumulate PFCs, perhaps increasing their effectiveness as sensitizers by attracting macrophages to the tumor site [31]. Animal studies have shown that PFCs do not sensitize healthy tissues to radiation therapy [32].

There have been at least six clinical trials of Fluosol DA reported: three as an adjunct to radiation therapy and three as an adjunct to chemotherapy. The radiation studies involved head and neck, lung, and glioma, whereas the chemotherapy studies treated non-small-cell lung, colorectal, and glioma.

The first trial involved [33] Stage III/IV patients with head and neck cancer. These patients were given 7–9 ml kg^{-1} of Fluosol DA (0.7–0.9 ml PFC kg^{-1}) on the first day of every week of therapy. They were given 5 weeks of therapy or 25 fractions of radiation and a total dose of 35–45 ml kg^{-1} of Fluosol DA (3.5–4.5cc PFC kg^{-1}). Patients breathed carbogen before and during each daily radiation treatment. There was some response to this therapy, but the total number of patients was only 15. There were four cases of acute complement reactions all of which were controlled with diphenhydramine. Eight of 15 patients exhibited serum enzyme elevations two to three times normal. These returned to normal 3 months posttherapy. Coagulation times, BUN, creatinine, serum albumin, and bilirubin were all normal. White cell counts and hematocrits were slightly depressed, but these changes were attributed to the radiation.

There was some evidence of increased incidence of mucositis in the Fluosol DA-treated patients.

A larger trial of head and neck patients was completed by Lustig et al. [34] in 1989. Forty-six patients were enrolled and 37 completed the protocol. Eleven experienced complement side effects and 17 of 46 exhibited serum enzyme elevations (ALT, AST, and alkaline phosphatase) of two to five times normal. One-year survival is 67% compared to the Radiation Therapy Oncology Group (RTOG) average of 62%.

A Phase I/II study of non-small-cell lung cancer patients treated with Fluosol DA and oxygen was conducted in 1985 [35]. Patients were enrolled in Stages II–IV without distant metastases. Fluosol dosing and radiation fractions were similar to the head and neck protocol discussed above with total doses of Fluosol DA ranging from 42 to 49 ml kg^{-1} (4.2–4.9 mL kg^{-1}). Forty-five patients were enrolled and 34 completed the program. There was a high frequency of complement responses in these patients with 44% reacting to the test dose or showing postinfusion reactions. Typical abnormalities in serum enzymes were noted with all returning to normal 3–6 months posttherapy. Seventeen of 34 patients had a complete response and 11 had a partial response. Thirteen patients remained alive 12–20 months posttreatment.

A Phase I/II trial of Fluosol DA as an adjuvant to high-grade brain tumors has been reported [36]. In this study, 18 patients were enrolled with a tumor grade of III to IV. Patients were given five to seven doses of 8 ml kg^{-1} of Fluosol DA. Once again, a high percentage of the patients, 66%, experienced side effects upon infusion and 11 of 18 patients who had elevated serum enzyme levels had normalized 3 months posttreatment. Mean survival time in this small patient population was >64 weeks compared to historical survival on radiation therapy of 54 weeks.

Of the studies evaluating Fluosol DA and oxygen breathing as an adjunct to chemotherapy, the most advanced application involves Carmustine (BCNU) in high-grade glioma patients. The Phase I/II trial enrolled 51 patients who received between 150 and 600 ml m^{-2} of Fluosol DA and 200 mg m^{-2} BCNU every 6 weeks for a total of three treatment regimes [37]. Of the 34 evaluable patients, 12 had a partial response (35%), 13 patients had stable disease, and 9 had disease progression. A Phase III trial is currently underway at the University of Kansas Medical Center.

Oxygent CA, a 90% w/v emulsion of PFOB, has been tested in animal models for adjuvant effect on cancer radiation therapy and chemotherapy with positive results [38]. No clinical data have been published.

Interstitial injection of an oxygenated, 35% v/v emulsion of perfluorophenanthrene has been reported to cause a significant tumor growth delay and increased survival in a animal model [39]. The perceived advantage of this approach is that PFC is localized in the tumor, reducing the liver toxicity seen in all intravenously administered trials of Fluosol DA. The limitation of this approach is that tumors must be palpable or otherwise visualizable to be injected. A Phase I study in head and neck cancer patients used this emulsion, trade-named Oncosol, at doses up to 3 ml per tumor without observable toxicity. Additional clinical studies are being planned.

VII. IMAGING

Perfluorocarbon emulsions have been studied as contrast agents for x-ray, ultrasound, and magnetic resonance imaging (MRI) and as agents for direct ^{19}F MRI.

Long and co-workers investigated brominated PFCs, both neat and emulsified in the 1970s because of the known x-ray opacity of the bromine atom [40–42]. They found that neat brominated PFCs could be administered intratrachially or orally and afforded excellent images without the side reactions. PFOB was subsequently developed by Alliance as a gastrointestinal (GI) MRI contrast agent trade-named Imagent MR. PFOB contains no protons and as such appears as a dark void in MR images. Clinical trials demonstrated the efficacy of Imagent MR [43,44] in darkening the bowel and allowing the recognition of bowel from adjacent structures. In the Phase III clinical trial, 127 subjects were studied. They were imaged before and 2–60 min after they had ingested 500–1000 ml of PFOB. There were no acute or subacute GI symptoms. The product was rectally eliminated rapidly without uptake even by diseased mucosa. The product was approved by the FDA in 1993. Clinical usage has been limited, not for lack of efficacy, but due to expense. Alliance announced in September 1994 that they would no longer support marketing of Imagent MR.

X-ray contrast with PFOB emulsions has been investigated for blood pool, liver, and lymph node imaging. PFOB emulsions (100% w/v) produced prolonged blood enhancement in animal models, and because the particles were taken up by the macrophages of the liver and spleen, these organs also showed x-ray enhancement [45]. Clinical trials have been conducted with this emulsion, trade-named Imagent BP [46,47]. In the former study, the emulsion was administered to 18 cancer patients, 14 of whom had liver metastases. The emulsion doses ranged from 0.5 to 3.0 ml kg^{-1}. Computed tomography (CT) of the liver and spleen was performed before and immediately after infusion and again 24 h later. PFOB increased the density of the blood, liver and spleen by 55, 39, and 317 Hounsfield units (HU), respectively. Tumor visualization was increased because the metastases enhanced minimally, 7 HU or lower, compared to the surrounding liver tissue. Peak enhancement of the liver and spleen occurred 24 h post-infusion. Previously undetected metastases were found in two of the patients. Adverse events occurred in 14 of the 18 subjects. These included lower back pain, fever, and malaise. The author concluded that the side effects could restrict the use of the product to a selected clinical population. In September 1994, Alliance announced it would stop development of this product.

Ultrasound enhancement by perfluorocarbon emulsions has been extensively studied by Mattrey et al. [48–50] using Fluosol DA and perfluorooctyl bromide. These studies have shown enhancement of the liver, spleen, tumor, kidney, and some of the vasculature. There has been one report of a clinical trial using Fluosol DA as an ultrasound contrast agent [50]. Administration of Fluosol DA at 8–16 ml kg^{-1} caused rim or diffuse enhancement developed in liver metastases. Echogenic enhancement by Fluosol DA allowed visualization of nonenhancing lesions which could not be seen before infusion. Side effects were minimal. Despite these seemingly positive results, there was no further clinical development of Fluosol DA for ultrasound contrast enhancement. Alliance is currently developing an ultrasound contrast agent based on PFOB called Imagent US. This product is still in preclinical development.

Recently, Beppu et al. [51] and DeMaria et al. [52] reported the use of emulsions of low-boiling perfluorocarbons such as perfluoropentane, which phase shifts from liquid at room temperature to gas particles when injected intravenously. Because ultrasound contrast is dependent on the density and particle size of the contrast media, these larger gas particles are inherently superior contrast agents compared to liquid emulsions. Preclinical studies in dogs with a PFP emulsion called QW3600 have generated excellent images. Unlike traditional microbubble contrast agents, QW3600 produces an intense and long-lasting enhancement of the myocardium after intravenous injection. The myocardial contrast effect continued long after the effect had washed out of the ventricular chambers. Ligation of coronary arteries clearly defined the area at risk. Although some loss of cardiac function was observed in this study. There was a dose-related decrease in arterial pO$_2$, increase in pulmonary artery pressure and decrease in cardiac output, blood pressure, and dP/dt. The authors claimed these effects were not clinically significant. A Phase I clinical trial has been completed, and at doses up to 0.1 ml kg^{-1} of emulsion, there were no significant sides effects [53]. Mild flushing and light-headedness was observed in some of the subjects. These effects all resolved within 2 min. Dense and complete left ventricle opacification was seen at 0.1 ml kg^{-1} for an average of 2.8 min, but there was no myocardial opacification at this dose. The blood half-life was 2 min. PFP was found in the expired air for 10–16 min postinfusion.

The use of PFCs for MRI has a great deal of potential because of ^{19}F nucleus is magnetically active and highly sensitive. In addition, there is only a trace of naturally occurring ^{19}F, so there is no natural background signal to interfere with the diagnostic signal. Early attempts to use ^{19}F imaging were successful but not clinically useful [54–56] because most PFCs have multiple signals causing misregistration of signals and diluting the amount of ^{19}F signal per molecule infused. Confounding this is the short T_2 time of most PFCs (<6 ms). Thus, the signal is not only dilute but short-lived.

Schweighardt cleverly solved the problem of multiple signals by using perfluoro-15-crown-5 ether (PFCE) in which all 22 of the fluorine atoms are identical and form a single peak [57]. In addition, the T_2 relaxation time of PFCE is 200 ms compared to 6 ms for PFOB and PFDCO, respectively. Using a 40% v/v emulsion of PFCE, excellent liver, spleen, tumor, and vascular images have been obtained at doses of 3 ml PFC kg^{-1} in animals [58].

Clark et al. were the first to discover that the T_1 relaxation times of PFCs change with oxygen tension [59]. They postulated that the dependence of T_1 on oxygen tension could be used to construct oxygen maps of the tissues in which the PFC was in contact. Eidelberg reduced this concept to practice in cats where 30% of the blood was replaced with an emulsion of perfluorotripropyl amine (PFTPA) and oxygen maps of the brain were obtained [60]. However, the large dose required coupled with the software modifications required to suppress the misregistered signals precluded the clinical utility of this emulsion. Sotak has recently used emulsions of PFCE to construct oxygen maps of the liver, spleen, and tumors in mice. When animals were treated with 3 ml PFC kg^{-1} of PFCE, significant differences in oxygen tension could be measured in the liver, spleen, and tumors of mice when they were breathing air compared to carbogen [58].

VIII. LIQUID BREATHING

Since Clark's initial liquid-breathing experiment, most of the research on PFCs has focused on the development of emulsified forms suitable for intravenous administration. However, a small but steady effort to establish liquid breathing as a therapeutic for lung lavage and respiratory distress has been continuing. Modell and co-workers worked out the mechanics of long-term liquid breathing of animals for evaluation of PFC toxicity, adsorption, and pulmonary physiology [61,62]. Schaffer and Moskowitz have made significant progress in both the physiology of liquid breathing and in developing respirators adapted for liquid breathing [63]. This work culminated in the first human clinical trial in preterm neonates. A 19-week-old neonate with respiratory distress syndrome was instilled with 30 ml kg^{-1} of an oxygenated PFC. Liquid ventilation was initiated as two 3–5-min cycles followed by conventional ventilation. During liquid ventilation cycles, tidal volumes of 15 ml kg^{-1} were delivered [64]. The infant survived for 19 h with markedly improved pulmonary parameters and expired from other causes.

Alliance has conducted a Phase I study in similar neonates using PFOB (LiquiVent™). In this protocol, the infants had exhausted conventional therapy, including two regimes of surfactant. The infants were liquid ventilated with PFOB for 15 min. Of seven neonates treated, two survived (R. Hopkins, personal communication, 1994). Alliance announced in October 1994 that they would begin enrolling patients for a Phase I study of liquid ventilation in adult respiratory distress syndrome.

IX. CONCLUSIONS

Significant progress has been made in many areas in the field of medical perfluorocarbon research. Many of the problems which prevented the use of PFC emulsions for oxygen transport have been overcome but limitations will still be encountered at some dose because of thrombocytopenia and bioaccumulation. Research to solve these problems is underway. The use of PFCs in MI failed in a large clinical trial and there seems to be no compelling reason to pursue additional formulations for this indication. Cancer therapy still holds great promise but has been set back by the withdrawal of the leading company in the field. Advances in the use of PFCs for ultrasound contrast and direct ^{19}F MRI have been made but other imaging uses such as CT and MRI contrast seem to have toxicity, efficacy or cost limitations. Liquid breathing has seen an exciting rejuvenation due to the results in neonates, however, it remains to be seen if the far more complex ARDS can be successfully treated with liquid ventilation.

REFERENCES

1. J. G. Riess and M. Leblanc, Perfluoro compounds as blood substitutes. *Angew. Chem. 17*:621 (1978).
2. J. G. Riess, Reassessment of the criteria for the selection of perfluorochemicals for second generation blood substitutes: Analysis of structure-property relationships. *Artif. Organs 8*:44 (1984).

3. R. J. Kaufman, Medical oxygen transport using perfluorochemicals, in *Biotechnology of Blood* (J. Goldstein, ed.), Butterworths–Heinemann, Boston, 1991, pp. 127–162.
4. R. J. Kaufman, Perfluorochemical emulsions as blood substitutes, in *Emulsions—A Fundamental and Practical Approach* (J. Sjöblom, ed.), Kluwer Academic Publishers, Boston, 1992, pp. 207–226.
5. L. C. Clark and F. Gollan, Survival of mammals breathing organic liquids equilibrated with oxygen at atmospheric pressure. *Science 152*:1755 (1966).
6. H. Sloviter and T. Kamimoto, Erythrocyte substitute for perfusion of brain. *Nature 216*:458 (1967).
7. R. P. Geyer, R. C. Monroe, and K. Taylor, Survival of rats totally perfused with a fluorocarbon-detergent preparation, in *Organ Perfusion and Preservation* (J. Norman, ed.), Appleton Century and Crofts, New York, 1968, pp. 85–97.
8. S. A. Gould, A. L. Rosen, L. R. Sehgal et al., Fluosol DA as a red-cell substitute in acute anemia. *N. Engl. J. Med. 314*:1653 (1986).
9. K. K. Tremper, G. M. Vercellotti, and D. E. Hammerschmidt, Hemodynamic profile of adverse clinical reactions to Fluosol DA 20%. *Crit. Care Med. 12*:428 (1984).
10. R. J. Kaufman, T. H. Goodin, and T. J. Richard, Efficacy of perfluorochemical emulsions in surgical anemia and shock resuscitation, in *Fifth International Symposium on Blood Substitutes: New Frontiers* (T. Chang, J. G. Riess, and R. M. Winslow, eds.), 1992. [Abstract.]
11. T. H. Goodin, E. G. Grossbard, R. J. Kaufman et al., A perfluorochemical emulsion for prehospital resuscitation of experimental hemorrhagic shock: a prospective, randomized controlled study. *Crit. Care Med. 22*:680 (1994).
12. R. J. Kaufman, The results of a Phase I clinical trial of a 40 v/v% emulsion of HM351 (Oxyfluor™) in healthy human volunteers, *Artificial Cells, Blood Substitutes and Immobilization Biotechnology 22*:A112 (1994).
13. A. C. Cernaianu, R. K. Spence, T. Vasilidze et al., Transfusion triggers with perflubron (Oxygent™) in a canine model of surgical hemodilution. In *Fifth International Symposium on Blood Substitutes: New Frontiers* (T. Chang, J. G. Riess, and R. M. Winslow, eds.), 1992. [Abstract.]
14. P. Keipert, Use of Oxygent™, a perfluorochemical (PFC) emulsion, as an anti-hypoxic agent to improve tissue oxygenation during acute blood loss, in *Blood Substitutes and Related Products: Advances in Development, Trial Design and Clinical Development* (E. Scatchard and M. McBride, eds.), International Business Communications, Southborough, MA, 1994.
15. R. K. Spence, Use of Oxygent™, perfluorocarbons in the twenty-first century: Clinical applications as transfusion alternatives, in *Blood Substitutes and Related Products: Advances in Development, Trial Design and Clinical Development* (E. Scatchard and M. McBride, eds.), International Business Communications, Southborough, MA, 1994.
16. S. F. Flaim, D. R. Hazard, J. Hogan, and R. M. Peters, Characterization and mechanism of side-effects of Imagent BP (highly concentrated fluorocarbon emulsion) in swine. *J. Invest. Radiol. 26*:S122 (1991).

17. P. J. Shaw, D. Bates, N. E. Cartlidge et al., Neurologic and neuropsychological morbidity following major surgery: Comparison of coronary artery bypass and peripheral cascular surgery. *Stroke 18*:700 (1987).
18. C. I. Blauth, P. L. Smith, J. V. Arnold et al., Influence of oxygenator type on the prevalence and extent of microembolic retinal ischemia during cardiopulmonary bypass. *J. Thorac. Cardiovasc. Surg. 99*:61 (1990).
19. D. M. Moody, M. A. Bell, V. R. Challa et al., Brain microemboli during cardiac surgery or aortography. *Ann. Neurol. 28*:477 (1990).
20. B. D. Spiess et al., Treatment of decompression sickness with a perfluorocarbon emulsion (FC-43). *Undersea Biomed. Res. 15*:31 (1988).
21. J. V. Arnold, D. Wagner, J. Fleming et al., Cerebral protection during CPB using a perfluorocarbon: A preliminary report, in *The Brain and Cardiac Surgery, Second Internation Conference*, 1992. [Abstract.]
22. R. Spears, J. Serur, D. Baim et al., Myocardial protection with Fluosol-DA during prolonged coronary balloon occlusion in the dog. *Circulation 73*:II-245 (1983). [Abstract]
23. M. Cleman, C. C. Jaffe, and D. Wholgelernter, Prevention of ischemia during percutaneous transluminal coronary angioplasty by transcatheter infusion of oxygenated Fluosol DA-20%. *Circulation 74*:555 (1986).
24. C. C. Jaffe, D. Wohlgelernter, and H. Cabin, et al., Preservation of left ventricular ejection fraction during percutaneous transluminal coronary angioplasty by distal transcatheter coronary perfusion of oxygenated Fluosol DA 20%. *Am. Heart J. 6*:1156 (1988).
25. D. H. Glogar, R. A. Kloner, J. Muller et al., Fluorocarbons reduce myocardial ischemic damage after coronary occlusion. *Science 211*:1439 (1981).
26. G. R. Nunn, G. Dance, J. Peters, and L. H. Cohn, Effect of fluorocarbon exchange transfusion on myocardial infarction in dogs. *Am. J. Cardiol. 52*: 203 (1983).
27. A. K. Bajaj, M. A. Cobb, R. Virmani et al., Limitation of myocardial reperfusion injury by intravenous perfluorochemicals. *Circulation 79*:645 (1989).
28. T. C. Wall, R. M. Califf, J. Blamkenship et al., Intravenous fluosol in the treatment of acute myocardial infarction. *Circulation 90*:114 (1994).
29. B. Teicher and S. Rockwell, Increase efficacy of radiotherapy in mice treated with perfluorochemical emulsions plus oxygen. *Am. Assoc. Cancer Res. 25* (1983). [Abstract.]
30. C. W. Song, I. Lee, T. Kasegawa et al., Increase in pO2 and radiosensitivity of tumors by Fluosol-DA (20%). *Cancer Res. 47*:442 (1987).
31. D. M. Long, F. K. Multer, A. G. Greenburg et al., Tumor imaging with x-rays using macrophage uptake of radioopaque fluorocarbon emulsions. *Surgery 84*:104 (1978).
32. T. P. Mate and S. Rockwell, Perfluorochemical emulsions do not affect bone marrow radiosensitivity, in *Abstracts American Society Therapy Radiation Oncologists Meeting*, Washington, DC, 1984.
33. C. M. Rose, R. Lustig, N. McIntosh, and B. Teicher, A clinical trial of Fluosol DA 20% in advanced squamous cell carcinoma of the head and neck. *Int. J. Radiat. Oncol., Biol. Phys. 12*:1325 (1986).

34. R. Lustig, N. McIntosh-Lowe, C. Rose et al., Phase I/II study of Fluosol-DA and 100% oxygen as an adjuvant to radiation in the treatment of advanced squamous cell tumors of the head and neck. *Int. J. Radiat. Oncol. Biol. Phys.* *16*:1587 (1989).

35. R. Lustig, N. Lowe, L. Prosnitz et al., Phase I/II study of fluosol and 100% oxygen breathing as an adjuvant to radiation in the treatment of unresectable non small cell carcinoma of the lung. *Int. J. Radiat. Oncol. Biol. Phys.* *17*: 202 (1989).

36. R. G. Evans, B. F. Kimler, R. A. Morantz, and R. A. Batnitzky, Lack of complications in long-term survivors after treatment with Fluosol® and oxygen as an adjuvant to radiation therapy for high-grade glioma. *Int. J. Radiat. Oncol. Biol. Phys.* *26*:649 (1989).

37. M. Gruber, M. Prados, C. Russell et al., Phase I/II study of Fluosol® and oxygen in combination with BCNU in malignant glioma. *Proc. Am. Assoc. Cancer Res.* *31*:190 (1990).

38. B. A. Teicher, S. A. Holden, G. Ara et al., A new concentrated perfluoro-chemical emulsion and carbogen breathing as a adjuvant to treatment with antitumor alkylating agents. *J. Cancer Res. Clin. Oncol.* *118*:509 (1992).

39. F. K. Schweighardt and D. Woo, *U.S. Patent 4*, 781,676 (1988).

40. D. M. Long, M. S. Liu, P. S. Szanto, and P. Alrenga, Efficacy and toxicity studies with radioopaque perfluorocarbon. *Radiology* *105*:232 (1972).

41. D. M. Long, M. S. Liu, P. S. Szanto et al., Forefront: Preliminary communication initial observations with a new x-ray contrast agent–radioopaque perfluorocarbon. *Rev. Surg.* *29*:71 (1972).

42. D. M. Long, F. K. Multer, A. G. Greenburg et al., Tumor imaging with X-rays using macrophage uptake of radioopaque fluorocarbon emulsions. *Surgery* *84*:104 (1978).

43. J. J. Brown, J. R. Duncan, J. P. Heiken et al., Perfluorooctyl bromide as a gastrointestinal contrast agent for MR imaging: use with and without glucagon. *Radiobiology* *181*:455 (1991).

44. R. F. Mattrey, M. A. Trammert, J. J. Brown et al., Oral contrast agents for magnetic resonance imaging. *Invest. Radiol.* *26*:S65 (1991).

45. R. F. Mattrey, D. M. Long, R. A. Slutsky, and G. B. Higgins, Perfluorooctyl Bromide as a blood pool contrast agent for liver, spleen and vascular imaging. *J. Comput. Assist. Tomogr.* *8*:739 (1984).

46. M. Behan, D. O. O'Connell, R. F. Mattrey, and D. N. Carney, Perfluoroctyl bromide as a contrast agent for CT and sonography: preliminary clinical results. *Am. J. Radiol.* *160*:399 (1993).

47. J. N. Bruneton, M. N. Falawese et al., Liver, spleen and vessels: preliminary clinical results of CT with perfluorooctyl bromide. *Radiology* *170*:179 (1989).

48. R. F. Mattrey, F. W. Scheible, B. B. Gosink et al., Perfluorooctyl bromide: a liver and spleen specific and a tumor imaging ultrasound contrast material. *Radiology* *145*:759 (1982).

49. R. F. Mattrey, G. R. Leopold, E. vanSonneberg et al., Perfluorochemicals as a liver and spleen seeking ultrasound contrast agent. *J. Ultrasound Med.* *2*:173 (1983).

50. R. F. Mattrey, G. Strich, R. E. Shelton et al., Perfluorochemicals as ultrasound contrast agents for tumor imaging and hepatosplenography: preliminary clinical results. *Radiology 163*:339 (1987).
51. S. Beppu, H. Matsuda, T. Shishido et al., Success of myocardial contrast echocardiography by peripheral venous injection method: Visualization of area at risk. *Circulation 88*:I-401 (1993). [Abstract]
52. A. N. DeMaria, H. Dittrich, O. L. Kwan et al., Myocardial opacification produced by peripheral venous injection of a new ultrasonic contrast agent. *Circulation 88*:I-401 (1993). [Abstract]
53. B. Cotter, O. L. Kwan, B. Kimura et al., Evaluation of the efficacy, safety and pharmacokinetics of QW3600 (Echogen) in man. *Circulation 90*:I-555, 1994. [Abstract]
54. P. Joseph, Y. Yuasa, H. Kundel et al., Magnetic resonance imaging of fluorine in rats infused with artificial blood. *Invest Radiol. 20*:504 (1985).
55. E. R. McFarland, J. A. Koucher, B. R. Rosen et al., In vivo ^{19}F NMR imaging. *J. Comput. Assist. Tomagr. 9*:8 (1985).
56. H. Longmaid III, D. Adams et al., In vivo ^{19}F NMR imaging of liver, tumor, and abscess in rats. Preliminary results. *Invest. Radiol. 20*:141 (1985).
57. F. K. Schweighardt, *U.S. Patent* 4,838,274 (1989).
58. B. J. Darzinski and C. H. Sotak, Rapid tissue oxygen tension mapping using ^{19}F inversion recovery echo planar imaging of perfluoro-15-crown-5-ether. *Magn. Reson. Med. 32*:88 (1994).
59. L. C. Clark, J. Ackerman, S. Thomas et al., Perfluorinated organic liquids and emulsions as biocompatible NMR imaging agents for ^{19}F and dissolved oxygen. *Adv. Exp. Med. Biol. 180*:835 (1985).
60. D. Eidelberg, G. Johnson, D. Barnes et al., ^{19}F NMR imaging of blood oxygenation in the brain. *Magn. Res. Med. 6*:344 (1988).
61. J. H. Modell, E. J. Newby, and B. C. Ruiz, Long term survival of dogs after breathing oxygenated fluorocarbon liquid, *Fed. Proc. 34*:312 (1970).
62. J. H. Modell, H. W. Calderwood, B. C. Ruiz et al., Liquid ventilation of primates. *Chest 69*:79 (1976).
63. T. H. Shaffer and G. D. Moskowitz, Demand-controlled liquid ventilation of the lungs, *J. Appl. Physiol. 36*:208 (1974).
64. J. S. Greenspan, M. R. Wolfson, D. Rubenstein, and T. H. Shaffer, Liquid ventilation of human preterm neonates *J. Pediatr. 117*:106 (1990).

8

NMR Self-Diffusion Studies of Emulsions

OLLE SÖDERMAN   Physical Chemistry 1, University of Lund, Lund, Sweden

BALIN BALINOV   Department of Physical and Analytical Chemistry, Exploratory Research, Nycomed Imaging AS, Oslo, Norway

I. INTRODUCTION

Emulsions are dispersions of one liquid in another, where the two liquids are not miscible. Emulsions are ubiquitous and of great technical importance, as they appear in a number of different applications which may be so diverse as in the pavement of roads and in pharmaceutical applications. Because of this fact, a number of different techniques have been developed to investigate many different properties of emulsions [1].

A class of systems which in many ways are related to emulsions are thermodynamically stable surfactant systems such as micellar solutions, microemulsions, and liquid crystals. It is fair to state that our understanding of such systems has increased rather dramatically during later years. This is to a large extent due to the development of suitable techniques by which such systems can be investigated. Here, different scattering methods have been important, but also nuclear magnetic resonance (NMR) spectroscopy has played an important role. In particular, the NMR methods which determine molecular self-diffusion have been central in increasing our understanding of surfactant systems, in general, and of surfactant solution systems, in particular.

This being the case, one might expect that NMR techniques should be able to provide important insight into emulsion properties. Therefore, it is somewhat surprising that very few NMR studies on emulsions are actually to be found in the literature.

The aim of this chapter is to point out a few application of the NMR self-diffusion technique on emulsion systems, applications which we deem as being able to provide important information with regard to central properties of emulsions, and for which information is difficult to obtain with other methods. These applications deal with the determination of the microstructure of the continuous phase and the relation of the structure to emulsion stability, the determination of emulsion droplet sizes, and, finally, the study of concentrated emulsions.

The disposition of this chapter is as follows. First, we introduce the NMR self-diffusion method, and then we describe in three separate sections the themes described above.

II. THE NMR SELF-DIFFUSION METHOD

A. Theoretical Background

Measurements of self-diffusion coefficients and flow by means of pulsed field gradient (PFG) techniques have evolved to become a very important tool in the study of a multitude of different problems. There are essentially two reasons for this, one of which has to do with the measuring technique as such. Thus, the PFG approach offers a convenient method of determining self-diffusion coefficients. The technique as such has recently been described in a number of review articles [2–4], so here we will merely state that the technique requires no isotopic labeling (avoiding possible disturbances due to addition of probes) and that it gives component-resolved self-diffusion coefficients with great precision in a minimum of measuring time. The main nucleus studied is the proton, but other nuclei, such as Li, F, Cs, and P, are also of interest.

The second reason for the success of the NMR self-diffusion technique rests on the fact that the method monitors transport over macroscopic distances (typically in the micrometer regime). Therefore, when the method is applied to the field of surface and colloid chemistry, the determined diffusion coefficients reflect aggregate sizes and obstruction effects for colloidal particles. This is the origin of the success the method has had in the study of microstructures of surfactant solutions and also forms the basis of its applications to emulsion systems.

The fact that the information is obtained without the need to invoke complicated models, as is the case for the NMR relaxation approach, is particularly important. In this context, it should be stressed that the PFG approach measures the self-diffusion rather than the collective diffusion coefficient, which is measured, for instance, by light-scattering methods.

The foundation for the use of NMR to monitor diffusion and flow was laid in 1950 in the seminal paper by Hahn [5]. Another milestone occurred around the middle of the sixties, when Stejskal and Tanner [6] demonstrated the use of pulsed field gradients following a suggestion by McCall et al. [7]. Finally, toward the end of the sixties, the method became component resolved when the Fourier transform (FT) approach was introduced [8].

In its simplest version, it consists of two equal and rectangular gradient pulses of magnitude g and length δ, sandwiched on either side of the 180° rf (radio frequency) pulse in a simple Hahn echo experiment. The experiment is outlined in Fig. 1. For molecules undergoing free (Gaussian)

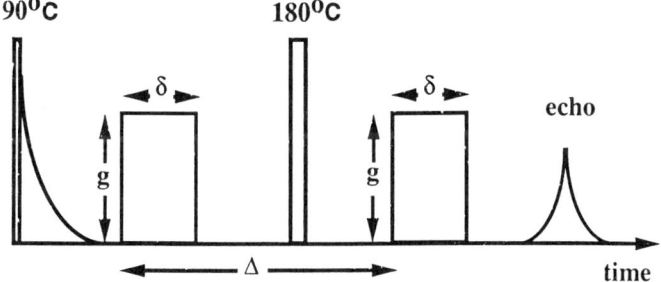

FIG. 1 The basic combination of rf and field gradient pulses used in the NMR self-diffusion experiment. Δ denotes the time between the leading edges of two pulsed field gradients of duration δ and magnitude g ($g = |\mathbf{g}|$).

diffusion characterized by a diffusion coefficient of magnitude D, the echo attenuation due to diffusion is given by [6,9]

$$E(\delta, \Delta, g) = E_0 \exp\left(-\gamma^2 g^2 \delta^2\left(\Delta - \frac{\delta}{3}\right)D\right),\tag{1}$$

where Δ represents the distance between the leading edges of the two gradient pulses, γ is the magnetogyric ratio of the monitored spin, and E_0 denotes the echo intensity in the absence of any field gradient. By varying either g, δ, or Δ (at the same time keeping the distance between the two rf pulses constant), D can be backed out by fitting Eq. (1) to the observed intensities.

As mentioned above, one of the key foundations of PFG diffusion experiments is the notion that the transport of molecules is measured over a time Δ (cf. Fig. 1) which we are free to choose at our own will in the range of from a few milliseconds to several seconds. This means that the length scale over which we are measuring the molecular transport is in the micrometer regime for low-molecular-weight liquids. When the molecules experience some sort of boundary with regard to their diffusion during the time Δ, the molecular displacement is lowered as compared to free diffusion, and the outcome of the experiment becomes drastically changed [10–12]. In fact, for restricted diffusion, the PFG experiment can be used for structure characterization.

One situation where this is encountered is for the case of restricted motion inside an emulsion droplet. In this case, the molecular displacements do not exceed the droplet size, which is indeed in the micrometer regime. In fact, the exact equations governing the echo amplitude as a function of the relevant parameters [corresponding to Eq. (1) for free diffusion] are not known for any other case than for free diffusion and free diffusion superposed on flow. Therefore, one has to rely on various levels of approximations.

In one of these, one considers gradient pulses which are so narrow that no transport during the pulse takes place. This has been termed the short gradient pulse (SGP) (or narrow gradient pulse, NGP) limit. This case leads to a very useful formalism whereby the echo attenuation can be written as

$$E(\delta, \Delta, g) = \int\int \rho(\mathbf{r}_0)P(\mathbf{r}_0|\mathbf{r}, \Delta) \exp[i\gamma g\delta\cdot(\mathbf{r} - \mathbf{r}_0)]\, d\mathbf{r}\, d\mathbf{r}_0,\tag{2}$$

where $P(\mathbf{r}_0|\mathbf{r}, \Delta)$ is the propagator which gives the probability of finding a spin at position $\mathbf{r}$ after a time Δ if it was originally at position $\mathbf{r}_0$. For free diffusion, $P(\mathbf{r}_0|\mathbf{r}, \Delta)$ is a Gaussian function and if this form is inserted in Eq. (2), Eq. (1) with the term $(\Delta - \delta/3)$ replaced with Δ is obtained, which is the SGP result for free diffusion. For cases other than free diffusion, alternate expressions for $P(\mathbf{r}_0|\mathbf{r}, \Delta)$ have to be used. Tanner and Stejskal [13] solved the problem of reflecting planar boundaries, whereas the case of interest to us in the context of emulsion droplets, (i.e., that of molecules confined to a spherical cavity of radius R) was presented by Balinov et al. [14]. The result is

$$
\begin{aligned}
E(\delta, \Delta, g) &= \frac{9[\gamma g \delta R \cos(\gamma g \delta R) - \sin(\gamma g \delta R)]^2}{(\gamma g \delta R)^6} + 6(\gamma g \delta R)^2 \\
&\times \sum_{n=0}^{\infty} [j_n'(\gamma g \delta R)]^2 \sum_m \frac{(2n + 1)\alpha_{nm}^2}{\alpha_{nm}^2 - n^2 - n} \\
&\times \exp\left(-\frac{\alpha_{nm}^2 D\Delta}{R^2}\right) \frac{1}{[\alpha_{nm}^2 - (\gamma g \delta R)^2]^2},
\end{aligned}
\tag{3}
$$

where $j_n(x)$ is the spherical Bessel function of the first kind and α_{nm} is the mth root of the equation $j_n'(\alpha) = 0$. D is the bulk diffusion of the entrapped liquid, and the rest of the quantities are as defined above. The main point to note about Eq. (3) is that the echo decay does indeed depend on the radius, and thus the droplet radii can be obtained from the echo decay for molecules confined to the sphere, provided that the conditions underlying the SGP approximation are met.

The second approximation used is the so-called Gaussian phase distribution. Originally introduced by Douglass and McCall [15], the approach rests on the approximation that the phases accumulated by the spins due to the action of the field gradients are Gaussian distributed. Within this approximation and for the case of a steady gradient, Neuman [16] derived the echo attenuation for molecules confined within a sphere, within a cylinder and between planes. For spherical geometry, Murday and Cotts [17] derived the equation for pulsed gradients according to Fig. 1. The result is

$$
\begin{aligned}
\ln[E(\delta, \Delta, g)] &= -\frac{2\gamma^2 g^2}{D} \sum_{m=1}^{\infty} \frac{\alpha_m^{-4}}{\alpha_m^2 R^2 - 2} \\
&\times \left(2\delta - \frac{\begin{array}{c} 2 + \exp[-\alpha_m^2 D(\Delta - \delta)] - 2\exp(-\alpha_m^2 D\delta) \\ -2\exp(-\alpha_m^2 D\Delta) + \exp[-\alpha_m^2 D(\Delta + \delta)] \end{array}}{\alpha_m^2 D}\right),
\end{aligned}
\tag{4}
$$

where α_m is the mth root of the Bessel equation $[1/(\alpha R)] J_{3/2}(\alpha R) = J_{5/2}(\alpha R)$. Again, D is the bulk diffusion coefficient of the entrapped liquid.

Thus, we have at our disposal two equations with which to interpret PFG data from emulsions in terms of droplet radii, neither of which are exact for all values of experimental and system parameters. As the conditions of the SGP regime are technically demanding to achieve, Eq. (4) (or limiting forms of it) have been used in most cases to determine the droplet radii. A key question is then under what conditions Eq. (4) is valid. That it reduces to the exact result in the limit of $R \to \infty$ is easy to show and also obvious from the fact that we are then approaching the case of free diffusion, in which the Gaussian phase approximation becomes exact.

Recently, Balinov et al. [14] performed accurate computer simulations aimed at further testing its applicability over a wide range of parameter values. An example is shown in Fig. 2. The conclusion reached in Ref. 14 was that Eq. (4) never deviates by more than 5% in predicting the echo attenuation. Thus, it is a useful approximation and we shall use it in the work presented in this chapter.

As pointed out above, the NMR echo signal E depends on the droplets radius which can be estimated by measuring E at different durations δ

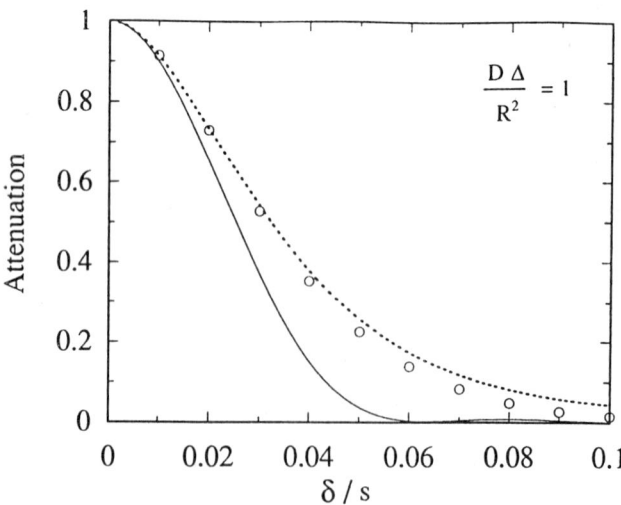

FIG. 2 Results of a simulation of the diffusion of water molecules in an emulsion droplet of radius R, given as the echo amplitude versus the duration δ of the field gradient pulse. The ratio $D\Delta/R^2$ is 1. The dotted line is the prediction of the Gaussian phase approximation [Eq. (4)], whereas the solid line is the prediction of the short gradient pulse [Eq. (3)]. (From Ref. 14.)

of the pulse gradient. A typical $E(\delta)$ plot is presented in Fig. 3 which demonstrates the sensitivity of the NMR self-diffusion experiment to resolve micrometer droplet sizes.

In conclusion, the PFG method puts at our disposal a method by which the diffusion coefficients for molecules can be measured. The values of these diffusion coefficients carry important information with regard to solution microstructure in surfactant systems which often constitute the continuous phase in emulsions. For molecules which do not undergo Gaussian diffusion but whose diffusion processes are perturbed by the presence of barriers which impede their diffusion, the echo attenuation will depend on the geometry of the these barriers and, as we shall see, on such properties such as the lifetime of the dispersed phase in the emulsion droplets, and thus important aspects of the enclosed phase may be studied.

B. Technical Details

There are a number of excellent review articles concerned with the technical details of performing PFG work [2–4], so we will only make a few comments concerning this topic here. One requires two additions to a

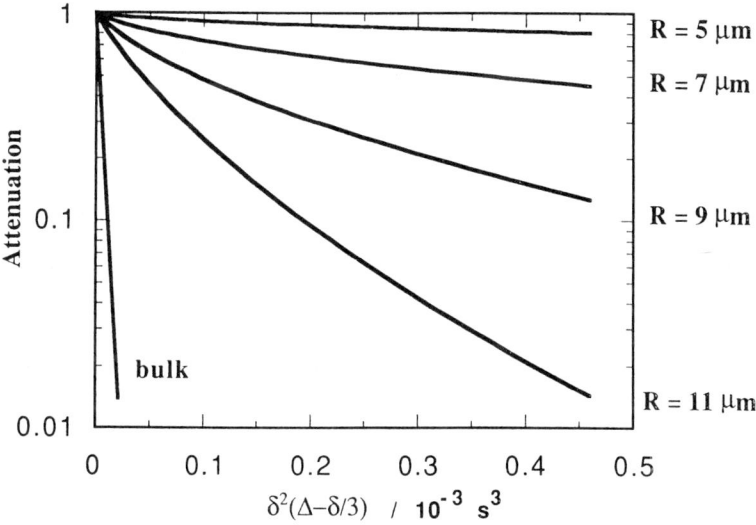

FIG. 3 The echo attenuation as a function of $\delta^2(\Delta - \delta/3)$ for different radii of emulsion droplets [according to Eq. (4)] with $\Delta = 0.100$ s, $\gamma g = 10^7$ rad m^{-1} s^{-1}, and $D = 2 \times 10^{-9}$ m^2 s^{-1}.

standard NMR spectrometer for performing the experiments. The first of these is a probe capable of delivering pulsed magnetic field gradients in addition to the ordinary rf pulses used in NMR work. The second addition is a gradient driver which generates the current necessary to produce the gradients. Both of these additions can nowadays be purchased at quite reasonable costs for virtually any type of spectrometer from the main spectrometer companies as well as from independent manufacturers of NMR equipment. An extensive discussion with regard to the specific requirements needed to perform high-quality PFG work can be found in Ref. 4.

Finally, one should note that any diffusion process depends on temperature. Thus, one has to use a proper control system to control the temperature when performing the measurements. As most modern spectrometers are equipped with temperature controllers, this fact poses no real problems.

III. APPLICATIONS

A. The Microemulsion Structure of the Continuous Medium in O/W and W/O Emulsion and Its Relation to Creaming and Coalescence

1. Background

The usefulness of self-diffusion coefficients in studying surfactant solutions lies in the direct information they provide about the microstructure of microheterogeneous systems [4,18–21]. To illustrate this fact, consider the two idealized situations depicted in Fig. 4. In Fig. 4a, an ordinary oil-swollen (or a reversed) micellar solution is shown, whereas an idealized bicontinuous structure is shown in Fig. 4b. In the former case, the diffusion coefficients of molecules confined to the micelles will be the same as those of the surfactant and will be given by the diffusion of the micelle as such, whereas the diffusion of molecules comprising the continuous medium will be rapid. For the situation in Fig. 4b, both molecules in the hydrophobic environment and in the water will diffuse rapidly and will be only slightly reduced from the values of the diffusion in the corresponding bulk liquids because of obstruction effects [22].

2. Phase Diagram

In order to study the influence of the microstructure of the continuous phase on the stability of emulsions, we have investigated the system sodium dodecyl-sulfate (SDS)/glycerolmono(2-ethyl-hexyl)ether/decane/brine (3 wt% NaCl) [23]. In this system, emulsions can be made of the

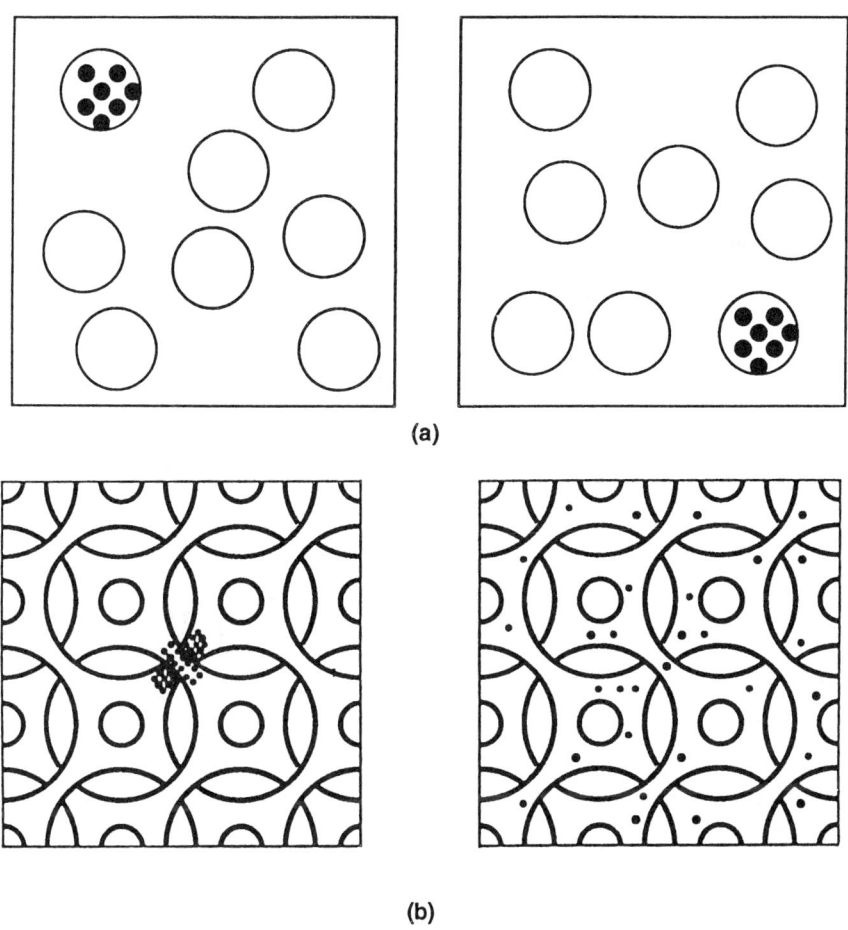

(a)

(b)

FIG. 4 (a) Illustration of an oil-swollen (or reversed) micelle; (b) Illustration of a bicontinuous structure. The filled points represent spin-labeled molecules at an initial time (left) and the same molecules after a certain time interval (right). The closed and bicontinuous structures will clearly give very different diffusion coefficients for the labeled molecules.

O/W or W/O type depending on the ratio between surfactant and cosurfactant. The total amount of surfactant and cosurfactant is kept constant at 5 wt%. The phase diagram is presented in Fig. 5. It is a cross section in a five-component phase diagram and it is difficult to predict the directions of the tielines in the two-phase areas where the emulsions are made. The

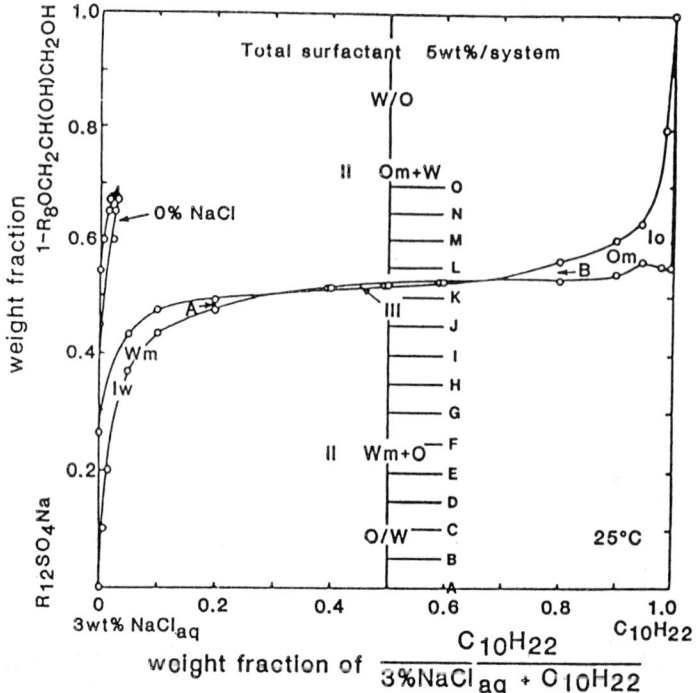

FIG. 5 The phase diagram for the SDS/glycerolmono(2-ethyl-hexyl)ether/decane/ brine (3 wt% NaCl) system at 25°C. Total surfactant concentration is 5 wt%. (From Ref. 40 with permission.)

samples are made with equal weights of brine and decane and, thus, their compositions are given by the vertical line in the middle of the diagram. The samples are prepared by weighing the components and then mixing them by gentle shaking by hand (keeping the time and amplitude of shaking as similar as possible among different samples). As emulsion stability depends critically on the droplets' size distribution, the emulsification process is critical. Nevertheless, some quantitative comparison can be made between the different samples prepared in this way.

3. NMR Diffusion in the Continuous Phase

The emulsions in this system are made in a two-phase area which for the O/W emulsions consists of an oil-rich phase and a phase of normal micelles, whereas for the W/O emulsions, it consists of a water-rich phase and a micellar phase of reversed micelles. The micellar phase is the continuous medium for both types of emulsions. In order to determine the structure of this continuous medium, we let the emulsion samples cream (or sediment) and separated the clear continuous medium from each sample. The clear medium was then transferred to a 5-mm NMR tube and the diffusion coefficients for both the oil and the water were determined with the NMR self-diffusion method as described above. The result is shown in Fig. 6, where the reduced diffusion coefficients (D/D_0, where D is the actual diffusion coefficient and D_0 is the diffusion coefficient of the neat liquid at the same temperature) for the oil and the water are plotted versus the ratio between cosurfactant and surfactant. For the O/W emulsion where SDS is the only surfactant, one finds that the continuous medium consists of small spherical micelles, that the water diffusion is fast (slightly lowered relative neat water due to obstruction effects [24], and that the oil diffusion is low and corresponds to a hydrodynamic radius of the oil-swollen micelle of about 50 Å (according to Stokes' law). When the cosurfactant is introduced and its amount is increased, the size of the micelle is increased, as can be inferred from the diagram by the lowered value of the reduced diffusion coefficient of the oil. Close to the three-phase area, the continuous medium is bicontinuous as the value of the reduced diffu-

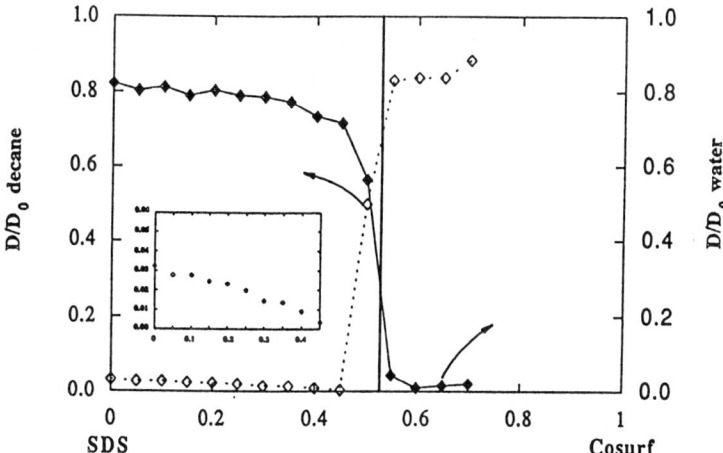

FIG. 6 Microemulsion structure in the continuous phase studied by the diffusion coefficients (D) divided by the diffusion coefficients (D_o) for the neat liquid versus the ratio of cosurfactant: SDS. The vertical line in the middle of the diagram refers to the three-phase area (cf. Fig. 5). The inset gives the reduced diffusion coefficient for decane with the scale on the abscissa enlarged. (From Ref. 40 with permission.)

sion coefficient is almost the same for both the oil and the water and equal to approximately 50% of the value for the bulk liquids. When the amount of cosurfactant is increased further, one passes over to the W/O emulsion area where the continuous medium is bicontinuous near the three-phase area and then changes to closed reversed micellar aggregates as can be seen from the now changed values of oil and water diffusion coefficients. Here, the hydrodynamic radius of the inverse micelles is about 70 Å.

4. Creaming and Coalescence Rate

It is difficult to measure an absolute value for the creaming rate of an emulsion, as there is often a broad size distribution of the emulsion droplets. For an isolated droplet in a continuous medium, the creaming rate u is

$$u = f\left(\Delta\rho, R^2, \frac{1}{\eta}\right), \tag{5}$$

where $\Delta\rho$ is the difference in density between the droplet and the continuous medium, R is the radius of the droplet, and η is the viscosity of the continuous medium. It is obvious that different droplet sizes will give different creaming rates. However, this expression does not take into account any obstruction effects which will lower the creaming rate when the droplets become more closely packed. Thus, we have measured the creaming mean "rate" as the time for three-fourths of the clear phase to appear ($t_{3/4}$); see Fig. 7. Inserted as comparison in the same diagram is the time of coalescence for the W/O emulsions, measured as the time for one-half of the water-rich phase to appear. The coalescence for O/W emulsions starts simultaneously throughout the creamed layer (after a much longer time, on the order of years) and cannot be measured in this way.

The creaming time for O/W emulsions increases up to the three-phase area where it suddenly decreases for the sample closest to this area. The increase of the creaming time can be ascribed to a decrease in the emulsion droplet size. NMR diffusion studies of the liquid inside the emulsion droplets show that the mean droplet size decreases toward the three-phase area. The experimental data for this diffusion are shown in Fig. 8, where a steeper initial slope indicates a larger droplet mean size [25] (cf. Fig. 3 and the discussion below). The sudden drop in the creaming time near the three-phase area corresponds well with the change in the continuous medium from closed micellar aggregates to a bicontinuous microemulsion. In fact, this sample is difficult to emulsify, and both creaming and coalescence occur fast here. On the W/O side of the three-phase area the same pattern is repeated but now in the other direction.

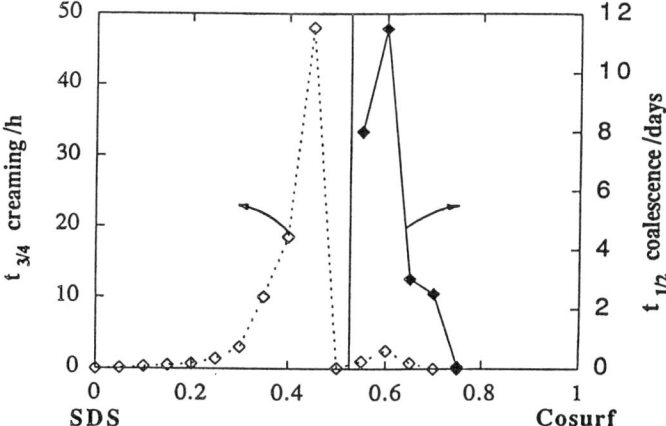

FIG. 7 The time for one-half of the coalesced emulsion layer to appear in W/O emulsions and the time for three-fourths of the clear phase to appear in a creamed emulsion versus the ratio of cosurfactant: SDS. The vertical line in the middle of the diagram refers to the three-phase area (cf. Fig. 5). (From Ref. 40 with permission.)

The surface tension will be the governing factor of the obtained emulsion droplet size in these emulsions gently shaken by hand. The general tendency for similar systems [26] is that the surface tension between the two phases that are emulsified decreases toward the three-phase area from both sides. The free energy which is needed to create the surface of an emulsion droplet is a function of the surface tension; thus the droplet size decreases with decreasing surface tension.

The creaming time as well as the coalescence time have peaks in both the O/W and W/O areas. The maximum in the creaming time is related to the emulsion droplet size and the presence of the bicontinuous microemulsion close to the three-phase area. The maximum in coalescence time is more difficult to interpret. The coalescence for O/W emulsions cannot be measured directly, but a careful study reveals that the coalescence first starts in both ends of the O/W area. There is a great difference in the coalescing time between the O/W and W/O emulsions, which is due to the fact that the former has a charged droplet interface, which hinders coalescence. The fast coalescence near the three-phase area can be explained by the bicontinuous microemulsion between the emulsion droplets that offers a good possibility for the molecules inside the emulsion droplets to migrate to another emulsion droplet. The fact that the coalescence is

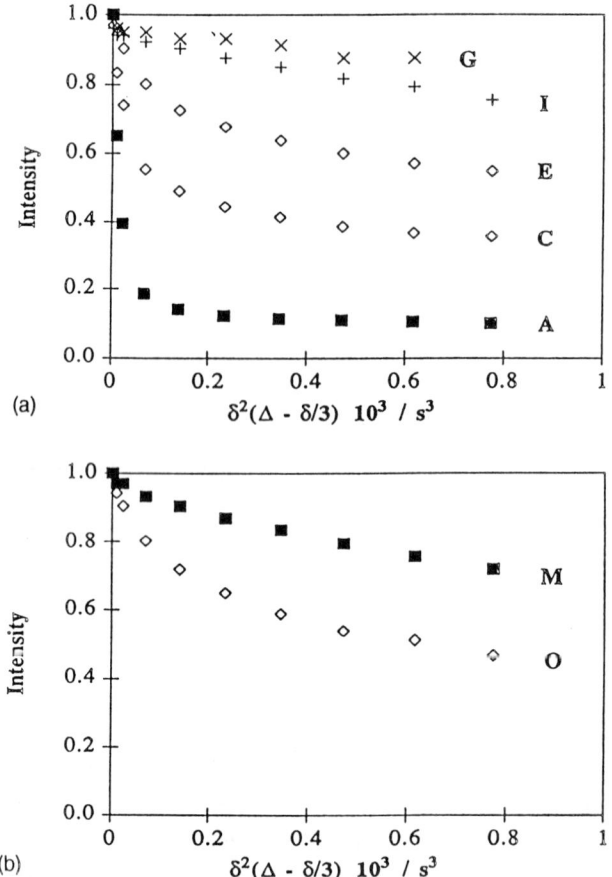

FIG. 8 The normalized NMR intensities (E/E_o) versus $\delta^2(\Delta - \delta/3)$ for the oil (a) and water (b) inside emulsion droplets in the studied samples. The letters refer to the sample compositions indicated in the phase diagram in Fig. 5. (From Ref. 40 with permission.)

more rapid at the high and low fractions of the cosurfactant could be explained by the large difference in curvature between the thermodynamically stable micelle and the unstable emulsion droplet. However, coalescence is of a complex nature and needs more research to be fully understood.

We anticipate that studies of this type will become increasingly important in the future, as there are several questions pertaining to the structure of the continuous phase and emulsion properties that still remain to be answered [27].

B. The Determination of Emulsion Droplet Radii by Means of the NMR PFG Method

As pointed out earlier, the echo attenuation curve for the PFG experiment when applied to molecules entrapped in an emulsion droplet is a signature of the size of the emulsion droplet. As a consequence, droplet sizes can be determined by means of the PFG experiment. The NMR sizing method, which was apparently first suggested by Tanner [9] has been applied to a number of different emulsions ranging from cheese to crude oil emulsions [25,28–33].

When applied to a real emulsion, one has to consider the fact that the emulsion droplets in most cases are polydisperse in size. This effect can be accounted for if the molecules confined to the droplets are in a slow-exchange situation, meaning that their lifetime in the droplet must be longer than Δ. For such a case, the echo attenuation is given by

$$E_{\text{poly}} = \frac{\int_0^\infty R^3 P(R) E(R) \, dR}{\int_0^\infty R^3 P(R) \, dR}, \tag{6}$$

where $P(R)$ represents the droplet size distribution function and $E(R)$ the echo attenuation according to Eq. (4) [or within the SGP approximation, Eq. (3)] for a given value of R. From NMR data it is difficult (although, in principle, not impossible) to determine the actual form of $P(R)$. However, given an analytical expression for $P(R)$, we may determine the parameters of that distribution function. A frequently used form is the log-normal function as defined in Eq. (7) as it appears to be a reasonable description of the droplet size distribution of many emulsions. In addition, it has only two parameters which makes it convenient for modeling purposes.

$$P(R) = \frac{1}{2R\sigma\sqrt{2\pi}} \exp\left(-\frac{(\ln 2R - \ln d_0)^2}{2\sigma^2}\right). \tag{7}$$

In Eq. (7), d_0 represents the diameter median and σ a measure of the width of the size distribution.

To illustrate the method and also discuss its accuracy we will use as an example some recent results for margarines (or low-calorie spreads) [28]. This system highlights some of the definite advantages of using the

NMR method to determine emulsion droplet sizes, because other nonperturbing methods hardly exist for these systems.

Given in Fig. 9 is the echo decay for the water signal of a low-calorie
spread containing 60% fat. These systems are W/O emulsions and, as can
be seen the water molecules do experience restricted diffusion (in the
representation of Fig. 9, the echo decay for free diffusion would be given
by a Gaussian function). Also given in Fig. 9 is the result of fitting Eqn.
(4), (6) and (7) to the data. As is evident, the fit is quite satisfactory and
the parameters of the distribution function obtained are given in the figure
caption. However, one might wonder how well determined these parameters are, given the fact that the equations describing the echo attenuation
are quite complicated. To test this matter further, Monte Carlo error investigations were performed in Ref. 28. Thus, random errors were added to
the echo attenuation and the least squares minimization was repeated 100
times, as described previously [34]. A typical result of such a procedure
is given in Fig. 10. As can be seen in Fig. 10, the parameters are reasonably
well determined, with an uncertainty in R (and σ, data not shown) of about
$\pm 15\%$.

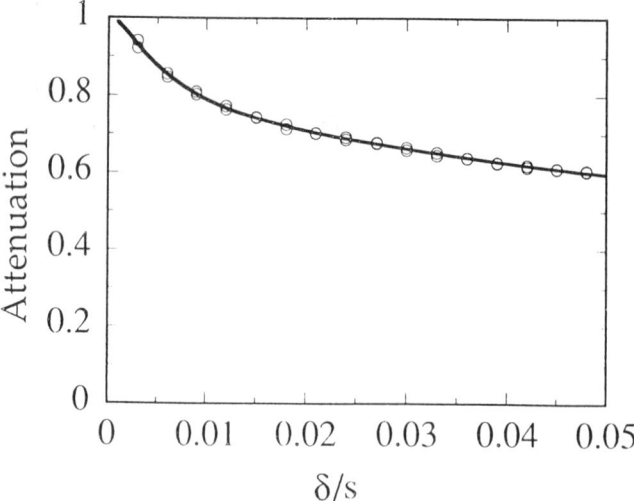

FIG. 9 Echo intensity for the entrapped water in droplets formed in a low-calorie
spread containing 60% fat versus δ. The solid line corresponds to the predictions
of Eq. (4), (6), and (7). The results from the fit are $d_0 = 0.82$ μm and $\sigma_0 = 0.72$.
(Adapted from Ref. 28.)

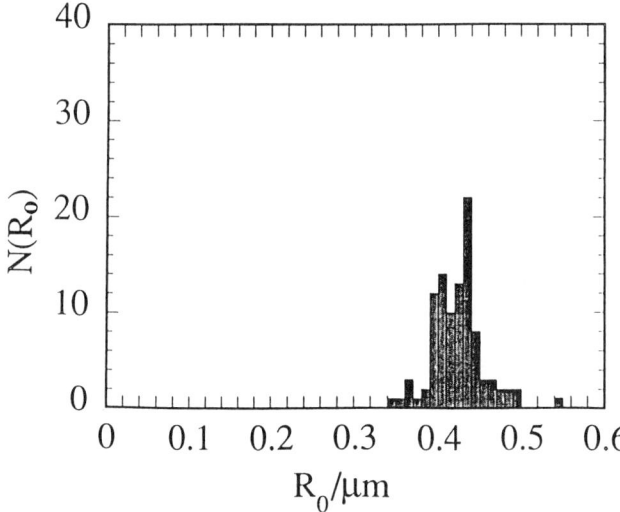

FIG. 10 A Monte Carlo error analysis of the data in Fig. 9. The value of the parameter d_0 in Eq. (7) is $d_0 = 0.82 \pm .044$ μm (note that $R_0 = d_{0/2}$ is plotted). (Adapted from Ref. 28.)

Creaming or sedimentation of emulsions with droplet sizes above 1 μm causes some experimental difficulty because of the change of the total amount of spins in the NMR active volume of the sample tube during the experiment. This can be accounted for by extra reference measurements with no gradient applied before and after each NMR scan. In addition, such reference measurements may provide information on the creaming rate, which is a useful characteristic of emulsions. Creaming or sedimentation is not a problem in the study of most food emulsions (such as low-calorie spreads), highly concentrated emulsions or viscous water-in-crude-oil emulsions [29].

Emulsion droplet sizes in the range from 1 μm up to 50 μm can be measured with rather modest gradient strengths of about 1 T m^{-1}. Note that the size determination rests on the molecular motion of the dispersed phase, so the method cannot be applied to dispersed phases with low molecular mobility. In practice, oils with self-diffusion coefficients above 10^{-12} m^2 s^{-1} is required for sizing of O/W emulsions. Of course, W/O emulsions with most conceivable continuous media can be sized.

Emulsion droplets below 1 μm can often be characterized by the Brownian motion of the droplet as such (exceptions are concentrated emulsions or other emulsions where the droplets do not diffuse). This is the approach taken in the study of microemulsion droplets, where the diffusion behavior of the solubilized phase is characterized by the droplets' (Gaussian) diffusion.

In conclusion, we summarize the main advantage of the NMR diffusion method as applied to emulsion droplet sizing. It is nonperturbing, requiring no sample manipulation (such as dilution with the continuous phase) and nondestructive which means that the same sample may be investigated many times, which is important if one wants to study long-time stability or the effect of certain additives on the droplet size. It requires small amounts of sample (typically on the order of a few hundred milligrams). Moreover, the total NMR signal from the dispersed phase in emulsions is usually quite intense because of the large amount of spins. This allows for quick measurement with a single scan per δ point.

C. PFG Studies of Concentrated Emulsions

In Sec. III.B, the discussion applies to the case where the molecules are confined to the droplets on the time scale of the experiment. This is a reasonable assumption for many emulsions, and it can, in fact, be tested by the NMR diffusion method by varying Δ. However, there are some interesting emulsion systems for which this is not always the case. These are the so-called highly concentrated emulsions (often termed high internal phase emulsions) [35,36], which may contain up to (and in some cases even more than) 99% dispersed phase. Here the droplets are separated by a liquid film which may be very thin (on the order of 100 Å) and which may in some instances be permeable to the dispersed phase.

It is useful to distinguish among three cases with regard to the lifetime of the molecules in the droplets. In the PFG experiment, the time scale is defined by the value of Δ. Thus, cases where the molecular lifetime is shorter, longer, or of the same order of magnitude as Δ give rise to very different behavior in the echo attenuation curves.

Starting with the first of these possibilities [i.e., that the lifetime is long in comparison to Δ (a situation often found in O/W concentrated emulsions)], the situation is of course similar to the one encountered in "ordinary" emulsion systems, and one can determine the droplet size distribution. But, in fact, one can obtain even more information about these interesting systems from PFG data.

Thus, the present authors have developed a technique by which the stability of such emulsions may be monitored. The background of the method is as follows. By determining the size of the emulsion droplets as a function of time for one particular sample (remember that the NMR method is nondestructive), we measure the increase in droplet size. This increase may then be interpreted in terms of the decrease in the total droplet area. Because the droplets are close packed, the area of the thin liquid film is approximately half of the total droplet area. There are two conceivable mechanisms for the decrease in droplet area. The first is rupture of the thin liquid film separating the droplets, which leads to a coalescence of two neighboring droplets. The second is a process akin to Ostwald ripening whereby dispersed molecules permeate the film.

For O/W emulsions (at least the ones we have investigated), the permeability across the thin liquid film is so slow that the stability is given by the film rupture mechanism. Given in Fig. 11 is the total droplet area as a function of time for a concentrated emulsion consisting of 98 wt% heptane, water, and CTAB. As can be seen there is a rather rapid initial decrease in the area, which levels out after about 12 h. From the initial part of the curve, the film rupture rate may be obtained, and for the data in Fig. 11 the value is 4×10^{-5} s^{-1}. At longer times, the emulsion becomes remarkable stable, and there is little or no further decrease in the total droplet area during a period of 1 year.

Turning to the cases where the lifetimes in the droplets are shorter or equal to Δ, these situations are at hand if the dispersed phase actually

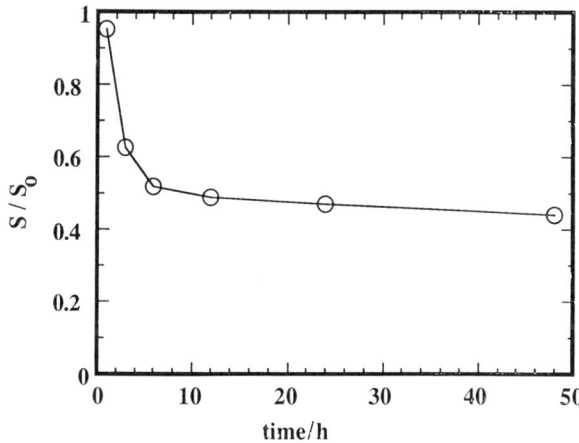

FIG. 11 Decrease of the relative droplet surface area (S/S_o) with time for a highly concentrated O/W emulsion containing 98 wt% heptane and 0.4 wt% CTAB as emulsifier. The S/S_o is calculated from the droplet size distribution as obtained by the NMR self-diffusion technique. The initial slope corresponds to a film rupture rate of $J = 4 \times 10^{-5}$ s^{-1}.

crosses the film by some mechanism, the detailed nature of which need not concern us here. We are then dealing with a system with permeable barriers (on the relevant time scale), and the system can now be regarded as belonging to the general class of porous systems.

For the case when the lifetime is short with respect to Δ, the PFG experiment enables one to determine the long-time diffusion coefficient for the dispersed phase (or for molecules dissolved in the dispersed phase). An example of such a case is given by an emulsion comprised of 96 wt% brine (of concentration 0.17 M with respect to NaCl), 2.3 wt% heptane, 1.4 wt% tetraethylene glycol dodecyl ether ($C_{12}E_4$), and 0.3 wt% soybean phosphatidyl choline. The echo attenuation for three different values of Δ for this system is presented in Fig. 12. The data set in Fig. 12 is not compatible with diffusion within a closed droplet. Rather, the data are compatible with a Gaussian diffusion, and by fitting Eq. (1) to the entire data set, one can back out one common diffusion coefficient, the value of which is 2.25×10^{-10} m^2 s^{-1}. This value should be compared with the bulk value of water diffusion (2.23×10^{-9} m^2 s^{-1}) at the relevant temperature. Thus, the water diffusion is Gaussian in nature, albeit with a reduced value of the diffusion coefficient, indicating that the droplets are semipermeable. There are essentially two parameters governing this process, namely, the lifetime in the droplets and the size of the droplets.

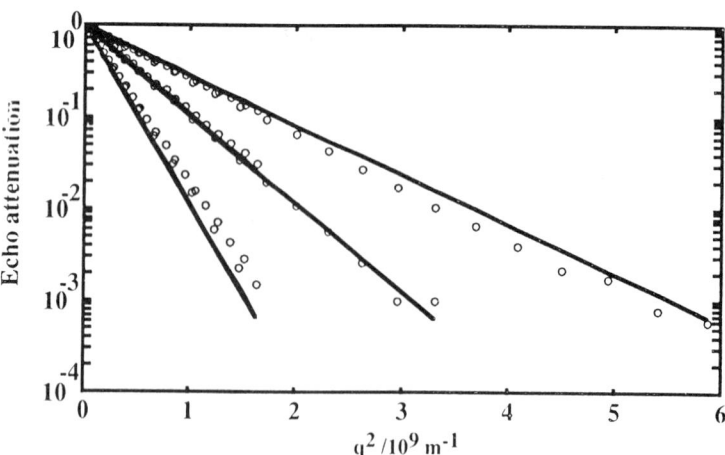

FIG. 12 Experimental echo attenuation curves versus q^2 ($q \equiv \delta\gamma g/2\pi$) for $\Delta =$ 140 ms (right curve), 250 ms (middle curve), and 500 ms (left curve). A global fit to the three attenuation curves gives $D = (2.25 \pm 0.022) \times 10^{-10}$ m^2 s^{-1}.

These are also parameters that are of great interest to determine for these systems. However, this is not possible when the lifetime is shorter than the value of Δ used.

There is a simple physical reason for this state of affairs. The diffusion process at long times is essentially a random walk of step length $2R$, where R is the droplet radius. For such a case, the diffusion coefficient is given by $(2R)^2/6\tau$, where τ is the lifetime of the walker in each droplet. Clearly, an infinite number of combinations of R and τ yield the same value for the long-time diffusion coefficient. An increase in the lifetime in the droplets can be compensated for by an increase in the step size. However if additional information is available, one may, of course, separate the parameters R and τ. The system described here has been studied by Kunieda and co-workers [37] and they report a value of the radius equal to 4 μm. Using this value for R, one obtains a value for the lifetime τ of 47 ms.

Finally, we turn to the case where the lifetime of the dispersed phase is of the same order of magnitude as Δ. For this case, one may actually separate the dynamic and structural information. Under these conditions, one may, in some cases, obtain a peak in the plot of the echo amplitude versus the gradient pulse duration, δ. This is a surprising result at first sight, as we are accustomed to observe a monotonous decrease of the echo amplitude with δ, but it is actually a manifestation of the fact that the diffusion is no longer Gaussian. Such peaks can be rationalized within a formalism related to the one used to treat diffraction effects [38], and the analysis of the data may yield important information regarding not only the size of the droplets but also the permeability of the dispersed phase through the thin films as well as the long-term diffusion behavior of the dispersed phase. We show in Fig. 13 an example of such a diffraction-like effect in a concentrated emulsion system [39]. The particular example pertains to a concentrated emulsion based on a fluorinated non-ionic surfactant, where the continuous medium is a perfluorinated oil.

The data in Figs. 12 and 13 are presented with the value of the quantity q on the abscissa. This quantity is defined as $q = \gamma g \delta/2\pi$ with the dimension of inverse length and it is, in fact, related to the scattering vector in scattering techniques. In fact, the inverse of the value of the position of the peak can be related to the center-to-center distance of the droplets. In the example given in Fig. 13, this value is 3.3 μm, which is in good agreement with twice the droplet radii as judged from microscope pictures taken of the emulsion. To obtain the droplet lifetime one needs access to a theory for diffusion in these porous systems. As far as we are aware, there does not at the present time exist such a theory. The present authors are in the process of developing a framework for interpreting experiments

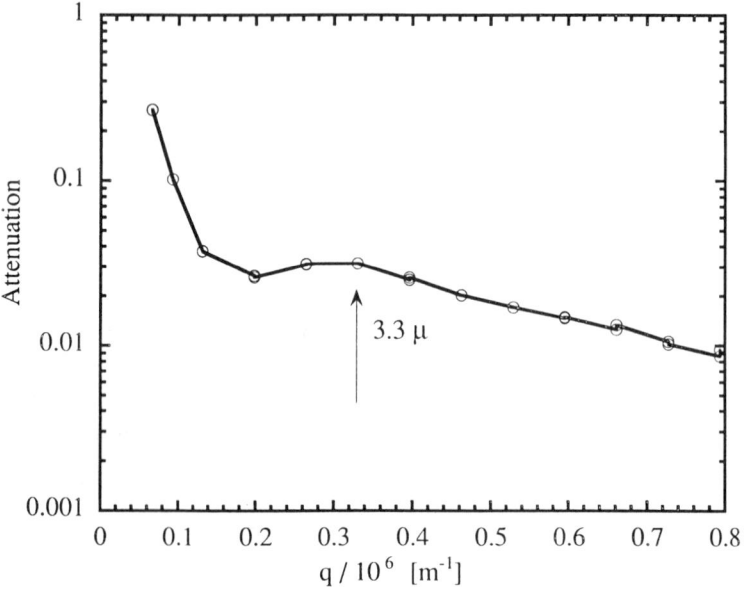

FIG. 13 The echo intensity versus the parameter q ($q \equiv \delta\gamma g/2\pi$) for the water in a concentrated W/O emulsion consisting of a partially fluorinated surfactant, perfluorodecaline, and water. (Adapted from Ref. 39.)

such as the one described above which is based on computer simulation of the diffusion of the molecules within in the droplets.

IV. CONCLUDING REMARKS

The purpose of this chapter has been to introduce, in our opinion, a very useful experimental technique, namely, the determination of flow and self-diffusion by means of the NMR PFG method. The key feature of this method is the fact that with it we can determine molecular mean displacements over distances which have lengths which are typically found in colloidal systems, in general, and in emulsion systems, in particular.

As a consequence, the method can be used to convey information of several very important aspects with regard to emulsion properties. We have focused on three such areas where we deem the method as particularly important in that the information obtained can scarcely be obtained by alternative methods. The first of these concerns the microstructure of the continuous phase. Knowledge of the structure is important when one is trying to gain insight into which processes determine emulsion stability. The second area deals with the determination of droplet size distributions, which is an important feature of emulsions. For some systems, conventional sizing techniques are not applicable. Example of such systems are often found in emulsions used in food technology. For these, the NMR technique is very well suited, as it does not depend on the physical status of the sample.

Finally, we have indicated that the technique is potentially very useful in the characterization of concentrated emulsions, where it may yield important information with regard to both structure and dynamic properties.

ACKNOWLEDGMENTS

The worked described here has been financially supported by the Swedish Board for Technical Development (NUTEC) and the Swedish Natural Science Research Council (NFR). Fruitful cooperation with Drs. Ingrid Lönnqvist, Bengt Jönsson, Per Linse, Ali Khan, and TorBjörn Wärnheim is gratefully acknowledged.

REFERENCES

1. P. Becher (ed.), in *Encyclopedia of Emulsion Technology, Vol. 2: Applications*, Marcel Dekker, Inc., New York, 1985.
2. P. T. Callaghan, C. M. Trotter, and K. W. Jolley, J. Magn. Reson. *37*:247 (1980).
3. P. Stilbs, Prog. Nucl. Magn. Reson. Spectrosc. *19*(1):1 (1987).
4. O. Söderman and P. Stilbs. Prog. Nucl. Mag. Reson. *26*:445 (1994).
5. E. L. Hahn. Phys. Rev. *80*:580 (1950).
6. E. O. Stejskal and J. E. Tanner, J. Chem. Phys. *42*:288 (1965).
7. D. W. McCall, D. C. Douglass, and E. W. Anderson, Ber. Bunsenges. Phys. Chem. *67*:336 (1963).
8. R. L. Vold, J. S. Waugh, M. P. Klein, and D. E. Phelps, J. Chem. Phys. *48*:3831 (1968).
9. J. E. Tanner, Thesis, University of Wisconsin, 1966.
10. P. T. Callaghan and A. Coy, in *NMR Probes of Molecular Dynamics* (P. Tycko, ed.), Kluwer Academic Publishers, Dordrecht, 1993.
11. P. T. Callaghan, *Principles of Nuclear Magnetic Resonance Microscopy*, Clarendon Press, Oxford, 1991.
12. P. T. Callaghan, A. Coy, T. P. J. Halpin, D. MacGowan, J. K. Packer, and F. O. Zelaya, J. Chem. Phys. *97*:651 (1992).
13. J. E. Tanner and E. O. Stejskal, J. Chem. Phys. *49*:1768 (1968).
14. B. Balinov, B. Jönsson, P. Linse, and O. Söderman, J. Magn. Reson. A. *104*:17 (1993).
15. D. C. Douglass and D. W. McCall, J. Phys. Chem. *62*:1102 (1958).

16. C. H. Neuman. J. Chem. Phys. *60*:4508 (1974).
17. J. S. Murday and R. M. Cotts, J. Chem. Phys. *48*:4938 (1968).
18. B. Lindman, O. Söderman, and H. Wennerström, in *Novel Techniques to Investigate Surfactant Solutions* (R. Zana, ed.), Marcel Dekker, Inc., New York, 1987. p. 295.
19. B. Balinov, U. Olsson, and O. Söderman, J. Phys. Chem. *95*:5931 (1991).
20. B. Lindman and P. Stilbs, in *Microemulsion Systems* (H. L. Rosano and M. Clausse, eds.), Marcel Dekker, Inc., New York, 1987, p. 129.
21. B. Lindman, K. Shinoda, U. Olsson, D. Andersen, G. Karlström, and H. Wennerström, Colloids Surf. *38*:205 (1989).
22. D. M. Anderson and H. Wennerström, J. Phys. Chem. *94*:8683 (1990).
23. K. Shinoda, H. Kunieda, T. Arai, and H. Saijo, J. Phys. Chem. *88*:5126 (1984).
24. B. Jönsson, H. Wennerström, P. Nilsson, and P. Linse, Colloid Polym. Sci. *264*:77 (1986).
25. I. Lönnqvist, A. Khan, and O. Söderman, J. Colloid Interface Sci. *144*(2): 401 (1991).
26. J. S. Zhou, M. Kamioner, and M. Dupeyrat, in *Microemulsion Systems* (H. L. Rosano and M. Clausse, eds.), Marcel Dekker, Inc., New York, 1987, p. 335.
27. D. H. Smith, G. K. Johnson, and D. B. Dadyburjor, Langmuir *9*:2089 (1993).
28. B. Balinov, O. Söderman, and T. Wärnheim, J. Am. Oil Chem. Soc. *71*:513 (1994).
29. B. Balinov, O. Urdahl, O. Söderman, and J. Sjöblom, Colloids Surf. *82*:173 (1994).
30. P. T. Callaghan, K. W. Jolley, and R. Humphrey, J. Colloid Interface Sci. *93*:521 (1983).
31. X. Li, J. C. Cox, and R. W. Flumerfelt. AIChE J. *38*(10):1671 (1992).
32. K. J. Packer and C. Rees, J. Colloid Interface Sci. *40*(2):206 (1972).
33. J. C. Van den Enden, D. Waddington, H. Van Aalst, C. G. Van Kralingen, and K. J. Packer, J. Colloid Interface Sci. *140*(1):105 (1990).
34. P. Stilbs and M. Moseley, J. Magn. Reson. *31*:55 (1978).
35. K. J. Lissant, J. Colloid Interface Sci. *22*:462 (1966).
36. H. M. Princen. J. Colloid Interface Sci. *91*:160 (1983).
37. H. Kunieda, N. Yano, and C. Solans. Colloids Surf. *36*:313 (1989).
38. P. T. Callaghan, A. Coy, D. MacGowan, K. J. Packer, and F. O. Zelaya, Nature (London) *351*(6326):467 (1991).
39. B. Balinov, O. Söderman, and J. C. Ravey, J. Phys. Chem. *98*:393 (1994).
40. O. Söderman, I. Lönnqvist, and B. Balinov, in *Emulsions—A Fundamental and Practical Approach* (J. Sjöblom, ed.), Kluwer Academic Publishers, Dordrecht, 1992, p. 363.

9

Flocculation and Coalescence in Emulsions as Studied by Dielectric Spectroscopy

JOHAN SJÖBLOM, HARALD FØRDEDAL, **and TORE SKODVIN** Department of Chemistry, University of Bergen, Bergen, Norway

I. INTRODUCTION

Dielectric spectroscopy has proved to be a powerful technique in analyzing heterogeneous systems of colloidal nature. In these studies, the time dependence of the processes occurring in the systems will determine the frequencies of immediate interest to the user. Low-frequency equipment has advantages when applied to aqueous solid suspensions, emulsions, or heterogeneous polymer blend systems. Especially in the first mentioned systems relaxation processes in the electrical double layer of the charged particles or droplets can be studied. For the reverse problem with electrically conducting cavities (aqueous domains) in nonpolar environments, the frequency dependence is highly dependent on the properties of the aqueous domains. The Maxwell–Wagner–Sillars relaxation in water-in-oil emulsions where the aqueous domains contain electrolyte is dependent on the overall mobility of charges inside the droplet and hence the electrolyte concentration. At higher concentrations of electrolyte (i.e., above 1% by weight), the time-dependent processes fall in the frequency region well suited for studies by the time-domain dielectric spectroscopy (TDS) principle. By making use of this principle, the measurements are made fast while preserving high accuracy.

Dielectric measurements on emulsified systems have been studied mainly by two research groups, that is, the group in Kyoto, Japan and the group in Pau, France. Recently, the groups in Chicago and Bergen have undertaken work in this field with the intention of developing the method to solve technological problems connected with the processing of the emulsified systems. Special interest has been devoted to crude-oil-based systems and problems related to petroleum exploitation. Wasan et al. [1,2] developed high-frequency dielectric measurements to determine emulsion type and water–oil content. Sjöblom et al. have utilized the time-domain dielectric spectroscopy principle to determine the level of the breakdown process in different kinds of emulsion. These processes involve both flocculation and coalescence. For the latter purpose, the TDS instrumentation was equipped with an external high electric field device.

In his review [3], Clausse gives an extensive background to the theory of electromagnetism and dielectric phenomena. He also gives a thorough derivation of many of the pertinent formula for different dielectric applications. Finally, he views the application of dielectric measurements to a variety of emulsified systems and microemulsions. As Clausse's treatment is very extensive, we have chosen in this chapter only to update the information lacking in his review. We will especially pay attention to association processes in emulsified systems as determined by means of time-domain dielectric spectroscopy (TDS).

II. BASIC THEORY

The permittivity of a substance is defined as $\epsilon^* = 1 + P/\epsilon_0 E$, where P is the polarization produced by an applied electric field E and ϵ_0 is the permittivity of free space. The polarization is due to charge movements and may have contributions from various mechanisms: for example, distortion polarization due to displacement of electrons and atoms within molecules, dipolar polarization due to reorientation of permanent molecular dipoles, or interfacial polarization in heterogeneous systems with boundaries between different components (i.e., the Maxwell–Wagner effect). (For general reviews, see Ref. 4.) At higher frequencies, the polarization may not be able to follow the applied ac field $E = E_0^{i\omega\tau}$, and a phase lag will appear between the polarization P and the field E. This is expressed by writing ϵ^* as a complex number:

$$\epsilon^*(\omega) = \epsilon'(\omega) - i\epsilon''(\omega), \tag{1}$$

where the imaginary part ϵ'' is sometimes called the loss factor.

On application of an electric field, the distortion polarization will be established instantly when compared to time intervals in which we are interested. If the development of the remaining part of the polarization is characterized by a time constant τ, its time dependence can be written $P(t) = P_0(1 - e^{-t/\tau})$. Insertion into Eq. (1) gives the following expression for the permittivity as function of the angular frequency, the Debye equation [5]:

$$\epsilon^*(\omega) = \epsilon_\infty + \frac{\epsilon_s - \epsilon_\infty}{1 + i\omega\tau} \tag{2}$$

or

$$\epsilon'(\omega) = \epsilon_\infty + \frac{\epsilon_s - \epsilon_\infty}{1 + \omega^2\tau^2}, \qquad \epsilon''(\omega) = \frac{(\epsilon_s - \epsilon_\infty)\omega\tau}{1 + \omega^2\tau^2}. \tag{3}$$

Here ϵ_s is the static permittivity, ϵ_∞ the limiting permittivity at high frequencies, and τ the dielectric relaxation time. The dielectric dispersion covers a wide frequency range, with a maximum in the loss factor at $\omega = 1/\tau$.

In the simple model above, it is assumed that the dielectric relaxation is characterized by a single relaxation time. In more complex systems, the spectrum may be characterized by several relaxation times. If these are well separated, the spectrum may be resolved and the different relaxation times determined. For systems with a broader dispersion than predicted by the Debye equations and which may not be resolved as a sum of a few such functions, Cole and Cole [6] have suggested the model function

$$\epsilon^*(\omega) = \epsilon_\infty + \frac{\epsilon_s + \epsilon_\infty}{1 + (i\omega\tau)^{1-\alpha}}. \tag{4}$$

The Cole–Cole equation corresponds to a symmetrical distribution of relaxation times τ with α as a parameter for the width of their distribution around a mean relaxation time τ.

In dielectrics that show a dc conductivity, the measured total permittivity will also include a contribution from the conductivity. In this case, the total permittivity can be written

$$\epsilon^*_{tot}(\omega) = \epsilon^*(\omega) - \frac{i\sigma}{\omega\epsilon_0} = \epsilon'(\omega) - i\epsilon''(\omega) - \frac{i\sigma}{\omega\epsilon_0}, \tag{5}$$

where σ is the conductivity. A high conductivity will lead to a large loss factor at low frequencies and difficulties in establishing the dipolar part of the spectrum at these frequencies.

III. DIELECTRIC TIME-DOMAIN SPECTROSCOPY

Traditionally, the static permittivity is measured by filling a capacitor with the dielectric under study and measuring its capacitance. To obtain the whole spectrum, up to the microwave region, the sample may be placed into a coaxial line or waveguides and measurements done at a large number of frequencies to cover a wide frequency range. In recent years, an alternative measurement technique in the time domain (TDS) has been developed [7,8], which makes it possible to cover a large frequency range in a single measurement. TDS is based on the study of the change in the shape of a fast-rising pulse propagating in a coaxial line when the sample is inserted into the line. From Fourier analysis it is known that the response pulse will contain information on the influence of the dielectric (i.e., its permittivity) over the entire frequency range up to some limiting frequency determined by the rise time of the step pulse. A necessary criterion for the characterization of a dielectric dispersion by means of time-domain dielectric spectroscopy is that the pulse shapes are monitored in a time window that is longer than the relaxation time studied.

Figure 1 shows the principle of a TDS spectrometer based on the total reflection method, that is, the study of reflected pulse shapes. The pulse generator (HP54123A) repetitively generates a step pulse with a short rise time, <25 ps, which propagates along a coaxial line. The pulse is reflected from a sample placed at the end of an open-ended coaxial line. The influence of the dielectric sample on the shape of the reflected pulse is registered by a sampling oscilloscope (HP54120B) and transferred to a com-

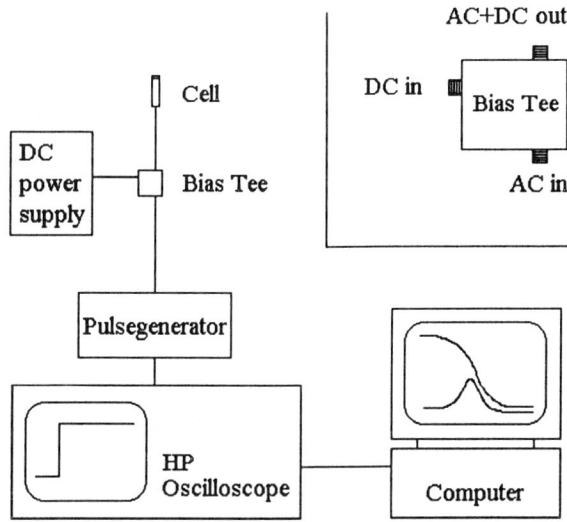

FIG. 1 Setup for the dielectric measurements according to the TDS principle at high external electric fields.

puter, which calculates the Fourier transforms and solves the reflection coefficient equation. If the pulse reflected by the empty cell is denoted $v(t)$ and the pulse reflected by the dielectric $r(t)$, Fourier transformation of these pulses gives

$$V(\omega) = \int_{-\infty}^{\infty} v(t)e^{-i\omega t}\, dt, \qquad R(\omega) = \int_{-\infty}^{\infty} r(t)e^{-i\omega t}\, dt. \qquad (6)$$

The ratio $S(\omega) = R(\omega)/V(\omega)$ is the reflection coefficient for electromagnetic waves of angular frequency ω reflected from the sample. This coefficient is a function of the permittivity of the sample:

$$S(\omega) = \frac{R(\omega)}{V(\omega)} = \frac{(\rho + \exp(-2i\omega l/c)\sqrt{\epsilon^*})\exp(2i\omega l/c)}{1 + \rho \exp(-2i\omega l/c)\sqrt{\epsilon^*}}. \qquad (7)$$

Here c is the speed of light in free space, l is the effective cell length, and $\rho = (1 - \sqrt{\epsilon^*})/(1 + \sqrt{\epsilon^*})$. This equation cannot be solved analytically, but the numerical solution will give the permittivity at a chosen frequency. The pulse shapes contain the necessary information to determine the dielectric spectrum. The influence of deviations from the ideal signal reflected from the sample cell, Eq. (7), due to connector mismatches and so forth can be reduced by measuring relative to a standard dielectric with known permittivity spectrum, not too different from that of the unknown.

For a dielectric with a dc conductivity σ, the reflected pulse $r(t)$ will not reach the same final level as the incident step pulse $v(t)$. If a sufficiently long-time window is used to allow the complete decay of all dielectric processes, the conductivity can be calculated from the time-domain data. This difference in final levels, r_∞ and v_∞, respectively, is a direct indication of conductivity, which gives

$$\sigma = \left(\frac{1 - r_\infty/v_\infty}{1 + r_\infty/v_\infty} \right) \frac{\epsilon_0 c}{l}. \tag{8}$$

Thus, the TDS measurement gives the information required to correct the total permittivity ϵ_{tot}^* for the conductivity contribution to obtain the dipolar part of the spectrum as illustrated in Eq. (5).

The TDS method described is based on a study of the pulse reflected from the sample [8,9]. Alternative TDS methods have been developed that utilize the pulse transmitted through the sample to obtain the dielectric spectrum [10,11]. An equation analogous to Eq. (7) will give the transmission coefficient as function of permittivity.

The different TDS methods have their pros and cons regarding sample volume, frequency range, and so on, but present techniques offer a fast method to determine the dielectric spectrum up to 20 GHz.

We have recently extended the dielectric setup to include measurements of systems that are exposed to high external electric fields, in order to monitor the electrocoalescence mechanism of w/o emulsions. In addition to the fast-rising ac step pulse, an external electric dc field is applied through a bias tee (Picoseconds Pulse Labs, 5530A), as shown in Fig. 1. The bias tee includes a dc block which allows the application of a voltage over the sample cell while it passes the fast-rising step pulse used to measure the permittivity without appreciably degrading the pulse.

IV. DIELECTRIC PROPERTIES OF EMULSIONS

Fundamental aspects of destabilization processes in emulsions are covered in previous chapters of this book. Basically, one can divide these processes in sedimentation, flocculation, and coalescence. Figure 2 gives a schematic view of these processes and the resulting emulsions. It is obvious that different volume elements in the samples in Fig. 2 will give very different dielectric responses. Hence, different ways to prepare the emulsions (i.e., the mechanical energy input), different surfactants, different viscosities, different salinities, and different storage times will all give

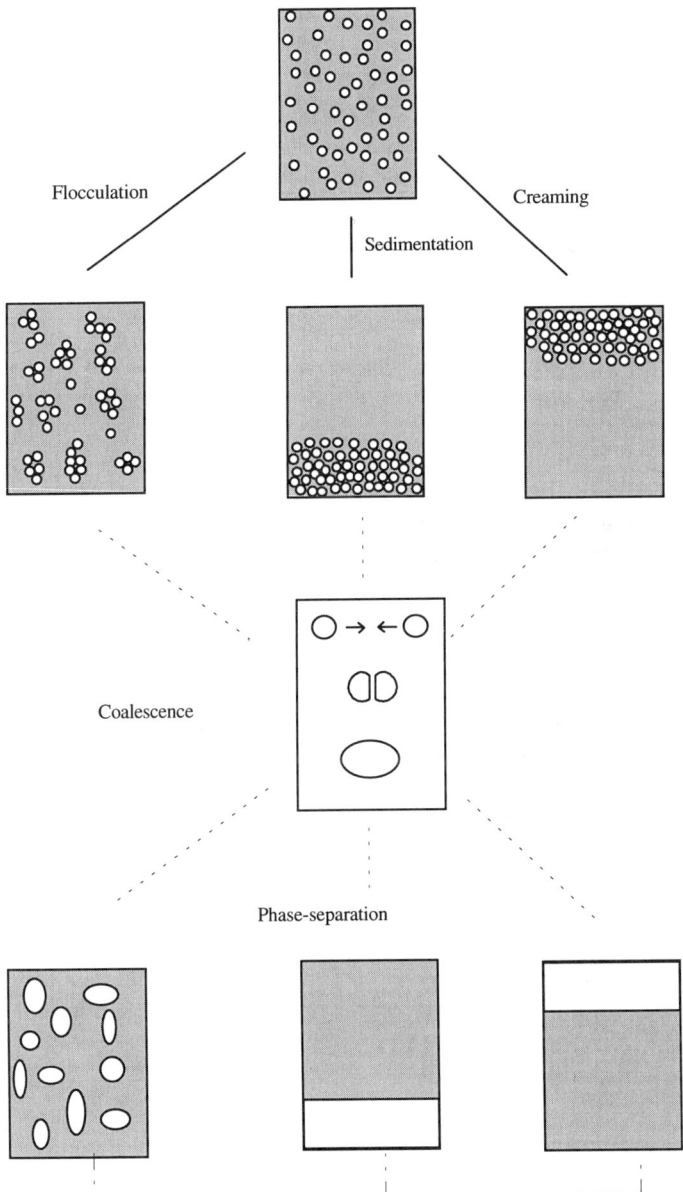

FIG. 2 A schematic representation of the processes in an emulsion (top) that eventually lead to a phase separation (bottom).

different stabilities and different dielectric responses. Figure 2 reveals the necessity of dielectric models covering both dilute and concentrated systems as well as flocculated systems with a large polydispersity in droplet/floc size. The following sections are devoted to these topics.

Wagner [12] presented in 1914 a theory of the dielectric properties of a dilute dispersion of spherical inclusions in a continuous medium. Wagner's work, together with the work of Wiener [13], makes up the basis for the theories of the dielectric properties of heterogeneous mixtures of the matrix-inclusion type. With the emulsion droplets as the inclusions and the continuous phase as the surrounding matrix, these theories have been shown to be applicable to emulsified systems as well. Other mixture models [14–16] may be better in predicting some special features of emulsions (for instance, percolation effects) but fail in other fields. Thus, we will concentrate on the extensions of the Wagner models. For a more thorough theoretical treatment of the dielectric properties of emulsions, the reader is referred to the reviews by Clausse [3], van Beek [17], and Hanai [18]. Numerical analysis and comparison of various mixture formulas have been comprised by Bánhegyi [19,20]. The Wiener formula, given by

$$\frac{\epsilon_s - \epsilon_{2s}}{\epsilon_s + 2\epsilon_{2s}} = \frac{\epsilon_{1s} - \epsilon_{2s}}{\epsilon_{1s} + 2\epsilon_{2s}} \phi_1, \tag{9}$$

expresses the effective static permittivity ϵ_s of a system consisting of spherical particles (sparsely) dispersed in a continuous medium. ϵ_{1s} and ϵ_{2s} are the static permittivities of the disperse and continuous phase, respectively. ϕ_1 denotes the volume fraction of dispersed particles. This equation is valid only for small values of ϕ_1 because the effect of particle interactions have been neglected.

Equation (9) has been extended to higher volume fractions by Bruggeman [21], and in later amendments by Hanai [22–24], the complex frequency-dependent permittivities have been introduced. The Hanai formula for the effective complex permittivity ϵ^* of a dispersion of spherical particles reads

$$\left(\frac{\epsilon^* - \epsilon_1^*}{\epsilon_2^* - \epsilon_1^*}\right)\left(\frac{\epsilon_2^*}{\epsilon^*}\right)^{1/3} = 1 - \phi_1, \tag{10}$$

with ϵ_1^* and ϵ_2^* representing the complex permittivities of the disperse and continuous phases, respectively. In principle, Eq. (10) is valid for volume fractions in the range $0 < \phi_1 < 1$.

For deviations from the spherical shape of the emulsion droplets, the equations given by Boyle [25] or Boned and Peyrelasse [26] may be applied. The Boyle equation, valid for spheroidal inclusions with a parallel alignment, reads

$$\left(\frac{\epsilon^* - \epsilon_1^*}{\epsilon_2^* - \epsilon_1^*}\right)\left(\frac{\epsilon_2^*}{\epsilon^*}\right)^{A_i} = 1 - \phi_1, \tag{11}$$

where A_i is a shape factor (or depolarization factor) along the axis of symmetry assumed to be parallel to the external field. In the case of a random orientation of dispersed spheroidal particles, Boned and Peyrelasse stated

$$\left(\frac{\epsilon_1^* - \epsilon^*}{\epsilon_1^* - \epsilon_2^*}\right)\left(\frac{\epsilon_2^*}{\epsilon^*}\right)^{3d}\left[\frac{\epsilon_2^*(1 + 3A) + \epsilon_1^*(2 - 3A)}{\epsilon^*(1 + 3A) + \epsilon_1^*(2 - 3A)}\right]^{3K} = 1 - \phi_1. \tag{12}$$

In Eq. (12), $d = A(1 - 2A)/(2 - 3A)$ and $3K = 2(1 - 3A)^2/[(2 - 3A)(1 + 3A)]$. Here, $A = A_1 = A_2 = (1 - A_3)/2$, where A_3 is the depolarization factor along the symmetry axis and A_1 and A_2 are the depolarization axes normal to the symmetry axis. For spheroids, the shape factors are determined by the ratios between the major and minor axes. When all the axes are equal, the shape factors are $\frac{1}{3}$ and Eqs. (11) and (12) are reduced to the Hanai formula [Eq. (10)]. In the case of prolate (needle-shaped) particles, $A < \frac{1}{3}$, whereas for oblates (disklike particles) $A > \frac{1}{3}$.

A. Relaxation Modes in Emulsions

Depending on the type of emulsion, the time-dependent processes vary to a large extent. In oil-in-water emulsions, one usually employs an ionic surfactant as the stabilizer of the oil droplets. The basic stabilization is electrostatic where the double layers of each droplet will overlap and provide protection. However, the ionic processes in the Stern layer and the diffuse layer are also the background to the experimentally well-known phenomenon that suspensions of particles in an electrolyte medium display a strong low-frequency dielectric dispersion. Pioneering work in this field has been carried out by Schwan et al. [27,28]. Important theoretical contributions have been given by Overbeek [29,30], Schwarz [31], Dukhin and Shilov [32], and Lyklema et al. [33–35].

The treatment of the polarization of the double layer is complex. Significant contributions originate from the counterions in the Stern layer and in the diffuse layer. A rough model for the system splits these mechanisms into a surface conductivity term and a freer diffusion term of the coions in the diffuse part. The relaxation time τ for very thin surface layers ($\kappa a \gg 1$, where κ is the reciprocal Debye length and a is a particle radius) is, according to Schwarz, proportional to a^2/D, where D is the diffusion coefficient of the counterions in the thin layer. This model neglects the influence of the diffuse layer, and Lyklema et al. [33–35] have accounted for different polarization situations in this layer in an extensive study. These studies show that the low-frequency dispersion tends to be due mainly to polarization in the diffuse part of the double layer and that the contribution of the bound counterions closest to the colloidal particle has a maximum at a charge density half of that at the surface. For fixed values of κa, the polarization of the diffuse layer increases more strongly with increasing surface charge densities than the polarization due to bound counterions in the Stern layer. For fixed surface charge densities, the polarization increases with κa.

An understanding of the relaxation mechanisms in the double layer is important, as it will provide insight into the ion-exchange kinetics among various parts of the double layer. In this way, dielectric spectroscopy may contribute to the interpretation of the dynamics of colloid particle interaction.

In water-in-oil emulsions, the Maxwell–Wagner–Sillars (MWS) equation predicts that dielectric relaxation due to an interfacial polarization will take place. The nature of this effect is that, depending on the frequency applied, the movement of the ions inside the aqueous droplet will give different responses to the field. At low frequencies, the situation inside the droplet will be that of permanent polarization (steady state) where the localization of the ions will be perceived as dependent on the applied outer, weak ac field. In a way the whole droplet will act as a dipole, hence increasing the polarization. At high frequencies, the random diffusion of the ions inside the aqueous cavity will not give rise to any permanent polarization and, hence, no permanent contribution to the permittivity is observed. Hence, the permittivity is substantially reduced. At intermediate frequencies, there will be a dielectric dispersion. According to the classical MWS theory, the relaxation time of such a process should obey

$$\tau = \epsilon_0 \frac{\epsilon_2 + (\epsilon_1 - \epsilon_2)(1 - \phi_1)A}{\sigma_2 + (\sigma_1 - \sigma_2)(1 - \phi_1)A}. \tag{13}$$

The level of τ and, hence, the localization of the dielectric dispersion is governed mainly by the conductivity, σ, in the denominator, together with the volume fraction disperse phase and the shape factor A. If the salinity level is low and the emulsion is stabilized by means of an ionic surfactant, there may be an overlap between the double-layer contributions and the MWS effect in the permittivity spectrum.

In our model emulsions, we have used nonionic surfactants and higher salinities, so that the relaxation process inside the aqueous droplets can be studied in the TDS region.

B. Some Special Features of the Model Functions

Theoretical dielectric spectra of emulsions may be calculated according to the Hanai equation by insertion of ϕ_1 and the complex permittivities ϵ_1^* and ϵ_2^* into Eq. (10). The theoretical spectra thus obtained show the features expected for emulsified systems; that is, in addition to the relaxations of the continuous and disperse phases, the Maxwell–Wagner–Sillars relaxation is displayed [12,36–38].

The effect of nonsphericity of the dispersed particles on the permittivity may be investigated by solving the Boyle and the Boned–Peyrelasse equations, that is Eqs. (11) or (12). For given ϕ_1, ϵ_1^*, and ϵ_2^*, Eqs. (11) and (12) yield different results when the shape factors are changed. The Boyle equation for parallel-oriented prolates ($A < \frac{1}{3}$) gives a higher low-frequency permittivity and a slower MWS relaxation as compared to the spherical case. As A is increased through $A = \frac{1}{3}$ and further to $A > \frac{1}{3}$ (oblates) the permittivity and relaxation time decrease to values below those of the spherical case. When the dispersed particles take on a random orientation we find a similar trend as for the parallel orientation when $A < \frac{1}{3}$; that is, the permittivities attain higher values and the relaxation time decreases as the deviation from sphericity becomes larger. However, due to the averaging over all directions, the permittivity and relaxation times for a given A is not affected to the same degree as in the case of parallel orientation. If A is increased from $\frac{1}{3}$, the permittivity and relaxation times increase. This behavior, which is contrasted by the decrease displayed for the parallel orientation, is also due to the averaging over all orientations.

C. Special Features for Real Emulsions Versus Monodisperse Model Systems

In the derivation of Eqs. (9)–(12), some basic assumptions on the systems are made. First, there is a demand that the dispersed particles be homogeneously distributed throughout the system. Second, any mutual interaction between the particles is neglected. The latter may be granted for Eq. (9) through the limited range of ϕ's for which Eq. (9) is valid. To arrive at Eqs. (10)–(12), an iterative procedure based on Eq. (9) is employed. For each iteration step, a small volume fraction $\Delta \phi$ of the dispersed particles is added to a system with an effective permittivity $\epsilon_{\text{eff}}(\phi)$, and no interaction between the particles already present (at volume fraction ϕ) and the particles being added is assumed.

In real emulsions, some of these demands may be fulfilled to some degree, but more often, deviations from the ideal models will be present. Because emulsions are thermodynamically unstable, one must also keep in mind that the state of the emulsion will change with time. The main processes in emulsions that lead to dielectric properties different from those predicted for ideal systems can be summarized as follows.

Gravity forces lead to sedimentation or creaming of the dispersed phase and, hence, a concentration gradient of the droplets through the system. The dielectric properties of such a system may be widely different from those of a system with an even droplet distribution [39].

New aggregates may be created due to the attractive forces between the emulsion droplets. This process is called flocculation, and the degree of flocculation in an emulsion plays an important role in the dielectric properties of the dispersed system [40–44]. The flocs represent sub-volumes where the concentration (or volume fraction) of the disperse phase is higher than the overall concentration. Because some of the continuous phase is incorporated in the floc structure, the effective volume of the flocs exceeds the volume of the disperse phase. The contribution to the total permittivity from the flocs will, in most cases, be larger than the sum of contributions from the individual droplets building up the flocs. The effect of flocculation is, thus, to increase the permittivity of the emulsions, especially at lower frequencies. This is illustrated in Table 1. The static permittivity of w/o emulsions ($\phi_{water} = 0.5$) was measured both during stirring (using two different devices) and at rest. The values of ϵ_s obtained during the stirring process closely resemble those predicted for a system with independent (or unflocculated) spherical droplets [Eq. (10)]. As soon as the stirring is stopped, ϵ_s increases by as much as 60% due to flocculation.

By calculating an average permittivity of the flocs and replacing the volume fractions of the disperse phase with the volume fraction of the

TABLE 1 The Static Permittivity, ϵ_s, of w/o Emulsions During Stirring and At Rest

Stirring device	Rotation frequency	ϵ_s During stirring	At rest
Emulsor screen Head	Maximum ($\sim$ 2500 rpm)	15.6	24.2
	Minimum ($\sim$ 500 rpm)	16.9	20.9
Vortex mixer		15.4	23.9

Note: The emulsions contained 50% (by volume) water and were stabilized by 1% NP-4.

flocs, Eqs. (10)–(12) may be used to estimate the permittivity of floccu-lated emulsions [32,41].

Flocculation may be followed by a coalescence (rupture of the thin film separating the droplets) of the droplets. This process alters the properties of the floc and, eventually, the dielectric properties of the total system.

In the following sections we will analyze flocculation and coalescence in more detail.

D. The Influence of Flocculation on Dielectric Parameters

The permittivity of flocculated emulsions, ϵ_{em}^*, can be expressed as a function of the volume fraction disperse phase, the permittivity of the individual phases, and a shape factor describing the flocs; that is,

$$\epsilon_{em}^* = f(\phi_1, \epsilon_1^*, \epsilon_2^*, A_f). \tag{14}$$

The approach depicted in Eq. (14) gives a very simplified view of the dielectric properties of an emulsion. Here it is assumed that any dissimilar-ity between the permittivity of a true emulsion and the permittivity calcu-lated from, for instance, Eq. (10) (valid for spherical-independent droplets) can be accounted for by a shape factor, A_f. This shape factor, however, cannot be interpreted as due to an asymmetric extension of the individual droplets but by the formation of new constellation of droplets, where a ratio between a minor and a major axis can be introduced. This situation is viewed in Fig. 3. This model corresponds to the formation of rather

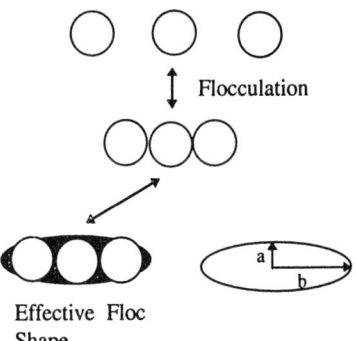

Effective Floc
Shape

FIG. 3 Schematic visualization of linear flocculation. The flocculated aggregates can be characterized by the shape factor A, which is related to the axial ratio $a:b$.

small linear flocs which can be called primary flocs. Primary flocs can occur in emulsions during the preparation when vigorous mixing is applied. If so, this situation will not explain the level of the flocculation after mixing when the droplets have a free Brownian movement. Under these conditions, larger secondary flocs will occur and their properties will now dictate the dielectric properties of the w/o emulsion. In order to gain more information on the floc structures, we have adopted a concept suggested by Dukhin and Shilov [32]. By defining the volume fraction occupied by flocculated aggregates, ϕ_f, as

$$\phi_f = \frac{\phi}{\phi_{df}}, \tag{15}$$

where ϕ_{df} is the volume fraction occupied by the droplets in the floc, the permittivity of the floc, ϵ_f, can be introduced as

$$\epsilon_f^* = f(\phi_{df}, \epsilon_1^*, \epsilon_2^*). \tag{16}$$

Thus, the total permittivity for the flocculated emulsion depends on the degree of flocculation [from Eqs. (15) and (16)]; that is,

$$\epsilon_{em}^* = f(\phi_f, \epsilon_f^*, \epsilon_2^*). \tag{17}$$

In Ref. 41, we generalized this model by also taking into account the shape of the flocs and of the droplets in the flocs, that is,

$$\epsilon_f^* = f(\phi_{df}, \epsilon_1^*, \epsilon_2^*, A_{df}) \tag{18}$$

and

$$\epsilon_{em}^* = f(\phi_f, \epsilon_f^*, \epsilon_2^*, A_f), \tag{19}$$

where A_{df} and A_f are the shape factors for the droplets in the flocs and for the floc, respectively.

The complex permittivity ϵ_f^* and the resulting ϵ_{em}^* is calculated by numerically solving, for instance, Eq. (12).

E. Dielectric Spectroscopy on Flocculated w/o Emulsions

Numerous articles on the dielectric properties of w/o emulsions have been published. However, in the main part of those investigations, the emulsions have been used as models for heterogeneous systems or just as a matrix for finely dispersed droplets [3]. Only in a few cases have the processes in or the structure of the emulsion as such been of main interest [1,2,45–47]. In recent articles, we have investigated the dielectric properties of several w/o emulsions by using the TDS technique. We have evaluated the feasibility of this technique as an instrument for probing the structural properties of the emulsions [41–44].

A common trend in all these investigations is that the static permittivity of the emulsions is much higher than would be expected from an analysis of Eq. (10). Also, the relaxation times deviate from the calculated values. Thus, a theoretical model based on spherical-independent droplets does not seem to be appropriate in these cases. In order to interpret the experimental data, several approaches have been made. In all these approaches we assume that the deviation from Eq. (10) can be traced to the flocculation of the emulsions. Our challenge has, thus, been to incorporate the flocculated state into the existing dielectric models on heterogeneous mixtures.

In Ref. 43, the dielectric properties of water-in-triglyceride emulsions were measured. The emulsions were stabilized by monoglycerides. The constituents of these emulsions mimic those of low-fat margarines. In order to explain the high measured permittivities, an effective floc shape was introduced (Fig. 3). By assuming that the flocs take on a more or less spheroidal shape, the permittivities were calculated from Eq. (11) with the shape factor as an adjustable parameter. In this manner, the measured permittivities could be explained if the flocs were spheroids with axial ratios ranging from 10 to 15. The relaxation times calculated based on the same assumptions were, however, not in good agreement with the experimental values. In this simple approach, only the static permittivity was utilized when extracting the flocculation parameters. Thus, any changes in the volume fraction disperse phase (i.e., taking into account the volume of continuous phase inside the flocs) were neglected.

In our series on water-in-crude-oil emulsions from the Norwegian continental shelf [42,48,49], one of the articles gives a dielectric spectroscopic characterization of authentic as well as model emulsions [42]. In [42], a similar model as that above for the flocculated aggregates is adopted. However, to find the shape factors of the aggregates, the Boned–Peyrelasse equation [Eq. (12)] was employed. In addition, not only the static permittivity but also the permittivity at high frequencies and the relaxation times were modeled. There was a remarkably good internal agreement among the shape factors that were extracted from the different dielectric parameters (Table 2). Basically, the flocs were characterized by axial ratios in the range from 1:3 to 1:10. Another interesting result from this investigation was the effect of the addition of hexylamine to the emulsions. Hexylamine is known to promote coalescence of the emulsion droplets. This coalescence will, as mentioned earlier, alter the dielectric properties

TABLE 2 Experimental Dielectric Parameters for Different Crude-Oil Emulsions Together with the Shape Factors That Give the Same Dielectric Parameter When Inserted in Eq. (12)

	Experimental values			Oblate ellipsoids $(0 < A < \frac{1}{3})$			Prolate ellipsoids $(\frac{1}{3} < A < \frac{1}{2})$		
ϕ_{water}	ϵ_s	ϵ_∞	τ (ns)	A^{ϵ_s}	A^{ϵ_∞}	A^τ	A^{ϵ_s}	A^{ϵ_∞}	A^τ
0.17	17.5	6.5	0.8	0.06	0.06	0.06	0.48	0.49	0.48
0.23	22.4	7.4	1.0	0.08	0.09	0.07	0.48	0.48	0.48
0.28	28.2	9.6	1.0	0.09	0.09	0.08	0.47	0.48	0.47
0.38	35.7	13.2	1.3	0.12	0.12	0.10	0.46	0.47	0.47
0.44	46.0	13.5	1.2	0.14	0.14	0.12	0.46	0.45	0.46
0.50	60.0	14.0	1.3	0.15	0.16	0.13	0.45	0.43	0.46

of the emulsion. As the amount of hexylamine was increased a considerably lower permittivity of the emulsions was measured (Table 3).

This first rather simple approach to analyze flocculated emulsions shows that frequency-dependent dielectric techniques can be developed with success for describing ongoing processes in emulsions. However,

TABLE 3 The Influence of Hexylamine on the Dielectric Parameters of w/o Emulsions

System	Conc. hexylamine (vol%)	ϵ_{st}	ϵ_{inf}	τ (ns)
NP-4	0	29.5	13.9	0.3
	1	25.0	13.9	0.2
	3	19.9	12.4	0.1
	5	18.8	12.2	0.1
Crude oil	0	31.0	11.3	0.3
	0.5	20.5	8.8	0.2
	1.0	20.2	8.4	0.2
	1.5	15.0	6.6	0.2
	2.0	12.3	5.7	0.2
	2.5	14.0	6.5	0.2

Note: The volume fraction water (5% NaCl solution) was 0.5. The emulsions are stabilized either by NP-4 or prepared from a crude oil.

for a detailed investigation of the emulsion state, the models should be refined. To explain the permittivity of a flocculated emulsion only by a single shape factor is an oversimplification. A polydispersity in floc sizes and shapes should be accounted for by a distribution of shape factors, A. In this way, a realistic distribution of droplets in a single state as well as in flocs of different sizes can be modeled. However, before this can be accomplished, reliable models for the Maxwell–Wagner-based relaxation time must be established.

As noted, additives may promote coalescence, and thus change the dielectric properties of the flocculated structures. There are also other means to change these structures. By applying a shear force, the flocs may be torn apart and eventually be totally disintegrated. To investigate this effect, a small flow system was set up, in which it was possible to record dielectric spectra of emulsions both at rest and during flow (Fig. 4). As seen from Fig. 5 the permittivities (both at low and high frequencies)

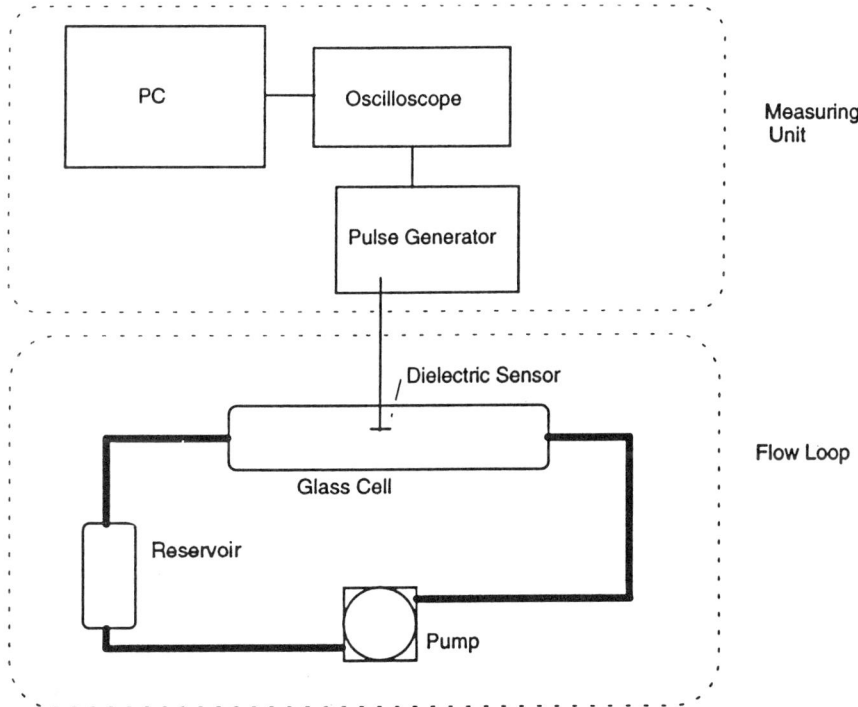

FIG. 4 Setup for the dielectric flow measurements.

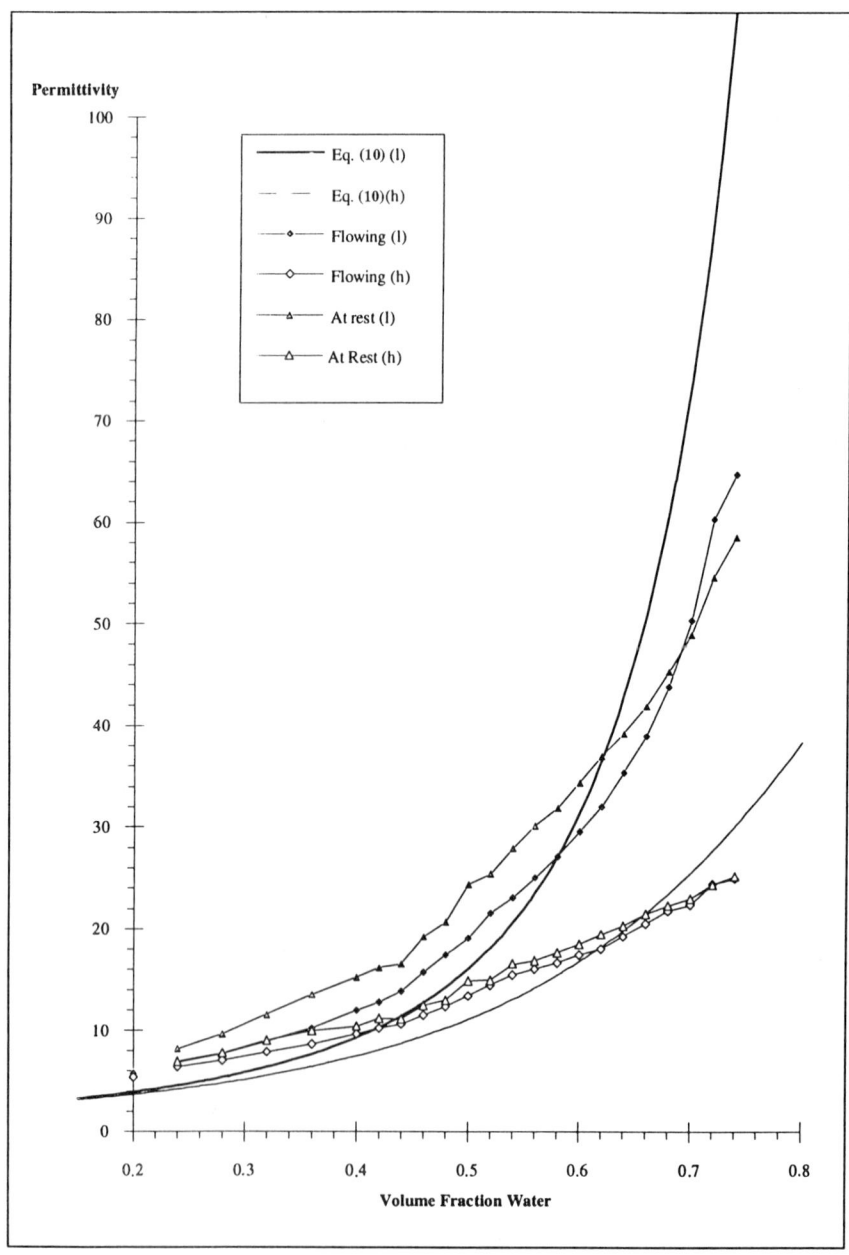

FIG. 5 The static [denoted (1)] and high frequency [denoted (h)] permittivities of w/o emulsions during flow and at rest versus volume fraction water (4% NaCl). The emulsions are stabilized by NP-4 (2% by volume in the oil phase). The permittivities calculated from Eq. (10) are also included.

decreased as the emulsions were subject to flow conditions. Even when the emulsions were flowing, the permittivities were higher than those predicted for an emulsion with individual spherical droplets. In this case, the applied forces under flow were not high enough to cause a complete disintegration of the flocculated state. Hence, the estimations of the floc sizes will suffer from a poorly defined reference state and are thus not very precise.

V. COALESCENCE

A commonly used method to break water-in-oil (especially water-in-crude-oil) emulsions is to apply an electric field over the emulsion. The principle behind the electrically induced coalescence is that the electric field acts to enlarge the droplets of the disperse phase by inducing coalescence. After this initial step, the water phase settles out under gravity. There is a variety of factors influencing the electrically induced coalescence, such as the dielectric properties of the disperse and the continuous phase, the volume fraction of the dispersed phase, conductivity, size distribution of the dispersed droplets, electrode geometry, electric field intensity, the nature of the electric field (ac or dc, uniform or nonuniform), and so forth [50–52]. The breakdown process is connected with a rupturing of the stabilizing bilayer surrounding the interacting aqueous droplets. As such, this process is difficult to study, but it has both important scientific and technological aspects. In biotechnology, this method has been used in order to create a broader understanding of cell fusion and cell hybridization. In the oil industry, it is common to rupture water-in-crude-oil emulsions by applying a high external ac or dc electric field [53].

The application of high electric fields to emulsified systems has been utilized since the pioneering work initiated by Cottrell in 1907 [54]. The first commercial electrical treaters were established at a production field in California in 1909, and Cottrell's first patents were granted in 1911 [55,56] Ever since, numerous improvements have been presented, resulting in a manifold of patents.

In order to investigate the electrical-induced breakdown of emulsions, several different techniques have been used, such as conductivity [57], light scattering [58], different microscopy techniques [59–61], dielectrophoretic [62], Coulter counter [63], capacitance [64], and different electrical measurements [65,66]. Much effort has been focused on the design of new coalescers, in order to obtain optimal conditions for the resolution of the dispersed water with respect to the electric field, such as pulse frequency and electric field intensity. Although many authors report on the demulsification of emulsions utilizing the electrostatic coalescence method, only a limited number of articles are related to the emulsion characteristics and the processes taking place at the interfaces. However, more recent publications have contributed to a deeper knowledge about the interfacial properties of the membranes separating the water domains and the actual processes involved in breaking the oil-continuous emulsion in electric fields [57,60,67–70].

From microscopy investigations and examinations of the forces acting on conducting spheres in a convergent electric field, Pearce suggested two effects as possible mechanisms for the electrical-induced coalescence of w/o emulsions [61]. The first is due to attractive forces which originate from the potential difference between the droplets, aligning the droplets into chains, and the second is the electrical breakdown of the thin oil film separating the droplets.

Brown and Hanson [71] studied the effect of oscillating electric fields on the coalescence of single water droplets at water–oil interfaces. It was observed that high-frequency oscillating fields promoted the coalescence of the water droplets. They also observed that at a critical electric field of 0.39 kV cm^{-1} the coalescence became instantaneous.

Williams and Bailey used laser light scattering to determine the changes in the droplet size distribution [58], during electrostatic resolution of w/o emulsions. This method has a limitation with regard to water content, but the technique exposed an increase in the mean droplet size.

Taylor reports on the resolution of water from water-in-crude-oil emulsions [57,60], using conduction current profiles and optical microscopy. The coalescence behavior is highly dependent on the properties of the interfacial membranes, and two distinct types of coalescence were identified. One where incompressible interfacial films led to an increase in the emulsion conductivity due to droplet chain formation, and one where mobile interfacial films led to low emulsion conductivity due to immediate droplet–droplet coalescence. Computer simulations reveal agreement between the microscopic observations and the simulations, based on dipole induction [70].

Hirato et al. conducted a fundamental study in which factors affecting the demulsification rate of w/o emulsions in high electric fields were studied in order to recycle the organic phase recovered after an extraction procedure [72]. The influence of important extraction parameters such as the level of the electric field, phase ratios, organic impurities, amount of surfactant, and electrolyte concentration on the demulsification rate were examined. They observed that an increase in the applied voltage, a reduction in amount of surfactant, an increase in electrolyte concentration, and an increase in the amount of dispersed phase all improved the demulsification rate.

A. Dielectric Spectroscopy at High Electric Fields

Dielectric spectroscopy has for some time proved to be a promising experimental technique for investigating w/o emulsions. Recently, we reported on dielectric investigations of w/o emulsions in high external electric fields [73,74] in order to systematically study the coalescence step, characterized by the MWS dispersion, as presented in Fig. 6.

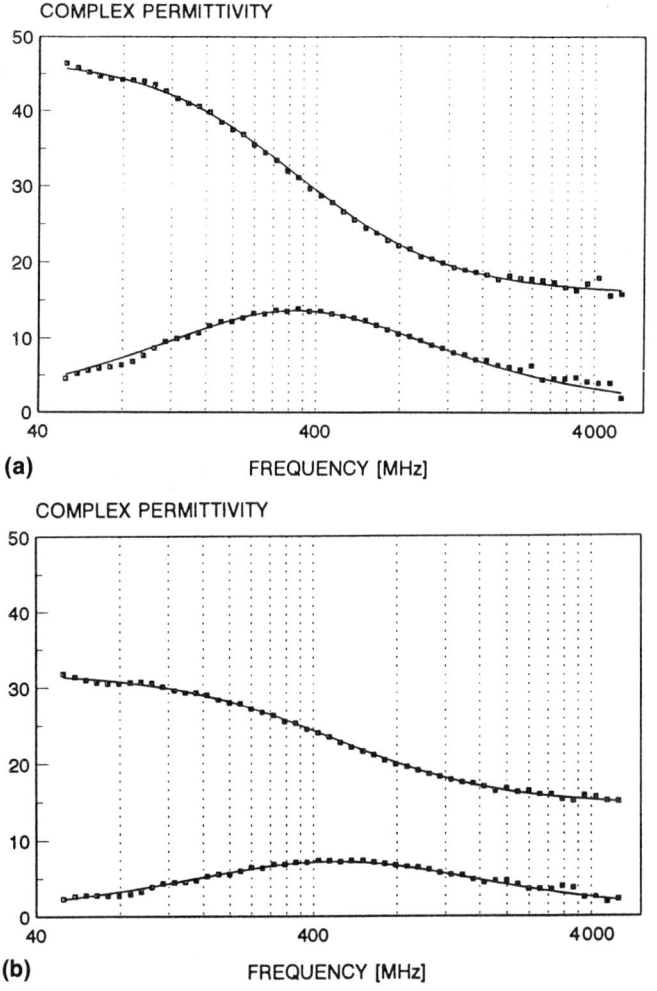

FIG. 6 Dielectric spectra of a w/o emulsion at different electric fields: 0 kV cm^{-1} cm (a), 0.32 kV cm^{-1} (b), and 0.56 kV cm^{-1} (c). The full solid lines represent a Cole–Cole model function.

COMPLEX PERMITTIVITY

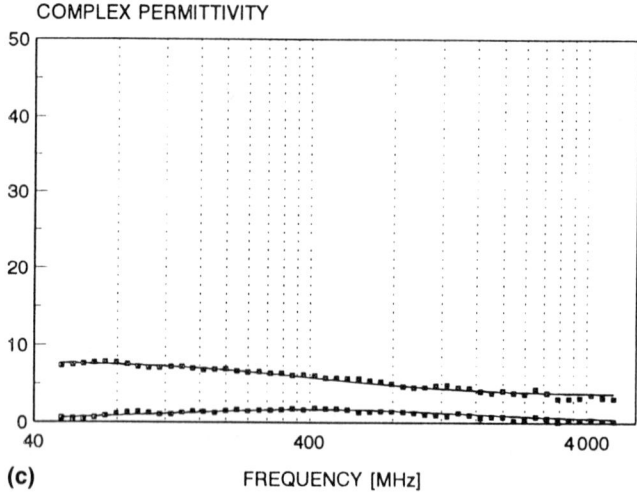

(c)   FREQUENCY [MHz]

FIG. 6 Continued.

Figure 6 shows the dielectric spectra of a w/o emulsion at three different electric dc fields. As an electric field of 0.33 kV cm^{-1} is applied to this emulsion, both the static permittivity as well as the increment decreases about 15 units. A further increase in the electric field to 0.56 kV cm^{-1} results in the static permittivity diminishing to 7 and the dielectric dispersion hardly discernible.

Figure 7 shows the time dependence of the dielectric parameters after the application of the electric field. An initial decrease is followed by a leveling out, and the final level of the plateau varies, depending on the magnitude of the electric field. The dielectric increment and the relaxation time show the same trend as the static permittivity. However, if the electric field is switched off when the static permittivity has leveled out, a reversibility of the permittivity is observed. The permittivity levels out at a value which can somewhat exceed the initial value of the permittivity, as seen in Fig. 8.

The changes in the dielectric parameters, as a high electric field is applied, are directly correlated to changes taking place in the measured sample. The effect of the electric field as a function of time after a high electric field is applied to the emulsified system clearly demonstrates that two distinct different behaviors are observed. At low electric fields, the

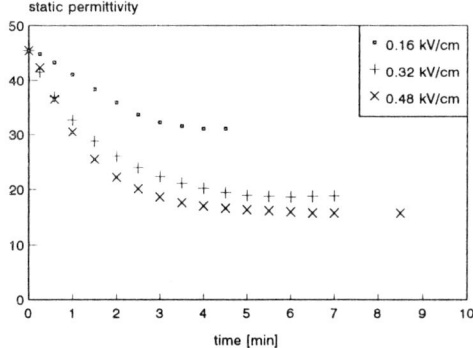

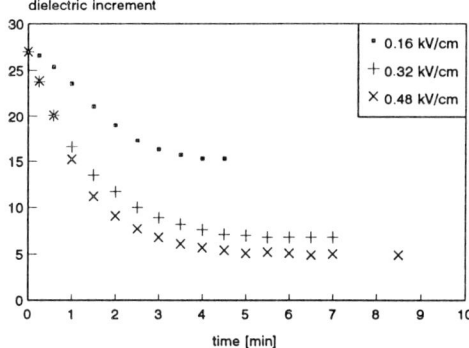

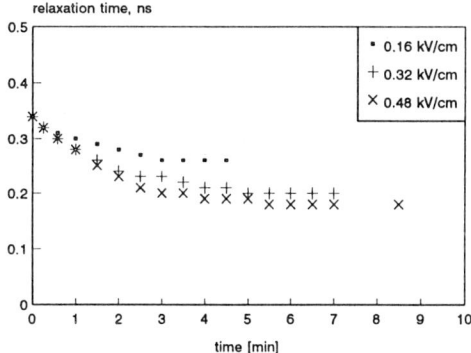

FIG. 7 The dielectric parameters, static permittivity (top), dielectric increment (center), and relaxation time in nanoseconds (bottom), as a function of the time after applying the external field.

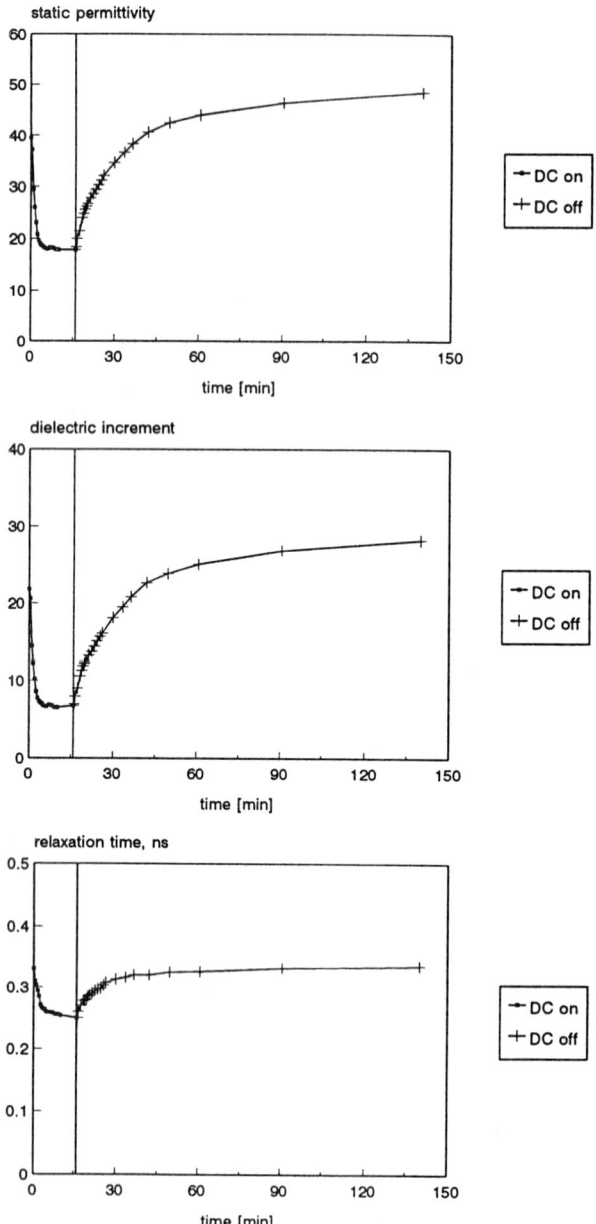

FIG. 8 The dielectric parameters static permittivity (top), dielectric increment (center), and relaxation time in nanoseconds (bottom), as a function of the time after applying the external field. The dc field is turned off after 15 min.

experimental findings can be explained by means of reversible mechanisms. However, as the value of the electric field increases, a breakthrough of the macroscopic conductivity occurs and an irreversible destabilization process takes place.

When a high electric field is applied to a w/o emulsion, the ions inside the emulsion droplets tend to orient themselves in the direction of the electric field. This can result in a droplet–chain formation between the electrodes, as the maximum value of the potential drop over the separating film will be found if the droplets orient themselves parallel to the electric field. At these low field strengths, the applied field is not high enough to induce coalescence. When the electric field is switched off, the system returns to a random distribution of the aqueous droplets, which explains the reversibility of the dielectric parameters. However, when the external electric field exceeds a critical value, the membrane protecting the aqueous droplets will rupture and a coalescence of the droplets will take place, giving rise to an irreversible process.

For w/o emulsions in the precoalesced state, the droplets no longer have a point contact, but they are separated by a thin film which is built up by both the monolayers from the aqueous droplets and some oil from the continuous phase. An exact thickness of the interdroplet distance cannot be given, but a good estimate is twice the length of the fully extended hydrocarbon chain of the surfactant molecule, that is, $\sim$40 Å. If the value of the potential drop over the stabilizing film is of the order of 0.5–1 V, the field in the film will be of the order of 10^3 kV cm^{-1}. At field strengths of this magnitude, different effects are produced over the thin membrane. However, in our emulsions we have noncharged monolayers separating the aqueous droplets which contain electrolyte. Hence, the monolayer, as such, contains no charges, only interacting dipoles. In addition, there might occur an ionic flux through the membrane, promoting the breakdown. As a consequence, the whole emulsified system will collapse and the measuring cell will finally be filled with two separated layers: one aqueous and one hydrocarbon. The total permittivity of this state can be calculated by a series coupling of capacitances. An estimate of Φ_{H_2O} = 0.65 and Φ_{oil} = 0.35 gives a value of 4–5 for the static permittivity, which is in good agreement with the value from Fig. 7. Discrepancies between the theoretical value for this final state and the experimentally obtained one can be due to the occurrence of a residual emulsion in the sample.

The behavior of concentrated emulsions in an external electric field is of course more complex. In this case, one can imagine a situation as described in Fig. 9. The figure describes the breakdown process as measured by means of static permittivity. This process seems to be rather gradual and monotonous. There are several underlying reasons for this.

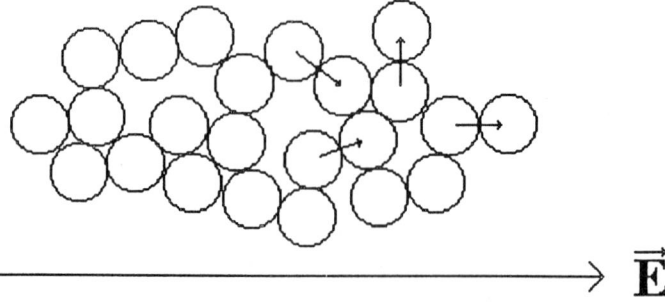

$$\longrightarrow \vec{E}$$

FIG. 9 Schematic illustration of the angular field dependence in a flocculated emulsion.

First, the membrane potential will have an angular dependence, for which the highest probability for coalescence is for droplets with contact points parallel with the applied field. The gradual coalescence will cause a redistribution of droplets and symmetries leading to a lower permittivity as calculated from the Boned–Peyrelasse equation. In addition to this effect, the membrane potential is also directly proportional to the size of the droplet. In the emulsions investigated, the droplet size will vary from 0.1 to 10 μm. At lower fields, a coalescence of the largest droplets will be promoted due to the highest membrane potentials. When the applied voltage is gradually increased, all droplets will sooner or later coalesce and a complete separation of water and oil will occur.

B. Multivariate Analysis of w/o Emulsions in High Electric Fields

With the dielectric instrumentation presented in Fig. 1 the emulsion stability can be precisely characterized and quantified. This will open possibilities of analyzing both different surfactant stabilizers as well as the behavior of emulsions in different processes.

In a real process where the emulsions are causing problems, as in the exploitation or separation stage in offshore petroleum production, a crucial point is what operational parameters can be adjusted without too large modifications of the whole process. Under offshore conditions, normal parameters in a situation like this are pressure, water cut, salinity, temperature, gas : oil ratio, and so forth. In order to model the behavior of a w/o emulsion under operational conditions, we undertook a multivariate study in which, we defined operational conditions as variables and emulsion stability (critical voltage) as response surface.

In our design, seven variables are investigated and a complete characterization of these variables would require $2^7 = 128$ experiments. However, in accordance with a classical two-factor design, as the number of variables increases, the desired information can often be obtained by performing only a fraction of the full factorial design, because higher-order interactions tend to become negligible and usually have no effect at all [75].

A 2^{7-3} reduced factorial design is used, which reduces the number of experiments from 128 to 16. The design is obtained by setting the level of the variables 5, 6, and 7 as combinations of levels of variables 1, 2, 3, and 4 [75]:

$$X_5 = X_1 X_2 X_3, \qquad X_6 = X_2 X_3 X_4, \qquad X_7 = X_1 X_3 X_4. \qquad (20)$$

Variables 1, 2, 3, and 4 are varied to include all combinations of the levels, whereas the levels of 5, 6, and 7 are determined from the generators above. Table 4 lists the variables chosen and the values of the high and low levels as well as the center points. The complete matrix for the variation of the variables is shown in Table 5. By using the critical electric field as a response variable, and using all variables, including cross-terms in the regression, a complete response model is obtained:

$$\hat{Y} = 0.96 - 0.410 X_1 + 0.255 X_2 - 0.145 X_3 - 0.100 X_4 + 0.460 X_5$$
$$+ 0.130 X_6 - 0.095 X_7 - 0.235 X_1 X_2 + 0.055 X_1 X_3 + 0.08 X_1 X_4 \qquad (21)$$
$$- 0.240 X_1 X_5 - 0.080 X_1 X_6 + 0.075 X_1 X_7 + 0.065 X_2 X_4.$$

TABLE 4 Defined Variables and Their Intervals Used in the Multivariate Analysis

Variables		Levels		Center points
		$-$	$+$	
wt% disperse phase	X_1	40	80	60
Salinity wt% NaCl	X_2	1	4	2.5
Temperature (°C)	X_3	4	23	13.5
Time (h)	X_4	0.25	24	12
Surfactant (wt%)	X_5	2	4	3
Ratio, surfactant ($S : T$)	X_6	0.25	0.75	0.5
Laponite, wt% of Φ_2	X_7	0.5	2.0	1.25

TABLE 5 2^{7-3} Factorial Design with Cross-Terms and Response

	X_1	X_2	X_3	X_4	X_5	X_6	X_7	X_1 X_2	X_1 X_3	X_1 X_4	X_1 X_5	X_1 X_6	X_1 X_7	X_2 X_4	Y (kV cm^{-1})
1	−	−	−	−	−	−	−	+	+	+	+	+	+	+	0.64
2	+	−	−	−	+	−	+	−	−	−	+	−	+	+	0.80
3	−	+	−	−	+	+	−	−	+	+	−	−	+	−	3.20
4	+	+	−	−	−	+	+	+	−	−	−	+	+	−	0.48
5	−	−	+	−	+	+	+	+	−	+	−	−	−	+	1.72
6	+	−	+	−	−	+	−	−	+	−	−	+	−	+	0.32
7	−	+	+	−	−	−	+	−	−	+	+	+	−	−	0.64
8	+	+	+	−	+	−	−	+	+	−	+	−	−	−	0.68
9	−	−	−	+	−	+	+	+	+	−	+	−	−	−	0.12
10	+	−	−	+	+	+	−	−	−	+	+	+	−	−	0.88
11	−	+	−	+	+	−	+	−	+	−	−	+	−	+	2.32
12	+	+	−	+	−	−	−	+	−	+	−	−	−	+	0.40
13	−	−	+	+	+	−	−	+	−	−	−	+	+.	−	1.04
14	+	−	+	+	−	−	+	−	+	+	−	−	+	−	0.12
15	−	+	+	+	−	+	−	−	−	−	+	−	+	+	1.28
16	+	+	+	+	+	+	+	+	+	+	+	+	+	+	0.72
17	0	0	0	0	0	0	0	0	0	0	0	0	0	0	1.08
18	0	0	0	0	0	0	0	0	0	0	0	0	0	0	1.08
19	0	0	0	0	0	0	0	0	0	0	0	0	0	0	1.04

Figure 10 shows the regression curve from the regression analysis. In Fig. 11, the regression coefficients are presented as histograms for internal comparison. The negative regression coefficients reduce the value of the critical electric field, whereas the positive ones increase the value of the critical electric field. A model based on main variables only will fail to predict the critical electric fields; hence, a model that takes into account interaction terms must be proposed. The prediction of the center samples shows that the model is slightly nonlinear. The bias helps us to extract the significant coefficients to be used in a revised model, where the significant coefficients are restricted to X_1, X_2, X_5, X_1X_2, and X_1X_5, giving the model

$$\hat{Y} = 0.96 - 0.410\,X_1 + 0.255\,X_2 + 0.460\,X_5 - 0.235\,X_1X_2$$
$$- 0.240\,X_1X_5. \tag{22}$$

Figure 12 shows a revised model based on three main variables and two interaction terms. This model explains 86.7% of the variation in the dependent variable, and the degrees of freedom increases from 1 to 10.

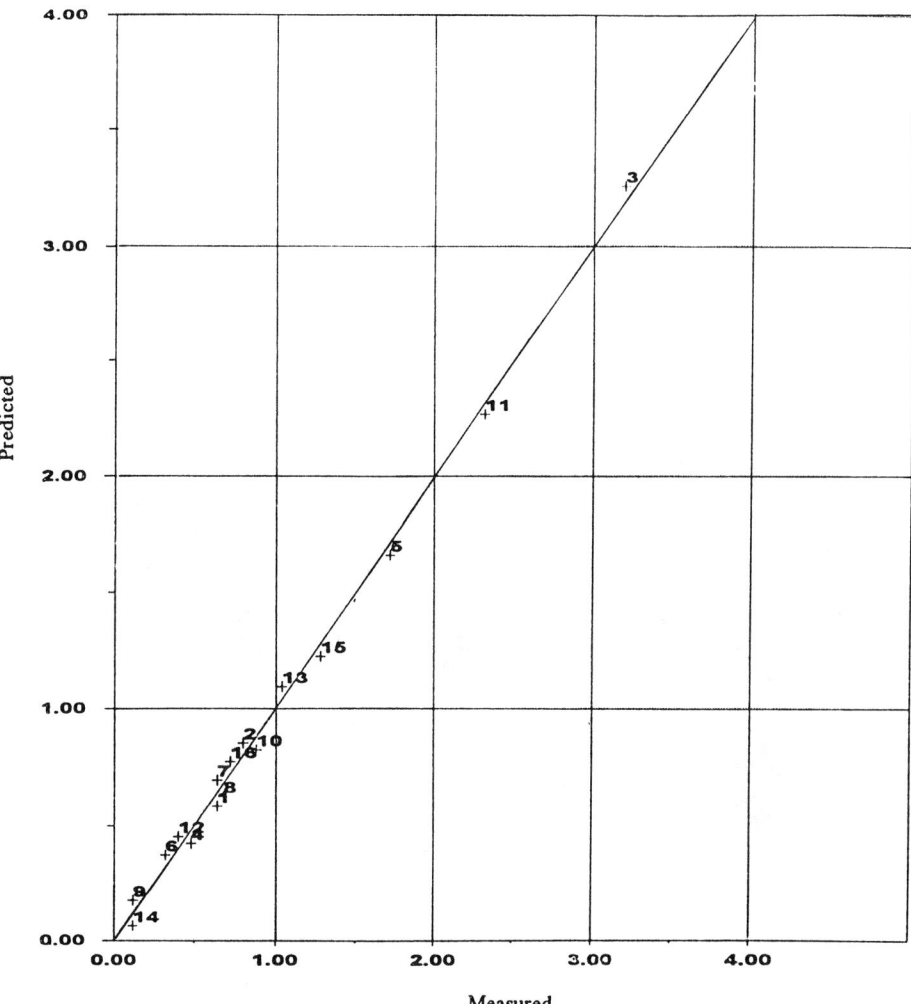

FIG. 10 Measured versus predicted values of the critical electric field needed to break the model emulsions. The solid line represents the fitted multivariate regression curve. Its mathematical expression is shown in Eq. (21).

In Ref. 62, all variables are examined and their importance for the stabilization/destabilisation of the emulsions is discussed.

In Fig. 13 the three most important variables are presented in a graph. The eight corners represent all possible combinations of the levels of these variables. The values in the corners are the mean value of the electric

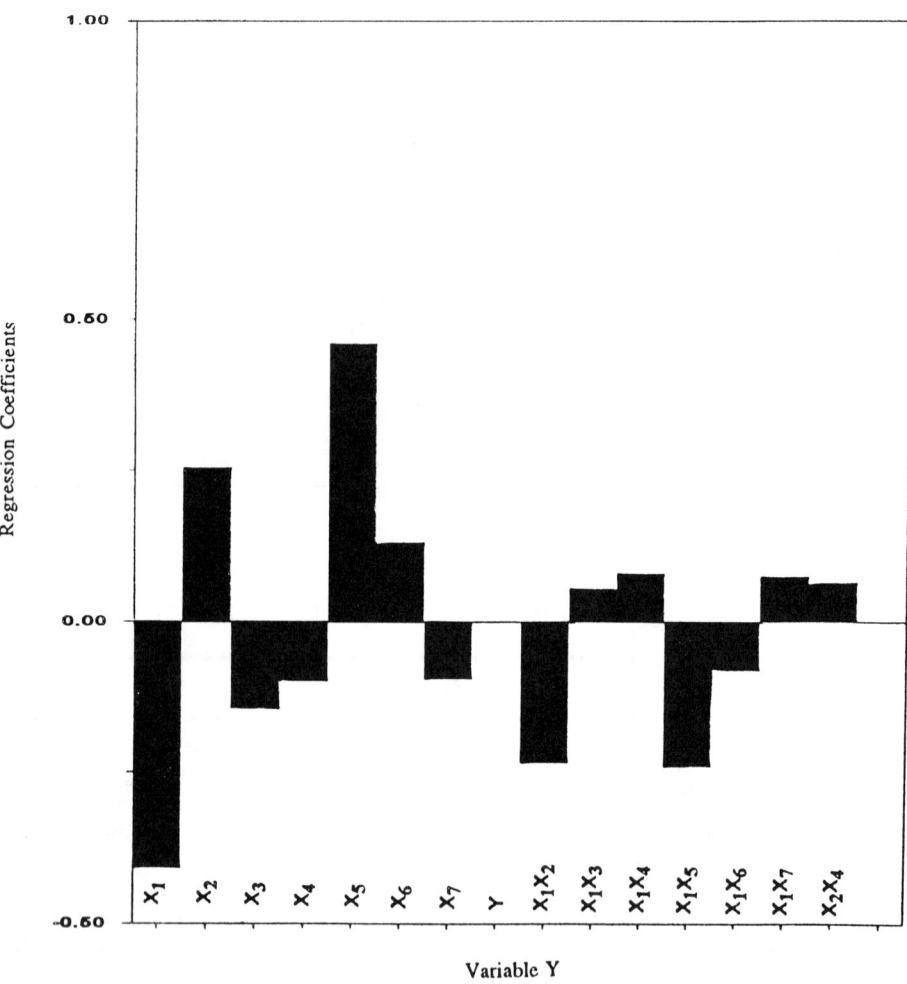

FIG. 11 The regression coefficients from Eq. (21). The negative values tend to reduce the critical electric field needed to break the emulsions, whereas the positive values increase it.

field, as there will be two measurements with identical sets of levels for X_1, X_2, and X_5. The lowest value of the electric field is measured as X_1 is high and X_2 and X_5 are low. The highest value of the electric field is in the diametrical opposed corner (e.g., as X_1 is at a low level and X_2 and X_5 are high).

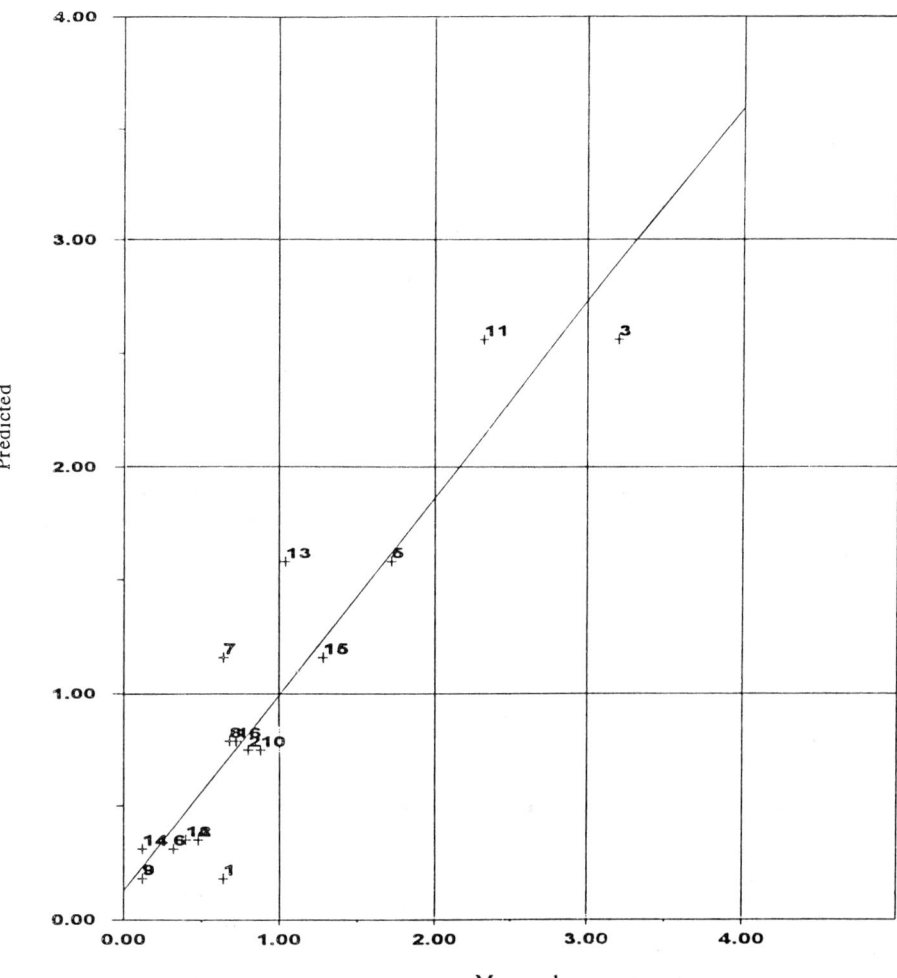

FIG. 12 A revised model based on three main variables and two interaction terms. Otherwise, the same notation as in Fig. 10. The mathematical expression of this model is given in Eq. (22).

The main purpose of such an investigation is to use reduced factorial design as a screening technique to extract information about the general trends of the variables investigated. All our findings are supported by previously published works [50,72,76–78]. However, our conclusions are, of course, limited to the setup of variables and experimental levels in

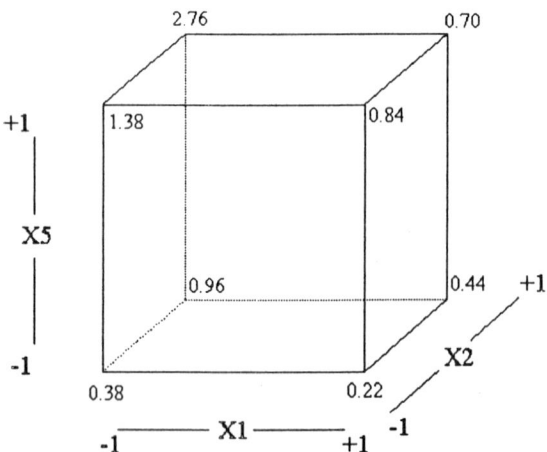

FIG. 13 A graphical presentation of the three most important variables. The highest and the lowest values of the electric field are observed in the diametrically opposed corners.

these. Any significance of an extrapolation outside this set cannot be justified by the present model.

C. Water-in-Crude-Oil Emulsions

In the previous sections, we have shown that the critical electric field where the breakthrough of the macroscopic conductivity occurs is a good quantitative parameter on emulsion stability. In the following, we extend the study to include water-in-crude-oil emulsions, and model emulsions stabilized by interfacially active fractions extracted from different crude oils.

Water-in-crude-oil emulsions are a special case of water-in-oil emulsions. In the exploitation of crude oil onshore or offshore, these emulsions can give rise to severe technical problems. In the North Sea, the hydrocarbon reserves include a variety of marginal fields which are either small or deep water fields. An efficient exploitation of these fields requires lower-cost options in subsea production or alternatively centralized or land-based processing units involving offshore platforms with low processing levels of the fluids. As a consequence, one can anticipate a transport in pipelines of unprocessed or minimum-processed fluids over considerable offshore subsea distances. In this multiflow transport pattern, water-in-crude-oil emulsions can cause severe problems by distancing the flow pattern of the aqueous, oil, and gas phases. These emulsions can be formed due to a normal turbulence in the pipes during flow. The existence of joints and valves in these pipe lines will further benefit a formation of emulsions under flow. However, at platforms, one encounters directly the problem with produced water-in-crude-oil emulsions from the wells. These emulsions consist of either formation water or injection water combined with the crude. When the production line is brought over chokes with a pressure gradient, the emulsions are formed. Also, for other reasons, it is of interest to dehydrate the crude oils. Water can under the right conditions form gas hydrates and is also responsible for scale and corrosion problems. A stable water-in-crude-oil emulsion also increases the viscosity of the emulsified system tremendously in comparison with the neat crude, which increases the pumping costs.

Crude oils are mixtures of numerous aliphatic and aromatic hydrocarbons, and oxygen, nitrogen, and sulphuric compounds. Some of these compounds are surface active in nature; that is, they can adsorb to water–crude interfaces and stabilize the system against coalescence. Two such classes of compounds are asphaltenes and resins. The nature and the relative amount of asphaltenes and resins from a crude vary, depending on the origin of the crude. The differences depend on several factors such as the nature of the source rock, the thermal evolution of the sedimentary organic material, the migration processes, and the efficiency of the petroleum trap [79]. Some production fields contain oil with a high degree of light components, whereas other fields contain a high content of heavy fractions, waxes, and sand, depending on the combinations of all the factors listed above.

In Bergen, we have during the past years developed procedures to separate and isolate different kinds of interfacially active components from different categories of crude oils. The basic idea behind the separation of the interfacially active fractions is to simulate interfacial conditions of relevance for true crude-oil systems.

The occurrence of wax particles is revealed by means of microscopy studies. Hence, the film built up by indigenous components should consist of nonspecified, nonpolar components, waxes (as lipids), and asphaltenes. Small wax particles are also incorporated into the film. However, inorganic particles such as clays have not been found in our samples [80]. Previous studies on the pH dependence of the interfacial films have revealed a low amount of ionizable and polar groups in the interfacial film.

In order to undertake more detailed studies of the stabilizing film, we have separated the components responsible for the emulsion stability [48]. Chemical analysis reveals the fractions to contain relatively low-molecular-weight, highly polydisperse, hydrocarbon polymers. The carbon:hydrogen ratio and infrared data [81] indicate the polymers to contain considerable amounts of aromatic groups. They can be dissolved in nonpolar solvents but not in water.

Interfacial tensions [82] and monolayer studies [83,84] show that the components in the film are not extremely surface active. Their surface pressures do not exceed 15 mN m^{-1} before film fracture and the minimal interfacial tension between crude oil and water is 25–30 mN m^{-1}. In a previous study when only asphaltenes were analyzed, we concluded that high interfacial pressure, π, was a necessity for stabilizing model water-in-oil emulsions by means of an asphaltenic fraction. A value of π of the order of 10 mN m^{-1} or less is insufficient in order to achieve stable emulsions. However, our previous studies also reveal that the nature of the organic phase is very decisive for the solution chemistry of these fractions. When benzene is added in sufficient quantities, the interfacial tension between the oil and the water phase is depressed; hence, the originally interfacially active fractions will no longer accumulate at the w/o interface. Hence, the whole emulsified system loses its stability.

Interfacial tensions of aqueous/organic phases where the latter contains the asphaltenes have been investigated by several groups. Sheu et al. [85] measured the dynamic interfacial tensions between alkaline aqueous phases and asphaltene/toluene organic phases. The asphaltene fraction was precipitated from a heptane solution. The interfacial tension was found to obey an exponential law for which the level of the equilibrium data depended on the alkalinity. For 2 M NaOH, γ_{eq} was close to 10 mN m^{-1} for 0.01% of asphaltene. When the amount of asphaltene was raised to 0.1% in 1 M NaOH, similar γ_{eq} values were obtained. The authors found that the dynamic interfacial tensions reflected the diffusion-controlled adsorption/desorption of asphaltene molecules. Equilibrium values reflect more a reaction-controlled situation with rearrangements of the asphaltene molecules.

Singh and Pandey [86] characterized the interfacially active fractions in Indian crudes. They also concluded that a high interfacial pressure correlated very well with a high w/o emulsion stability.

Interfacially active fractions from North Sea crude oils have also been exposed to aging under normal atmospheric and ultraviolet conditions. Fourier transform–infrared spectroscopy reveals that the carbonyl peak grows markedly because of the C=C mode. At the same time, the spectral region between 900 and 700 cm^{-1} reveals that a condensation process takes place upon aging. The interfacial activity increases in all fractions as the aging proceeds. For two of the crude oils, this is accompanied by an increase in the w/o emulsion stability [87]. A parallel study of the whole crude oil showed similar results upon aging [88].

In-depth studies of thin films, their rupture, rheology, and drainage effect on demulsification have been performed by Wasan et al. and are summarized in an excellent way in Chap. 4.

In this part of the section we report on the stability of different water-in-crude-oil emulsions. These have been true crude-oil-based emulsions, model emulsions stabilized by indigenous components separated from crude oil, and finally emulsions based on completely synthetic components. The study also includes interfacial tension measurements and gravity tests on the emulsions and molecular weight (M_w) characterizations on the extracted interfacially active fractions [89]. In the petroleum industry, numerous additives are added in the production, transportation, and separation steps in order to inhibit different processes, like crystallization, foaming, emulsion formation, and so forth. The additives have a specific purpose but can, in many cases, interact with ionic as well as nonionic compounds in the crude oils. In order to study the effect of different additives on the emulsion stability we have chosen to introduce different surface-active molecules.

Three different crudes were investigated. B1 is from an Elf production field in France, whereas B2 (Elf) and G (Statoil) are from the North Sea. The procedure for extracting the interfacially active fractions from the crude oils is presented in Fig. 14. The fractions are divided into a precipitated fraction (asphaltenes) and an adsorbed fraction (resins). The molecular weight of the different fractions were measured with size exclusion chromatography (HPSEC), calibrated on polystyrene M_w standards.

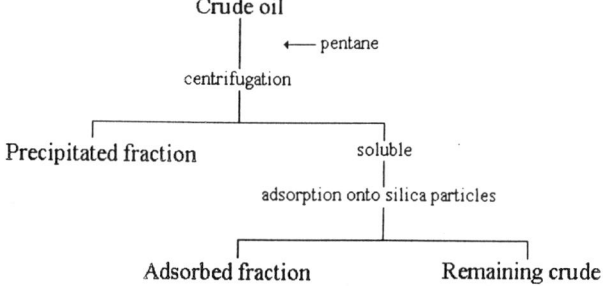

FIG. 14 Process for separation of indigenous surfactants from crude oils.

The interfacially active fractions extracted from crude oils B1, B2, and G have different amounts and ratios between asphaltenes and resins, as observed from Fig. 15. The molecular weights of the asphaltenes extracted from B1 are as much as three times higher than those extracted from B2 and G. The corresponding molecular-weight measurements of the resins are shown in Fig. 15 and show the same consecutive order. From the figure, it is clearly seen that the resins from B1 have both a smaller portion of light compounds as well as a higher portion of heavier compounds than B2 and G. On the other hand, these two crude oils show small differences in M_w between the asphaltene and the resin fractions. This indicates, in good accordance with our previous work on similar constituents, that the distinction between the asphaltene and the resin fractions is not so well defined [90]. Because the resins from B1 have approximately the same M_w as the asphaltenes from B2 and G, these asphaltenes can more or less be defined as heavy resins, which is also in accordance with Refs. 91 and 92. This fact will influence the ability of these fractions to stabilize the emulsions.

Based on the behavior of the crude oil or model emulsions in external electric fields one can classify the ability of natural surfactants to stabilize and prevent coalescence. For crude-oil emulsions, the highest E_{cr} is obtained for crude B1. For this crude, no visual separation of the aqueous phase is observed within 12 h, and a relatively high electric field in order to break this emulsion is required. The corresponding values of the electric field for crude oil G and B2 emulsions are distinctly lower.

Emulsions stabilized by 2% of asphaltenes show stabilities corresponding to the original crude oil emulsions. An emulsion stabilized by 2% of asphaltenes breaks at an electric field close to the value required to break the original crude-oil emulsion. This seems reasonable because the actual amount of asphaltenes in the crudes often lie around 2–3%. This is observed for both the B1 and G crude oils. The level of the critical electric field is much smaller for the G emulsions than for B1 emulsions. Analogously, 1% of asphaltenes from B1 are capable of stabilizing the emulsion, whereas 1% of asphaltenes from G are not. This can again be explained by the difference in polarity of the compounds, as revealed from the M_w measurements. The interfacially active fractions extracted from B2 are poor stabilizers. Even 2% of asphaltenes are not capable of stabilizing the emulsion. Of the systems investigated, it is only the emulsion stabilized by 1% of asphaltenes and 1% resins from B2 that is stable, indicating that the interplay between the asphaltenes and the resins is of importance. However, a relatively low electric field to irreversibly break this emulsion is required. A general trend observed for all crudes is that the resins alone

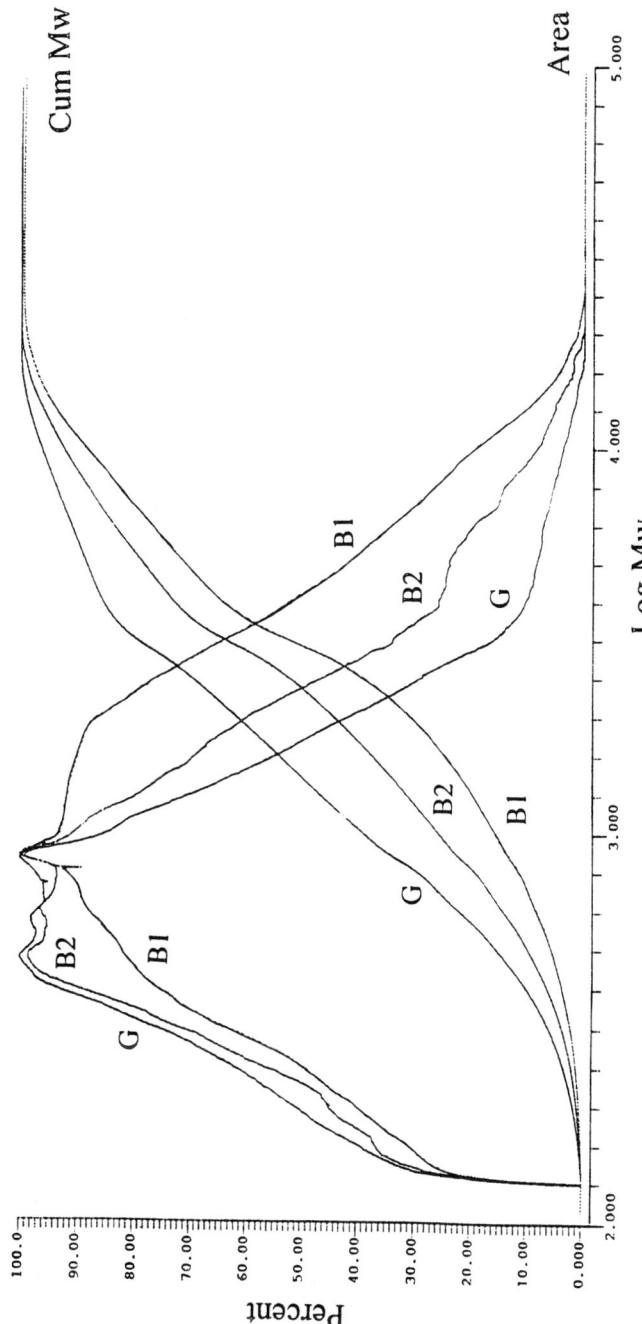

FIG. 15 Molecular-weight distribution of the resins extracted from the crude oils B1, B2, and G.

are not capable of stabilizing the emulsions. After being agitated, the water and oil phases separate within seconds.

The experimental data clearly indicate that the stabilizing fraction is the asphaltenes, not the resins, although the resins can show a high interfacial activity as judged from interfacial tension measurements in Table 6. However, when these fractions are combined, the result is in most cases a stabilization of the water-in-crude-oil emulsion. When 1% of asphaltenes are combined with 1% of the resins, the resulting emulsions show a lower stability than the original crude emulsions and those stabilized by 2% of asphaltenes. This also rules out the possibility that these emulsions should be stabilized by the asphaltenes alone. Hence, the active interaction between asphaltenes and resins will give rise to stability. Obviously, the resins, due to their higher interfacial activity, will be the first species to reach the w/o interface. After this, the asphaltene molecules will accumulate at the interface building up a complex macromolecular constellation at the interface. In order to preserve the stability level of the emulsions, this new interface must have some essential properties with regard to rigidity of the original asphaltene interface. Hence, there should be a critical ratio between asphaltenes and the resins for the interface to have necessary properties.

An investigation for the commercial surfactants similar to that for the crude-oil-based systems also reveals large differences in the level of the critical electric field. The E_{cr} required to break an emulsion stabilized by Span 80 (mono-oleate sorbitan ester) is distinctly higher than for NP-4 (ethoxylated nonylphenol). The difference in stabilizing behavior can, to some extent, also be observed from the decantation/sedimentation data. Emulsions stabilized by NP-4 sediment until only 10% of the continuous phase separates the water droplets in the emulsion. The percentage of water in the rest-emulsion thus increases from 50% to approximately 90%. Span 80, on the other hand, stabilizes emulsions that do not readily sediment, as only 5% of the continuous phase separates. Decantation/separation is mainly a result of flocculation processes taking place in the emulsion, and the decantation data in Table 6 clearly show that there are distinct differences among the commercial surfactants and among the different interfacially active fractions. Microscopy pictures show that Span 80 gives rise to rather small monodisperse droplets under our mixing conditions, while for NP-4 substantially larger and more polydisperse droplets are observed.

When two or more stabilizers are present in the same mixture, a competitive adsorption between the different emulsifiers can occur, which can be of crucial importance for the emulsion stability. A stabilizer can be removed from the interface if another stabilizer is added in sufficiently

TABLE 6 Interfacial Tensions, Decantation Data, and E_{cr} for Emulsions Stabilized by Indigenous as Well as Commercial Surfactants

50:50 emulsion (by volume) + additives to the oil phase	w/o or o/w	Separation dispersed phase (%) after 1 h	after 12 h	Separation continuous phase (%) after 1 h		γ (mN m^{-1})	E_{cr} (kV cm^{-1})
2% NP-4	w/o	0	0	50	90	0.3	0.34
2% Span 80	w/o	0	0	5	5	5.0	0.87
Crude B1	w/o	0	0	0	0	20.1	0.84
2% asph.	w/o	0	0	5	20	26.5	0.81
1% asph.	w/o	0	0	15	25	29.3	0.32
2% res.	u	100	100	100	100	28.0	—
1% res.	u	100	100	100	100	32.7	—
1% asph. + 1% res.	w/o	0	0	15	30	30.3	0.60
1% asph. + 1% NP-4	u	100	100	100	100	0.3	—
1% res. + 1% NP-4	u	100	100	100	100	0.3	—
0.75% asph. + 0.75% res. + 0.5% NP-4	u	100	100	100	100	2.1	—
1% asph. + 1% Span 80	w/o	0	0	0	2	3.7	0.37
1% res. + 1% Span 80	w/o	0	0	0	2	4.2	0.37
0.75% asph. + 0.75% res. + 0.5% Span 80	w/o	0	0	0	5	3.6	0.34
Crude B2	w/o	0	0	0	0	27.8	—
2% asph.	u	100	100	100	100	30.9	—
1% asph.	u	100	100	100	100	36.1	—
2% res.	u	100	100	100	100	26.4	—
1% res.	u	100	100	100	100	28.4	—
1% asph. + 1% res.	w/o	0	0	50	60	28.2	0.08
Crude G	w/o	0	0			10.6	0.26
2% asph.	w/o	0	0	50	90	24.7	0.24
1% asph.	u	100	100	100	100	32.7	—
2% res.	u	100	100	100	100		—
1% res.	u	100	100	100	100	31.5	—
1% asph. + 1% res.	w/o	0	0	45	90	28.5	0.13
1% asph. + 1% NP-4	u	100	100	100	100	2.6	—
1% res. + 1% NP-4	u	100	100	100	100	0.3	—
0.75% asph. + 0.75% res. + 0.5% NP-4	u	100	100	100	100	3.2	—
1% asph. + 1% Span 80	w/o	0	0	0	8	3.5	0.26
1% res. + 1% Span 80	w/o	3	25	2	10	4.7	0.05
0.75% asph. + 0.75% res. + 0.5% Span 80	w/o	0	0	0	8	4.3	0.13

u: unstable emulsion

high concentration, resulting in a loss of functionality for the removed stabilizer. The addition of Span 80 to an emulsion stabilized by asphaltenes and/or resins still gives stable w/o emulsions, but the level of E_{cr} is distinctly lower than the value observed for the reference emulsions stabilized by commercial as well as the natural surfactants. However, a conspicuous behavior is observed when NP-4 is added to an emulsion stabilized by interfacially active fractions. This nonionic surfactant completely destabilizes the emulsion, independent of how the system is stabilized. At the same time, the interfacial tension is at a very low level, which indicates a high interfacial activity.

VI. CONCLUSIONS

Dielectric time-domain spectroscopy has proved to be an interesting technique for studying water-in-oil emulsions. From accurate low-frequency, relaxation-time, and high-frequency measurements it is possible to determine shape factors essentially describing the flocculation level. When the dielectric instrumentation is equipped with an external electric field, information about the coalescence can be obtained when exceeding a critical voltage. With a reliable number on emulsion stability, different kinds of model or commercially based emulsified systems can be compared. Also, emulsions under operational conditions can be examined by means of a multivariate calibration.

ACKNOWLEDGMENTS

Harald Førdedal and Tore Skodvin would like to acknowledge the STP program "Surface and Colloid Chemistry for the Oil Industry" financed by The Norwegian Research Council (NFR) and the oil companies Elf Aquitaine, Saga Petroleum a/s and Statoil a/s for Ph.D. grants. The dielectric instrumentation was also financed by NFR. Valuable discussions with Prof. S. S. Dukhin and Dr. Bo Gestblom are acknowledged.

REFERENCES

1. C. Thomas, J. P. Perl, and D. T. Wasan, J. Colloid Interface Sci. *139*:1 (1990).
2. J. P. Perl, H. E. Bussey, and D. T. Wasan, J. Colloid Interface Sci. *108*:525 (1985).
3. M. Clausse, in *Encyclopedia of Emulsion Technology*, Vol. 1 (P. Becker, ed.), Marcel Dekker, Inc., New York, 1983, p. 481.
4. N. E. Hill, W. E. Vaughan, A. H. Price, and M. Davies, *Dielectric Properties*

and Molecular Behaviour, Van Nostrand, London, 1969; J. B. Hasted, *AqueousDielectrics*, Chapman & Hall, London, 1973; E. H. Grant, R. J. Sheppard, and G. P. South, *Dielectric Behaviour of Biological Molecules in Solution*, Reinhold, New York, 1978.

5. P. Debye, *Polar Molecules*, Reinhold, New York, 1929.
6. K. S. Cole and R. H. Cole, J. Phys. Chem. *9*:341 (1941).
7. R. H. Cole, Annu. Rev. Phys. Chem. *28*:283 (1977).
8. R. H. Cole, J. G. Berberian, S. Mashimo, G. Chryssikos, A. Burns, and E. Tombari, J. Appl. Phys. *66*:793 (1989).
9. R. H. Cole, S. Mashimo, and P. Winsor, J. Phys. Chem. *84*:786 (1980).
10. B. Gestblom and E. Noreland, J. Phys. Chem. *81*:782 (1977).
11. B. Gestblom and H. Elmgren, Chem. Phys. Lett. *90*:412 (1982).
12. K. W. Wagner, Arch. Electrotechnol. *2*:371 (1914).
13. O. Wiener, Abh. Sächs. Akad. Wiss. *32*:509 (1912).
14. C. J. F. Böttcher, Rec. Trav. Chim. *64*:47 (1945).
15. Hsu and T. D. Gierke, J. C. Molnar, *Macromolecules 16*:1945 (1983).
16. H. Looyenga, Physica *31*:401 (1965).
17. L. K. H. van Beek, Prog. Dielectr. *7*:69 (1967).
18. T. Hanai, in *Emulsion Science* (P. Sherman, ed.), Academic Press, London, 1968, Chap. 5.
19. G. Bánhegyi, Colloid Polym. Sci., *264*:1030 (1986).
20. G. Bánhegyi, Colloid Polym. Sci., *266*:11 (1988).
21. D. A. G. Bruggeman, Ann. Phys. Leipzig *24*:636 (1935).
22. T. Hanai, Kolloid Z. *171*:23 (1960).
23. T. Hanai, Kolloid Z. *175*:61 (1961).
24. T. Hanai, Bull. Inst. Chem. Res. Kyoto Univ. *32*:341 (1961).
25. M. H. Boyle, Colloid Polym. Sci. *263*:51 (1985).
26. C. Boned and J. Peyrelasse, Colloid Polym. Sci. *261*:600 (1983).
27. H. P. Schwan, G. Schwarz, J. Maczuk, and H. Pauly, J. Phys. Chem. *66*: 2626 (1962).
28. H. P. Schwan, Adv. Biol. Med. Phys. *5*:147 (1957).
29. J. Th. G. Overbeek, Kolloid Beih. *54*:287 (1943).
30. J. Th. G. Overbeek, Adv. Colloid Sci. *3*:97 (1950).
31. G. Schwarz, J. Phys. Chem. *66*:2636 (1962).
32. S. S. Dukhin and V. N. Shilov, *Dielectric Phenomena and the Double Layer in Disperse Systems and Polyelectrolytes*, John Wiley & Sons, New York, 1974.
33. J. Lyklema, S. S. Dukhin, and V. N. Shilov, J. Electroanal. Chem. *143*:1 (1983).
34. M. M. Springer, A. Korteweg, and J. Lyklema, J. Electroanal. Chem. *153*: 55 (1983).
35. J. Lyklema, M. M. Springer, V. N. Shilov, and S. S. Dukhin, J. Electroanal. Chem. *198*:19 (1986).
36. J. C. Maxwell, *A Treatise on Electricity and Magnetism*, Clarendon Press, Oxford, 1892.

37. K. W. Wagner, Arch. Electrotechnol. *3*:100 (1914).
38. R. W. Sillars, Proc. Inst. Electr. Eng. *80*:378 (1937).
39. K. Asami, and T. Hanai, Colloid Polym. Sci. *270*:78 (1992).
40. T. Hanai, Kolloid Z. *177*:57 (1961).
41. J. Sjöblom, T. Skodvin, T. Jakobsen, and S. S. Dukhin, J. Dispersion Sci. Technol. *15*:401 (1994).
42. T. Skodvin, J. Sjöblom, J. O. Saeten, O. Urdahl, and B. Gestblom, J. Colloid Interface Sci. *166*:43 (1994).
43. T. Skodvin, J. Sjöblom, J. O. Saeten, T. Wärnheim, and B. Gestblom, Colloids Surfaces A *83*:75 (1994).
44. T. Skodvin, T. Jakobsen, and J. Sjöblom, J. Dispersion Sci. Technol. *15*: 449 (1994).
45. B.-M. Sax, G. Schön, S. Paasch, and M. J. Schwuger, Progr. Colloid Polym. Sci., *77*:109 (1988).
46. R. M. Hill, E. S. Beckford, R. C. Rowe, C. B. Jones, and L. A. Dissado, J. Colloid Interface Sci. *138*:521 (1990).
47. R. M. Hill and J. Cooper, J. Mater. Sci. *27*:4818 (1992).
48. J. Sjöblom, O. Urdahl, H. Høiland, A. A. Christy, and J. Johansen, Prog. Colloid Polym. Sci. *82*:131 (1990).
49. J. Sjöblom, H. Söderlund, S. Lindblad, E. J. Johansen, and I. M. Skjärvö, Colloid Polym. Sci. *268*:389 (1990).
50. V. B. Menon and D. T. Wasan, *Encyclopedia of Emulsion Technology*, Vol. 2 (P. Becker, ed.), Marcel Dekker, Inc., New York, 1983, Chap. 1.
51. P. J. Bailes and S. K. L. Larkai, Trans. Inst. Chem. Eng. *59*:229 (1981).
52. P. J. Bailes and S. K. L. Larkai, Chem. Eng. Res. Des. *62*:33 (1984).
53. L. C. Waterman, Chem. Eng. Proc. *61*:51 (1965).
54. B. Speed, J. Ind. Eng. Chem. *11*:153 (1919).
55. F. G. Cottrell, US Patent No. 987114 (1911).
56. F. G. Cottrell and J. B. Speed, US Patent No. 987115 (1911).
57. S. E. Taylor, Inst. Phys. Conf. Ser. *118*:185 (1991).
58. T. J. Williams and A. G. Bailey, IEEE Trans. Ind. Appl. *IA-22*:536 (1986).
59. D. C. Chang, Biophys. J. *56*:641 (1989).
60. S. E. Taylor, Colloids Surf. *29*:29 (1988).
61. C. A. R. Pearce, Br. J. Appl. Phys. *5*:136 (1954).
62. U. Zimmermann, J. Membrane Biol. *67*:165 (1982).
63. U. Zimmermann, G. Pilwat, and F. Riemann, Biophys. J. *14*:881 (1981).
64. U. Zimmermann, Biochim. Biophys. Acta, 694, 227 (1982).
65. P. J. Bailes and S. K. L. Larkai, Trans. Inst. Chem. Eng. *60*:115 (1981).
66. C. P. Galvin, Inst. Chem. Eng. Symp. Ser. *88*:101 (1984).
67. R. A. Mohammad, A. I. Bailey, P. F. Luckham, and S. E. Taylor, Colloids Surf. *80*:223 (1993).
68. R. A. Mohammad, A. I. Bailey, P. F. Luckham, and S. E. Taylor, Colloids Surf. *80*:237 (1993).
69. R. A. Mohammad, A. I. Bailey, P. F. Luckham, and S. E. Taylor, Colloids Surf. *83*:261 (1994).

70. T. Y. Chen, R. A. Mohammad, A. I. Bailey, P. F. Luckham, and S. E. Taylor, Colloids Surf. *83*:273 (1994).
71. A. H. Brown and C. Hanson, Trans. Faraday Soc. *61*:1754 (1965).
72. T. Hirato, K. Koyama, T. Tanaka, Y. Awakura, and H. Majima, Mater. Trans. JIM *32*:257 (1991).
73. B. Gestblom, H. Førdedal, and J. Sjöblom, J. Dispersion Sci. Technol. *15*: 449 (1994).
74. H. Førdedal, E. Nodland, J. Sjöblom, and O. M. Kvalheim, J. Colloid Interface Sci. *173*:396 (1995).
75. G. E. P. Box, W. G. Hunter, and J. S. Hunter, Statistics for Experimenters, An Introduction to Design, Data Analysis and Model Building, John Wiley & Sons, New York, 1978.
76. R. Aveyard, B. P. Binks, P. D. I. Fletcher, X. Ye, and J. R. Lu, in *Emulsions—A Fundamental and Practical Approach* (J. Sjöblom, ed.), NATO ASI Series C363, Kluwer, Dordrecht, 1992, p. 97.
77. J. Sjöblom, O. Urdahl, K. G. N. Børve, L. Mingyuan, J. O. Saeten, A. A. Christy, and T. Gu, Adv. Colloid Interface Sci. *41*:241 (1992).
78. J. Sjöblom, L. Mingyuan, H. Höiland, and E. J. Johansen, Colloids Surf. *46*:127 (1990).
79. R. Pelet, F. Behar, and J. C. Monin, Organ. Geochem. *10*:481 (1986).
80. E. J. Johansen, I. M. Skjärvö, T. Lund, J. Sjöblom, H. Söderlund, and G. Boström, Colloids Surf. *34*:353 (1990).
81. Li Mingyuan, A. A. Christy, and J. Sjöblom, in *Emulsions—A Fundamental and Practical Approach* (J. Sjöblom, ed.), NATO ASI Series C363, Kluwer, Dordrecht, 1992, p. 157.
82. J. Sjöblom, Li Mingyuan, T. Gu, and A. A. Christy, Colloids Surf. *66*:55 (1992).
83. K. G. Nordli, J. Sjöblom, J. Kizling, and P. Stenius, Colloids Surf. *57*:83 (1991).
84. K. G. Nordli, J. Sjöblom, and P. Stenius, Colloids Surf. *63*:241 (1992).
85. E. Y. Sheu, M. M. De Tar, and D. A. Storm, Fuel *71*:437 (1992).
86. B. P. Singh and B. P. Pandey, Indian J. Technol. *29*:443 (1991).
87. J. Sjöblom, Li Mingyuan, A. A. Christy, and H. P. Rønningsen, Colloids Surf. *96*:261 (1995).
88. H. P. Rønningsen, J. Sjöblom, and Li Mingyuan, Colloids Surf. (in press).
89. H. Førdedal, Y. Schildberg, J. Sjöblom, and J.-L. Volle, Colloids Surf. (in press).
90. Y. Schildberg, J. Sjöblom, A. A. Christy, J. L. Volle, and O. Rambeau, J. Dispersion Sci. Technol. (in press).
91. K. J. Leontaris and G. A. Mansoori, Soc. Petrol. Eng. J. *16258*:149 (1987).
92. H. Lian, J. R. Lin, and T. F. Yen, Fuel *73*:423 (1994).

10

Ultrasonic Characterization
of Emulsions

KJELL-EIVIND FRØYSA and ØYVIND NESSE   Department of
Industrial Instrumentation, Christian Michelsen Research AS,
Bergen, Norway

I. INTRODUCTION

Acoustical methods can be used as a characterization of emulsions. Both particle size and concentration may be achieved by accurate acoustical measurements. Acoustical methods are applicable in optically opaque media. They are nonintrusive in many situations. In addition, low-intensity acoustic waves leave the medium unchanged after propagating through it. High-power acoustics may, on the other hand, change the constitution of the mixture and may, therefore, in some cases be used actively to speed up desired processes. In this chapter, we will pay attention to low-intensity ultrasonic waves only. Ultrasonic waves are acoustic waves with frequencies above the audible ones. That means that the frequency is above 20 kHz. The choice of frequency depends on the application, as we will see below. Generally, we can say that the instrumentation gets more complex the higher the frequency. The highest ultrasonic frequencies used are above 1 GHz. Much more frequent is the use of frequencies below 100 MHz.

In this chapter, we will describe some of the potentials of acoustical methods for the characterization of emulsions, in addition to describing some previous works. The most frequently used techniques in this context are the measurements of the speed of sound and the measurements of the attenuation. We will, therefore, pay most of our attention to these techniques. We will present both simplified and more comprehensive theoretical models, present simulated results, describe how to set up proper experiments, and present experimental results. The goal is to extract droplet size distribution and/or volume fraction of the suspended droplets from the measured acoustical data.

II. PREVIOUS WORK

In the literature, there are many articles on ultrasonics in connection with two-phase fluids like emulsions. These articles have appeared in many journals. For nonacousticians, it may, therefore, be difficult to orientate among these works. Two review articles may be useful when ultrasonics in emulsions is considered. Pal [1] considers several main techniques for measuring the oil and water content of emulsions. One of the main techniques described in that article is ultrasonics. Three ultrasonic techniques have been mentioned there, and they will be presented below. McClements [2] considers one of the ultrasonic techniques described by Pal in more detail. This is the measurement of the speed of sound and attenuation. In that article, both the mathematical/physical principles and a possible experimental technique are discussed.

The first ultrasonic technique described by Pal is the sound-scattering technique. This technique considers the fact that liquid droplets suspended in another immiscible liquid will scatter sound in all directions. By sending a coherent (for example, sinusoidal) sound wave through the emulsion, we will, in addition to the coherent sound field going straight through the emulsion, get a diffuse or incoherent sound field scattered in all directions from the emulsion. For a fixed frequency, this scattered field depends on the concentration of the suspended fluid and on the size of the suspended fluid spheres. By measuring the intensity of the scattered field in several directions, it may be possible to extract information both on the average radius of the fluid spheres and on the concentration of the emulsion. These techniques have been used on solid/liquid mixtures. To our knowledge, little has been done on applying such techniques on emulsions. Therefore, in order to use such techniques, a considerable job has to be done on development of the techniques on emulsion. When the fluid spheres have radii of a few micrometers, this technique would probably demand quite high frequencies, typically in the range 100 MHz to 1 GHz.

The second method described by Pal is the measurement of the speed of sound in the fluid mixture. This method, combined with the measurement of the attenuation of acoustic waves, was the main object in the article of McClements [2]. This method is, in our opinion, the method which is simplest to use and the method which has been most widely studied. We will, therefore, pay most of our attention to this method and postpone the presentation of the method to the next sections of the chapter.

The ultrasound vibration potential (UVP) technique is the last acoustical technique mentioned by Pal. In this technique, we assume that the suspended liquid appears as droplets. When a sound wave is incident on a droplet, it will cause the droplet to move. Typically, a compressibility contrast between the continuous and the suspended fluid will cause the droplet to pulsate (monopole oscillation), and a density contrast will cause the droplet to oscillate back and forth relative to the surrounding fluid (dipole oscillation). Depending on the frequency, higher order modes like quadrupole and octupole oscillations may be generated. At present, this is not important. When the fluid droplets are charged, the dipole oscillation produces an electric field. The net result of these small electric fields is a macroscopic electric field denoted as the UVP signal. This signal depends on the concentration of the suspended droplets, and it is, therefore, possible to use it for estimates of the concentration of the emulsion. The technique has been used both for O/W and W/O emulsions. However, the UVP signal also depends on parameters other than the concentration, and therefore the method may be uncertain. For more information and references on this method, we refer to Pal.

In addition to the above-mentioned techniques, we mention another possible technique for measuring concentration. When an emulsion is flowing through a pipe bend, the flow will produce noise. When droplets of the suspended fluid are present, we expect this noise to be different from the noise in a single phase flow. This principle has been investigated at SINTEF/SI for measuring the oil content in dilute O/W emulsions [3]. At present, the concept is still under development, and it is difficult for us to discuss the potential of the method.

III. SIMPLIFIED PHASE VELOCITY MODELS

The speed of sound in water is generally higher than the speed of sound in oil. Both the speed of sound in water and in oil depend on the temperature. The speed of sound in water is 1402 m s^{-1} at 0°C, increases to a maximum of 1555 m s^{-1} at around 74°C, and is 1543 m s^{-1} at 100°C. At temperatures of around 20°C, the speed of sound in water changes with around 3 m s^{-1} for each degree Celsius. At this temperature, most oils will have a decreasing speed of sound with temperature. An example is Exxsol D80 which is a model oil produced by Exxon. The speed of sound at 21.9°C was measured to around 1315 m s^{-1}. The increase of temperature of 1°C gives a reduction in speed of sound of around 4 m s^{-1} for temperatures around 20°C. Mixtures of oil and water will have a speed of sound which is somewhere between the speed of sound in the two components. This means that by measuring the speed of sound and the temperature in an emulsion, it may be possible to extract information concerning the volume fraction of the two components when knowing which oil is present in the emulsion. In media where a suspended fluid appears as droplets in a continuous fluid (like emulsions), the speed of sound typically depends on frequency. Emulsions are, thus, dispersive media and they have, therefore, both a group velocity and a phase velocity at a given frequency. We will consider the phase velocity, and this term will consequently be used instead of the speed of sound when referring to emulsions. As we will see below, not only the absolute level of the phase velocity but also the change in phase velocity as a function of frequency may provide useful information both on the particle sizes present and on the concentration of the emulsion. The attenuation of sound waves in emulsions may be much stronger than the attenuation of sound waves in pure water and in pure oil. It increases with concentration and can be used for concentration estimates when the particle sizes are known.

In order to compute the concentration from the measured sound speed, we need a model for the sound speed as a function of concentration. The simplest of such models do not depend on the droplet size and predict a phase velocity which is independent of frequency. Probably the most widely used of such models is Wood's formula [4], also denoted as Urich's formula [5]. This formula can be explained by first considering the formula

$$c = \sqrt{\frac{1}{\kappa_c \rho}} \tag{1}$$

for the speed of sound in a single-phase fluid. κ_c is the adiabatic compressibility of the liquid and ρ is the density of the liquid. We now introduce a two-phase medium where the indexes 1 and 2 correspond to the two homogeneous fluids. In addition, let φ_1 and φ_2 denote the volume fractions of the two media. Consequently, $\varphi_1 + \varphi_2 = 1$. The average density of the two-phase medium is $\rho_{av} = \varphi_1 \rho_1 + \varphi_2 \rho_2$. If we similarly denote the average compressibility as $\kappa_{av} = \varphi_1 \kappa_1 + \varphi_2 \kappa_2$, we obtain the following formula for the phase velocity in the two-phase mixture:

$$c_{mix} = \sqrt{\frac{1}{\kappa_{av}\rho_{av}}} = \sqrt{\frac{1}{(\varphi_1/\rho_1 c_1^2 + \varphi_1/\rho_1 c_1^2)(\varphi_1 \rho_1 + \varphi_2 \rho_2)}}. \tag{2}$$

This formula is Wood's formula for a two-phase medium, and it can be used if we know the speed of sound and the density of the two media which are mixed. Wood's formula can be generalized to multifluid mixtures in a straightforward manner. It is expected to be approximately valid for emulsions, especially in the low-frequency regime. In this case, the wavelength is much larger than the size of the individual suspended spheres, and the sound wave will experience average values for the acoustical properties of the two media.

An alternative simple model for the phase velocity in emulsions is the formula of Wyllie [6]. In this formula, we can think that the emulsion is replaced by a layered medium (two layers) in which all of the oil is placed in one layer and all of the water is placed in the other layer. In this way, the effective phase velocity of the emulsion will be given by the equation

$$\frac{1}{c_{mix}} = \frac{\varphi_1}{c_1} + \frac{\varphi_2}{c_2}. \tag{3}$$

Note that in this equation, the density does not appear. When Wyllie's formula can be used, it is expected to be a high-frequency approximation, due to the fact that the sound changes between propagating with the phase velocities of the two media. The wavelength must then be so small that there are several wavelengths within each appearance of the suspended fluid. However, we have no guarantee that Wyllie's formula is applicable in this way for emulsions.

Pal mentions these models as his model 1 and model 2. In addition, there are other simplified models. Pal mentions the formula

$$c_{mix} = \varphi_1 c_1 + \varphi_2 c_2, \tag{4}$$

which is a simple averaging of the speed of sound in the two media. This is Pal's model 3. This formula is not motivated physically. In Ref. 7, these three models have been compared.

There is also a fourth model mentioned by Pal. This model has been proposed by Tsouris and Tavlarides [8]. The formula is claimed to be superior to other models but can, in some cases, give large errors. In addition to the density and the phase velocity of the two media, the formula takes into account the fact that there is one continuous and one suspended phase.

IV. MULTIPLE-SCATTERING MODELS

Common for all these formulas is that they do not give any frequency dependence on the phase velocity on emulsions. In addition, they do not give the attenuation in the emulsions. Certain models exist where the attenuation is given by similar formulas. These formulas will depend on the size of the suspended fluids and are, therefore, more complicated than the formulas above. Emulsions contain fluid droplets, and the acoustical behavior depends on the wavelength. This has been demonstrated in the literature both theoretically and by measurements; see, for example, Refs. 2 and 9. We will now briefly present the type of models which also describe the frequency dependency of the phase velocity in emulsions. These are the multiple-scattering theories. The theories consist of two parts. The first part is the behavior of the sound close to a single fluid sphere, and the next part is to use this information to compute the average macroscopic behavior of the sound wave in the emulsion, including the phase velocity and attenuation of the sound wave. Readers who are not interested in the mathematical/physical discussion, but just in the implementation of a multiple-scattering model, can go to the end of this section, after Eq. (21).

A plane sound wave incident on a fluid sphere suspended in another immiscible fluid, will, as described above, cause the fluid to oscillate in several modes. For emulsions, the monopole and dipole will be the only significant modes when the wavelength of the incident wave is much larger than the size of the suspended spheres. As an example, the wavelength in water is around 1.5 mm when the frequency is 1 MHz, and around 15 μm when the frequency is 100 MHz. When the wavelength becomes comparable to or less than the size of the fluid spheres, higher order modes have to be taken into account also. The solution becomes more complicated and the exact number of oscillation modes now depends on the frequency.

Generally, the reflection from a single fluid sphere has been described by Allegra and Hawley [10] and by Epstein and Carhart [11]. We will now briefly outline this phenomenon. When a plane wave of frequency ω is incident on the sphere, it can be represented as

$$p_i = Ce^{ikz - i\omega t}, \tag{5}$$

where C is a constant, k the wave number of the wave in the continuous fluid, ω the angular frequency $(= 2\pi f)$, z the distance along the propagation direction, and t the time. If we let the suspended fluid sphere with radius a be located with its center in the origin and let r be the distance from the origin to an observation point, it can be shown that a reflected wave of the following form will be generated [10,11]:

$$p_r = Ce^{-i\omega t} \sum_{n=0}^{\infty} i^n A_n (2n + 1) h_n(kr) P_n(\cos \theta). \tag{6}$$

For further explanation of the geometry, see Fig. 1. Far from the fluid sphere $(kr \gg 1)$, this formula reduces to

$$p_r \approx Ce^{-i\omega t} \frac{e^{ikr}}{r} \left(\frac{1}{ik} \sum_{n=0}^{\infty} A_n (2n + 1) P_n(\cos \theta) \right) \overset{\text{def}}{=} Ce^{-i\omega t} \frac{e^{ikr}}{r} f(\theta). \tag{7}$$

P_n is the Legendre polynomial and h_n is the spherical Bessel functions; see, for example, Ref. 12 for definitions. $f(\theta)$ is the far-field amplitude of the scattering from a single sphere. In this sum, the $n = 0$ term represents

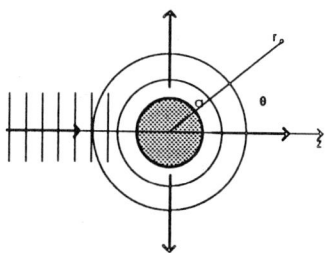

FIG. 1 Geometry in the reflection from a single sphere.

the monopole oscillation mode. The next terms are the dipole mode, quadrupole mode, octupole mode, and so on. In addition, a sound wave will be generated in the interior of the sphere. Thermal and rotational (viscous) waves will also be generated in boundary layers close to the fluid-sphere surface. These waves will take energy from the sound wave propagating through the emulsion, and it is, therefore, important to be able to describe these waves properly. Thus, the coefficients A_n depend generally on the thermal and viscous properties of the two media. The computation of the coefficients is numerically rather complex. However, as stated above, in the long-wavelength limit, where $ka \ll 1$, the two first terms in the series in Eqs. (6) and (7) are the only significant ones, and Eqs. (6) and (7) simplify to

$$p_r = Ce^{-i\omega t}\frac{e^{ikr}}{r}\left[\frac{1}{ik}\left(A_0 + 3A_1\left(1 + \frac{i}{kr}\right)\cos\theta\right)\right] \tag{8}$$

and

$$p_r \approx Ce^{-i\omega t}\frac{e^{ikr}}{r}\left[\frac{1}{ik}(A_0 + 3A_1\cos\theta)\right], \qquad kr \gg 1. \tag{9}$$

In order to simulate the propagation of sound waves through emulsions, we have to combine the information concerning the reflection close to a single sphere to the overall behavior of the sound wave. The scattered sound field close to a single sphere can be rescattered by a neighboring sphere. This happens with the scattered waves from all fluid spheres, and it leads to a very complicated system of equations to be solved in order to achieve the overall behavior. However, we are not interested in the multiple scattered sound field close to a specific sphere but only the average macroscopic effects of the multiple scattering. Within several approximations, there are several so-called multiple-scattering theories which provide the phase velocity and attenuation from the information of the reflection of a sound wave close to a single fluid sphere. Probably the most widely used model is the so-called Waterman–Truell multiple-scattering theory [13] first published by Urich and Ament [14]. As most other theories, this theory is limited to low concentrations of the suspended fluid. It is applicable not only for emulsions but also, for instance, for bubbly liquids. For emulsions, comparisons between measurements and theoretical simulations indicate that the model is valid for volume fractions of the suspended fluid up to about 50%; see Ref. 15 and below. For emulsions with monosized spheres, the Waterman–Truell multiple-scattering theory generally gives the wave number K of a plane wave in the emulsion by the following equation:

$$\left(\frac{K}{k}\right)^2 = \left(1 + \frac{3\varphi}{2ik^3a^3} \sum_{n=0}^{\infty} (2n + 1)A_n\right)^2 - \left(\frac{3\varphi}{2ik^3a^3} \sum_{n=0}^{\infty} (-1)^n(2n + 1)A_n\right)^2.$$

(10)

In the long-wavelength region ($ka \ll 1$), this expression reduces to

$$\left(\frac{K}{k}\right)^2 = \left(1 + \frac{3\varphi}{i(ka)^3} A_0\right)\left(1 + \frac{9\varphi}{i(ka)^3} A_1\right).$$

(11)

K is generally a complex quantity, and the phase velocity (c_{mix}) and the attenuation (α_{mix}) can be computed from this quantity. They can be found as

$$c_{mix} = \frac{\omega}{\text{Re}\{K\}} \quad \text{and} \quad \alpha_{mix} = \text{Im}\{K\}.$$

(12)

The phase velocity c is then given in meters per second and the attenuation α is given in Nepers per meter (Np m). For an explanation of the unit Np m^{-1}, see below where experimental techniques are described. In the long-wavelength region, the coefficients A_0 and A_1 can be given explicitly [9,10]:

$$A_0 = \frac{i(ka)^3}{3}\left(\frac{\rho k'^2}{\rho'k^2} - 1\right) - k^2acT\rho\kappa H\left(\frac{\beta}{\rho c_p} - \frac{\beta'}{\rho'c_p'}\right)^2$$

(13)

and

$$A_1 = -i(ka)^3(\rho - \rho')\left[3\left(\frac{2(\rho' - \rho)}{1 + 3(1 + i)\delta_v/2a + \frac{3i\delta_v^2}{2a^2}} + 3\rho\right)\right]^{-1},$$

(14)

where

$$H = \left(\frac{1}{1 - iZ} - \frac{\kappa \tan Z'}{\kappa'(\tan Z' - Z')}\right)^{-1},$$

(15)

$$Z = \frac{(1 + i)a}{\delta_t},$$

(16)

$$\delta_t = \sqrt{\frac{2\kappa}{\omega\rho c_p}},$$

(17)

$$\delta_v = \sqrt{\frac{2\mu}{\omega\rho}}.$$

(18)

In the primed variables, we use data for the suspended fluid, whereas in the unprimed variables, we use data for the continuous fluid. The data needed for each phase are

c Speed of sound
ρ Density
μ Shear viscosity
β Thermal expansion factor
κ Thermal conductivity
c_p Specific heat capacity at constant pressure

In addition, we need the wave number

$$k = \frac{\omega}{c} + i\alpha, \tag{19}$$

where α is the attenuation (Np m^{-1}) in the pure fluid for the actual frequency. This can either be found tabulated or computed from the formula

$$\alpha = \frac{\omega^2}{2\rho c^3}\left(\mu_b + \frac{4}{3}\mu + \frac{(\gamma - 1)\kappa}{c_p}\right). \tag{20}$$

Here, μ_b is the bulk viscosity of the fluid and γ the ratio of the specific-heat capacities. For gases, γ is often tabulated. For both gases and liquids, γ can be computed from the formula

$$\gamma = 1 + \frac{\beta^2 c^2 T}{c_p}. \tag{21}$$

From Eqs. (11)–(21), the predicted phase velocity and attenuation can be computed by the Waterman–Truell multiple-scattering theory when the wavelength is much larger than the suspended fluid spheres ($ka \ll 1$). The presentation of the Waterman–Truell multiple-scattering theory given here is for emulsions where the suspended fluid droplets are of one size only. In reality, there is a distribution of droplet sizes. If we consider Eq. (11), the size-dependent parts in that formula are A_0/a^3 and A_1/a^3. When there is a distribution of sizes, we use a volumetric average of these quantities in Eq. (11). For more details, we refer to Refs. 9 and 13.

The drawback of the multiple-scattering theories is that we need several physical constants of each of the two fluid components in the mixture. This is, of course, because the sound propagation in emulsions is a quite complex phenomenon when described to the desired precision including correct description of the dispersion of the sound waves. The density and speed of sound of each medium are of great importance, and so are the thermal and viscous coefficients of the continuous medium. The predicted phase velocity and attenuation do not vary so much with the viscous and thermal coefficients of the suspended medium. Water is an especially well-studied liquid, and it is possible to obtain reliable data for the coefficients for this medium. Thus, it is at least possible to predict reliably the phase velocity and attenuation in water-continuous emulsions. The precision of simulations in oil-continuous emulsions depends much more on the precision of the physical data for the actual oil at the actual temperature. As a help in simulations, we give values for the relevant physical data for pure water at 20°C at atmospheric pressure:

c = 1482.3 m s.
ρ = 998.1 kg m^{-3}
μ = 0.00100 Pa s
μ_b = 0.0029 Pa s
β = 2.07 $\times$ 10^{-4} K^{-1}
κ = 0.597 W/m K
c_p = 4182.7 J/kg K
γ = 1.007
α = 6.54 $\times$ 10$^{-16}\omega^2$ Np m^{-1}

In concluding the theoretical presentation of multiple-scattering theories, we have to say that the Waterman–Truell multiple-scattering theory presented here is just one of several multiple-scattering theories. Common for multiple-scattering theories is that they are valid for low concentrations of the suspended medium. As stated above, for emulsions, this means that when the volume fraction is above approximately 50%, the Waterman–Truell multiple-scattering theory will not be valid. In this case, new acoustical phenomena will appear because of the stronger interaction between neighboring droplets. Other multiple-scattering theories where these interactions to some extent have been accounted for have been published. Here we mention the theories by Ma et al. [16] and by Berger and Twersky [17]. A complete implementation of these theories is quite complex, and we refer to the respective articles for details. For a simplified implementation of the multiple-scattering theory by Ma et al. when the wavelength is much larger than the fluid spheres ($ka \ll 1$), we refer to Ref. 9. However, the measurements at a higher concentration of the suspended fluid presented later in this chapter show behaviors which cannot be described even by these more sophisticated multiple-scattering models.

V. PREDICTED ACOUSTICAL PROPERTIES OF EMULSIONS

We will now use the Waterman–Truell model to present some general properties of the propagation of ultrasonic waves through emulsions. In these examples, as in the rest of this chapter, we have used Exxsol D80 as the oil. The density of Exxsol D80 at 20°C and 1 bar is approximately 800 kg m^{-3} and the speed of sound is approximately 1320 m s^{-1}. In the theoretical examples shown here, the temperature is 20°C and the pressure is atmospheric.

In Fig. 2, we have computed the phase velocity in 40% oil-in-water emulsion as a function of frequency. The computation has been performed for monosized droplets in the emulsions, and in the two cases shown, the droplet radii are 1 μm and 10 μm, respectively. The frequency range is from 1 Hz to 1 GHz, that is, from far below the ultrasonic regime up to very high ultrasonic frequencies. We see that in both cases, there are large changes in phase velocity as a function of frequency. Typically, at the lowest frequencies, the phase velocity is almost constant at a value close to 1360 m s^{-1}. This is the region in which an effective medium model like Wood's formula [Eq. (2)] should be expected to be valid, because the wavelength is very large compared to the suspended droplets, and an averaging of density and compressibility might be expected. However, Wood's formula in this case predicts a phase velocity of about 1394 m s^{-1}, which is far above the low-frequency limit in Fig. 2. The difference between the two estimates can be shown to be mostly due to a thermal effect. This thermal effect is accounted for in the Waterman–Truell multiple-scattering theory but not in Wood's formula. As the frequency in-

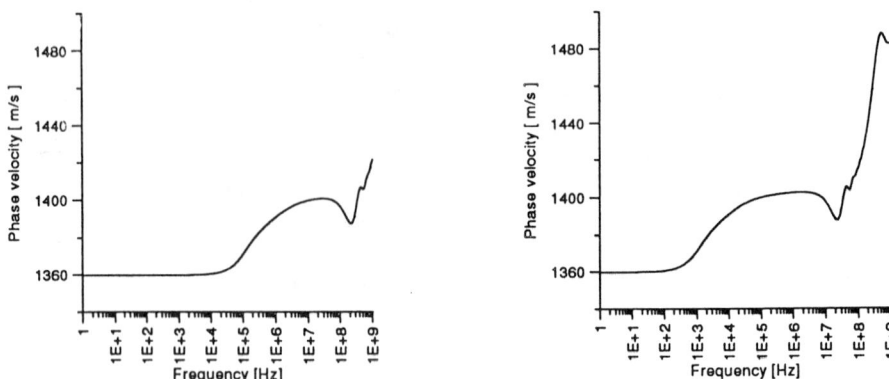

FIG. 2 Predicted phase velocity as a function of frequency for a 40% oil (Exxsol D80) in water emulsion. Temperature: 20°C; pressure: 1 bar; droplet radius: 1 μm (left curve) and 10 μm (right curve). The simulations are based on the Waterman–Truell multiple-scattering theory.

creases, there is a significant increase in the phase velocity in Fig. 2, up to a value of roughly 1400 m s^{-1}. In the two examples, this increase happens at different frequencies. In the 1-μm emulsion, it starts close to 100 kHz, whereas in the 10-μm emulsions, it starts close to 1 kHz. An increase in radius of a factor of 10 thus reduces the frequency where the rise starts with a factor of $100 = 10^2$. If we understand this effect, it can be used for particle size measurements. The rise appears to be due mostly to differences in the thermal properties of the two media. Because of this difference, thermal waves will be generated at the boundary of each droplet. These waves are highly attenuated, and at the frequencies where the rise appears, they propagate a distance comparable to the size of the droplets. In addition to this effect, the difference in density between the two media contributes to some extent similarly to the thermal waves. The long-wavelength version of the Waterman–Truell multiple-scattering theory presented in Eqs. (11)–(21) describes this rise, and the approximative equations can be used to study the origin of this effect.

When increasing the frequency further, we see that the phase velocity first goes through a minimum and thereafter increases to the value of the phase velocity in water. We see that this effect also depends on the particle size, but this time, an increase in radius of a factor of 10 gives a reduction in frequency by the factor of 10. At the frequencies where this happens, the wavelength of the acoustical wave is comparable to the size of the suspended droplets; thus, ka is no longer much smaller than 1, and the simplified version of the Waterman–Truell multiple-scattering theory described in Eqs. (11)–(21) cannot be used. We see again that here is an effect which depends on the particle size, and there should be possibilities of using this effect for particle size measurements. In addition to the detection of particle sizes, such plots can be used for the concentration measurements through measuring the absolute level of the phase velocity in frequency regions where it does not change much with frequency and droplet radius.

To complete the discussion of Fig. 2, three comments are necessary. The first comment is from the fundamentals of acoustics. When we measure as high as 1 GHz, we can expect relaxation phenomena to appear in the water and in the oil. These phenomena are not taken into account in the present theory. Thus, for the highest frequencies (100 MHz–1 GHz, say), the theoretical predictions may have to be corrected due to such phenomena. The second comment is that at the highest frequencies, the acoustic waves are predicted to be strongly attenuated and, therefore, the phase velocity measurement can be difficult, especially when the concentration of the emulsion is as high as in this example (40%). The third comment is also related to measurements. The measurements of phase velocities for all frequencies from 1 Hz to 1 GHz would require numerous different sound sources and experimental setups. In addition, the electronics needed would be quite complicated. The precision of the measurement could also vary significant over the various frequency bands. Therefore, in practice, we have to pay attention to a much narrower frequency region than presented here and see if there are properties in such frequency regions which can be related to droplet size and to the concentration of the emulsion.

In Fig. 3, we have paid most of the attention to the frequency region from 100 kHz to 1 MHz. In this region, we see from Fig. 2 that the rise in phase velocity from the low-frequency value close to 1360 m s^{-1} to the intermediate-frequency value close to 1400 m s^{-1} takes place for droplets both of radius 1 μm and 10 μm. Therefore, this frequency region contains a lot of information and is worth being studied further. In addition, this frequency region is commonly used in ultrasonic measurements in liquids, and the instrumentation is reasonably simple compared to higher frequencies.

In Fig. 3, we show the behavior of the phase velocity and the attenuation as a function of frequency, droplet, size and concentration (volume fraction). The medium is still an Exxsol D80 oil in water emulsion. In the upper row of the figure, the phase velocity and attenuation are shown as a function of frequency. The volume fraction is 40% and the droplet radius is 1 μm. We see the rise in phase velocity as a function of frequency is similar to the one shown in Fig. 2. The attenuation is quite high compared to pure water. It increases almost linearly with frequency in contrast to water, where it increases as the square of the frequency. In the next row, we see the phase velocity and the attenuation as a function of droplet radius. Here, the frequency is 500 kHz and the volume fraction is still 40%. We see that here is a large dependency on the radius, especially for the attenuation. The attenuation maximum is due to the same thermal effect as was described above in the discussion of Fig. 2. In the last row, we see the phase velocity and attenuation as a function of volume fraction from 0 (pure water) to 0.5 (= 50% oil in water). We see that the phase velocity has an almost linear decrease with volume fraction and the attenuation has a similar increase. The plots in Fig. 3 indicate that it is possible to predict an average droplet radius by measuring the phase velocity and attenuation in the frequency region from 100 kHz to 1 MHz. When the volume fraction is above 50%, the theory on which these simulation is based breaks down. However, measurements presented below indicate that it is also possible to measure droplet size and/or concentration in that case. In Fig. 3, we have discussed oil-in-water emulsions. Water-in-oil

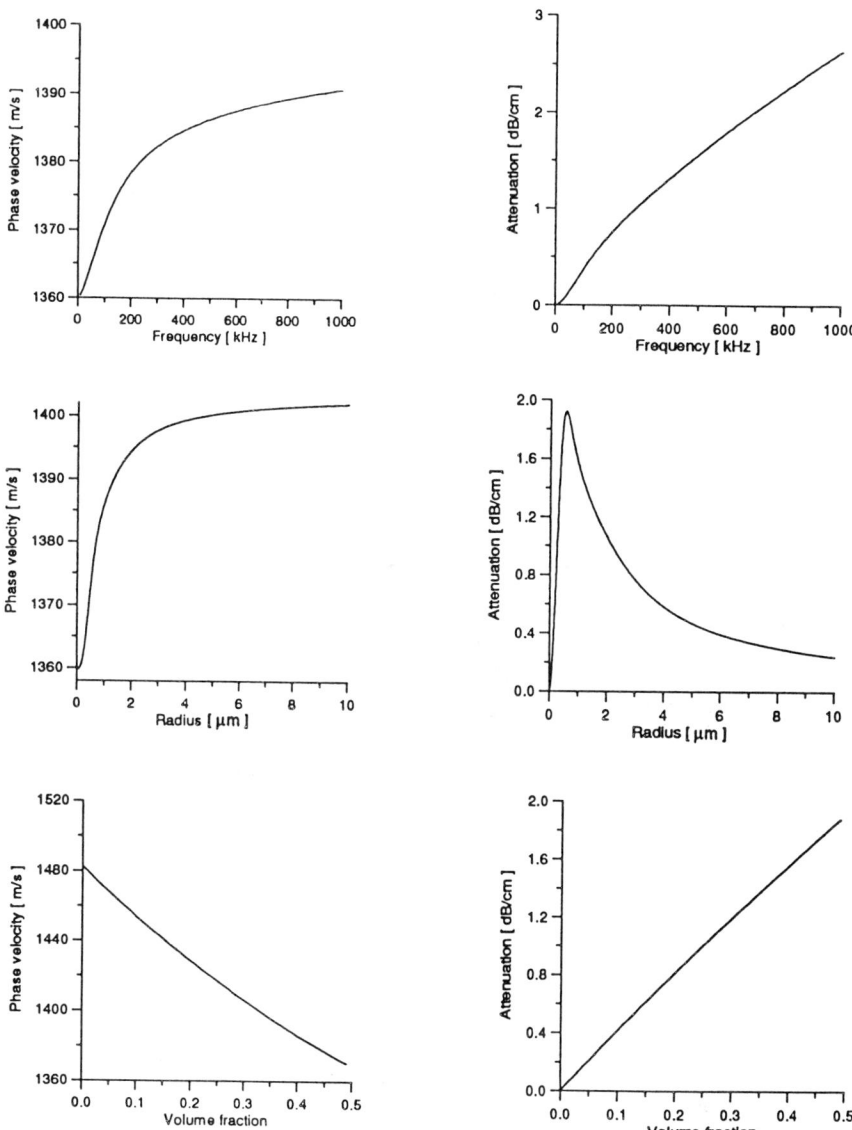

FIG. 3 Predicted phase velocity and attenuation in an oil (Exxsol D80)-in-water emulsion from the Waterman–Truell multiple-scattering theory. Temperature: 20°C; pressure: 1 bar. Upper curves: droplet radius 1 μm, volume fraction 40%; middle curves: frequency 500 kHz, volume fraction 40%; lower curves: frequency 500 kHz, droplet radius 1 μm.

emulsions will behave similarly and can be characterized similar to oil-in-water emulsions.

Finally, we have to say that the precision of these simulations will be decreased if we do not know exactly which components the emulsion contains. In such a case, the qualitative behaviors are not expected to change much, but the quantitative results may be different. A study is needed to see how good the measurements can be in this case.

VI. EXPERIMENTAL TECHNIQUES

There are numerous ultrasonic measurement techniques which relate acoustical and physical parameters of the medium [18]. In this section, we describe a pulse-echo technique with a buffer rod to measure phase velocity and attenuation. This technique can be used successfully in comparisons between experiments and the multiple-scattering theory presented above.

A. Pulse-Echo Technique with Buffer Rod

A pulse-echo measurement cell with buffer rod is shown in Fig. 4. A single transducer which acts as both transmitter and receiver is placed in front of the buffer. The acoustical signal travels through the buffer, and at the buffer–sample interface, the signal is partly reflected and partly transmitted. The reflected part yields the first echo while the signal which is reflected at the sample–reflector interface travels back through the sample and buffer yields the second echo. A wave propagation diagram of this situation is shown in Fig. 5. Evaluation of the first and second echo forms

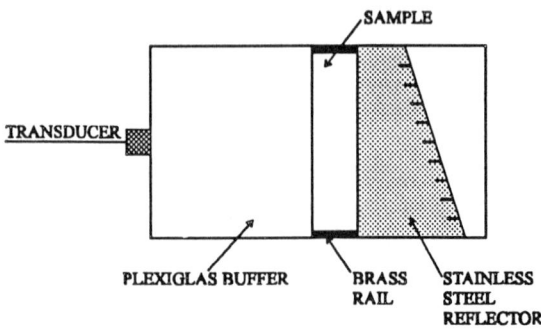

FIG. 4 A measurement cell.

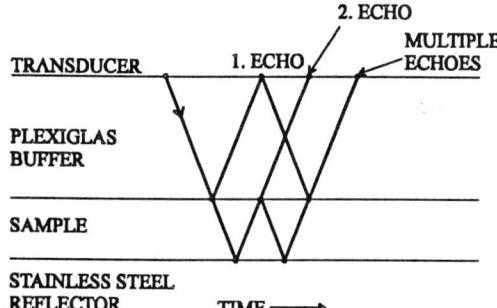

FIG. 5 Wave propagation diagram in the measurement cell of Fig. 4.

the basis in the pulse-echo technique. In principle, the phase velocity of the sample is determined by taking the time difference between the first and second echo, and the attenuation is found by measuring the amplitude of these two echoes. But in order to achieve accurate measurements, one must correct for diffraction (beam spreading) and consider the design, hence the limits, of the measurement cell. Several investigators have studied the effect of diffraction on acoustic wave propagation [19–22], and pulse-echo techniques with buffer rod in particular [23–26]. A brief discussion is given below.

B. Diffraction of Acoustical Waves

Diffraction arises from the finite dimension of the transducer compared with the wavelength. Due to diffraction, the acoustic wave fronts are not plane. Also, the radiated acoustic wave field is not confined to a region defined by the area of the transducer but spreads out. This diffracted wave will generally have a lower amplitude than the amplitude of a plane wave. The apparent attenuation which is due to beam diffraction affects low-frequency (long-wavelength) signals more than high-frequency signals because low-frequency signals spread out more than high-frequency signals. In order to measure absorption properly, a correction for diffraction thus has to be made. The diffraction also alters the phase of the measured signal. When using the phase inside a burst, diffraction effects have to be corrected for in order to measure the phase velocity properly.

Papadakis et al. [23] have outlined a procedure for correction of amplitude and thus also phase velocity in a measurement cell, Fig. 4, where each echo is corrected to its undiffracted value *before* evaluation of attenuation and phase velocity. The procedure is general and one can apply the tabulated correction values given by Khimunin [20,21] or apply directly the correction factor given by Bass [19].

First, the normalized distance S is calculated for each echo:

$$S = \frac{z\lambda}{a^2} = \frac{zc}{a^2 f},$$ (22)

where z is the propagation distance, λ is the wavelength, a is the transducer radius, c is the phase velocity, and f is the signal frequency. Using index b for the buffer and index s for the sample (emulsion), the S value for each echo is

1st echo: $S_b = \dfrac{2l_b c_b}{a^2 f},$ (23)

2nd echo: $S_s = \dfrac{2l_b c_b}{a^2 f} + \dfrac{2l_s c_s}{a^2 f},$ (24)

where l_b and l_s are the lengths of buffer and sample, respectively. If the measurement technique is based on multiple echoes in the sample (reverberations), S is found as above by adding the propagation distance in each medium. For instance, a double echo in the sample yields

3rd echo: $S_{s2} = \dfrac{2l_b c_b}{a^2 f} + \dfrac{4l_s c_s}{a^2 f}.$ (25)

Second, amplitude and phase velocity, hence phase shift, are corrected, for instance, by using the tabulated values by Khimunin [20,21] for the nondimensional distance S for a set of parameter values ka where k is the wave number, $k = 2\pi/\lambda$.

C. Calibration

Calculation of attenuation requires knowledge of the reflection and transmission coefficients at the interfaces between sample and buffer and reflector. In other words, the characteristic impedance $Z = \rho c$ (i.e., the product of density and phase velocity) of each medium must be known. In this section, we describe how the impedance and the length of the sample can be measured. Water is a convenient calibrant fluid because its phase velocity [27], density, and absorption coefficient [28] are well known. Pulse-echo technique has been used by Kline [24] to measure attenuation and dispersion in solids. He used water as buffer and reflector, and the impedance of the solid with known thickness was then indirectly calculated via the transmission and reflection coefficients. In a measurement cell for fluids, the length l_s of the sample volume can be determined by measuring the phase velocity of water (which is known) at a controlled temperature. The acoustical travel time across the sample volume should be repeated for several different temperatures in order to improve the estimate of the sample length. One must compensate for the thermal expansion of the measurement cell. The acoustic impedance Z of the buffer can be determined by the use of the formula for the reflection coefficient R [see Eq. (28)] and the amplitude ratio of the first echo with air and with water as the sample:

Echo air is sample $\quad A_{\text{air}} = A_i e^{-2l\alpha} R_{ba} \approx -A_i e^{-2l\alpha}$ (26)

Echo water is sample $\quad A_{\text{water}} = A_i e^{-2l\alpha} R_{bw}$ (27)

Thus, we get

$$R_{bw} = -\frac{A_{\text{water}}}{A_{\text{air}}} = -\frac{Z_w - Z_b}{Z_w + Z_b}$$ (28)

Here, the index bw indicates the propagation direction, which here means an acoustic wave propagating through the buffer toward the buffer–water interface. The last equality in Eq. (28) comes from standard reflection theory in acoustics; see, for instance, Ref. 28. In the expression above, one assumes that the reflection coefficient is approximately equal to -1 when air is the sample. This is because $Z_{\text{air}} \ll Z_{\text{buffer}}$.

The impedance Z is defined as the ratio of the acoustic pressure p in a medium to the particle speed u, $Z = p/u$. This expression is a complex quantity. For a plane wave, this ratio is a real quantity, $Z = \rho c$, which, as stated above, is called the characteristic impedance of the medium. Z being complex means that the pressure p is not in phase with the particle speed u. In pulse-echo techniques with a buffer rod, we may use $Z = \rho c$; thus, Z is real.

Once R_{bw} has been measured, the characteristic impedance of unknown samples can be determined by

$$R_{\text{sample}} = \frac{A_{\text{sample}}}{A_{\text{water}}} R_{bw} = \frac{Z_s - Z_b}{Z_s + Z_b}.$$ (29)

In addition, if the phase velocity has been measured, the density can now easily be found. Further details can be found in the work by McClements and Fairley [25]. They use this impedance technique, Eqs. (28) and (29), combined with the second echo, Fig. 5, to measure attenuation in emulsions. The impedance technique is particular useful when the attenuation in a sample is to high to permit the second echo. Because R is calculated from the quotient of two amplitude measurements, the problem of measuring absolute signal amplitudes is overcome. The reference amplitudes, A_{sample} and A_{water}, should be measured in the whole frequency range because the transducer does not have a constant frequency response and Z_{buffer} may also depend on frequency. However, this measurement technique requires a constant magnitude of the signals incident on the buffer–sample interface, $A_i(f)e^{-\alpha l}$. On a practical basis, one must take into account that the excitation signal fluctuates, both in amplitude and frequency. This source of error can be reduced by averaging many signals.

A more robust measurement technique which allows fluctuation in the excitation signal is accomplished by using the ratio of the 1 and 2 echoes as reference signals with water as sample, that is, as calibrant. The method requires that the impedances of buffer and reflector are known. Reference signals must be measured and stored for each frequency component. The two echoes with emulsion as sample are then compared with the two reference signals. Furthermore, if the differences of the effect of diffraction on phase velocity and attenuation in water and oil are small, one may consider the correction for diffraction to be implicitly included in the calibration measurements. Thus, the calculations to correct for diffraction can be abandoned. Details are given below.

D. Phase Velocity

Phase velocity which is based on the difference in travel time between the 1st and 2nd echoes and the length of the sample volume is determined from the following equation:

$$c_{sample} = \frac{2l_s}{t_{sample}} = \frac{c_{calibrant}t_{calibrant}}{t_{sample}}. \tag{30}$$

The time difference is found from corresponding zero crossings of the first and second echoes, Fig. 6, which yields the two-way travel time across the sample volume. In principal, this is the same as PEO (pulse-echo overlap) which is widely used; see, for example, Tardajos et al. [29]. Note the 180° phase shift the first echo has undergone relative to the second echo. This is because the impedance of emulsion is lower than the impedance of the Plexiglas buffer. The average of zero crossings located within the steady-state region of the burst is used, in this example about 35–75% of the burst length. Linear interpolation between the sampling points may be used in the detection of the various zero crossings. Finally, the phase velocity is corrected for diffraction effects. This rather cumbersome but robust method is repeated for each frequency component, hence each tone burst.

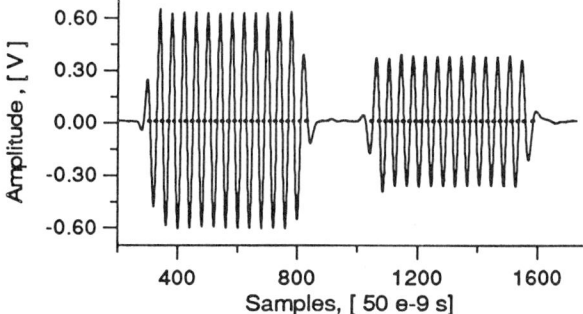

FIG. 6 Time sequence of the 1 and 2 echoes. Buffer: Plexiglas; sample: 40% water-in-oil; reflector: stainless steel; signal: tone burst of 500 kHz; sampling rate: 20 MHz. The zero crossings are marked.

E. Attenuation

The attenuation in emulsion may be determined by comparing the amplitudes of the first and second echoes. The amplitude may be determined by use of the Hilbert transform [30]. The Hilbert transform is an all-pass filter which introduces approximately a constant 90° phase shift of all frequency components of the signal. The Hilbert transform of cosine is sinus. As an example, let us denote the Hilbert transform $H(f)$ and let $f(t) = B \cos \omega t$, then

$$H(f) = iB \sin \omega t = ig, \quad i = \sqrt{-1}.$$

The magnitude M is

$$M = [(B \cos \omega t + iB \sin \omega t)(B \cos \omega t - iB \sin \omega t)]^{1/2}$$
$$= (f^2 + g^2)^{1/2} = B.$$

Similarly, we will find the amplitude of a general burst $f(t)$ as $|f(t) + H(f(t))|$. Figure 7 shows the magnitude of the signal in Fig. 6, as estimated by this algorithm. An estimate of the amplitude is found in the same manner as the phase velocity by taking the average of the digitized magnitude estimates located in a steady-state region of the burst. Alternatively, one may use a peak detector.

Because both echoes must travel through the buffer, the principal effect of the buffer on the echoes is the change in amplitude due to the impedance difference between sample and buffer. The 2 echo will also be modified due to the impedance difference between sample and reflector. Thus, in

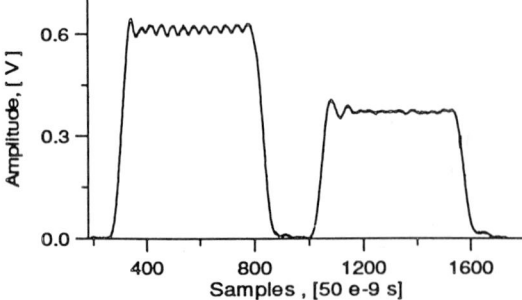

FIG. 7 Magnitude of the signal in Fig. 6. The 1 and 2 amplitudes are the average signal level in the sample range 500–700 and 1200–1400, respectively.

addition to correction for diffraction, the attenuation must also be corrected for transmission and reflection losses at the interfaces. This correction may be accomplished through a plane-wave, normal incidence model. The reflection and transmission coefficients are then based on the characteristic impedance of the respective media:

$$A_1 = R_{bs}A_i e^{-\alpha_b 2l_b}, \tag{31}$$

$$A_2 = T_{bs}R_{sr}T_{sb}A_i e^{-2(\alpha_b l_b + \alpha_s l_s)},$$

where $T = 1 + R$. By correcting the amplitudes for diffraction and then rearranging the expressions, we get the attenuation in the sample

$$\alpha_s = \frac{1}{2l_s} \ln \left(\frac{A_1 T_{bs} R_{sr} T_{sb}}{A_2 R_{bs}} \right) \left[\frac{Np}{m} \right]. \tag{32}$$

The units Np m^{-1} and dB m^{-1} are related by $e^{\alpha_{Np}} = 10^{\alpha_{dB}/20}$, thus 1 Np m$^{-1} \approx 8.7$ dB m^{-1}. R_{sr} can be determined in the same manner as R_{bs}.

A slightly alternative method is to measure the amplitudes in the sample relative to the calibrant. The method is based on the assumption that the attenuation in the calibrant can be neglected and that the difference of the effect of diffraction on attenuation in the calibrant and in sample is small; that is, the phase velocities are roughly of the same magnitude. In oil–water emulsions this may be achieved by using water as the calibrant. The attenuation in water is about 0.05 dB m^{-1} at 500 kHz, whereas the emulsions have attenuation typically in the order of 100 dB m^{-1} = 1 dB cm^{-1}. Thus, we set the attenuation in water to 0 dB m^{-1}. The preliminary attenuation in water and emulsion due to geometrical spreading, diffraction, and absorption (and scattering in emulsion) is

Water:

$$Wa = 20 \log\left(\frac{A_2^w R_{bw}}{A_1^w T_{bw} R_{wr} T_{wb}}\right),$$ (33)

Emulsion:

$$Ea = 20 \log\left(\frac{A_2^e R_{be}}{A_1^e T_{be} R_{er} T_{eb}}\right).$$ (34)

Thus the attenuation, α, in emulsion is

$$\alpha = Ea - Wa = 20 \log\left(\frac{A_2^e R_{be} A_1^w T_{bw} R_{wr} T_{wb}}{A_1^e T_{be} R_{er} T_{eb} A_2^w R_{bw}}\right) \left[\frac{dB}{2l_{\text{sample}}}\right].$$ (35)

F. Experimental Considerations

1. Temperature

Like water, most materials have acoustical properties which are dependent on the temperature. The experiments should therefore be carried out at a controlled and stable temperature. Usually the measurement cell is immersed into a thermostatically regulated water bath, and in most applications, a temperature stability of $\pm 0.1°C$ is sufficient to achieve an accurate acoustical characterization of emulsion.

2. Length of Buffer and Sample

In order to fulfill the wave propagation diagram in Fig. 5. one must adapt the travel time in the buffer to the travel time in the sample. The maximum number of periods in a tone burst is achieved by having a two-way travel time in the sample being equal to one-way travel time in the buffer. For example, the buffer and sample lengths of the measurement cell in Fig. 4 are 10 cm and 2.5 cm, respectively and the corresponding phase velocities are about 2700 m s^{-1} in Plexiglas and 1300–1500 m s^{-1} in the oil–water emulsion. The measurement technique requires a signal detection in the steady-state region of the tone burst. This region is reached within some number of periods, for example, 10 periods. Frequency, the inverse of period, is related to the wavelength by the phase velocity, $c = \lambda f$. Thus, a tone burst of 10 periods is shorter when applying a high-frequency signal relative to a low-frequency signal. In other words, a measurement cell for high-frequency measurements may have shorter buffer and sample lengths than a low-frequency measurement cell.

At the buffer–sample interface, the incident wave generates a shear wave which propagates back toward the transducer but at a much lower amplitude and velocity than the compressional wave. The shear wave may interfere with the 1 and/or 2 echos and thus distort the amplitude and time measurements. In many experiments, interference can be avoided by a proper choice of buffer material and length and sample length. The shear wave, c_s, is found via Poisson's ratio σ:

$$c_s = c_c \sqrt{\frac{2\sigma - 1}{2\sigma - 2}}, \tag{36}$$

where c_c is the compressional wave. In Plexiglas, $\sigma = 0.35$ which yields $c_s \approx 1300 \text{ m s}^{-1}$. The shear-wave amplitude in the Plexiglas buffer, Fig. 4, has been measured at 500 kHz with air as the sample. The shear-wave amplitude was about 33 dB below the amplitude of the compressional wave. Its disturbance, Δt, on time measurements in the worst case, that is when the compressional and shear wave is 90° out of phase, is

$$\Delta t = \frac{-\tan^{-1}(A)}{2\pi f} \approx \frac{-A}{2\pi f}, \qquad A = \frac{A_s}{A_c}, \tag{37}$$

where A_s is the shear-wave amplitude and A_c is the compressional-wave amplitude. In our experiment, Fig. 4, $20 \log(A_s/A_c) = -33$. This means that Δt at most is about 7 ns at 500 kHz, which yields a change in phase velocity of 0.3 m s^{-1} relative to 1400 m s^{-1} (about 0.02%).

Side wall reflections from the buffer or sample may also interfere with the direct signal if the angle of beam spreading is large enough. The effect is analogous with interference of shear waves. The angle of beam spreading, θ, for a circular transducer is given by [28]

$$\theta = \sin^{-1}\left(\frac{3.83}{ka}\right). \tag{38}$$

To prevent interference from side wall reflections, one must adjust the length and lateral dimensions of the measurement cell.

3. Nonparallelism

Misorientation of the transducer with respect to the interfaces can introduce phase cancellations across the transducer which cause errors in measurements of travel time and amplitude. Parallel interfaces are of particular importance in making accurate attenuation measurements. A misorientation angle φ between transducer and interface reduces the signal amplitude by a factor γ given by [31]

$$\gamma = \frac{2J_1(2kan\varphi)}{2kan\varphi}, \tag{39}$$

where J_1 is Bessel function of 1. order and n is number of echoes. In our pulse-echo technique, $n = 1$. To achieve a high degree of parallelism at low cost, one may use two high-precision rails mounted between buffer–sample and sample–reflector interfaces; see Fig. 4.

4. Reflector

Ideally, the reflector should reflect all acoustical energy incident on the front. However, some energy is partly transmitted into the reflector and

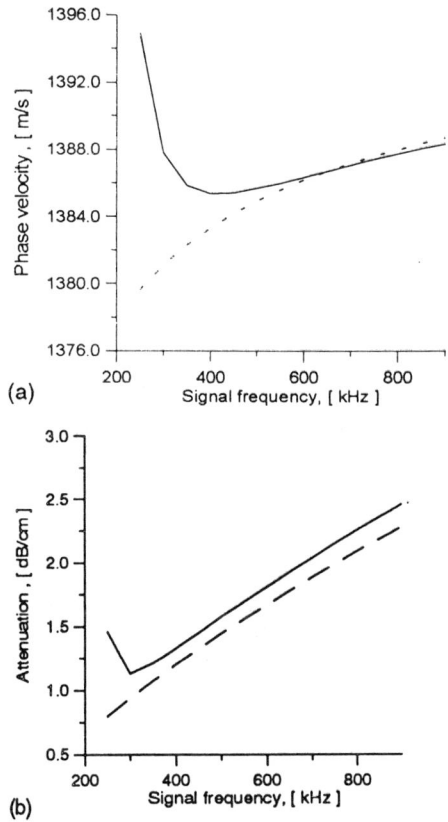

FIG. 8 (a) Phase velocity and (b) attenuation in 40% oil-in-water emulsion. Solid line: 0.0001% air in emulsion; dashed line: no air in emulsion.

can be retransmitted into the sample. One may overcome this problem by making a long reflector with a conical end embedded by an absorbing material. An other alternative is to use a thin stainless-steel reflector with air backing [32]. In Fig. 4, the reflector has an angled back part that guides the signals toward the side wall. In addition, the cuts in the reflector distort the signals and make the reflector less sensitive to reverberations.

5. Gas Bubbles

The presence of gas bubbles in emulsion may significantly alter the ultrasonic measurements [33]. Figure 8 shows the effect of 0.0001% gas bubbles in an 40% oil-in-water emulsion. The calculations are carried out by use of the Waterman–Truell [13] multiple-scattering model. In this example, the bubble Sauter [34] radius is about 15 μm and so the resonance is located around 200 kHz. Experimental work done by Cheyne et al. [35] verifies the behavior of the phase velocity below and above the bubble resonance. In emulsions with low viscosity, the gas bubbles introduced by mixing the components reach the surface within minutes. However, in high-viscosity emulsions, for instance, a 70% oil-in-water emulsion, the emulsion should be placed in a vacuum vessel before the ultrasonic measurements take place in order to remove gas bubbles.

VII. EXPERIMENTAL RESULTS

An experimental investigation of oil–water emulsions has been carried out with volume fractions of dispersed phases ranging from 30% to 80% [15]. Here, we present some of these measurements. The measurement cell in Fig. 4 was used to measure phase velocity and attenuation at frequencies from 250 to 900 kHz. For this purpose, we used a wide-band Panametrics V302 transducer with center frequency at about 500 kHz. The measurements in emulsion were assessed relative to the calibrant which was distilled water, and therefore no compensations for diffraction effects were applied. The uncertainty of the phase velocities is estimated to ± 1 m s^{-1}, except at the lowest frequencies where the uncertainty is slightly higher. The uncertainty of the attenuation measurements is estimated to about ± 0.1 dB cm^{-1}. The emulsions were made by mixing distilled water, Exxsol D80 (a model oil), and surfactant. The amount of surfactant, Berol 26 and Berol 02 which yield oil and water continuous emulsions respectively, was 1% of the oil–water volume. The mixture was homogenized for 2 min. This yields a typical droplet Sauter [34] radius of about 1 μm. In our frequency range, the wavelengths are much longer than the droplet radius and thus the long-wavelength limit of the multiple-scattering theories is valid [Eqs. (11)–(21)]. However, we do not compare

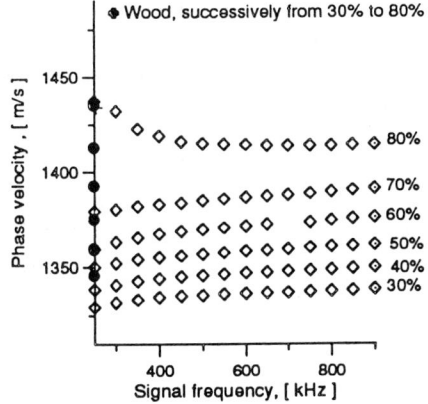

FIG. 9 Measured phase velocities in water-in-oil emulsions from 30% to 80%.

water-in-oil emulsions with theory because accurate values of heat capacity and thermal conductivity of Exxsol D80 are not available. When Exxsol D80 is the continuous phase, these parameters are among the most important ones. The experiment was carried out at 21.9°C.

An overview of phase velocity measurements versus frequency in water-in-oil emulsions is shown in Fig. 9. The dispersed volume fractions are from 30% to 80%. We see that up to 70%, the phase velocity increases with increasing frequency and increasing concentrations. However, at

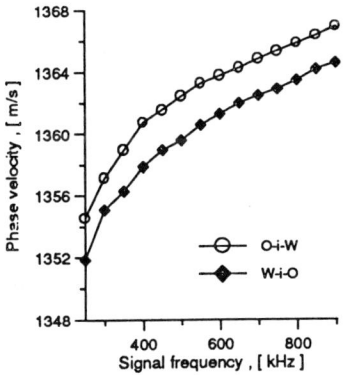

FIG. 10 Measured phase velocity in 50% oil-in-water and water-in-oil emulsions of about equal sized distributions.

80%, this trend is partly reversed and the highest phase velocity is found at the lower end of the frequency range. Compared to Fig. 8a, this result suggests that air is entrapped in the 80% emulsion with resonance frequency near 200 kHz, that is, a bubble diameter of 15 μm. In Fig. 9, we also see that Wood's formula [Eq. (2)], which does not account for thermoviscous losses, overpredicts the phase velocities. The water-in-oil and oil-in-water emulsions exhibit roughly the same trend as far as the phase velocity versus frequency is concerned, but the phase velocity in oil-in-water versus concentration is (of course) reversed because Exxsol D80 has a lower phase velocity than distilled water. The 50% oil-in-water

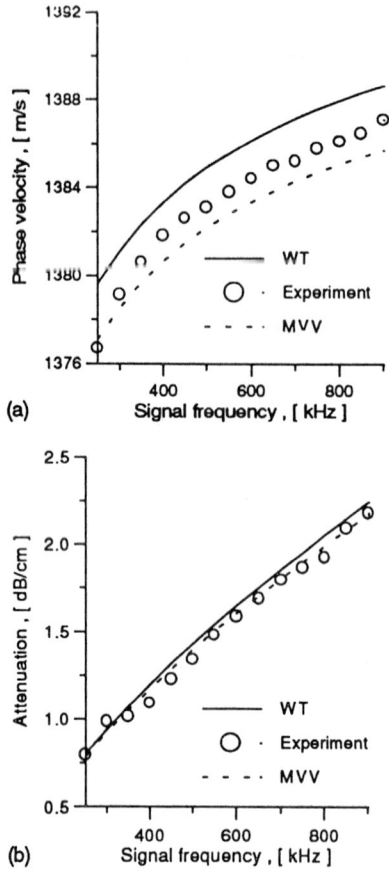

FIG. 11 Comparison of measured and theoretical (a) phase velocity and (b) attenuation in 40% oil-in-water emulsion.

and water-in-oil emulsions of about equal droplet size distribution are shown in Fig. 10. We see that the water-continuous emulsion has a slightly higher phase velocity than the oil-continuous emulsion.

In Fig. 11, we compare measured phase velocity and attenuation to theoretical predictions. In this case, we have a 40% oil-in-water emulsion. The effective mean radius (Sauter [34]) of the emulsion is 0.95 μm, and the actual size distribution is accounted for in the theoretical predictions. We see that the measured phase velocity is between the predicted velocities from the Waterman–Truell multiple-scattering theory (WT) and the multiple-scattering theory by Ma, Varadan, and Varadan (MVV) [16]. As in McClements' work [9] (who used n-hexadecane in water), we find that the WT theory predicts a slightly higher phase velocity and a slightly higher attenuation than the MVV theory. The agreement between the measured phase velocity and attenuation and the theories is better in our case than in Ref. 9. Comparisons between experiment and theories in Ref. 9 show that WT and MVV slightly underpredict the phase velocities and significantly overpredict the attenuation. The attenuation measurements in Ref. 9 for volume fractions of 35.7% and 56.4% were about 30% below the theoretically predicted values.

Attenuation at 500 kHz in oil-in-water and water-in-oil versus concentration is shown in Fig. 12. The theories of WT and MVV do not predict the nonmonotonic behavior at about 50–60% where the attenuation reaches a maximum. A feasible explanation is that at high concentrations the thermal and viscous waves close to each droplets now propagate further than

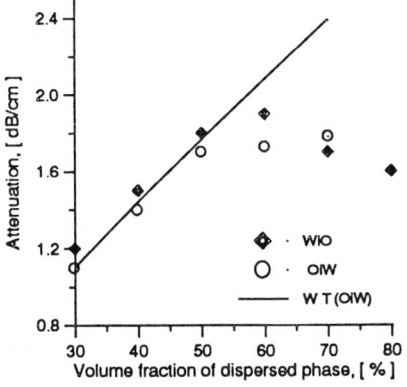

FIG. 12 Attenuation versus volume concentration of dispersed phase for a 500-kHz signal in oil-in-water and water-in-oil emulsions.

the interdroplet distance. The theories do not account for interaction effects between thermal and viscous boundary layers of neighboring droplets and, thus, the theories break down in this region. The water-in-oil and oil-in-water emulsions exhibit about the same characteristics. The attenuation measurement at 80% oil-in-water is most probably influenced by air embedded in the emulsion because the attenuation at 250 kHz is greater than at 300 kHz, about 1.7 dB cm^{-1} and 1.2 dB cm^{-1}, respectively. This agrees with a bubble resonance near 200 kHz; see Fig. 8b.

VIII. SUMMARY

Experimental work shows that there is an excellent agreement between theory and measurements at concentrations below 50%. Ultrasonic characterization of emulsion is feasible and is, in particular, suitable for determining concentration and droplet size distribution in emulsions. Its main drawback is the requirement of detailed knowledge of the physical properties of the components in the emulsion, which is needed in the theoretical model(s). Especially, the properties of the continuous phase must be known. However, the first generation of ultrasonic instrumentation for characterization of emulsion is already on the market. Below, we give a short presentation of two such products.

IX. COMMERCIAL ANALYZERS

There are not many commercial ultrasonic emulsion analysers available. In this section, we give a brief description of two ultrasonic instruments, Cygnus UVM1 [36] and OPUS [37] which are available from Cygnus Limited, UK or Nusonics, USA and Sympatec, Germany, respectively.

Cygnus UVM1 [36] (Ultrasonic Velocity Meter) operates in the pulse-echo mode. The instrument consists of a single 2.25-MHz transducer mounted in a cylindrical stainless-steel cell. This measurement cell is, in principle, equal to the measurement cell shown in Fig. 4. The ultrasonic travel time and temperature are measured and fed into a personal computer where calculation of the volume fraction of the dispersed phase takes place. For food materials, the concentration is found from a simple relationship between the ultrasonic velocity and physical properties. The travel time and temperature accuracy is ±10 ns and ±0.1°C. The Cygnus UVM1 [36] is primarily designed to monitor solid-fat content in emulsions.

OPUS [37] (On-line Particle size analysis by Ultrasonic Spectrometry) measures the attenuation in emulsions and suspensions (solid particles in liquid) over a broad frequency range. The instrument consists of two transducers: one transmitter and one receiver. The sample is placed between the transducers. In the short-wavelength limit where the wavelength is less than the particle radius, the acoustical signal is reflected at continuous phase–particle interface. In this region, also called geometrical scattering, most of the signal is reflected and the attenuation across the cell is relatively high. At larger wavelengths, the geometrical scattering is not fully established, and the attenuation is much weaker. To some extent, this is related to the increase in phase velocity at high frequencies in oil-in-water emulsions, as shown in Fig. 2. OPUS [37] takes advantage of these relationships. The particle size is determined by investigating the transition zone between high and low attenuation. The concentration of the dispersed phase is obtained from a set of equations that are linear at low and medium concentrations. OPUS [37] measures concentration of dispersed phase up to 25% and the particle/droplet diameter range is 5–3000 μm.

ACKNOWLEDGMENTS

The work presented in this chapter has been partially supported by The Research Council of Norway, Norsk Hydro, and STATOIL through the strategic technology program TOP at Christian Michelsen Research AS, and by The Research Council of Norway through the PROSIT program. Professor Halvor Hobæk, Department of Physics, and Professor Johan Sjøblom, Department of Chemistry, at the University of Bergen have assisted in the experimental work.

REFERENCES

1. R. Pal, Colloid Surf. A: Physicochem. Eng. Aspects *84*:141 (1994).
2. D. J. McClements, Adv. Colloid Interface Sci. *37*:33 (1991).
3. B. Vigerust, K. Esbensen, and B. Hope, SINTEF SI Report No. STF27 A93010, SINTEF SI, Oslo, Norway, 1993 [in norwegian].
4. A. W. Wood, *A Textbook of Sound*, Bell, London, 1911.
5. R. J. Urich, J. Appl. Phys. *18*:983 (1947).
6. M. R. J. Wyllie, A. R. Gregory, and G. H. F. Gardner, Geophysics *23*:459 (1957).
7. J. G. Berryman, J. Acoust. Soc. Am. *93*:2666 (1993).
8. C. C. Tsouris and L. L. Tavlarides, Ind. Eng. Chem. Res. *32*:998 (1993).
9. D. J. McClements, J. Acoust. Soc. Am. *91*:849 (1992).
10. J. R. Allegra and S. A. Hawley, J. Acoust. Soc. Am. *51*:1545 (1972).
11. P. S. Epstein and R. R. Carhart, J. Acoust. Soc. Am. *25*:553 (1953).
12. P. M. Morse and U. Ingard, *Theoretical Acoustics*, McGraw-Hill, New York, 1968.
13. P. C. Waterman and R. Truell, J. Math. Phys. *2*:512 (1961).

14. R. J. Urich and W. S. Ament, J. Acoust. Soc. Am. *21*:115 (1949).
15. Ø. Nesse, *18th Scandinavian Symposium in Physical Acoustics* (H. Hobæk, ed.), University of Bergen, Bergen, Norway, 1995.
16. Y. Ma, V. K. Varadan, and V. V. Varadan, J. Acoust. Soc. Am. *87*:2779 (1990).
17. N. E. Berger and V. Twersky, J. Acoust. Soc. Am. *89*:604 (1991).
18. L. C. Lynnworth, *Ultrasonic Measurements for Process Control: Theory, Techniques, Applications*, Academic Press, Boston, 1989.
19. R. Bass, J. Acoust. Soc. Am. *30*:602 (1958).
20. A. S. Khimunin, Acoustica *27*:173 (1972).
21. A. S. Khimunin, Acoustica *32*:192 (1975).
22. X. M. Tang, M. N. Toksöz, P. Tarif, and R. H. Wilkens, J. Acoust. Soc. Am. *83*:453 (1988).
23. E. P. Papadakis, K. A. Fowler, and L. C. Lynnworth, J. Acoust. Soc. Am. *53*:1336 (1973).
24. R. A. Kline, J. Acoust. Soc. Am. *76*:498 (1984).
25. D. J. McClements and P. Fairley, Ultrasonics *29*:58 (1991).
26. J. C. Adamowski, F. Buiochi, C. Simon, E. C. N. Silva, and R. A. Sigelmann, J. Acoust. Soc. Am. *97*:354 (1995).
27. N. Bilaniuk and G. S. K. Wong, J. Acoust. Soc. Am. *93*:1609 (1993).
28. L. E. Kinsler, A. R. Frey, A. B. Coppens, and J. V. Sanders, *Fundamentals of Acoustics*, 3rd ed., John Wiley & Sons, New York, 1982.
29. G. Tardajos, G. G. Gaitano, and F. R. M. de Espinosa, Rev. Sci. Instrum. *65*:2933 (1994).
30. J. G. Proakis and D. G. Manolakis, *Introduction to Digital Signal Processing*, MacMillan, New York, 1989.
31. R. Truell, C. Elbann, and B. B. Chick, *Ultrasonic Methods in Solid State Physics*, Academic Press, New York, 1969.
32. R. E. Lacey and F. A. Payne, Am. Soc. Acoust. Eng. *37*:1583 (1994).
33. K.-E. Frøysa, in *17th Scandinavian Symposium in Physical Acoustics, Compliation of Abstracts* (M. Vestrheim and H. Hobæk, eds.) University of Bergen, Bergen, Norway 1994).
34. C. Orr, in *Encyclopedia of Emulsion Technology, Vol. 1: Basic Theory* (P. Becher, ed.), Marcel Dekker, Inc. New York, 1983, pp. 369–405.
35. S. A. Cheyne, C. T. Stebbings, and R. A. Roy, J. Acoust. Soc. Am. *97*:1621 (1995).
36. Cygnus Instruments ltd., Cygnus House, 30 Prince of Wales Road, Dorchester, Dorset DT1 1PW, England.
37. Sympatec GmbH, Burgstätter-Strasse 6, D 3392 Clausthal-Zellerfeld, Germany.

Index